Biometric-Based Physical and Cybersecurity Systems

Mohammad S. Obaidat • Issa Traore
Isaac Woungang
Editors

Biometric-Based Physical and Cybersecurity Systems

Springer

Editors
Mohammad S. Obaidat
ECE Department
Nazrabayev University
Astana, Kazakhstan

Isaac Woungang
Department of Computer Science
Ryerson University
Toronto, ON, Canada

Issa Traore
Department of Electrical
and Computer Engineering
University of Victoria – UVIC
Victoria, BC, Canada

ISBN 978-3-030-07526-2 ISBN 978-3-319-98734-7 (eBook)
https://doi.org/10.1007/978-3-319-98734-7

This Springer imprint is published by the registered company Springer Nature Switzerland AG
The registered company address is: Gewerbestrasse 11, 6330 Cham, Switzerland

To Our Families

Preface

Biometrics represents one of the most robust and reliable forms of human identification in physical and cyber security. The last decade has witnessed tremendous advances in sensor technologies and data processing techniques and algorithms. This has led to the strengthening of traditional biometrics technologies (e.g., fingerprint, face, iris, retina, keystroke dynamics, and voice) and the emergence of several new technologies, which are showing great promises. The confluence of the consumer markets and national security needs have led to a growing demand for biometrics products and services. For instance, the integration of biometric sensors in smartphones and the use of these technologies for online banking have boosted the adoption of biometric technologies for the masses.

Biometrics carries a strong ability to establish whether an individual is genuine or an impostor. Such an ability to reliably identify individuals coupled with the extreme difficulty of forging biometric data has made it the most trusted form of human identification.

According to Research and Markets Ltd., a market research firm, the global biometrics market will reach \$42.4 billion by 2021. Due to advancements in the technological landscape and better understanding by people, biometrics technologies are now used in various sectors including government, defense, law enforcement, finance, e-commerce, and education. Some countries have started using biometrics to prevent identity fraud during elections or in welfare management and provision.

Several technological advancements have occurred in this area in the last 15 years or so. Now it is possible to collect high-quality data adapted to specific context. For instance, new optical fingerprint scanners are able to sense the presence of a finger and start the scanning process automatically; they can adapt to the capture environment and filter out some of the imperfection and noise. Breakthrough in webcam technologies has improved face detection effectiveness; increasing the accuracy of face biometric technologies.

There are several emerging technologies such as DNA analysis, keystroke recognition, gait analysis, EEG/ECG analysis, mouse dynamics, odor, and touchscreen dynamics that are showing great promises in industry.

As part of this upward trend, we have seen the appearance of multimodal biometric systems that leverage in hybrid platforms the strength of several technologies to improve identification accuracy.

However, despite the positive outlook and trends in biometric technologies, there are still several open or unresolved challenges that researchers are still grappling with. Providing adequate responses to those challenges is essential for the biometrics discipline in order to maintain its reputation and sustain its preponderance in cyber and physical security. Some of the challenges include, but are not limited to, the following:

- Robustness and reliability of biometric scanning technologies
- Reduced dependencies on underlying platforms
- Security of biometric templates and samples
- Privacy of biometric users
- User acceptability
- Biometrics forgery
- Biometrics systems testing
- Biometrics hardware and middleware
- Mobile biometrics devices and platforms
- Cloud-based biometric systems
- Scalable and dependable biometric system architectures
- Integration of biometrics into cryptosystems

The book presents the state of the art in biometrics technologies and reports on new approaches, methods, findings, and technologies developed or being developed by the research community and the industry to address the aforementioned challenges.

The book focuses on introducing fundamental principles and concepts of key enabling technologies for biometric systems applied for both physical and cyber security, disseminates recent research and development efforts in this fascinating area, investigates related trends and challenges, and presents case studies and examples.

It also covers the advances, and future in research and development in biometric security systems, by reviewing the fundamental techniques in the design, operation, and development of these systems.

It is intended for researchers, developers and managers, and students of computer science, information science and technology and electrical and computer engineering, in particular, graduate students at the master and PhD levels, working or with interest in the aforementioned areas.

The book consists of a selection of 21 peer-reviewed chapters contributed by leading experts in biometrics. Chapter 1 is an introduction that gives general perspective on biometrics, and the remaining chapters are structured around the following five major themes: advances in legacy or traditional biometric technologies, emerging biometric technologies, hybrid biometric technologies, enabling technologies or platforms, and the interaction between biometric technology and society. Chapter 1 introduces the concept of biometrics. The categorization of biometric technologies is presented, along with some discussion on biometrics requirements and characteristics and the uses of biometrics technologies. This chapter also presents well-established biometric systems along with operation modes. Finally, some perspectives in terms of contemporary developments that keep up with the promises and expectations of the biometric discipline are discussed.

Part I: Advances in Legacy Technologies

This part consists of four chapters (Chaps. 2, 3, 4, and 5) covering advances and improvements in traditional biometrics technologies, e.g., fingerprint, hand geometry, and face.

Chapter 2 discusses advances in fingerprint technologies and illustrates application of this technology for solving modern problems such as test takers identification.

Chapter 3 discusses biometric identification using eye biometrics, with a focus on iris and retina, and their fusion in a bimodal eye biometric framework.

Chapter 4 discusses the benefit and challenges with 3D hand geometry compared to simple hand geometry biometrics.

Chapter 5 presents fundamental aspects and design of 3D face recognition systems. Discussion of practical issues, such as benchmarking and standardization, and underlying privacy and security challenges, is conducted as well.

Part II: Emerging Technologies

This part consists of six chapters (Chaps. 6, 7, 8, 9, 10, and 11) that present progress and success stories related to recent biometric technologies, e.g., gait, keystroke dynamic, online signature.

Chapter 6 presents advances in keystroke dynamics biometrics. Discussion of conventional and nonconventional feature models is provided as well.

Chapter 7 introduces a new approach for human identification by extracting and analyzing behavioral biometrics in Wi-Fi signals and discusses potential application of this technology.

Chapter 8 discusses research challenges involved in using stylometry for continuous authentication and introduces an approach to tackle some of those challenges.

Chapter 9 presents the different forms of gait biometrics and discusses their applications and perspectives.

Chapter 10 compares online and offline signature verification and presents in detail online signature-based authentication methods.

Chapter 11 explores the state of the art of EEG biometrics, presents the underlying infrastructure and applications, and discusses its promise and prospects.

Part III: Hybrid Technologies

This part consists of five chapters (Chaps. 12, 13, 14, 15, and 16) covering improvements and advances in developing hybrid biometric systems and devices.

Chapter 12 introduces a multimodal biometric invariant fusion authentication system based on fusion of Zφ invariant moment of fingerprint and face features.

Chapter 13 explores healthcare sensors that have the capability collectively to capture different biometrics such as heart beat rate, blood pressure, and iris, enabling the generation of runtime secret key for secure communications.

Chapter 14 presents the state of the art and uses of wearable devices for biometric authentication and explores multimodal biometrics using wearable technologies.

Chapter 15 discusses cognitive features that can be extracted from three different biometrics, namely, mouse dynamics, eye tracking, and keystroke dynamics, and then presents a hybrid model for integrating and synchronizing the features collected from each of these biometrics.

Chapter 16 discusses the design of a multi-biometric authentication based on various characteristics of the finger knuckle.

Part IV: Enabling Technologies

This part involves three chapters (Chaps. 17, 18, and 19) that cover technological advances in developing, deploying, and operating biometric systems from cloud computing and IoT perspectives.

Chapter 17 reviews the state of the art in leveraging cloud resources to deliver Biometrics-as-a- Service (BaaS).

Chapter 18 highlights the authentication challenges for the cloud computing environment and the limitation of traditional solutions and then discusses a number of recent proposals to improve security while maintaining user privacy.

Chapter 19 explores secure solutions for IoT using biometric features of users as well as end users.

Part V: Technology and Society

This part consists of two chapters (Chaps. 20 and 21) covering ethical, legal, and sociological issues related to biometric deployment and operation in the real world.

Chapter 20 explores how to ensure the integrity of election using e-voting by identifying the requirement of e-voting system, discussing the underlying security challenges, and outlining possible solutions.

Chapter 21 discusses the ethical, legal, and social implications of biometrics technologies for society and illustrates case studies such as the Unique Identity Authority of India (UIDAI) project.

The above chapters represent a good coverage of recent advances in biometrics that we believe will lead the readers to great understanding of the state of the art of biometrics technologies. We hope that this will represent a good resource for readers from academia and industry in pursuing future research and developing new applications.

Amman, Jordan — Mohammad S. Obaidat
Victoria, BC, Canada — Issa Traore
Toronto, ON, Canada — Isaac Woungang

Acknowledgments

We would like to thank all the authors who have contributed quality chapters to this book. Special thanks to all our editorial advisory board members and the reviewers who invested a lot of efforts and time in selecting the highest quality chapters possible. We would like also to thank the *Springer* team who helped and advised us in conducting this book project to term.

Finally, we would like to thank our families for their tireless support throughout this project.

Contents

1 **Introduction** 1
Issa Traore, Mohammad S. Obaidat, and Isaac Woungang

Part I Advances in Legacy Technologies

2 **Advances in Fingerprint Technology** 13
Rupa Patel and Soodamani Ramalingam

3 **Recognition-Based on Eye Biometrics: Iris and Retina** 37
Josef Hájek and Martin Drahanský

4 **3D Hand Geometry Recognition** 103
Michal Dvořák and Martin Drahanský

5 **Fundamentals and Advances in 3D Face Recognition** 125
Soodamani Ramalingam, Aruna Shenoy, and Nguyen Trong Viet

Part II Emerging Technologies

6 **Advances in Key Stroke Dynamics-Based Security Schemes** 165
Mohammad S. Obaidat, P. Venkata Krishna, V. Saritha, and Shubham Agarwal

7 **Behavioral Biometrics Based on Human-Computer Interaction Devices** 189
Chi Lin and Mohammad S. Obaidat

8 **Continuous Authentication Using Writing Style** 211
Marcelo Luiz Brocardo, Issa Traore, and Isaac Woungang

9 **Facets and Promises of Gait Biometric Recognition** 233
James Eric Mason, Issa Traore, and Isaac Woungang

10 Online Signature-Based Biometric Recognition 255
Sudeep Tanwar, Mohammad S. Obaidat, Sudhanshu Tyagi, and Neeraj Kumar

11 EEG-Based Biometrics 287
Florian Gondesen, Matthias Marx, and Dieter Gollmann

Part III Hybrid Technologies

12 Multimodal Biometric Invariant Fusion Techniques 321
P. Viswanatham, P. Venkata Krishna, V. Saritha, and Mohammad S. Obaidat

13 Biometrics Based on Healthcare Sensors 337
Mohammad S. Obaidat, Tanmoy Maitra, and Debasis Giri

14 Biometric Authentication for Wearables 355
Harin Sellahewa, Nasiru Ibrahim, and Sherali Zeadally

15 Cognitive Biometrics for User Authentication 387
Ahmed Awad and Yudong Liu

16 Finger Knuckle-Based Multi-Biometric Authentication Systems 404
Aditya Nigam and Phalguni Gupta

Part IV Enabling Technologies

17 Leveraging Cloud-Based Resources for Automated Biometric Identification 437
Wei Lu, John Hancock, Benjamin Osowiecki, Aram Taft, and Wei Li

18 Automated Biometric Authentication with Cloud Computing 455
Hisham Al-Assam, Waleed Hassan, and Sherali Zeadally

19 Biometric Security and Internet of Things (IoT) 477
Mohammad S. Obaidat, Soumya Prakash Rana, Tanmoy Maitra, Debasis Giri, and Subrata Dutta

Part V Technology and Society

20 Protecting the Integrity of Elections Using Biometrics 513
Mohammad S. Obaidat, Tanmoy Maitra, and Debasis Giri

21 Ethical, Legal, and Social Implications of Biometric Technologies 535
Sudeep Tanwar, Sudhanshu Tyagi, Neeraj Kumar, and Mohammad S. Obaidat

Index 571

About the Editors

Mohammad S. Obaidat (Fellow of IEEE and Fellow of SCS) is an internationally well-known academic/researcher/scientist. He received his PhD and MS degrees in Computer Engineering with a minor in Computer Science from the Ohio State University, Columbus, Ohio, USA.

Among his previous positions are Advisor to the President of Philadelphia University for Research, Development and Information Technology; President of the Society for Modeling and Simulation International, SCS; Senior Vice-President of SCS; Dean of the College of Engineering at Prince Sultan University; Chair and Professor at the Department of Computer and Information Science and Director of the MS Graduate Program in Data Analytics at Fordham University; and Chair and Professor of the Department of Computer Science and Director of the Graduate Program at Monmouth University. Dr. Obaidat is currently a Full Professor at the King Abdullah II School of Information Technology, University of Jordan, Nazarbayev University, Kazakhstan, PR of China Ministry of Education Distinguished Professor at University of Science and Technology Beijing, China, and Honorary Distinguished Porfessor at The Amity University.

He has received extensive research funding and has published to date over 65 books, over 60 book chapters and over 850 refereed technical articles in scholarly international journals and proceedings of international conferences – about half of them are journal papers. Professor Obaidat has served as a consultant for several corporations and organizations worldwide. Dr. Obaidat is the Founding Editor in Chief of the Wiley's *Security and Privacy* journal. He is also the Editor in Chief of the Wiley's *International Journal of Communication Systems* and the FTRA's *Journal of Convergence*. He served as the Editor in Chief of the KSIP's *Journal of Information Processing*. He is also an Editor of *IEEE Wireless Communications, IEEE Systems Journal, Simulation: Transactions of the Society for Modeling and Simulations International (SCS)*, Elsevier's *Computer Communications journal*, Springer's *The Journal of Supercomputing*, *IET Wireless Sensor Systems*, SCS *Journal of Defense Modeling and Simulation*, *International Journal of Communication Networks and Distributed Systems*, *The Academy Journal of*

Communications, *International Journal of BioSciences and Technology*, *International Journal of Information Technology*, and *ICST Transactions on Industrial Networks and Intelligent Systems*, among others.

He also served as editor of *IEEE Transactions on SMC*-Parts A, B, and C, the *Wiley Security and Communication Networks journal, Journal of Networks*, *International Journal of Information Technology, Communications and Convergence, IJITCC*, Inderscience, Elsevier's *Computers and Electrical Engineering* journal, and *International Journal of Wireless Networks and Broadband Technologies*, among others. He has guest edited numerous special issues of scholarly journals such as *IEEE Transactions on Systems, Man and Cybernetics*, SMC, *IEEE Wireless Communications*, *IEEE Systems Journal, IEEE Access, Simulation: Transactions of the Society for Modeling and Simulation International (SCS)*, Elsevier's *Computer Communications* journal, *Computers and Electrical Engineering* journal, Wiley's *Security and Communication Networks*, *Journal of Networks*, and *International Journal of Communication Systems*, among others. Obaidat has served as the Steering Committee Chair, Advisory Committee Chair, and Program Chair of numerous international conferences.

He is the founder of two well-known international conferences: the International Conference on Computer, Information, and Telecommunication Systems (CITS) and the International Symposium on Performance Evaluation of Computer and Telecommunication Systems (SPECTS). He is also the co-founder of the International Conference on Data Communication Networking (DCNET) and SIMULTECH.

Between 1994 and 1997, Obaidat has served as distinguished speaker/visitor of *IEEE* Computer Society. Since 1995, he has been serving as an *ACM* Distinguished Lecturer. He is also an *SCS* Distinguished Lecturer. Between 1996 and 1999, Dr. Obaidat served as an IEEE/ACM Program Evaluator of the Computing Sciences Accreditation Board/Computing Sciences Accreditation Commission (CSAB/CSAC). Obaidat is the Founder and first Chairman of SCS Technical Chapter (Committee) on PECTS (Performance Evaluation of Computer and Telecommunication Systems). He has served as the Scientific Advisor for the World Bank/UN Digital Inclusion Workshop – The Role of Information and Communication Technology in Development. Between 1995 and 2002, he has served as a member of the board of directors of the Society for Computer Simulation International. Between 2002 and 2004, he has served as Vice-President of Conferences of the Society for Modeling and Simulation International (SCS). Between 2004 and 2006, Prof. Obaidat has served as Vice-President of Membership of the Society for Modeling and Simulation International (SCS). Between 2006 and 2009, he has served as the Senior Vice-President of SCS. Between 2009 and 2011, he served as the President of SCS. Many of his recent papers have received the best paper awards from IEEE AICCSA 2009, IEEE GLOBECOM 2009, IEEE DCNET 2011, IEEE CITS 2013, IEEE CITS 2016, and IEEE ICC 2019 International Conferences. Professor Obaidat has been awarded a Nokia Research Fellowship and the distinguished Fulbright Scholar Award. He received the SCS *Outstanding Service Award* for his excellent leadership, services, and technical contributions. Dr. Obaidat received very recently the Society for Modeling and Simulation International's (SCS) prestigious *McLeod Founder's Award* in recognition of his outstanding technical and professional contributions to modeling and simulation. He received in December 2010 the IEEE ComSoc-GLOBECOM 2010 Outstanding Leadership Award for his outstanding leadership of Communication Software Services and Multimedia Applications (CSSMA) Symposium. He received the Society for Modeling and Simulation International's (SCS) prestigious *Presidential Service Award* for his outstanding unique, long-term technical contributions and services to the profession and society. He was inducted to the SCS *Hall of Fame – Lifetime Achievement Award* for his technical contribution to modeling and simulation and for his outstanding visionary leadership and dedication to increasing the effectiveness and broadening the applications of modeling and simulation worldwide. He was awarded in 2017 the IEEE CITS Hall of Fame Distinguished and Eminent Award. He has been awarded with the Amity University Distinguished Honorary Professor Award. He received recently the IEEE ComSoc TC on Communication Software award For his achievements and original technical contributions to cybersecurity, wireless networking, computer networks and modeling & simulation.

He has been invited to lecture and give keynote speeches worldwide. His research interests are cybersecurity, biometrics-based security, wireless communications and networks, IoT, telecommunications and networking systems, security of network, information and computer systems, security of e-based systems, performance evaluation of computer systems, algorithms and networks, green ICT, high-performance and parallel computing/computers, applied neural networks and pattern recognition, and adaptive learning and speech processing. During2004/2005, he was on sabbatical leave as Fulbright Distinguished Professor and Advisor to the President of Philadelphia University in Jordan, Dr. Adnan Badran. The latter became the Prime Minister of Jordan in April 2005 and served earlier as Deputy Director General of UNESCO. Professor Obaidat is a Fellow of the Society for Modeling and Simulation International (*SCS*) and a Fellow of the Institute of Electrical and Electronics Engineers (*IEEE*).

Issa Traore obtained a PhD in Software Engineering in 1998 from the Institute Nationale Polytechnique (INPT)-LAAS/CNRS, Toulouse, France. He has been with the faculty of the Department of Electrical and Computer Engineering of the University of Victoria since 1999. He is currently a Full Professor and the Coordinator of the Information Security and Object Technology (ISOT) Lab (http://www.isot.ece.uvic.ca) at the University of Victoria. His research interests include biometrics technologies, computer intrusion detection, network forensics, software security, and software quality engineering. He is currently serving as Area Editor for the Wiley's *Security and Privacy* journal. Dr. Traore is also co-founder of the Plurilock Security Solutions Inc. (http://www.plurilock.com), a network security company which provides innovative authentication technologies, and is one of the pioneers in bringing behavioral biometric authentication products to the market.

Isaac Woungang (Senior IEEE Member) received his PhD degree in Mathematics from the University of South, Toulon and Var, France, in 1994. From 1999 to 2002, he worked as Software Engineer in the Line Photonic Group at the Nortel Networks, Ottawa, Canada. Since 2002, he has been with Ryerson University, where he is now a Full Professor of Computer Science and Coordinator of the Distributed Applications Broadband and Network Laboratory (DABNEL) research Lab. His current research interests include radio resource management in wireless networks, cloud security, and routing in opportunistic networks. He has published 8 edited and 1 authored books and over 80 refereed journals and conference papers. He is currently a Full Professor and the Coordinator of the *International Journal of Communication Networks and Distributed Systems* (*IJCNDS*), Inderscience, UK. He has also served as Chair of the Computer Chapter, IEEE Toronto Section. He is currently serving as Area Editor for the Wiley's *Security and Privacy* journal.

Chapter 1
Introduction

Issa Traore, Mohammad S. Obaidat, and Isaac Woungang

The last decade has seen a dramatic increase in the adoption rate of biometric technologies. This is driven by significant improvement in the enabling technologies (e.g., sensors) and the data processing capabilities (i.e., computing power, algorithms, to name a few). Advances in mobile computing, with the appearance of smart devices, and the increasing needs of security in critical areas such as finance and government are major drivers of the biometric security field. In the last decade, the population operating, using directly, or impacted indirectly by biometric solutions has grown dramatically.

According to Research and Markets [1], a market research firm based in Ireland, with expertise in biometrics technologies, the global biometrics marketplace is forecast to grow at a compounded annual growth rate (CAGR) of a little over 17% in the next 10 years, reaching about $52.7 billion by 2025. The biometrics field is spearheaded by a bubbling and fast growing market segment which covers major geographical areas across the globe. The biometrics field is also marked by its diversity and innovation. From the inception of the earliest forms of biometrics dating back in the 1800s to until about 30 years, the field went through a slow evolution, where only a handful of technologies, mostly physiological, were available or in use. In contrast, in the last three decades, with the appearance of disruptive

I. Traore (✉)
Department of Electrical and Computer Engineering, University of Victoria – UVIC, Victoria, BC, Canada
e-mail: itraore@ece.uvic.ca

M. S. Obaidat
King Abdullah II School of Information Technology, Amman, Jordan

University of Jordan, Amman, Jordan

I. Woungang
Department of Computer, Ryerson University, Toronto, ON, Canada
e-mail: iwoungan@ryerson.ca

M. S. Obaidat et al. (eds.), *Biometric-Based Physical and Cybersecurity Systems*,
https://doi.org/10.1007/978-3-319-98734-7_1

technologies such as the Internet, smartphones, cloud computing, and so on, the biometric field has experienced an explosion with the appearance of a growing body of new and disruptive technologies. Advances in sensor technologies, data processing techniques and technologies, computational intelligence and data mining, and other related fields have helped achieving dramatic improvements in the reliability and robustness of biometric technologies. This has contributed to raising the confidence level and improving the perception of these technologies by the different stakeholders, e.g., the public, policy makers, entrepreneurs, etc.

1 What Is Biometrics?

According to the Oxford English Dictionary, published about two decades ago, biometrics is the "application of statistical analysis to biological data" [2]. While this definition broadly captures the practice of biometrics analysis, it misses out its essence by the lack of specificity. A more recent definition, available in the online dictionary *Dictionary.com*, bridges this gap, with the following definition:

> "the process by which a person's unique physical and other traits are detected and recorded by an electronic device or system as a means of confirming identity."

Biometrics consists of the measurement of biological signals for the purpose of human identification. There are many biological signals which can be used for human identification, but only a few of them can be measured. The different ways a human can effectively identify other humans is countless, for instance, based on sensory systems (e.g., vision, hearing, touch, taste, smell, etc.). However, only a few of those human identification systems qualify as biometrics. The main hurdle is the challenge related to collectability and measurability. The inability to measure many of these signals, due to the lack of adequate sensors, rules them out as acceptable biometrics.

1.1 Biometrics Classes

Biometric technologies can be categorized based on the type of signals they rely on, which consist primarily of the following kinds: physiological, behavioral, and cognitive. Physiological characteristics are inherent to the human physiology. Examples of physiological characteristics include hand geometry, finger minutia, and facial features, to name a few.

Behavioral characteristics are traits that are learned or acquired based on human actions. Examples of behavioral traits include keystroke dynamics, mouse dynamics, gesture dynamics, signature dynamics, voice, and gait features.

Cognitive biometrics relies on the cognitive, emotional, and conative state of an individual as the basis to recognize the individuals. In general, these states of mind are extracted by recording physiological or behavioral bio-signals, such as the electroencephalogram (EEG), electrocardiogram (ECG), and electro-dermal response (EDR) of the individual in response to the presentation of an authentication stimulus, e.g., an image portraying a memorable event.

1.2 Biometrics Requirements and Characteristics

Any human physiological, behavioral, or cognitive characteristic can be used as a biometric trait as long as it satisfies the following requirements [3]:

- *Universality*: The characteristic or trait must be applicable to each human being.
- *Distinctiveness*: Any two persons should be sufficiently different in terms of the characteristic.
- *Permanence*: The characteristic should be sufficiently invariant over a period of time.
- *Collectability*: The characteristic can be measured quantitatively.

As mentioned earlier, collectability is a key requirement that is often overlooked in favor of the other requirements (e.g., distinctiveness) which are more obvious. However, several potential technologies have failed to make the cut because of the lack of practical ways of measuring the bio-signals.

In addition to the aforementioned criteria, a practical biometric trait is required to address the following additional requirements:

- *Performance*: achievable recognition accuracy and the speed and resources required
- *Acceptability*: extent to which people are willing to accept the use of a particular biometric identifier in their daily lives
- *Resistance to circumvention*: reflects how easily the system can be fooled using fraudulent methods such as forgeries based on synthetic samples and other evasion techniques
- *Privacy preservation*: protection of private user information embedded in the biometric templates and underlying technologies

1.3 Uses of Biometrics Technologies

Biometric technologies provide the basis for an extensive array of highly secure human identification and verification solutions. These include:

- Physical access control and surveillance
- Authentication
- Digital forensics
- Time attendance
- Border security

- Passport integrity
- In-person and electronic voting

One of the largest areas of application is in automated fingerprint identification systems (AFIS), which are used in forensics investigation, criminal record checking, passport integrity checking, and boarder security, to name a few.

Biometrics are used in various industries, including government and law enforcement, commercial and retail, health care, travel and immigration, financial and banking, and so on.

Governmental applications cover national identity cards, passports, driving licenses, social security cards, voter registration, welfare registration, and many others. The technologies are used as reinforcement or replacement for some of these critical documentations or processes.

Multiple biometrics, combined in multimodal frameworks to deliver much improved accuracy and robustness, are used to secure restricted areas in airports, national security facilities, etc.

With the dramatic improvement in computational capabilities and progress made in developing smart sensors, the biometric technology landscape is also witnessing a shift from primarily hardware-based systems to software-based solutions powered by smartphones and cloud computing.

2 Biometric Systems

Among the well-established physiological biometrics solutions, with wide user populations, are fingerprint, iris recognition, facial recognition, hand geometry, vein recognition, signature recognition, and palm print recognition [4–7].

Well-established behavioral solutions include voice recognition and dynamic signature recognition [8, 9].

Besides the aforementioned technologies, there are several emerging biometric solutions such as body odor, ear pattern, and lip prints recognition in the physiological category [10, 11] and keystroke dynamics, touchscreen dynamics, mouse dynamics, stylometry, and gait recognition in the behavioral category [12, 13]. Several products have been released in the market related to these technologies that cover a variety of industries such as banking and finance, online fraud detection, health care, etc.

2.1 Operation Modes

A biometric system is basically a pattern recognition system that operates by acquiring raw biometric data (e.g., finger scan, typing rhythm, mouse movement, touchscreen interactions) from an individual. The captured data is processed by extracting identifying parameters also referred to as biometric features. The extracted features form the basis of the so-called biometric signature, also referred to as biometric profile or template. The templates are stored in a database and used later to check the user identity by matching them against new biometric samples.

Biometric systems involve three primary modes of operation depending on the application: enrollment, verification, and identification modes.

In the enrolment mode, a biometric template is built for the user by capturing sample data and constructing a mathematical representation of corresponding biometric traits.

In the verification mode, the user claims a certain identity (e.g., by supplying a user name). The verification process consists of validating the person's identity by comparing the captured biometric data with the biometric template(s) stored in the system's database for the claimed identity. It is a *1-to-1* matching process. Verification forms the basis of authentication and access control systems.

In the identification mode, the system is presented with an unattributed sample, e.g., a latent fingerprint collected in a crime scene. The identification process consists of recognizing the actual individual that produced the sample by searching the templates of all the individuals in the database for a match. It is a *1-to-n* matching process.

Identification is typically used for forensics investigation. But, it can also be used for fraud detection. For instance, to prevent fraudsters from enrolling in a biometric system using multiple assumed identities, an identification process can be executed against existing biometric databases, before completing each new enrollment. This can be used to prevent election frauds when voting relies on a biometric database.

2.2 Usage Models

Traditionally, applications of biometrics were for physical access control, e.g., controlling access to secured facility or buildings using a fingerprint scanner or through face recognition. Other common traditional applications are in forensics, of which automated fingerprint identification system (AFIS) is the most prominent example.

More recent applications in the physical security area include integrity check for passports, citizen identification for in-person voting, etc.

In the Internet era and advances in smartphone technologies, the use of biometrics for user authentication is becoming commonplace.

Authentication consists of proving that the claim made by an individual to exhibit certain identity is genuine. As mentioned above, biometric authentication is established by requiring the individual to provide biometric data sample and then checking this sample against the original template stored for the claimed identity.

There are three kinds of biometric authentication approaches:

– Static authentication
– Active authentication
– Continuous authentication

Static authentication consists of checking the identity of the individual once, typically at log-in time. While static authentication is crucial for gate keeping, it is not enough to ensure the security of the session. The remaining session can still be hijacked by a hacker for nefarious purpose.

Active authentication consists of re-authenticating the individual; this occurs typically once after log-in. Active authentication has the potential to catch session hijacking, if such occurrence has taken place before the re-authentication.

Continuous authentication is a more permanent form of active authentication. The user identity is checked repeatedly after log-in. Continuous authentication could happen periodically, after a certain amount of activity or at the expiration of some predefined time interval. It could be made more stringent by re-authenticating the individual at every action (e.g., mouse click or keystroke) performed by the individual. While a stringent approach reduces drastically the windows of vulnerability, it could increase the overhead on the underlying authentication system. A more balanced approach is recommended where the re-authentication window is kept relatively small but sufficiently high (e.g., greater than one action) to reduce the performance overhead.

2.3 System Performance

As a pattern recognition system, a biometric system compares a sample against an existing biometric template and then computes a similarity score also referred to as biometric (matching) score. A biometric score s is compared against a threshold, say t, in order to make a binary decision: accept the sample as genuine if the score is greater than the threshold ($s \geq t$), or reject the sample, otherwise. The quality of a biometric system is measured by evaluating its ability for accurately making the accept/reject decisions over a user population.

The performance of biometric technologies is commonly measured using a group of related metrics widely accepted in industry and academia. These include the following types of errors:

- False Acceptance Rate (FAR): error rate resulting from mistaking biometric measurements from two different persons to be from the same person also called false match rate (FMR)
- False Rejection Rate (FRR): error rate resulting from mistaking two biometric measurements from the same person to be from two different persons also called false non-match rate (FNMR)
- Failure to Capture Rate (FTCR): the rate at which the system is unable to process the captured raw biometric data and extract features from it

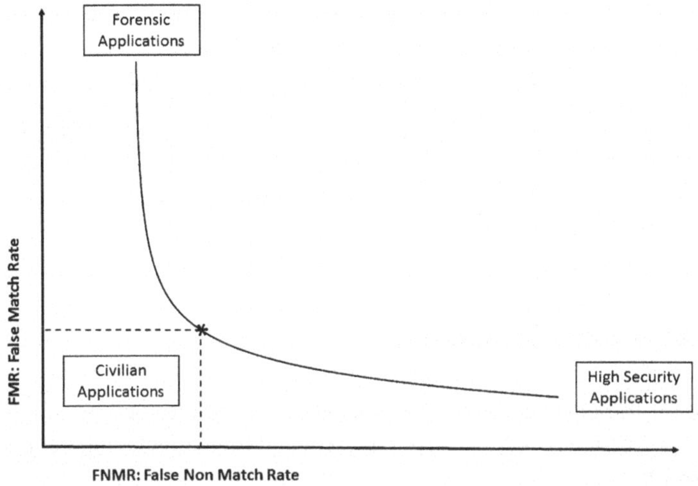

Fig. 1.1 ROC curve [14]

- Failure to Enroll Rate (FTER): the rate at which the system is unable to enroll a user's biometric features

Both FAR (or FMR) and FRR (or FNMR) are functions of the system threshold *t*. Dependent on the value of threshold *t*, different sets of (FAR, FRR) combinations will be achieved. These two measures must always be provided together when rating a biometric system; FAR alone is meaningless and vice versa. The combination (FAR, FRR, *t*) is referred to as an operating point. When designing and evaluating a biometric system, it is customary to provide the so-called Receiver Characteristic Curve (ROC). The ROC curve is a graphical plot of different operating points as shown in Fig. 1.1.

FAR and FRR vary in a contrarian way depending on the threshold *t*:

- FAR increases if *t* is decreased to make the system more tolerant to input variations and noise.
- If *t* is increased to make the system more secure, then *FRR* will increase accordingly.

The optimal operating point depends on the relative cost of a false acceptance versus a false rejection. In many cases, a false rejection is much less costly than a false acceptance; a false rejection is a source of frustration, whereas a false acceptance represents a security breach.

The ROC curve allows determining another important performance metric referred to as crossover error rate (CER) or equal error rate (EER). It is the (operating) point where the FAR is equal the FRR. Most existing biometric solutions performances are reported in terms of EER. The lower the crossover point, the more accurate the biometric system.

Additional performance metric that is used specifically to evaluate biometrics-based continuous authentication systems is the re-authentication delay also referred to as *Time-To-Alarm (TTA)* or *Time-To-Detect (TTD)* [15]. This corresponds to the duration after which the individual is re-authenticated; it could be expressed in terms of the amount of (primitive) actions (e.g., mouse click or keystrokes) performed by the individual. A trade-off must be made between TTA and accuracy metrics (i.e., FAR, FFR, and EER).

3 Summary and Perspectives

There are several contemporary developments which are testimony to the fact that the biometrics field has very promising present and future, including but not limited to the following:

- Fast growing market for biometric businesses
- Growing number of users and operators around the globe
- Diversity of use cases and applications: e-government, national security, border control, financial fraud detection and prevention, etc.

The field of information and system security relies on a few important building blocks. Cryptography is one such building block as it enables secrecy and integrity of data communication and storage. In the last three decades, the biometrics discipline has appeared to be a serious contender to be among those crucial building blocks as it provides sound mechanisms to support authentication and access control, which are central pillars of information and system security.

With the technological advances achieved in the recent years, and the significant increase in biometric system reliability and robustness, biometrics technologies have become pervasive. At the same time, like any fast growing field, the biometrics discipline has its own growing pains that will need to be addressed adequately, for it to be considered a viable building block of information and system security. Examples of critical issues that need adequate solutions include the following:

- Heterogeneity challenges and its impact on end-user accuracy and usability (in particular for physiological biometrics)
- Stability challenges and its impact on accuracy and robustness (in particular for behavioral biometrics)
- Privacy and security challenges
- Psychological acceptability
- Biometric evaluations: aging effects, confidence interval estimation, databases, environment impacts, performance modelling and prediction, protocol and benchmarking, social impacts, and usability studies
- Biometric system designs and standards: anti-spoofing, cancellable template and protection, encryption, large-scale identification and big data, mobile and remote

biometrics, security and privacy, smart card, standardization, template selection, and update

Among those issues, some of the most pressing ones pertain to the security and privacy of biometric systems. As a matter of fact, biometric systems have vulnerabilities of their own. A biometric system can be circumvented by bypassing the system altogether, and accessing whatever resources it is intended to protect, without triggering the biometric check. Attack can be directed at the biometric data, by intercepting and replaying transmitted biometric samples, or by breaching the template database. By hacking the template database, biometrics samples can be stolen and reused in some cases. The primary losers, in case of biometrics theft, are the owners of the samples. While it is easy to replace a stolen password or credit card, by design, biometric traits are not replaceable. This means, victims of biometric theft would permanently be exposed to possible identity fraud.

Concerns about the privacy implications of biometric technologies are tremendous [16, 17]. One of the common concerns is about the sharing of biometric data, i.e., whether the biometrics data you provide at specific outlet, for instance, in online transactions, could find their way into government databases, or used third-party organizations without your knowledge.

Furthermore, some biometric technologies contain far more information beyond their initial intent. For instance, DNA contains information about the genome, which could be used for various kind of genetic analysis.

The legal or regulatory framework around biometric data collection and retention is nonexistent in many countries, or still being written in others. This puts users and other stakeholders in the crosshair of potential abuse or misuse of biometric technologies.

In order to keep up with the promises and expectations of the biometric discipline, effective approaches must be developed to address the aforementioned challenges and many others. Fortunately, the biometric research community and industry are relentlessly working on developing innovative solutions to strengthen biometric technologies and make them more dependable.

References

1. Global Biometric System Market Analysis & Trends - Industry Forecast to 2025. (Research and Markets, Guinness Centre, Taylors Lane, Dublin 8, Ireland), http://www.researchandmarkets.com/publication/medpaeb/420980
2. J. Pearsall, B. Trumble (eds.), *The Oxford English Reference Dictionary*, 2nd edn. (Oxford University Press, Oxford, 1996)
3. A.K. Jain, A. Ross, S. Prabhakar, An introduction to biometric recognition. IEEE Trans. Circuits Syst. Video Technol. Spec. Issue Image- Video-Based Biomet. **14**(1), 4–20 (2004)
4. W.W. Bledsoe Man-machine facial recognition: report on a large-scale experiment. Technical report PRI-22. Panoramic Research Inc., California (1966)

5. R. Sanchez-Reillo, C. Sanchez-Avila, A. Gonzalez-Marcos, Biometric identification through hand geometry measurements. IEEE Trans. Pattern Anal. Mach. Intell. **22**(10), 1168–1171 (2000)
6. G. Aguilar, et al. Fingerprint recognition. Internet monitoring and protection, 2007. ICIMP 2007. Second international conference on. IEEE, 2007
7. G. Guo, M.J. Jones, Iris extraction based on intensity gradient and texture difference. Applications of computer vision, 2008. WACV 2008. IEEE workshop on. IEEE, 2008
8. J. Wu, J. Konrad, P. Ishwar. Dynamic time warping for gesture-based user identification and authentication with Kinect. Acoustics, speech and signal processing (ICASSP), 2013 I.E. international conference on. IEEE, 2013
9. J.A. Markowitz, Voice biometrics. Commun. ACM **43**(9), 66–73 (2000)
10. J. Kasprzak, B. Leczynska, Cheiloscopy, in *Human Identification on the Basis of Lip Trace (in Polish)*, (KGP, Warsaw, 2001)
11. Odinaka, Ikenna, et al. ECG biometrics: a robust short-time frequency analysis. Information forensics and security (WIFS), 2010 I.E. international workshop on. IEEE, 2010
12. A. Ahmed, I. Traore, A new biometric technology based on mouse dynamics. IEEE transactions on dependable and secure computing 4.3 (2007)
13. A.A.E. Ahmed, I. Traore, Free text recognition of keystrokes. IEEE Trans. Cybernet. **44**(4), 458–472 (2014)
14. E.M. Newton, L. Sweeney, B. Malin, Preserving privacy by de-identifying face images. IEEE Trans. Knowl. Data Eng. **17**, 232–243 (2005)
15. I. Traore, A. A. E. Ahmed (eds.), *Continuous Authentication Based on Biometrics: Data, Models, and Metrics* (IGI Global, Hershey. ISBN 978-1-61350-129, 2012)
16. W. Payne, *Biometrics: a Double Edged Sword - Security and Privacy* (SANS Institute, 2002). https://www.sans.org/reading.../biometrics-double-edged-sword-security-privacy-137
17. S. Prabhakar, S. Pankanti, A.K. Jain, Biometric recognition: security and privacy concerns. IEEE Secur. Priv. **99**(2), 33–42 (2003)

Part I
Advances in Legacy Technologies

Chapter 2
Advances in Fingerprint Technology

Rupa Patel and Soodamani Ramalingam

1 Fingerprint Basics

Biometric identifiers in general fall into two categories, namely, behavioural and physiological. Behavioural characteristics relate to pattern of behaviour such as voice, handwriting and gait. Physiological characteristics relate to shape of the body such as DNA, face recognition, iris, retina and fingerprints. In comparison to other identifiers, fingerprint technology has a reputation of having a very good balance of all the desirable properties: universality, distinctiveness, permanence, collectability, performance, acceptability and circumvention [1]. Table 2.1 shows a comparison of the biometric technologies and how each of these rate against the desirable properties. From this table, it is clear that fingerprint has distinctly a medium-high ranking of the desirable properties.

Fingerprint technology has been widely used in critical applications. From traditional ink and paper in the 1800s to introduction of sensors a century later and touchless swipe sensors manufactured in the 2000s, fingerprint technology has grown and since being used for various applications proving that this technology can be used to simplify processes. Long gone are the days when fingerprints were only used to solve crime investigation.

As shown in Fig. 2.1, following the traditional method of using ink and paper in 1800s, the 1900s introduced use of optical sensor which captures an optical image as used in mobile devices, the capacitive sensors which use tiny capacitor circuits to collect data about a fingerprint and swipe sensors which complete an image by

R. Patel (✉)
University of Hertfordshire, Hatfield, UK
e-mail: r.36.patel@herts.ac.uk

S. Ramalingam
School of Engineering and Technology, University of Hertfordshire, Hatfield, UK
e-mail: s.ramalingam@herts.ac.uk

M. S. Obaidat et al. (eds.), *Biometric-Based Physical and Cybersecurity Systems*,
https://doi.org/10.1007/978-3-319-98734-7_2

Table 2.1 Comparison of biometric technologies [1, 2]

Biometric identifier	Universality	Distinctiveness	Permanence	Collectability	Performance	Acceptability	Circumvention	Introduction cost
DNA	H	H	H	L	H	L	L	H
Face	H	L	M	H	L	H	H	L
Fingerprint	**M**	**H**	**H**	**M**	**H**	**M**	**M**	**L**
Hand geometry	M	M	M	H	M	M	M	L
Hand vein	M	M	M	M	M	M	L	M
Iris	H	H	H	M	H	L	L	H
Retina	H	H	M	L	H	L	L	H
Signature	L	L	L	H	L	H	H	L
Voice	M	L	L	M	L	H	H	M

H high, *M* medium, *L* low

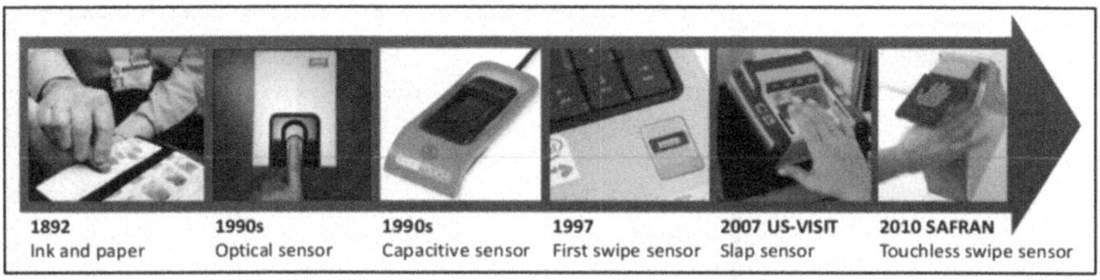

Fig. 2.1 Evolution of fingerprint recognition sensors from the 1800s to 2010 [3]

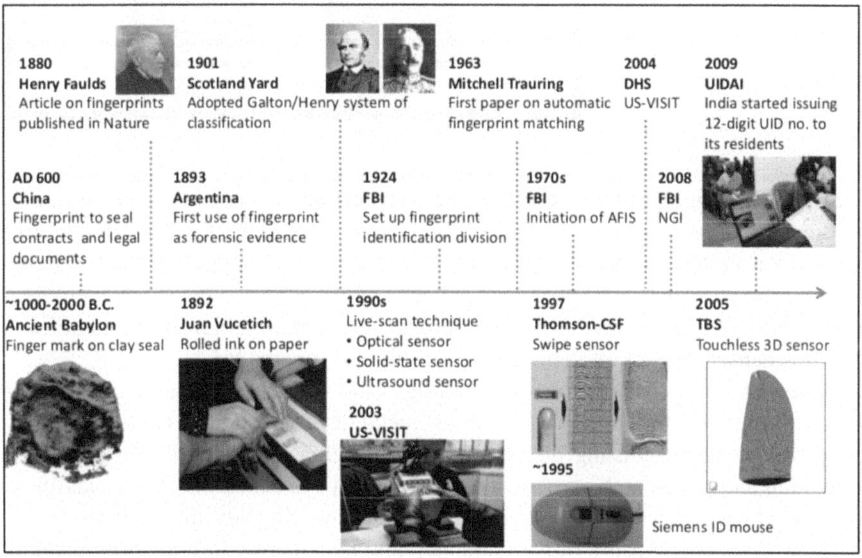

Fig. 2.2 Fingerprint recognition milestones [3]

joining pieces of images captured when a user swipes their finger as moved in different positions.

Figure 2.2 shows how the timeline of sensors compares with the actual implementation of fingerprint technology in contrast to the milestones of fingerprint recognition [3], from the introduction of Sir Henry's classification in the 1800s to Scotland Yard adopting this in 1901 and FBI initiating the Automated Fingerprint Identification System (AFIS) in the 1970s. This system uses a process of identifying one or many unknown fingerprints against a database. Two decades later, the swipe sensors were introduced with touchless sensors being invented in 2005 followed by the creation of Unique Identification Authority of India project in 2009. Its prime purpose is to provide a unique identity (Aadhaar) to all Indian residents.

Hygiene is always the most talked about subject when fingerprint technology is first discussed with new users. The introduction of touchless swipe sensor in 2010 was ground-breaking discovery within the field of biometrics where fingerprint technology is concerned. The touchless fingerprint technology uses digital camera

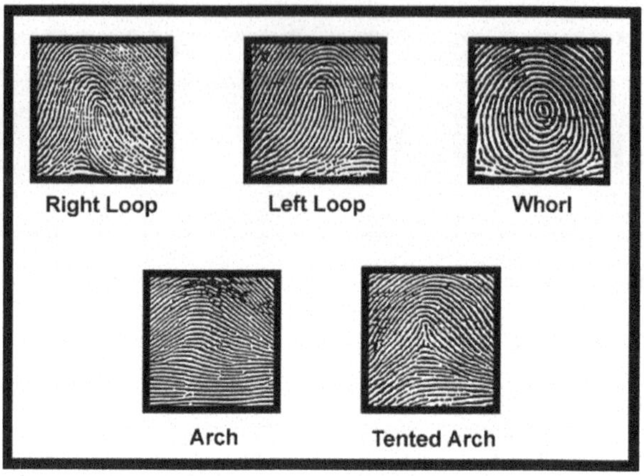

Fig. 2.3 Types of fingerprint patterns [6]

to acquire the fingerprint image and does not require any contact between the skin of the finger and the sensing area. One of the main modules of a touchless fingerprint recognition system is the preprocessing of a fingerprint image which can remove certain problems such as (i) low contrast between the ridge and the valley pattern on a fingerprint image, (ii) non-uniform lighting and (iii) motion blurriness and defocus which is caused by lack of depth of field of digital cameras [4].

A fingerprint is a pattern consisting of ridges and valleys on the surface of a finger formed during the third to fourth months of foetal development [5]. Each pattern can be identified by recognising the type of class it belongs to, whorl, left loop, right loop, arch and tented arch, as shown in Fig. 2.3. This is called the Henry classification system as introduced by a British policeman, Sir Edward Richard Henry [7]. The arch class consists of the plain arch and tented arch. In fingerprint patterns within the arch class, the ridges enter on one side, form an arch in the centre and exit on the other side [8, 9]. The loop class consists of left loop and right loop. Ridges enter, form a curve and exit on the same side [8, 9]. Whorls consist of circles, more than one loop or a mixture of pattern types forming plain whorl, double loop whorl, central pocket whorl and accidental whorl [8, 9]. Research shows that 65% of the population have loops, 30% have whorls and 5% have arches [5].

1.1 Fingerprint Enrolment and Related Technologies

Fingerprint enrolment is a process in which users present their finger to a fingerprint sensor; the system then extracts key features from the fingerprint using image processing techniques and converts these into a template. The final step involves storing the template into a database or a token (such as ID card or passport) which would then allow fingerprint authentication to be carried out for a user to be identified or verified.

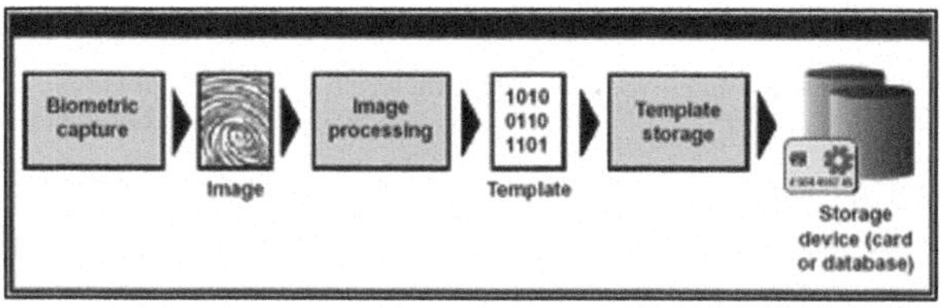

Fig. 2.4 Enrolment process of a fingerprint [10]

Figure 2.4 shows how the fingerprint enrolment process works. A fingerprint is captured with the use of finger print sensor and saved as a digital image. The quality of image captured is a critical criterion for effective recognition such as moisture of hand, placement of finger on the sensor including its position and pressure of thumbprint. Once the fingerprint is captured, extraction of key features of the image is a crucial stage in both enrolment and authentication stages. Feature extraction refers to a process of using image processing techniques to extract key characteristics of the fingerprint as a pattern.

The extracted features are concatenated to form a feature vector that is saved as a template for future identification or verification of an individual. This process consisting of a biometric capture, feature extraction and template formation constitutes the enrolment process.

Advanced fingerprint readers have the ability to be used in live environments such as in the airports with the facility to be powered over Ethernet. They also have the ability to store data on the device. Embedded intelligence on these readers enables deciding on the quality of image captured and reject if poor.

Enrolment procedures are evaluated through performance such as *failure to enrol rate (FER)* which is defined as the rate at which people are unable to enrol. Another metric is the *failure to capture (acquire) rate (FCR)* which is defined as the rate at which biometric information cannot be obtained during the use of a biometric system, even though the individual was previously able to enrol [11].

1.2 Feature Extraction with Minutiae

Minutiae are often referred to as points of interest in a fingerprint, such as bifurcation and ridge endings [5]. Other points of interest include island, crossing point, delta point and core point [12]. The various ways in which ridges can be discontinuous form characteristics of fingerprints called minutiae: Table 2.2 shows the definition of characteristics of different types of minutiae.

A representation of the key features in an actual fingerprint image is shown in Fig. 2.5. Most commercial fingerprint recognition systems utilise only ridge endings and bifurcations as these two types of minutiae alone are able to provide sufficient uniqueness. In order to locate the features accurately, the feature extraction process includes filtering and optimisation of images. Thin ridges present in a fingerprint image are often filtered to the width of a single pixel by fingerprint technologies

Table 2.2 Characteristics of minutiae

Type of minutia	Characteristics that define it
Termination/ridge endings	Ridge comes to an end
Bifurcation	A ridge divided into two ridges
Island/short ridges or independent ridge	A ridge that commences, travels a short distance and then ends [13]
Crossover or bridge	A short ridge that runs between two parallel ridges [13]
Core	Topmost point on the innermost upwardly curving ridgeline (approximately centre of the fingerprint) [12]
Delta	Triangular region located near a loop [14]
Ponds/lake	Empty spaces between two ridges [15]

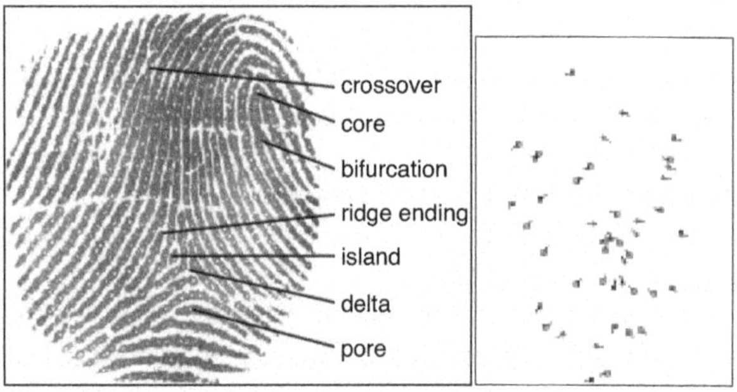

Fig. 2.5 (**a**) Types of minutiae [16]. (**b**) Location of minutiae

[17]. This is important to the performance of a biometric system as the quality of feature extraction directly affects a system's ability to generate templates. A good quality image has around 40–100 minutiae [18].

Minutiae extraction methods can be split into two categories: those that work on binarised fingerprint images and those that work directly on greyscale fingerprint images [18]. Binarised fingerprint images are those that have been converted into binary data via a binarisation process which transforms an enhanced grey-level image into a binary image [19]. A binary image consists of values 1s and 0s where 1 is assigned to all the ridge pixels and 0 is assigned to nonridged pixels [20]. The ones that work directly on greyscale images do not use the binarisation and thinning process [19].

Minutiae extraction on binarised fingerprint images can be further classified into unthinned and thinned binarised images. Thinned binarised images are obtained from a skeletonisation process that reduces the width of binarised ridgeline to one pixel [21]. This is based on the number of neighbouring pixels in the way the minutiae points are detected by location of end points on the thinned ridge skeleton. Bifurcation points are selected if they have more than two neighbouring pixels, and end points are selected if they have one neighbouring pixel [18]. Loss of information, being time-consuming and observation of spurious minutiae due to breaks, holes and undesired spikes are some of the problems identified with the binarisation and thinning process [18, 22].

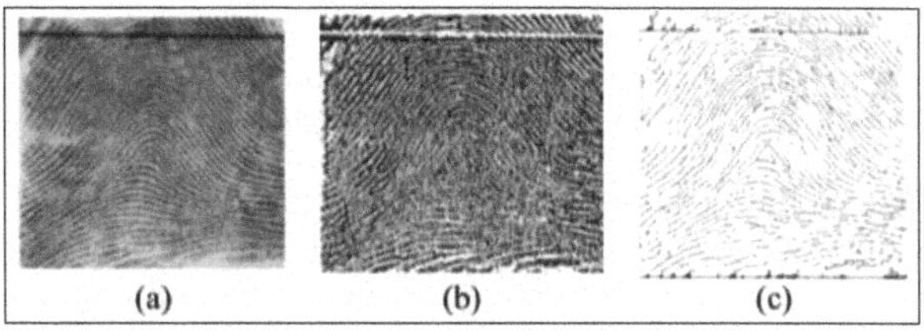

Fig. 2.6 Binarised and greyscale images. (**a**) Fingerprint image, (**b**) binarised image, (**c**) thinned image [23]

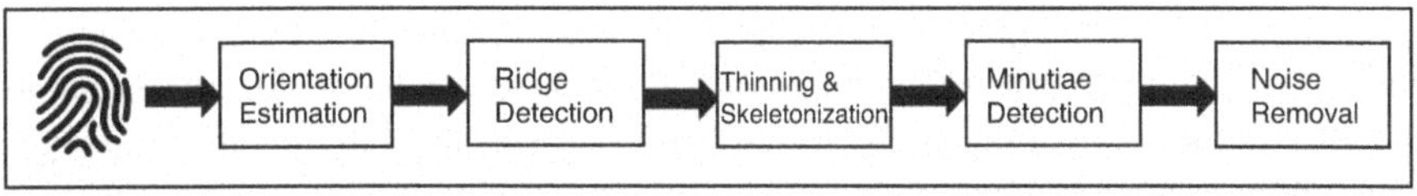

Fig. 2.7 Fingerprint feature extraction algorithm

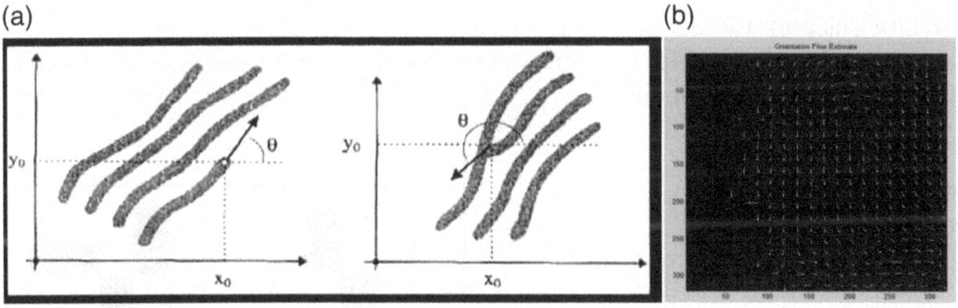

Fig. 2.8 (**a**) Minutiae orientation of ridge and bifurcation [24]. (**b**) Orientation flow estimate [21]

With greyscale fingerprint images, the sinusoidal shapes formed by the combination of ridges and valleys in a local neighbourhood have a well-defined frequency and orientation [20]. Figure 2.6 shows the result of binarisation and thinning process on a greyscale image.

A typical feature extraction algorithm involves five operations as shown in Fig. 2.7 [7]: (i) orientation estimation, to estimate local ridge directions; (ii) ridge detection, to separate ridges from the valleys by using the orientation estimation resulting in a binary image; (iii) thinning algorithm/skeletonisation, to reduce the ridge width to one pixel; (iv) minutiae detection, to identify ridge bifurcations, those with three ridge pixel neighbours and ridge endings and those with one ridge pixel neighbour; and (v) post-processing, to remove spurious minutiae.

Orientation Estimation: The minutiae pattern is characterised by the direction of the ridge ending and bifurcation points as shown in Fig. 2.8a. The bifurcation angle is the angle between the horizontal and the direction of the valley ending between the bifurcations at the minutiae location [25]. An orientation map provides ridge flow in an image that is partitioned into nonoverlapping blocks as shown in Fig. 2.8b. This map allows to identify the dominant direction of ridge lines and remove other spurious ridges.

Ridge Detection: The next stage is that of deciding if the ridge is a termination point or a bifurcation typically by examining a local neighbourhood (window size 3 × 3). In [26], the pixels are examined for their connectivity as shown in Fig. 2.9.

Thinning: Thinning or skeletonisation is a process of reducing thick fingerprints into one-pixel-wide ridge lines. At this stage, each minutia is described by a feature vector containing its (*x*,*y*) coordinates, type (ridge or termination) and orientation.

Minutiae Detection: Minutiae detection identifies ridges with three ridge pixel neighbours and termination as one ridge edge pixel. Typically a 3 × 3 window is convolved over the image. Often minutiae from thinned images do not correspond to minutiae in the original image. Spurious results are produced in the thinned images which need to be post-processed.

Post-processing: Any other unwanted patterns of minutiae are removed from the image. These include spikes, breaks and holes. Morphological operators may be used for this purpose.

Information recorded for each minutia include the coordinates, orientation of the ridge segment and type of minutiae of each minutiae point. Figure 2.10 shows a flow

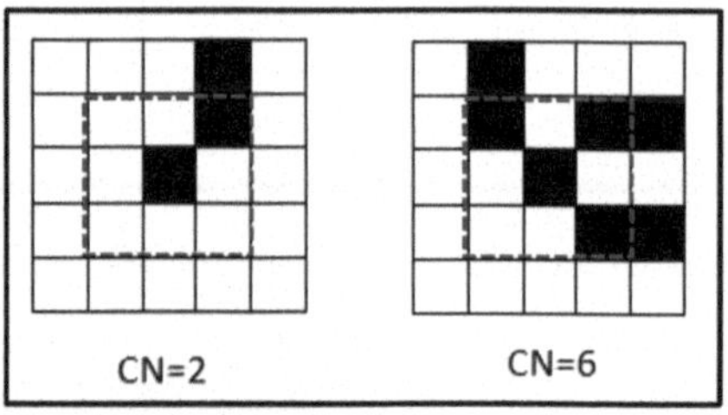

Fig. 2.9 CN value for ridge ending and bifurcation

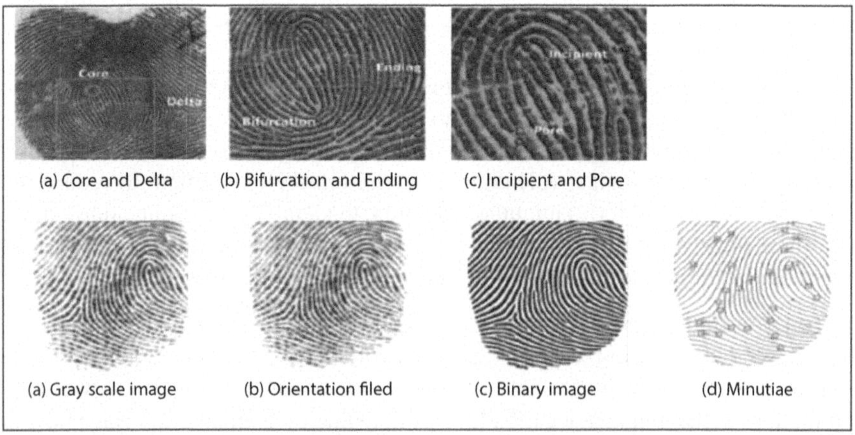

Fig. 2.10 Key feature extraction. (**a**) Feature levels in a fingerprint, (**b**) and (**c**) are magnified versions of the fingerprint regions indicated by boxes in (**a**). (**d**) represent the steps for a typical minutiae extraction [27]

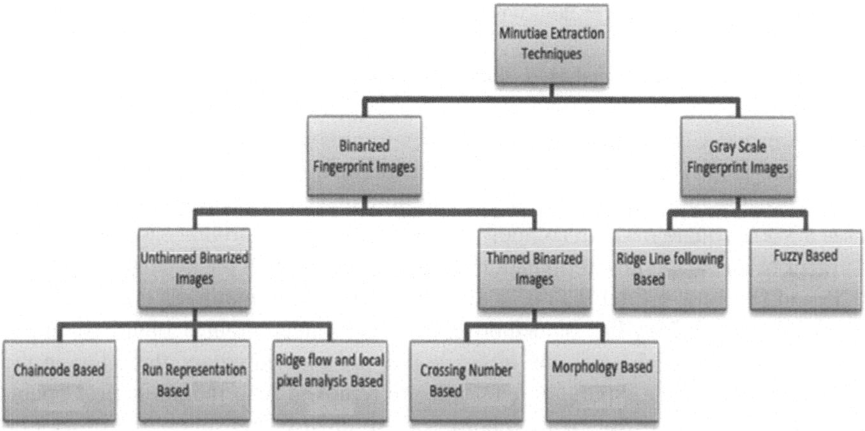

Fig. 2.11 Classification of key feature extraction techniques [18]

chart of a typical minutiae extraction algorithm. Following extraction of minutiae, a stored template could typically contain the minutiae position (*x*,*y*), minutiae direction (angle) and minutiae type (ridge ending or bifurcation) [28]. During the enrolment and authentication stages, the template is stored in the database to then being used in the matching process as a reference template or database template during enrolment or query template in the authentication process during fingerprint matching.

1.3 Feature Extraction Categorisation

The feature extraction algorithms differ variedly depending on whether binary or greyscale images are used and whether skeletonisation is applied or not. This categorisation is shown in Fig. 2.11. The binarised fingerprint images are split into thinned binarised images where each ridge is converted to one pixel width via the skeletonisation process. The end points and bifurcation points located on the thinned ridge skeleton then detect the minutiae points based on the neighbouring pixels. If they have more than two neighbours, the bifurcation points are selected, and if they have a single neighbouring pixel, the end points are selected.

As methods based on thinning are sensitive to noise, techniques such as chaincode processing, run-based methods and ridge flow and local pixel analysis-based methods which fall under the unthinned binarised images can be used [18]. The chaincode processing method uses object contours by scanning image from top to bottom and right to left to identify the transitions from white background to black foreground. This follows a representation of an array of contour elements which are traced counterclockwise, and each element denotes a pixel on the contour. A minutiae ending is located when the ridge makes a left turn by tracing a ridgeline along the boundary counterclockwise. If it makes a right turn, a bifurcation minutia is identified [18, 29].

The run representation-based method results in fast extraction of fingerprint minutiae that are based on the horizontal and vertical run-length encoding from binary images. Characteristic images are then found by checking the runs adjacency of the runs [29]. This minutiae extraction technique does not require a computationally expensive thinning process. The ridge flow and local pixel analysis technique uses a 3×3 square mask to compute the average of pixels around each pixel in the fingerprint image. A ridge termination minutia is identified if the average is less than 0.25 and for those greater than 0.75 determine a bifurcation minutia [18].

In the morphology-based method which falls under thinned binarised image feature extraction technique, the image is preprocessed to reduce the effort in the post-processing stage. Spurs, bridges, etc. are removed with morphological operators followed by the use of morphological hit-or-miss transform to extract true minutiae. Morphological operators are shape operators whose composition allows the natural manipulation of shapes for the identification and the composition of objects and object features [18].

In methods such as ridgeline following based and fuzzy-based techniques, minutia is extracted directly from greyscale fingerprint images without using binarisation and thinning processes [29]. The ridgeline following technique uses local orientation field by following ridge flow lines, and the fuzzy-based technique uses distinct levels of grey pixels: dark consisting of darker pixels of ridges and bright consisting of lighter pixels of valleys and furrows. These two levels are modelled by using fuzzy logic by applying appropriate fuzzy rules to extract minutiae accurately [18, 29].

1.4 Crossing Number

With fingerprint recognition systems, the concept of a crossing number (CN) as a feature extraction process is widely used. The CN method uses a skeleton image where the ridge pattern is eight-connected denoted by N8. This refers to an investigation of the eight neighbouring pixels of the central pixel, *P*. The minutiae are extracted by scanning the local neighbourhood of each ridge pixel in the image using a 3×3 window as shown in Fig. 2.12. It then assigns a CN value based on the type

Fig. 2.12 A 3×3 pixel window [12]

P4	**P3**	**P2**
P5	**P**	**P1**
P6	**P7**	**P8**

Table 2.3 Crossing number point system

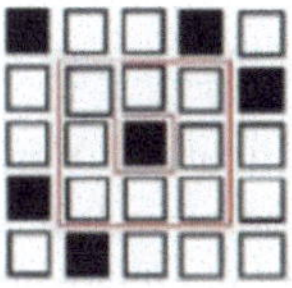	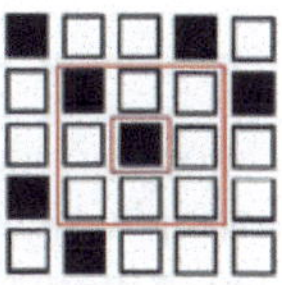	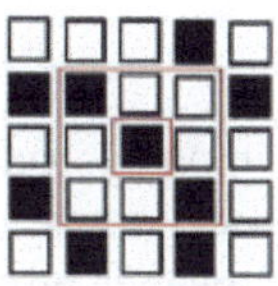	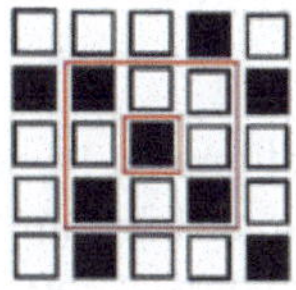	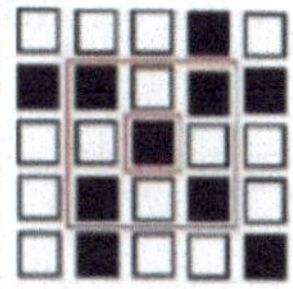
(a) Non-minutiae N8 = 0, CN = 0	(b) Ridge ending N8 = 1, CN = 1	(c) Continuing ridge N8 = 2, CN = 2	(d) Bifurcation point N8 = 3, CN = 3	(e) Complex minutiae N8 > 3, CN = 4

Table 2.4 Pseudocode for determining direction from CN

% Direction for ridge ending point
if CN=1 then
if P1=1 then direction =W
if P3=1 then direction =S
if P7=1 then direction =N
if P5=1 then direction =E
if P4=1 then direction =SE
if P2=1 then direction =SW
if P6=1 then direction =NE
if P8=1 then direction =NW
end if
	% Direction for ridge bifurcation point
	if CN=3 then
	if P1 and P3 and P7=1 then direction =W
	if P1 and P3 and P5=1 then direction =S
	if P1 and P7 and P5=1 then direction =N
	if P3 and P5 and P7 =1 then direction =E
	if P4 and P3 and P5=1 then direction =SE
	if P3 and P2 and P1=1 then direction =SW
	if P3 and P5 and P6=1 then direction =NE
	if P4 and P8 and P5=1 then direction =NW
	end if

of central pixel and its neighbouring pixels as shown in Table 2.3. The grid in the last column of the table shows a sub-image of size 5 × 5. In the first row, the central pixel highlighted in black has no neighbours in its 3 × 3 connectivity. Hence, the CN value returned is 0. In the remaining rows, there is an increasing value of CN due to the number of connected neighbours.

The CN values for a ridge and bifurcation are calculated using a formal definition for CN as follows:

$$CN = 0.5 \sum_{i=1}^{8} \mid P_i - P_{i+1} \mid , \quad P_9 = P_1 \ [12] \tag{2.1}$$

This is translated as half the sum of the differences between pairs of adjacent pixels in the eight neighbouring pixels. P_i is the pixel value which is in the neighbourhood of the central pixel, P where $P_i = (0 \text{ or } 1)$ and P9 = P1 [4]. This method involves recording of locations of the minutiae points and their considered directions: N, S, W, E, NE, NW, SE and SW. The most important minutiae points are the ridge endings and bifurcation; hence the angle and direction of these minutiae are very important. The directions can be determined for the conditions of CN = 1 and CN = 3 [10] using the following pseudocode in Table 2.4.

2 Pattern Matching Concepts for Fingerprint Identification and Verification

Biometric authentication can be split into two: verification and identification. Verification is a one-to-one matching process where the template may be stored in a token such as ID card or passport. The token is presented to the system and checked against relevant physiological characteristic presented at time of authentication. The other form is identification which carries out one-to-many match within a database of stored templates. Verification is much quicker than identification.

Fingerprint verification can be done by using a one-to-one matching process. This is a verification process whereby decision-making considers if the matching score exceeds a threshold. If so, it 'Accepts' the fingerprint. Otherwise it 'Rejects' the fingerprint. The matching is done via a token. In this method, the fingerprint template is stored on a token such as an ID card. A user is required to present their token which has the template stored on it, followed by presenting a finger to the fingerprint reader to verify if the two templates match. When a user presents their finger to a reader, the key features are extracted to then perform a verification check against the template stored on the token.

Fingerprint identification can be done using a one-to-many process. This is an identification process whereby decision-making considers two possibilities:

Close-set identification which returns the identity associated with the highest matching score

Open-set identification which returns the identity associated with this score or declares that the test biometric data does not belong to any of the registered individuals

The Automated Fingerprint Identification System (AFIS) has been developed for human identification in which the matching module computes a matching score between two fingerprints. A high score is assigned to fingerprints from the same finger and a low score to those that do not match. To claim that two fingerprints are from the same finger, the following factors are evaluated [1]:

- Global pattern configuration agreement – two fingerprints must be of the same type.
- Qualitative concordance – corresponding minutiae must be identical.
- Quantitative factor – at least a certain number of minutiae details must be found.
- Corresponding minutiae details – must be identically interrelated.

There are various fingerprint matching techniques discussed by researchers and experts, but the most commonly used in fingerprint recognition systems are [1, 30]:

- *Minutia-based:* matches local features such as bifurcations and ridge endings based on location and direction of each point as shown in Fig. 2.13a.
- *Correlation-based:* two fingerprint images are superimposed, and the correlation between corresponding pixels is computed for different alignments such as

Fig. 2.13 Fingerprint matching techniques. (**a**) Minutiae-based matching [31]. (**b**) Correlation-based matching [32]. (**c**) Pattern-based matching [33]

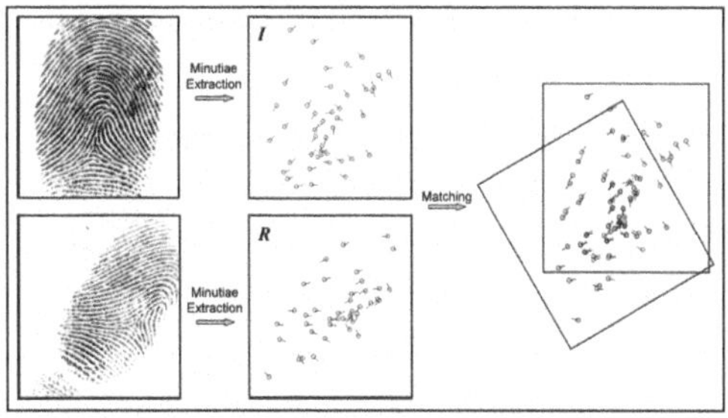

(**a**) Minutiae-based matching [31]

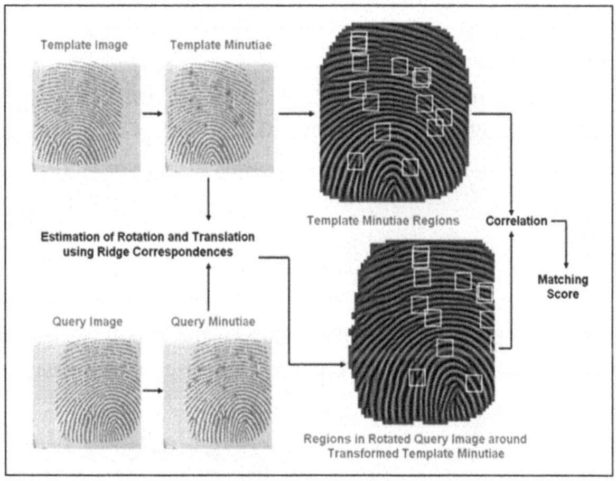

(**b**) Correlation-based matching [32]

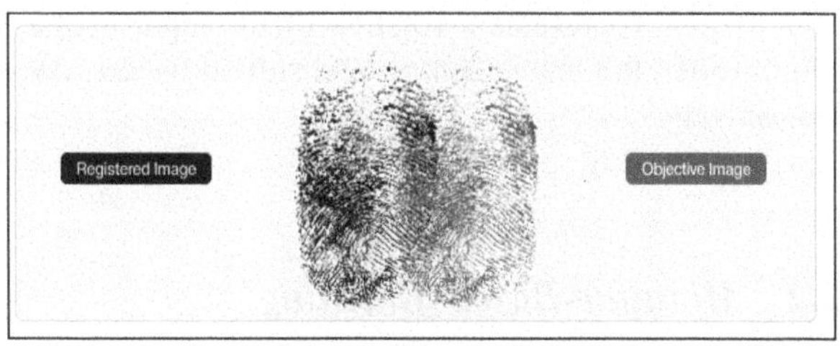

(**c**) Pattern-based matching [33]

various displacements and rotations as shown in Fig. 2.13b. This approach requires less computation but is less robust against image distortions [30].

- *Pattern-based or ridge feature-based:* compares two images for similarity such as shape, texture information, local orientation, ridge shape and frequency as shown in Fig. 2.13c.

2.1 Correlation-Based Matching

The correlation-based matching technique uses a sum of squares approach to determine the match between the test input and the templates in the database. This is given by:

$$\mathrm{SSD}(T,I) = |T - I|^2 = (T - I)^{\mathrm{T}}(T - I) = |T|^2 + |I|^2 - 2T^{\mathrm{T}}I \text{ [1]} \quad (2.2)$$

where SSD represents the sum of squared differences between the template and input images and is an indicator of the diversity between these two images, T represents the template image, I represents the input fingerprint and the superscript T represents the transpose of a vector.

If the terms $|T|^2$ and $|I|^2$ are constant, the diversity between the two images is minimised when the cross-correlation CC is maximised:

$$\mathrm{CC}(T,I) = T^{\mathrm{T}}I \text{ [1]} \quad (2.3)$$

where $\mathrm{CC}(T,I)$ is a measure of image similarity.

The similarity cannot be simply computed by simply superimposing T and I and applying the above equation due to the displacement and rotation. The similarity between two fingerprint images T and I can be measured as:

$$S(T,I) = \max_{(\Delta x,\ \Delta y,\theta)} \mathrm{CC}\left(T, I^{(\Delta x,\ \Delta y,\theta)}\right) \text{ [1]} \quad (2.4)$$

$I^{(\Delta x,\ \Delta y,\ \theta)}$ represents a rotation of the input image I by an angle θ around the origin (often the image centre) and shifted by Δx, Δy pixels in directions x and y, respectively.

2.2 Minutiae-Based Matching

In this algorithm, each minutia may be described by several attributes including its location in the fingerprint image, orientation, type (i.e. ridge termination or ridge bifurcation), a weight based on the quality of the fingerprint image in the neighbourhood of the minutia, etc. Let T and I represent the template and input fingerprints respectively, each represented as a feature vector as given by Eqs. (2.5) and (2.6) and whose size is determined by the number of minutiae m and n, respectively.

$$T = (m_1, \ldots, m_m), \quad m_i = (x_i, y_i, \theta_i), \quad i = 1 \ldots m \text{ [1]} \quad (2.5)$$

$$I = \left(m_{1,} \ldots, m'_{n}\right), m'_{j} = \left(x'_{j}, y'_{j}, \theta'_{j}\right),\ j = 1 \ldots n\ [1] \tag{2.6}$$

Minutia m'_j in the input image and m_i in the template image are considered matching if the spatial distance (sd) between them is smaller than a given tolerance r_0 and the *direction difference* (dd) between them is smaller than an angular tolerance θ_0:

$$\mathrm{sd}\left(m'_j,\ m_i\right) = \sqrt{\left(x'_j - x_i\right)^2 + \left(y'_j - y_i\right)^2} \leq r_0\ [1] \tag{2.7}$$

$$\mathrm{dd}\left(m'_j,\ m_i\right) = \min\left(\left|\theta'_j - \theta_i\right|, 360^\circ - \left|\theta'_j - \theta_i\right|\right) \leq \theta_0\ [1] \tag{2.8}$$

Equation (2.8) takes the minimum of $|\theta'_j - \theta_i|$ and $360° - |\theta'_j - \theta_i|$ because of the circularity of angles (e.g. the difference between angles of 2° and 358° is only 4°).

2.3 *Pattern Matching*

With the pattern matching technique, each fingerprint is represented by a feature vector of size $80 \times 8 = 640$ and is called the *finger code*. The formula for pattern matching is given by:

$$V_{ij} = \frac{1}{n_i}\left(\sum_{C_i}\left|g\left(x, y : \theta_j, \frac{1}{10}\right) - \overline{g_i}\right|\right)\ [1] \tag{2.9}$$

where:

- V_{ij} of the vector ($i = 1\ldots80$ is the cell index; $j = 1\ldots8$ is the filter index) denotes the energy revealed by the filter j in cell i and is computed as the average absolute deviation from the mean of the responses of the filter j over all the pixels of the cell i as above.
- C_i is the ith cell of the tessellation (tiling of fingerprint area of interest with respect to the core point).
- n_i is the number of pixels in C_i.
- The local texture information in each sector is decomposed into separate channels by using a Gabor filter bank (fingerprint enhancement method), so $g(.)$ is defined by the Gabor filter equation.
- g_i is the mean value of g over the cell C_i.

Research shows that due to the large variability in different impressions of the same finger, matching fingerprint images is an extremely difficult problem [1]. Some of these variations are due to:

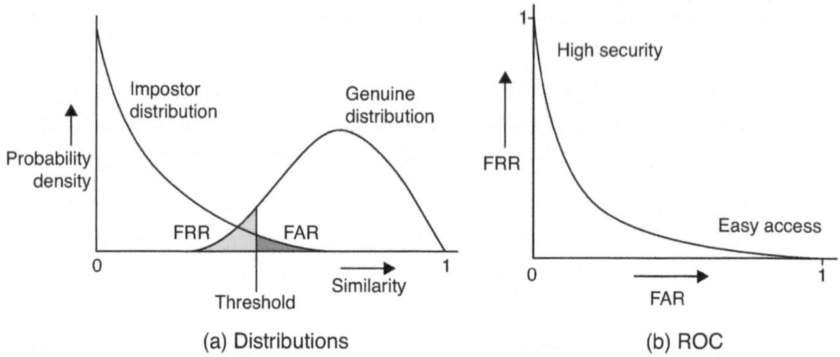

Fig. 2.14 Match and non-match distributions and receiver operating curve (ROC) [34]

- *Displacement*: The same finger may be placed in different locations on the fingerprint capture device sensor.
- *Rotation*: As per displacement, the same finger may be placed at different angles on the sensor.
- *Pressure and skin condition*: Finger pressure, dryness of the skin, skin disease, grease, dirt, sweat and humidity in the air can all result in a non-uniform contact on the sensor.
- *Noise*: Introduced by the fingerprint sensing system. For example, residues left over from the previous fingerprint capture.

2.4 Fingerprint Performance Metrics

With fingerprint biometric authentication, a similarity measure is estimated between an input and the template. The system accepts or rejects the claimed identity based on whether the similarity score is above a predetermined threshold. Such a threshold is determined through experimentation of the distributions of similarity measures for the specific database. In general, the distribution of similarity measures of genuine and imposter attempts does not have a crisp boundary of thresholds. An overlap exists between these two categories of attempts as shown in Fig. 2.14. The matching performance is thus measured by two error measures [27]:

- False accept rate (FAR) or false match rate (FMR) – the likelihood of a fingerprint system incorrectly matching an input pattern to a non-matching template in the database. This is shown by the dark grey-shaded area in the figure.
- False reject rate (FRR) or false non-match rate (FNMR) – the likelihood of the system failing to detect a match between the input pattern and a matching template in the database as shown by the light grey-shaded area in Fig. 2.14.

It is reported [34] that advanced systems have FAR = 10^{-4} and FRR = 10^{-2}. More realistic systems have error rates of have FAR = 10^{-2} and FRR = 5×10^{-2}. A receiver operating curve (ROC) provides insight into the system performance by plotting the FAR against FRR. This enables tuning of the threshold to meet the requirements of the application. Systems requiring high security such as airports can set up a high threshold where FAR is very low, whilst others as in a classroom access may have easy access and therefore lower thresholds with low FRR. Identification systems often use an equal error rate measure as a performance metric, and this defines the specific point on the ROC where FAR is equal to FRR.

2.5 Fingerprint Databases

Several databases are available for training and testing a fingerprint system largely due to the research community generating its own databases as well as international organisations that run competitions in the field. Some databases are listed below:

1. *NIST Fingerprint Database 4* [35]: This database consists of 2000 8-bit greyscale fingerprint image pairs. Each image is 512 × 512 pixels with resolution 19.7 pixels/mm and 32 rows of white space. There are five classes of images, namely, arch, left loop, right loop, tented arch and whorl.
2. *CASIA-FingerprintV5* [36]: There are two datasets, one with the subject's co-operation and the other without the co-operation. Dataset 1 has 17 users with fingerprints from middle fingers of both hands and a total of 4 samples. Three different sensors are used. The dataset dimensionality is 68 fingers × 4 samples × 3 sensors = 816 image samples. Similarly, the non-co-operative dataset dimensionality is 64 fingers × 4 samples × 3 sensors = 768 images.
3. *FingerDOS* [37]: This is a collection of fingerprint images acquired using an optical sensor. The database consists of 60 subjects with a total of 3600 fingerprint images acquired from thumb, index finger and middle finger of left and right hands. Each subject has 10 samples and saved as 8-bit greyscale images.
4. *Spoofed Fingerphoto Database* [38, 39]: This database consists of spoofed fingerprints where co-operative fingerprints were captured with two different sensors and the spoofed fingerprints were acquired using iPad, Laptop and printout. It consists of 64 subjects × 2 fingers = 128 classes. Further the database varies in illumination and backgrounds two different illumination conditions.

3 Practical Consideration of Fingerprint Technology During Enrolment and Matching

Practical considerations in deploying a fingerprint technology are explained with a use case scenario. This section, in particular, will be a useful practitioner's guide to enable such deployment.

Whilst considering use of fingerprint technology, protection of data, obtaining consent from end-user and selecting a well-tested system should be given importance when implementing such technology. Any fingerprints obtained should be stored as a template consisting of binary data only, and the actual fingerprint images should never be stored. In addition, as part of the de-identification process, all data should be stored on a secure database and encrypted to minimise misuse of data and risks of hacking.

The General Data Protection Regulation [40] implemented in May 2018 has recognised biometric information as sensitive information which was excluded in the Data Protection Act. Hence, getting consent from the end-user should also be a crucial part of the process. For added data security, the fingerprint database should not consist of any other personal information such as date of birth and should be independent of other systems. Certain systems may have specific licence restrictions based on number of users stored, but there are good fingerprint systems that do not have these restrictions.

In a scenario where fingerprint system is being used to speed up processes, i.e. student attendance, or for exam process, network speed should be tested to ensure that the system can still function in the event of loss of network. Most biometric terminals do have the ability to store secure templates on the device which would help speed up the matching process and enhance system accuracy especially when using one-to-many identification. In the event where a template is stored on a token such as an ID card, a one-to-one match can improve system performance as the matching process is carried out locally rather than via the network. Storing fingerprint templates on a token can pose a few challenges such as capacity on the token or no means of identification when token is lost. The highest risk would be misuse of the token if found by an unauthorised user.

Along with data storage and other factors, using a fool-proof system is highly recommended, and one of the ways in which this can be achieved is by enrolling at least two fingers and use more than one as means of identification. Enrolment of more than one fingerprint leads to added benefits in case of cuts, bruises or other obstructions to one of the fingers which would then allow continuity of service by permitting the user to use the additional enrolled finger. Taking these factors into account can help implement a robust fingerprint recognition system in a fast-paced environment yet providing secure means of identification.

4 Fingerprint Use Case in an Education Environment

In this section, we consider an application that is proposed as part of the first author's research work. The main aim of this work is to provide fingerprint identification in a higher educational context to avoid any hassle during examination periods. Details are as follows.

4.1 Requirement

This project stemmed up as a result of students often forgetting to bring along their student ID cards to an examination room. This leads to change in anxiety levels of students as they will then need to approach the examination office or ID Office who will issue a temporary receipt/ID to sit in the exam room after verifying their identity through other means. To circumvent this situation, this work provides a solution through the use of fingerprint recognition system.

4.2 Planning and Ethics Approval

The student population is distributed across different educational disciplines who have strong opinions on sharing their personal identities. Similarly, the deans of schools and other professional services including the Students Union and Examination Office needed to be approached for their approval. Hence, this planning stage was crucial to the project. Several brain storming sessions were conducted in arriving at a solution for appropriate usage of fingerprint system for the student community. An advisory group has been formulated which continues to monitor the progress of this work and advice at appropriate times.

An initial online survey was carried out with both staff and students which basically explained how the data would be captured, stored, saved and its life time explained. The survey also invited concerns that were further addressed on an individual basis. The survey revealed over 70% of acceptance rate in favour of use of this technology in a university environment for processes such as exams, attendance monitoring and door entry system.

Cross-disciplinary research on ethics also exists within the university. Hence, this work has been subject to such research query. This research work benefits from cross-disciplinary research as it only strengthens the case.

4.3 Fingerprint Sensors

Several commercial fingerprint readers were considered. With the help of one of the vendors, a trail run is planned to evaluate the suitability in a sensitive environment.

The chosen FBI PIV IQS certified fingerprint sensor features all-in-one integration: sensor + embedded feature extractor and matcher + internal user database with the capacity of up to 50,000. It is the first ever multimodal device capable of capturing and processing finger vein and fingerprint biometric data at the same time which is also resistant to spoofing a false acceptance rate of 10^{-4}, and false reject rate is ten times lower than with the best of the two modalities.

It boasts an average speed of 1 s for one-to-one authentication and 1.5 s for a one-to-many authentication (1:5000). Fingerprint images can be stored at 500 dpi, 256 greyscale, 400 × 400 pixels: RAW or WSQ compressed. The chosen fingerprint sensor can also generate a multimodal template (fingerprint and finger vein) or a fingerprint template. During initial experiments whilst carrying out amateur tests, it was quickly identified that the device generated small size templates (at 2 KB) and raw images (at 157 KB). An XML document consisting of user records of up to 23 users generated a file of up to 40 KB. This demonstrates that in this scenario, the maximum size of a database consisting of fingerprint data would require approximately 53 MB space to store such information.

4.4 Experiment: Subject with Knowledge of Biometrics

A cohort in the final year of the study was chosen for the trial run.

A module delivered by the second author was selected to carry out initial tests using the fingerprint/finger vein reader to enrol and authenticate users. Prior to capturing the data, students were given a brief presentation explaining the purpose of the project, benefits it will bring to them, details of what the enrolment process and authentication process would entail and the requirement to sign consent forms. Each student who agreed to take part in this pilot was given a 'Participant Information Sheet' which consisted of project details and contact details of the first author should they need to send withdrawal request at any point.

Once the student signed the consent form, the enrolment process commenced. In a class attended by 23 students, 13 students agreed to take part in the pilot and provided their fingerprints. It took 16 min to add record, enrol and test identify 13 students. This averages to 1.23 min per student taking 3 s to enrol and 2 s to identify the registered fingerprint. Along with time taken, identification scores were also noted once the student had been registered on the system and asked to present their finger again to test the identification process. During enrolment, it took one student four attempts to enrol their fingerprint, but the rest of the students were registered on the system within one attempt.

Each enrolled student was given a feedback form which consisted of 13 closed questions where they were required to comment on various aspects such as speed, accuracy, comfort, privacy and general statements which related to the initial survey carried out. The purpose of this was to compare results of survey without users having used or tested the system against feedback received once they had used/tested the system.

The final stage of this test required the enrolled students to present their fingers on the scanners during an in-class phase test 2 weeks later. During this process, the identification matching scores were noted to then compare with the ones noted down when initially tested during enrolment stage.

4.5 Inferences Drawn

Due to the quick identification and acceptance rate of over 70% in favour of this technology, a larger cohort of at least 200 students will be used for a pilot to test the system against speed and accuracy.

This will involve capturing fingerprint data in a working student registration environment. The total time it takes to register a student (along with the pilot fingerprint data collection process) will be measured and compared against standard registration times. Students will also be asked to provide feedback on the system, like the ones from the final year Biometrics students. They will also be asked to provide consent by completing a consent form.

Once the students have been enrolled on the system, they will be asked to present their finger on a fingerprint wall reader or tablet once when they enter the exam room and once when they exit the room. The system will note the date, time and location during the identification process. As this is a pilot, they will be asked to sign the manual register as per the current process to eliminate any gaps in exam registration.

If the pilot is successful, the fingerprint technology will have the scope to be rolled out for the rest of the exam sessions, attendance system and even door entry system where feasible. To encourage using the technology to its maximum potential, use of fingerprint system is also being considered for awards ceremonies where students can pick up their tickets using fingerprint rather than other means of ID.

This work has been supported by the University of Hertfordshire Proof of Concept Award and the Diamond Fund within the project, 'A Biometric Approach to Prevent False Use of IDs'.

5 Conclusion

In this chapter, the authors have provided an insight into the overall picture of how fingerprint technology works whilst narrating its evolution from 1800 to 2010 and how it achieved the medium ranking against desirable properties compared to other biometric modalities.

The classification system and various minutiae extraction methods have been demonstrated in Sect. 1 of this chapter. In summary, minutiae extraction methods are split into two categories: based on binarised fingerprint images consisting of values 1s and 0s and the ones that work directly on greyscale fingerprint images that do not use the binarisation and thinning process. As the crossing number feature extraction process is widely used, this has been widely discussed by providing pseudocode and formulas used to calculate values for ridge and bifurcation.

As discussed in Sect. 2, fingerprint matching and verification process form an important part of the pattern matching concepts. Fingerprint verification is carried out by one-to-one matching, whereas identification uses one-to-many matching. The identification process considers two possibilities when making decision: close-set and open-set. Various factors and matching techniques such as minutiae-based, correlation-based and pattern-based have been demonstrated along with a detailed look at the matching algorithms.

The final section gives a DOs and DON'Ts guidance to assist with decision-making and implementation process when selecting a fingerprint recognition system when deploying such system as fingerprint technology advances rapidly.

Glossary

Circumvention An unauthorised user gaining access illegitimately to the system and data, i.e. by spoofing method.

Collectability How easy it is to acquire fingerprints. In some environments where users would pose issues, i.e. washing hands or worn fingers, more expensive means might be used to acquire a useable fingerprint image.

Filtering Removing unnecessary noise in an image during preprocessing stage to help enhance quality of an image.

Threshold Biometric match threshold refers to a point at which it becomes reasonably certain that a biometric sample matches a particular reference template.

Feature extraction Built values called features derived from an initial set of measured data which is intended to be informative and nonredundant.

References

1. A.K. Jain, D. Maio, D. Maltoni, S. Prabhakar, *Handbook of Fingerprint Recognition* (Springer, New York, 2003)
2. Advantages and disadvantages of technologies, Available: http://biometrics.pbworks.com/w/page/14811349/Advantages%20and%20disadvantages%20of%20technologies
3. A. Jain, 50 Years of Biometric Research: Almost Solved, The Unsolved, and The Unexplored, Michigan State University (Ed), (ICB, Madrid, 2013), p. 44
4. J. Shah, U. Poshiya, Touchless fingerprint recognization. Asian J. Comput. Sci. Inf. Technol. **3** (5), 73–76 (2013)
5. M.S. Al-Ani, A novel thinning algorithm for fingerprint recognition. Int. J. Eng. Sci. **2**, 43–48 (2013)
6. Fingerprint recognition, Accessed on: 30/05/2018. Available: http://www.ece.uah.edu/biometric/fingerprint_recognition.htm
7. S.S. Ponnarasi, M. Rajaram, Impact of algorithms for the extraction of minutiae points in fingerprint biometrics. J. Comput. Sci. **8**, 1467–1472 (2012)
8. T. Trimpe, Fingerprint basics. (2009). Available: http://sciencespot.net/Media/FrnsScience/fingerprintbasicscard.pdf
9. A. Patel, V. Agrawal, V.H. Shah, Improve fingerprint and recognition using both minutiae based and pattern based method. Orient. J. Comput. Sci. Technol. **7**(2), 251–256 (2014)
10. Smart Card Alliance Secure Personal ID Task Force, Smart cards and biometrics in privacy-sensitive secure personal identification systems, **10**(5). Available: https://www.securetechalliance.org/newsletter/may_2005/feature_0505.html
11. A.S. Patrick, *Fingerprint Concerns: Performance, Usability, and Acceptance of Fingerprint Biometric Systems* (National Research Council of Canada, Ottawa, 2008). Available: https://www.andrewpatrick.ca/essays/fingerprint-concerns-performance-usability-and-acceptance-of-fingerprint-biometric-systems/
12. H.H. Ahmed, K.N. Hamdy, M.S. Tolba, M.A. ELRashidy, Comparative study of various techniques of fingerprint recognition systems. Int. J. Comput. Netw. Technol. **3**(3), 91–103 (2015)
13. Fingerprint identification, Available: http://www.realtimenorthamerica.com/download/Fingerprint_Identification.pdf
14. A.J. Bertino, Forensic science: fundamentals and investigations 2012 update, 2012. [Online]. Available: http://ngl.cengage.com/search/productOverview.do?N=201+4294918536&Ntk=NGL%7CP_EPI&Ntt=1367707560133261614320027647301534107526&Ntx=mode%2Bmatchallpartial
15. P. Verma, Y. Bahendwar, A. Sahu, M. Dubey, Feature extraction algorithm of fingerprint recognition. Int. J. Adv. Res. Comput. Sci. Softw. Eng. **2**(10) 292–297
16. S. Bhattacharya, K. Mali, Fingerprint recognition using minutiae extraction method, pp. 1–5. Available: https://www.researchgate.net/publication/257430785_Fingerprint_Recognition_Using_Minutiae_Extraction_Method
17. K.P. Tripathi, A comparative study of biometric technologies with reference to human interface. Int. J. Comput. Appl. **14**(5), 10–15 (2011)
18. R. Bansal, P. Sehgal, P. Bedi, Minutiae extraction from fingerprint images - a review. Int. J. Comput. Sci. Issues **8**(5), 74–85 (2011)
19. D. Kocharyan, H. Sarukhanyan, Feature extraction techniques and minutiae-based fingerprint recognition process. Available: https://archive.org/stream/FeatureExtractionTechniquesAndMinutiae basedFingerprintRecognitionProcess/34-39#page/n0/mode/1up
20. N. Zaeri, Minutiae-based fingerprint extraction and recognition. INTECH (2011). https://doi.org/10.5772/17527
21. S.Z. Li, A. Jain, *Encyclopedia of Biometrics* (Springer, 2009). [Online]. Available: http://link.springer.com/referenceworkentry/10.1007%2F978-0-387-73003-5_394

22. D. Maio, D. Maltoni, Direct gray-scale minutiae detection in fingerprints. IEEE Trans. Pattern Anal. Mach. Intell. **19**(1), 27–40 (1997)
23. S. Samavi, F. Kheiri, N. Karimi, Binarization and thinning of fingerprint images by pipelining. Presented at the third conference on machine vision, image processing & applications (MVIP 2005), 2005
24. U. Rajanna, A. Erol, G. Bebis, A comparative study on feature extraction for fingerprint classification and performance improvements using rank-level fusion. Pattern. Anal. Appl. **13**(3), 263–272 (2010)
25. K.B. Raja, J. Ravi, K.R. Venugopal, Fingerprint Recognition Using Minutia Score Matching. Int. J. Eng. Sci. Technol. **1**(2), 35–42 (2010)
26. I.G. Babatunde, A.O. Charles, A.B. Kayode, A modified approach to crossing number and post -processing algorithms for fingerprint minutiae extraction and validation. J. Manag. Comput. Sci. J. **6**(1), 1–7 (2011)
27. A.K. Jain, J. Feng, K. Nandakumar, Fingerprint matching. IEEE Comput. Soc. **43**(2), 36–44 (2010)
28. N. Singh, S. Rani, Fingerprint recognition and identification by minutiae detection phase spectrum analysis algorithm. Conference Proceedings, Int. J. Softw. Web Sci (IJSWS) **4**(2), 109–115 (2013)
29. D. Thakkar, Minutiae based extraction in fingerprint recognition. Available: https://www.bayometric.com/minutiae-based-extraction-fingerprint-recognition/
30. K. UCHIDA, Detection and recognition technologies fingerprint identification. NEC J. Adv. Technol. **2**(1) (2005)
31. N. Celik, N. Manivannan, W. Balachandran, S. Kosunalp, Multimodal biometrics for robust fusion systems using logic gates. J. Biom. Biostat. **6**(1), 1–6 (2015)
32. A.K. Jain, K. Nandakumar, Local correlation-based fingerprint matching. ICVGIP, 2004
33. Fingerprint verification. Available: http://www.dds.co.jp/en/fv/algorithm.html
34. A.M. Bazen, *Fingerprint Identification - Feature Extraction, Matching, and Database Search* (Universiteit Twente, Enschede, 2002)
35. NIST, NIST Special Database 4, in *NIST 8-Bit Gray Scale Images of Fingerprint Image Groups*, ed. by FIGS, S. R. D. Program (Ed), (National Institute of Standards and Technology, Gaithersburg)
36. National Laboratory of Pattern Recognition (NLPR). Center for Biometrics and Security Research. CASIA-FingerprintV5 [Online]. Available: http://biometrics.idealtest.org/dbDetailForUser.do?id=7
37. F. Francis-Lothai, D.B.L. Bong, FingerDOS: a fingerprint database based on optical sensor. WSEAS Trans. Inf. Sci. Appl. **12**(29), 297–304 (2015)
38. A.T.A. Taneja, A. Malhotra, A. Sankaran, M. Vatsa, R. Singh, Fingerphoto spoofing in mobile devices: a preliminary study, in *International conference on biometrics: theory, applications and systems*, 2016
39. A.M.A. Sankaran, A. Mittal, M. Vatsa, R. Singh, On smartphone camera based fingerphoto authentication, in *International conference on biometrics: theory, applications and systems*, 2015, pp. 1–7
40. Sensitive data and lawful processing. Available: https://www.twobirds.com/~/media/pdfs/gdpr-pdfs/25%2D%2Dguide-to-the-gdpr%2D%2Dsensitive-data-and-lawful-processing.pdf?la=en

Chapter 3
Recognition-Based on Eye Biometrics: Iris and Retina

Josef Hájek and Martin Drahanský

1 Introduction

In today's security world, biometrics is an interesting and important approach. Ideally, the user interacts with a simple interface, and in a matter of seconds, the biometric system scans the biometric characteristic, whether it is fingerprint or eye iris, and decides whether the user is allowed to pass or not.

This work is focused on the way of scanning and processing the features of a human eye for biometric and biomedical purposes. The features are iris and retina, as they include enough unique information that by its range covers a set bigger than the current human population on Earth.

However, such systems are not perfect, and there is always room for improvement. Recently, it has been discovered that a viable course of the future of biometrics may lie in multimodal biometric systems, which combine more than one source of biometric information for evaluation (such as fingerprint and palm veins), unlike unimodal biometric systems, which only use one.

Biometry applies to different parts of the human body. If the part is damaged or removed altogether, further identification using this biometric is not possible. From all possible affections of our senses, loss of vision has the biggest impact on a person's everyday life. This impact is higher than a loss of memory, voice, or hearing, because 80% of all sensory information that human brain receives comes through our vision. The interesting fact is that a vast majority of visual losses can be prevented by an early recognition and diagnosis of the illness and its treatment. Visiting ophthalmologist can be stressful for some people not only because of fear of the examination itself but also, for example, because of a lack of time. This could be

J. Hájek · M. Drahanský (✉)
Brno University of Technology, Centre of Excellence IT4Innovations,
Brno 62166, Czech Republic
e-mail: drahan@fit.vutbr.cz

M. S. Obaidat et al. (eds.), *Biometric-Based Physical and Cybersecurity Systems*,
https://doi.org/10.1007/978-3-319-98734-7_3

at least partially solved by automated devices, which could evaluate the current patient's condition without any human intervention and possibly even determine anamnesis.

Relatively, lots of iris scanning and recognition algorithms exist. Devices are working, usable in practice, with compact dimensions and quite a good price. Its disadvantage is its inability to detect life and thus its easy fraudulence by, e.g., specially tailored contact lens, etc. For an area of high-security level, the iris is not recommended. If we now focus on eye retina, there is currently no existing biometric system that would be based on this modality. Eye retina has its disadvantages in more difficult scanning form, which is, however, an advantage for liveness detection, in bigger degree user cooperation and higher cost of the device; on the other hand, the possibility of creating fake eye retina is very low (nearly impossible), and also eye retina possesses relatively lot of stable features that can be used for recognition of a big number of people. If we think about the possibility of mixing these two biometric characteristics into one multimodal biometric system, we come to a clear view that the combination has its purpose because the current trends of biometric systems lead toward multimodal biometrics. That kind of device, which would combine eye retina and iris, currently does not exist on the market neither does it exist in any functional concept which would enable using this technology in practice.

We also cannot omit medical (or biomedical) applications, where ophthalmoscopes and fundus cameras are used. These are very expensive; they also do not enable automatic eye retina scanning. The iris is not very interesting for medical use – in eye retina, much more diseases are visible. We can also say that there is no useful expert system on the market which would enable an automatic detection and recognition of diseases that show up in eye retina. When designing and constructing a solution that would fulfill the criteria mentioned, it is necessary to consider these:

- User acceptance (fear of users from its usage).
- Amount of cooperation of user and device necessary.
- Quality of captured images.
- Scanning and processing time.
- Overall price of device and associated software.

Creating a complex and sufficiently robust device is also possible covering the requirements of both biometric and medical segments. The same device with small modifications can be used in both areas, while the only significant change would be software which in biometric systems would extract biometric features with a task of comparing these features with a saved pattern in a database; and in medical systems, it would build an image of the eye retina and would also save the iris; however the expert system would focus mainly on the detection of diseases and suggest possible diagnoses, which would make the ophthalmologist's job easier.

2 Anatomy of the Eye

The human eye is a unique organ which reacts to light and similar to other parts of the body can be used for biometric recognition purposes such as fingerprints, hand veins and geometry, face, etc. In the eye, there are two crucial parts which have relatively high biometric entropy and thus are very suitable for recognition. The first one is the iris, i.e., the colored front part of the eye. The second one is the retina which is responsible for light sensing and cannot be observed by the naked eye because it is located on the back side of the eyeball. Both of them are very well protected against physical damage because they are inner structures of the eye. Iris and retina biometric patterns are unique for each individual although, e.g., the color of the iris is genetically dependent (also applies for monozygotic twins). The base anatomy of the human eye is described in Fig. 3.1.

The human eye consists from [2]:

- *Transparent crystalline lens* is located immediately behind the iris. It is composed of fibers that come from epithelial (hormone-producing) cells. It acts to fine-tune the focusing of the eye.
- *Ciliary body* extends from the iris and connects to the choroid. This annular structure produces aqueous humor, holds the intraocular lens in a place, and also has a supporting function during eye accommodation.
- *Iris* is colored annular structure regulating pupil diameter and thus amount of light coming into the eye. Eye color is defined by the iris.

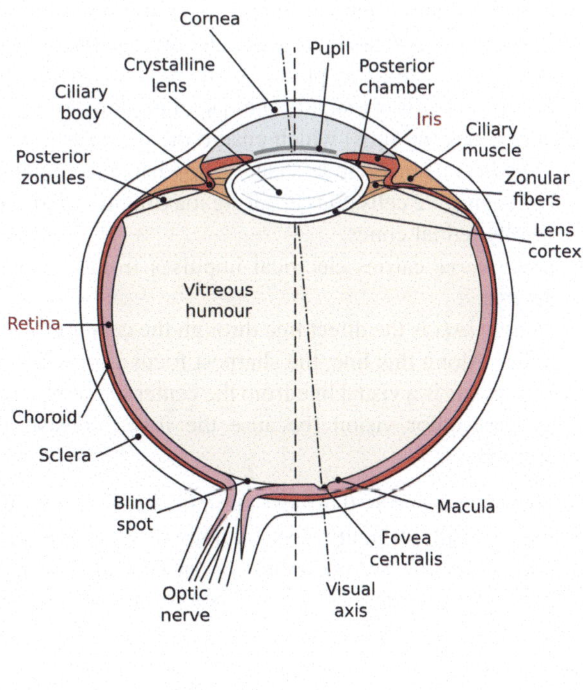

Fig. 3.1 Eye anatomy. (Modified from [1])

- *Pupil* is a hole located in the center of the iris. For an external observer, it seems to be a small black spot because light coming through the pupil into the eye is absorbed by biological tissues inside the eye.
- *Cornea* acts as a window of the eye. It is a transparent front surface of the eye covering the iris and the pupil admitting light into the eye.
- *Posterior chamber* contains aqueous humor and is located behind the peripheral part of the iris and in front of the suspensory ligament of the lens.
- *Ciliary muscle* is a ring of tiny muscles enabling changes in lens shape and thus controlling eye accommodation at varying distances. The muscles affect zonular fibers in the eye. When the muscles contract, they pull themselves forward and move the front lens part toward the axis of the eye.
- *Zonular fibers* are connections of the ciliary body and the crystalline lens which suspend in position during eye accommodation.
- *Lens cortex* is a part of lens comprising secondary lens fibers (long, thin, transparent cells, form the bulk of the lens).
- *Posterior zonules* pull the ciliary body, anterior zonules, and lens into the unaccommodated form.
- *Vitreous humor* is a transparent gelatinous mass filling the interior space of the eye. It contains no blood vessels, and more than 98% of its volume is water.
- *Choroid* is the vascular layer of the eye. It lies between the sclera and retina and provides nourishment for the back of the eye.
- *Sclera* is the protective outer layer of the eye enclosing the eyeball except the part covered by the cornea.
- *Retina* is a nervous tissue layer covering the back of the eyeball. It consists large amount of light-sensitive cells in which stimulation of electrochemical reactions is initiated and electrical impulses are transmitted to the brain.
- *Blind spot* is a small area where the optic nerve comes into the eye. On this area, there are no photoreceptors; i.e., there is no sensitivity to light.
- *Fovea centralis* is a small central spot or pit in the center of the macula containing only cones (no rods) which ensure the most sensitive color vision.
- *Macula* is a small circle shape yellowish area containing a maximum number of light-sensitive cells, thus ensuring maximum visual acuity. It is made up of almost wholly retinal cones.
- *Optic nerve* carries electrical impulses from visual stimuli in the retina out of the eye.
- *Optical axis* is the direct line through the center of the cornea, pupil, lens, and the retina. Along this line, the sharpest focus is drawn when we look at an object.
- *Visual axis* is a visual line from the center of the pupil to the fovea. This axis gives the best color vision (because the fovea consists of high-density cones and no rods).

The human eye is the most complicated and one of the most important sense organs from all. From the physical point of view, it is a transparent biconvex optical system, which focuses the light rays onto the surface of the retina by the help of the cornea and eye lenses. The *cornea* is a front elastic part of the eye. It is transparent,

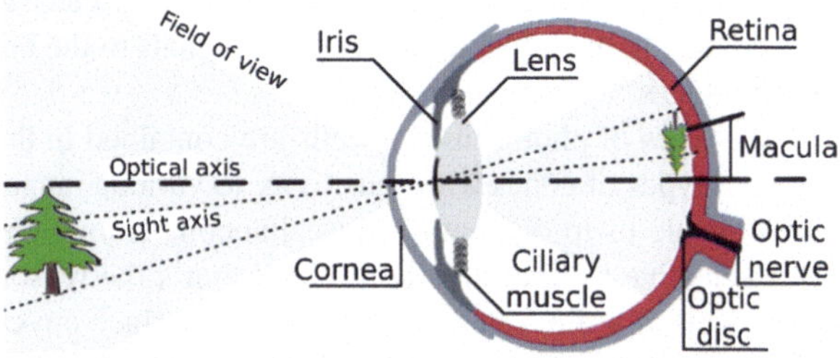

Fig. 3.2 Light projection into the eye

without vessels, and with the fixed optical power of more than 43 diopters represents approximately two-thirds of the eye's total power.

Another optic element of the eye is a *lens* with the ability to change its shape and thus optical power and focal length. It allows to focus on an object at various distances from the observer (eye accommodation). The minimal and maximal distance which the eye is able to focus on is given by two points [3]:

- *Near point* – the shortest distance between an object and the observer which still gives a sharp image on the retina. In this case, the optical system of the eye has the largest optical power of more than 60 diopters. A healthy eye is able to focus on the object at a distance of 25 cm without any exertion.
- *Far point* – the longest distance between an object and the observer which still gives a sharp picture on the retina. The eye lens has the lowest optical power. For a healthy individual, this distance is in infinity.

The light entering the eye is controlled by the *iris*. The surface of the iris has a shape of an annulus and is able to control the diameter of the pupil and thus the amount of light falling on the *retina*. It has a similar function as the aperture of a camera lens.

Fig. 3.2 shows the projection into the eye. The resulting image of the outside world is in the eye inverted (upside down), mirrored, and reduced. The captured image is subsequently transformed to electric signals which are directed to the brain via the optic nerve. In the brain, the image is processed, i.e., it is flipped upside down and mirrored according to the reality.

A thorough understanding of eye anatomy is closely related to a proposal of optical system for acquirement of digital copy of the eye that can be used for medical as well as for biometric purposes.

2.1 Anatomy of the Retina

As shown in Fig. 3.1, the *retina* is located on the back part of the inner surface of the eyeball. The retina is considered as part of the central nervous system and is the only one which can be observed noninvasively and directly. This light-sensitive tissue has a similar function like the film in a camera. Optical system within the eye focuses an image onto the retina surface, initiating several electrical and chemical reactions. Nerve fibers in the retina transport these signals to the brain which interprets them as visual images.

Two types of photosensitive cells are contained in the retina – the rods and three different types of cones with sensitivity to various ranges of wavelengths; hence the cones enable us to distinguish miscellaneous colors. Only one type of rod is much more sensitive to light than cones, and that's the reason why we cannot recognize colors well in a dim light. A whole retina surface covers approx. 70% of the inner surfaces of the eyeball and contains approximately seven million cones and about 75–150 million rods.

On the retina, there are two most conspicuous structures (see Fig. 3.3) – *optic disc* (blind spot) and a *macula* (yellow spot). The optic disc is actually the head of an optic nerve entering the eye and is also a point where the blood vessels, supplying the retina, come into the eye. On colored fundus images, it has bright yellow or even white color. It has more or less a circular shape, which is interrupted by protruding vessels. Sometimes the shape can be elliptical, which is caused by the non-negligible angle between the level of image and level of the optical disc. The diameter is approx. 1.8 mm and is placed from 3 to 4 mm to the nasal side of the fovea. Optic disc completely lacks any light-sensitive cells, so if the light reaches this place, it cannot be visible for the person. In this case, the missing part of the object is completed by the brain (that is why it is also called *blind spot*). We can easily convince ourselves about the existence of blind spot. A test is shown in Fig. 3.4.

When observing the cross with the right eye while the left eye is closed, at a certain distance from the image, the black circle disappears. This is exactly the moment when the image is depicted on the optic disc.

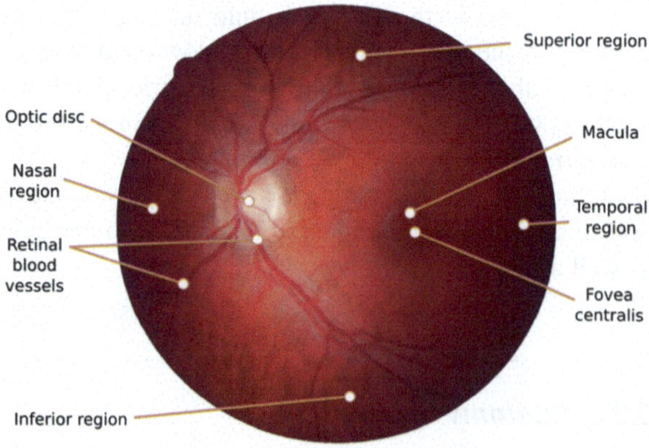

Fig. 3.3 Overview of the retina

Fig. 3.4 Blind spot test

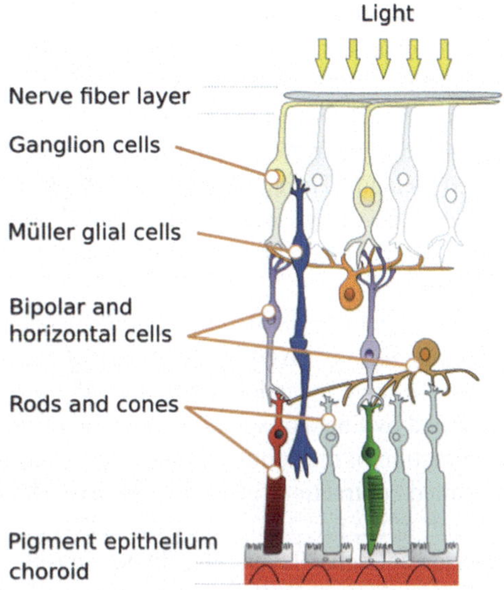

Fig. 3.5 Retinal layered structure. (Modified from [4])

On the other hand, the macula is an area of the sharpest vision. It has a circular shape with a diameter of approx. 5 mm and contains a high density of photosensitive cells, and cones predominate. In the fovea, there are cones only (no rods). Cells density decreases with distance from the macula to the rest surface of the retina. Interestingly, the name "yellow spot" is not derived from the color of the macula (which is rather dark red) but is according to its color observed in the eyes of dead people. The retina itself is a tissue with a thickness of 0.2–0.5 mm. It is described with several layers as shown in Fig. 3.5.

Light first passes through the optic fiber layer and the ganglion cell layer where the most amount of nourishing blood vessels is located. The light is transferred through the layer by Müller glial cells that act as optic fibers and then is received by the photoreceptor cells, cones, and rods which convert the light into the nerve impulses sent through the nerve fibers and optic nerve to the brain. The absorption of photons by the visual pigment of the photoreceptors is firstly translated into a biochemical message and then an electrical message stimulating all the appropriate neurons of the retina. Nourishment of the photoreceptor cells is ensured by the layer of retinal pigment epithelium cells which are fed by blood vessels in the choroid.

2.2 Anatomy of the Iris

The *iris* is a front colored part of the eye which is not hidden inside and can be observed by the naked eye. The annular shape with a central hole called pupil has the same functionality as an aperture of the photo camera – regulating the amount of light coming into the eye. The outer border of the iris is fixed, connected to the ciliary body, while the size of the pupil can vary depending on ambient light. The pupil is not located exactly in the middle but is a little bit moved down and separates the anterior chamber from the posterior chamber. The pupil appears to be black because no light is coming from the eye.

On the back side of the iris lies heavily pigmented layer – epithelium – preventing excessive light from entering the eye. The pigment that gives the iris its color is called melanin. The amount of melanin gives the unique color of the iris. If less, long wavelengths of light are absorbed and short wavelengths are reflected; thus, the eye seems to be blue. On the other hand, more amount of melanin causes brown color [5].

The iris can be divided into pupillary zone and outer ciliary zone. The size of the pupillary zone is given by the maximal size of the extended pupil. These areas are separated by a meanderingcircular ridgeline called *collarette*. Pupillary margin is encircled by fibers of sphincter papillae muscle lying deeply inside the stroma layer. Contraction of the sphincter causes pupil constriction which subsequently results in so-called contraction furrows in the iris. The depth of these furrows depends on the dilation of the pupil caused by the action of the dilator muscle which belongs to the anterior epithelial layer. The dilator muscle fibers are arranged in a radial pattern ending at the root of the iris. Another artifact giving the iris its typical structure is *crypts* occurring adjacent to the collaret, and smaller crypts are located on the iris periphery. The surface of the iris comprises a relatively high amount of structural and circular furrows, pits, and contraction folds. All the described features contribute to a highly detailed iris pattern that is very diverse across human population. While some biometric traits change with age, the iris stops developing around the age of 2 years [6]. In the case of twins, iris recognition has an advantage over face recognition. Monozygotic twins may be nearly identical in terms of facial features, but their irises are highly likely to display several differences in texture.

It is also interesting to note that it was claimed that each area of the body is represented by a corresponding area and patterns, colors, and other characteristics in the iris of the eye. This technique is called *iridology* [122] and can be used to determine information about general health of the person. However, this practice has never been proven as a reliable medical examination method because it makes no anatomic or physiological sense, and well-controlled scientific evaluation of iridology has shown entirely negative results (Fig. 3.6).

3 Iris Recognition

Iris recognition is currently one of the most secure technologies for access control systems. Due to the fine and unique texture of the iris, the probability of having the same iris texture is around 1 in 10^{78} [7], thus ensuring sufficient coverage of the population. However, the fact that the iris is located visibly and it is possible to take a photo from a distance of some meters, the risk of the iris pattern copy and subsequent counterfeit is relatively high. Hence, there should always be some additional security mechanisms (e.g., liveness detection) for high-secured access control.

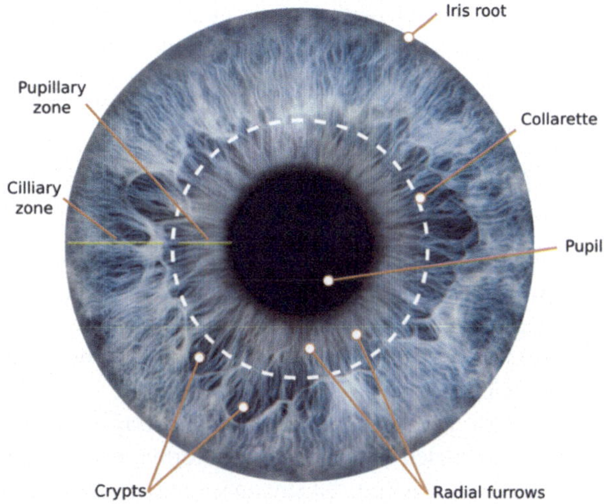

Fig. 3.6 Detailed iris anatomy

3.1 History of Iris Recognition

Although the patented and currently used algorithms have been introduced relatively recently, the concept behind iris recognition has a much longer history. The first idea about using pattern or coloration of the iris for recognition was published in 1886 by Alphonse Bertillon who mentioned in his work that "the features drawing of the areola and denticulation of the human iris" can be used for human recognition [8]. In 1936 the ophthalmologist Frank Burch proposed the concept of the method for individual recognition using iris patterns. However, since that time, it took more than six decades until two American ophthalmologists Dr. Leonard Flom and Dr. Aran Safir proposed and managed a patent for the iris identification concept in 1987. The concept was introduced, but they had no algorithms or implementation, and so their patent remained unrealized. Thus after 2 years, they approached Dr. John Daugman from Harvard University to develop algorithms for automatic and fast identification based on human iris. He formulated three characteristics which determine the iris as an ideal organ for recognition [9].

- It is an inner organ of the body very resistant to external influences.
- It is practically impossible to change its structure without causing eye damage.
- It is physiologically responsive to light, which allows to perform the natural liveness tests.

Dr. Daugman was awarded a patent for automated iris recognition in 1994. One year earlier, the Defense Nuclear Agency of the United States started to work on a prototype of iris recognition system which has been successfully completed and tested in 1995, thanks to the effort of doctors Flom, Safir, and Daugman. In the same year, the first commercial product became available [10]. The patented algorithms became widely licensed by several companies, and research on many aspects of this technology and alternative methods has exploded to rapidly growing academic literature on related topics – optics, sensors, computer vision, security, etc.

3.2 Daugman's Algorithm

The algorithm invented by Dr. John Daugman was the first patented and subsequently deployed approach for automatic iris recognition system. It was the most significant milestone in iris recognition technology. Although patented more than 10 years ago, its principles are still used by some current iris recognition technologies.

The algorithm uses 2D Gabor wavelet transform. Each particular pattern on the iris is demodulated to obtain phase information for features extraction. Information is encoded into *iris code* bit stream, stored in databases allowing search at speeds of millions of iris patterns per second on a single CPU.

The first step of Gabor demodulation is locating the iris in the scanned picture. The iris must be scanned with high quality so that it can be mapped into phase diagrams containing the information about the iris position, orientation, and the number of specific identification attributes. It is then possible to compare it with the database using a pattern once the extraction is done. The principle of Daugman's algorithm is depicted in Fig. 3.7.

Firstly, the iris and its radius are located in the picture. It is done using the following operator [11]:

$$\max_{(r,x_0,y_0)}\left|G_\sigma(r)\frac{\partial}{\partial r}\oint_{r,x_0,y_0}\frac{I(x,y)}{2\pi r}\,ds\right| \tag{3.1}$$

where $G_\sigma(r)$ is Gaussian function of smoothing according to σ, $I(x,y)$ is the rough input picture, and the operator searches for the maximum in blurred partial

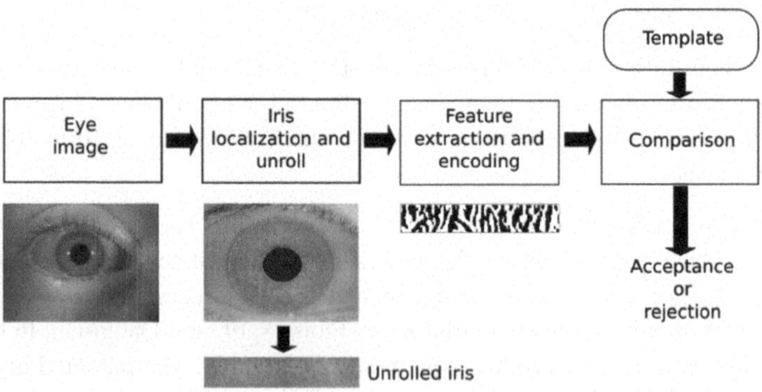

Fig. 3.7 Identification process according to Daugman's algorithm

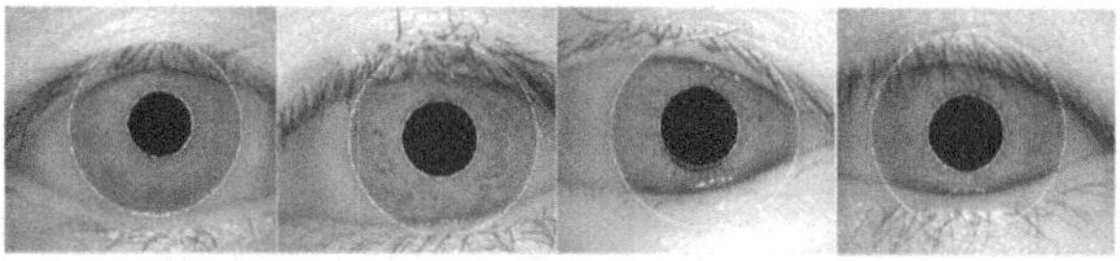

Fig. 3.8 Examples of located irises [12]

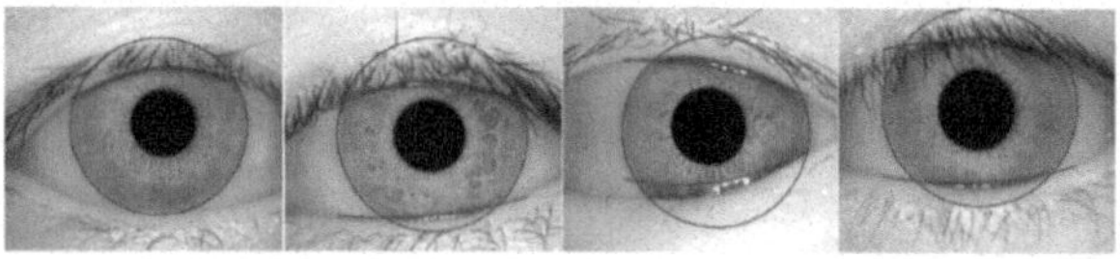

Fig. 3.9 Examples of located eyelids [13]

derivation with attention to radius r and middle coordinates (x_o, y_0). The operator is basically a circular edge detector and returns the maximum if the potential circle shares the middle of the pupil and the radius. Examples of located irises are shown in Fig. 3.8.

Another step is an eyelid localization. Using similar procedure as the one used when locating the iris – the exact position of the upper and the lower eyelid is found. Part of the previous formula used to detect the outline is changed from circular to arched, while parameters are set according to standard statistical methods of estimation so that they ideally correspond to each of the eyelid edges. Examples of located eyelids are shown in Fig. 3.9.

3.2.1 Daugman's Rubber Sheet Model

Daugman's rubber sheet model maps each point inside the iris into polar coordinates (r, θ), where r is in interval $\langle 0, 1 \rangle$ and θ is the angle in interval $\langle 0, 2n \rangle$.

This model compensates pupil dilatation and size inconsistency, thanks to the use of polar coordinates system which is invariant toward a size and translation. The model, however, does not compensate for rotational inconsistency, which is handled by the moving of the iris template in direction θ during comparison phase until both templates match together. Polar coordinate system usage is shown in Figs. 3.10 and 3.11.

3.2.2 Iris Features Encoding

Gabor filter in polar coordinate system is defined as [15]:

$$G(r, \theta) = e^{j\omega(\theta-\theta_0)} e^{-\frac{(r-r_0)^2}{\alpha^2}} e^{-\frac{(\theta-\theta_0)^2}{\beta^2}} \tag{3.2}$$

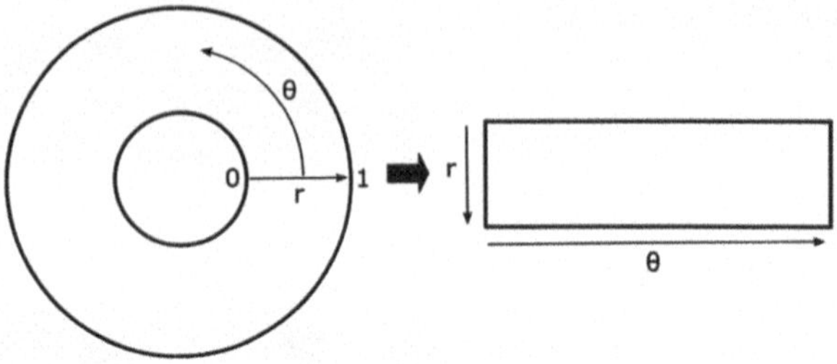

Fig. 3.10 Iris and pupil centers are coincident. (Modified from [14])

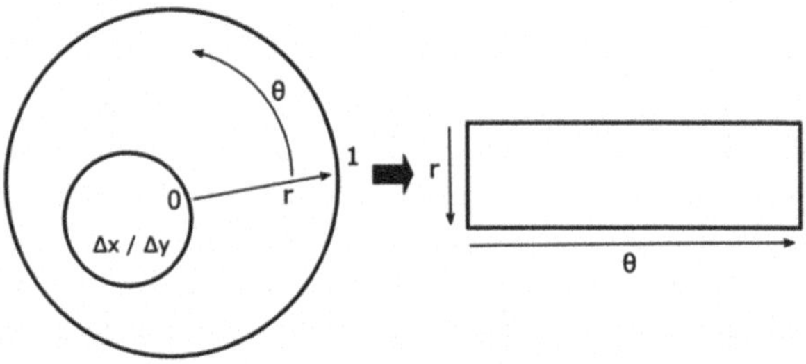

Fig. 3.11 Iris and pupil centers are not coincident. (Modified from [14])

where (r, θ) indicates the position in the picture, (α, β) defines effective height and length, and ω is filter's frequency. Demodulation and phase quantization are defined as [11]:

$$g_{\{Re,Im\}} = sgn_{\{Re,Im\}} \iint_{\rho\varnothing} I(\rho, \varnothing) e^{j\omega(\theta_0 - \varnothing)} e^{-\frac{(r_0-\rho)^2}{\alpha^2}} e^{-\frac{(\theta_0-\varnothing)^2}{\beta^2}} \rho d\rho d\varnothing \tag{3.3}$$

where $I(r, \phi)$ is the rough iris picture in polar coordinate system and $g_{\{\mathrm{Re,\,Im}\}}$ is a bit in the complex plane that corresponds to the signs of real and imaginary parts of the filter's responses. Illustration of the whole encoding process is shown in Fig. 3.12.

Iris' code contains 2048 bits – that is, 256 bytes. The size of input picture is 64×256 bytes, the size of iris' code is 8×32 bytes, and the size of Gabor filter is 8×8. An example of the encoded iris information contains Fig. 3.13.

3.2.3 Iris Codes Comparison

The comparison is done by calculating the Hamming distance between both 256-byte iris codes. Hamming distance between iris code *A* and *B* is defined as the amount of exclusive sums (XOR) between individual bits [9]:

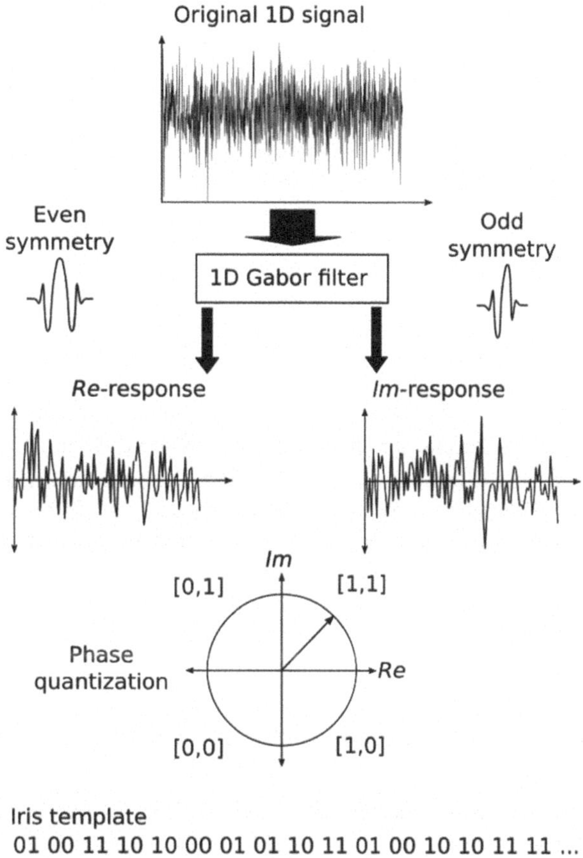

Fig. 3.12 Encoding process illustration. (Modified from [16])

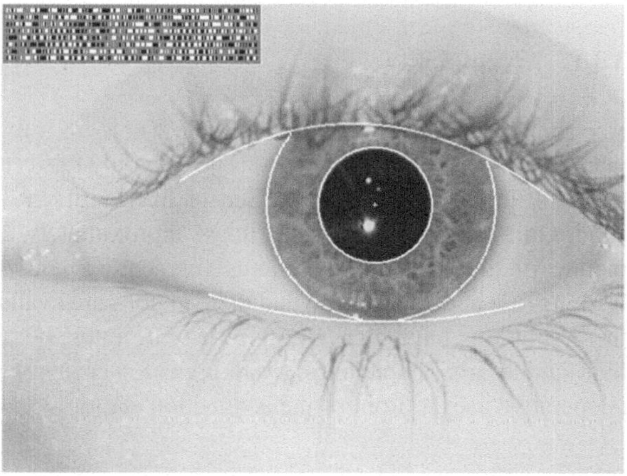

Fig. 3.13 Iris code example [11]

Fig. 3.14 Iris codes comparison

$$HD = \frac{1}{N}\sum_{j=1}^{N} A_j \otimes B_j \tag{3.4}$$

where $N = 2048$ (8×256), if the iris is not overshadowed by an eyelid. In that case, only valid areas are taken into consideration when calculating Hamming distance.

If both samples are obtained from the same iris, Hamming distance is equal or nearly equal to zero, thanks to high correlation of both samples. To ensure rotational consistency, one of the samples is shifted left or right, and each time the Hamming distance is calculated, the lowest Hamming distance is then considered as a final result of the comparison. An example of iris code comparison is shown in Fig. 3.14.

3.3 *Characteristic of the Iris Recognition Technology*

A selection of characteristics related to the suitability of iris recognition is listed below.

3.3.1 Acceptability

Acceptability of identification using the iris is on a middle level as no immediate interaction with the user is needed. The user only needs to stand in front of the device and look in the direction of the sensor from a given distance without moving his head. On the current devices, it takes approximately 2 s to scan and evaluate the picture of the user's iris. Systems for acquirement of irises on the fly (during walking) are in development, and first versions are available on the market. However these solutions have higher false to acquire rates, because the probability to get a high-quality iris sample during walking is lower in comparison with calmly staying cooperative user in front of the acquisition station.

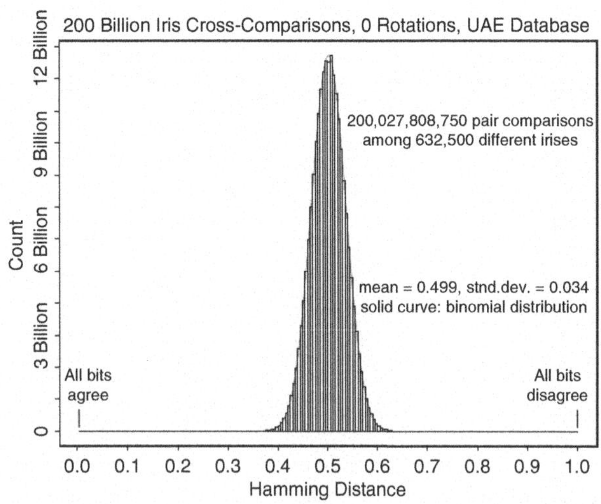

Fig. 3.15 Distribution of Hamming distance [17]

3.3.2 Reliability

During iris scan, insufficient information may be gotten because of ambient lighting, eyes not being open enough, and so on. However, it is relatively a reliable identification method.

The accuracy of the comparison of two iris samples is defined by the Hamming distance, that is, the number of bits that are different in both iris samples. For example, if the Hamming distance is 0.2, two irises differ by 20%. It is noted that a probability of incorrect comparison is 1:26,000,000, and Hamming distance of 0.32 (i.e., about 1/3 same bits from both samples) is sufficient.

Fig. 3.15 shows a distribution of Hamming distances when comparing big amount of irises [17]. The graph creates a binomial distribution with 50% probability (0.5). It also shows that it is highly improbable that two various irises could differ in less than 1/3 of information.

Table 3.1 shows the probabilities of false accept and false reject depending on the Hamming distance of two iris patterns. The Hamming distance value 0.342 is a point of equal error rate (ERR) where false accept and false reject rates are the same. This means that if the difference between currently scanned iris code record and one in the database is 34.2% or greater, they are considered to be from two different individuals.1.

3.3.3 Permanence

The iris is an internal organ and thus well protected but externally visible. Furthermore, the iris does not change with aging – one enrollment should be sufficient for a lifetime with the exception of damage due to accident or disease.

Table 3.1 Hamming distances and error rates probabilities [18]

Hamming distance	False accept probability	False reject probability
0.280	1 in 10^{12}	1 in 11,400
0.290	1 in 10^{11}	1 in 22,700
0.300	1 in 6.2 billion	1 in 46,000
0.310	1 in 665 million	1 in 95,000
0.320	1 in 81 million	1 in 201,000
0.330	1 in 11 million	1 in 433,000
0.340	1 in 1.7 million	1 in 950,000
0.342	1 in 1.2 million	1 in 1.2 million
0.350	1 in 295,000	1 in 2.12 million
0.360	1 in 57,000	1 in 4.84 million
0.370	1 in 12,300	1 in 11.3 million

3.4 Advantages and Disadvantages of Iris Technology

Iris recognition is relatively new among other usual biometric methods; however, it has attracted attention from industry, from government, and also from the army due to its highly desirable properties for personal identification.

External visibility of the human retina ensures a relatively easy scan of its structure. On the other hand, some civil liberties campaigners have voiced concerns about privacy because the iris pattern can be captured from relatively long distance (up to tens of meters) without any cooperation and knowledge of the person.

3.4.1 Advantages

- *Pattern permanence* – iris pattern is well protected and stable through the whole life of an individual. It is not prone to external influences, unlike the face, hand, or fingers. However, the iris can still be affected by eye diseases like diabetes or some other serious disease causing alternations in the iris.
- Uniqueness – remarkable uniqueness of the iris is given by richness of texture details – crypts, coronas, stripes, furrows, etc. Even genetically similar people have totally different iris texture.
- *User-friendliness* – iris is an externally visible organ and enables scanning from distance without any close interaction with a sensor. It requires minimal cooperation with the user. It also makes the recognition more hygienic in comparison to touch-based biometrics such as fingerprint recognition.
- *Speed and scalability* – iris region images can be normalized into rectangular regions of fixed size; thus, fixed-length feature codes can be extracted extremely fast, and matching can be performed easily by the XOR operation. For this reason, the iris recognition is very suitable for large deployments with databases of thousands of users.

- *Relative simple liveness detection (anti-spoofing)* – is given by natural physiology of changing pupil size depending on ambient light or by eye movement called hippus [19].
- *Robustness* – iris recognition system is well resistant to changes in the external environment. For example, voice recognition cannot be processed properly with excessive background noise.

3.4.2 Disadvantages

- *Fear of eye damage* – one of the main disadvantages of using iris recognition system is that the user has to trust the system because sometimes it is said to be a harmful system when using it for a longer period of time because the iris is constantly being scanned by infrared light.
- *Price* – e.g., widely used fingerprint recognition is much cheaper in general. However, iris recognition is still one of the most accurate biometrics, and the prices for devices are dropping down each year, because of increasing amount of installations worldwide.
- *Reliability* – iris recognition can be easily affected by the use of contact lenses or glasses, eyelashes, or reflection from the cornea, and this often results in false rejection of the user.
- *Security* – it can be quite easy to counterfeit an iris sensor. An attacker needs to have iris pattern obtained from a user (is possible to take iris picture from a distance without any user cooperation and awareness) and print the pattern or make a fake contact lenses. For better security, the system has to be equipped with liveness detection (anti-spoofing).

3.5 Related Standards

- *ANSI INCITS 379–2004: Information technology – Iris Image Interchange Format.* This standard describes the format for transmitting visual information about the iris. This includes attribute definition, data record and samples, and matching criteria.
- *ISO/IEC 19794–6:2011: Information technology – Biometric data interchange formats – Part 6: Iris image data.* This standard defines two alternative formats for data representation. The first is based on direct saving into uncompressed format, and the second one requires some preprocessing. However, data is compact and contains only information regarding the iris.

3.6 Commercial Applications and Devices

Many examples of practical usage do exist. These systems are most widespread in the United Arab Emirates, where they are used in airports and ports (c. 3.8 million comparisons daily). Another example can be the system at Schiphol Airport in the Netherlands used by people with high flight frequency. In Czech Republic, this system has not been deployed yet for a practical larger-scale application. Another noncritical example of use is at El Salvador sugar mill [20] where the clatter of time clocks has been replaced by a quiet time-and-attendance system based on iris recognition. The lines have been reduced and time fraud eliminated. In Afghanistan, UNHC (United Nations High Commission) uses iris recognition to control immigrants from surrounding countries.

Relatively large numbers of devices that are capable of iris recognition are currently available on the market. Some of them (just small selection) are mentioned below.

- **BM–ET200/BM–ET330 (Panasonic)**
Panasonic BM-ET200/BM-ET330 [21] offers small wall mounted recognition device able to enroll up to 5025 users depending on the mode the device is operating in. Templates are stored locally in ciphered internal memory. The number of users can be higher in the case of using Ethernet network deployment with database stored and maintained on the central server. Recognition time is 0.3 s with scanning distance of 30–40 cm. Supported operation modes are (a) 1:N – identification mode (one to many), (b) 1:1 – verification mode based on PROX Card or PIN input or iris data from local cache or server database, (c) 1:1 verification with smart card, and (d) standalone-built-in web server for enrollment (Fig. 3.16).

- **IrisAccess 4000 (LG Electronics)**
LG IrisAccess 4000 [22] offers iris recognition technology which can be easily integrated into current security systems through common Ethernet, USB, or serial connection, but local template database is not possible. The devices scan both irises, and a variety of authentication modes can be configured for the right, left, or both

Fig. 3.16 Panasonic BM-ET200 [21]

Fig. 3.17 IrisAccess 4000 [22]

eyes at the same time. Its optical system is also prepared for multi-biometric use allowing facial recognition. However, this has to be handled by external third-party application. Scanning distance is 25–37 cm and offers a standard "one-to-many" identification mode with identification time from 1 to 2 s (Fig. 3.17).

- **iCAM D1000 (Iris ID)**
Iris ID Systems, Inc., is the key developer and driver of the commercialization of iris recognition technology since 1997. The recognition system iCAM D1000 is the sixth generation, and thousands of them are deployed in different countries worldwide. It is a 1082 mm height wall mount device with autonomous positioning system for variously tall people. Straight interaction with the device is not required – scanning distance is 75 cm (±20 cm). It is able to capture the face and iris images in 2–3 s (dual capture) and transfer them to PC equipped with recognition software [23] (Fig. 3.18).

- **BO 2120 EVM (Iritech, Inc.)**
Another well-known player in the field of iris recognition technology is a company Iritech, Inc., offering monocular or binocular iris recognition devices and OEM modules ready to be integrated into existing product lines. An example of hardware module is an evaluation kit BO 2120 EVM [24]. It is a small monocular system working in near-infrared light spectrum with scanning distance of 5 cm and encrypted internal memory for up to 1000 iris templates. Matching query against full memory of templates is within less than 0.5 s. Available application programming interface for several programming languages allows fast integration into a new or into the current devices where is a requirement for iris recognition technology (Fig. 3.19).

Fig. 3.18 iCAM D1000 [23]

Fig. 3.19 Iritech BO 2120 [24]

4 Retinal Recognition

According to available information, the retina recognition is not currently used in practice at all. This is caused by several factors, e.g., complicated optical system, price, and low user-friendliness. Retinal recognition has clear advantages in uniqueness and a number of features compared to other biometrics (up to 400), and also, it is the only place in the human body with the possibility to observe the blood vessels noninvasively and directly. Also, no standards for retinal recognition currently exist; however, it is basically the image of the blood vessels similar to the one used in the recognition using hand or finger blood vessels. Tests of the first constructed device for retinal recognition reported no false accepts and three-attempt false reject error rate (the user is rejected if a match is not found in three trials) of less than 1% [25]. Any counterfeit of the retinal recognition system is very difficult, because, in the first step, the attacker should get a retinal image, which is not possible without user cooperation and awareness. In the second step, the attacker would have to imitate an optical system of the eye so that the biometric system could perform a scan of the retinal fake.

Human recognition which uses the pattern of blood vessels on the back of the eyeball is a very specific branch of biometrics. Although retina-based technique of identification is often perceived as a very secure biometric approach, this method is not widely used due to a few major disadvantages. One of them is related to security. The retina is located inside on the back of the eye, and it is technically difficult and inconvenient to acquire the retinal image. The retina is not a visible part of the human body in comparison to the iris or face, which allows being captured even from longer distances. In this case, it is necessary to have an appropriate optical system which is able to focus the whole eye through the pupil and take a picture of the retinal part. This process leads to complicated and expensive optic device.

On the other hand, it is very hard to counterfeit such recognition because the attacker would have to obtain a retinal image from the relevant individual and simulate an optical system of the eye.

4.1 History of Retinal Recognition

In 1935, ophthalmologists Carleton Simon and Isidore Goldstein who have been exploring eye diseases have found out that the pattern of blood vessels of the retina is different and unique for each individual. They have subsequently published a paper about using retinal blood vessels pattern as a unique pattern for identification [26]. Their research was also supported by Dr. Paul Tower who published a paper dealing with the study of twins in 1955 [8]. He found out that retinal vessels pattern gives the lowest similarity among other biometric patterns also for monozygotic twins. At that time, the idea of recognition based on retinal vascular pattern was relatively timeless.

With the concept of simple fully automated device that is able to acquire a retinal image and verify a user's identity came Robert Hill in 1975 – founder of EyeDentify, Inc., company, and gave immense effort for its development. However, a fully working device was introduced into the market after several years.

In that time, several other companies have been trying to use available medical fundus cameras and modify them for identification purposes. However, these cameras had several essential disadvantages, e.g., relatively complicated system for the eye and device optical axes alignment, illumination in visible spectrum which was uncomfortable for a user, and, last but not the least, high price of medical fundus cameras.

Another attempt led to using illumination in a non-visible spectrum (infrared). For these rays is an eye almost transparent except the choroid that reflects it back creating blood vessels pattern. Infrared illumination is not visible to human (thus is more user-friendly), and when the eye is illuminated in infrared range, the pupil is not contracted, and thus, it is possible to acquire larger surface of the retina thanks to a wider field of view.

The first working prototype was created in 1981. Camera optical system using infrared illumination was connected to a personal computer for pattern analysis. After intensive testing, an algorithm of simple correlation was chosen as the most suitable. After another 4 years, EyeDentify, Inc., introduced the standalone product EyeDentification System 7.5 into the market which has been able to identify a person based on the retinal image and the PIN code stored in the internal database. The device was performing a circular scan of the retina using low-intensity infrared light resulting in a contrast feature vector. The image was composed from 256 12-bit logarithmic samples reduced into the record of 40 bytes for each eye.

4.2 *EyeDentification System 7.5*

The development of retinal recognition device was led mainly by EyeDentify, Inc., company. This corresponds also with numerous obtained patents in this field. The first usable retinal recognition device was introduced in 1981. The product was called EyeDentification System 7.5, and continual improvement of the technology for another 5 years was finished by a system with three registration modes, verification, and identification. Functional principle of the device can be divided into three nontrivial subsystems [3]:

- *Picture, obtaining and processing signals* – optical system and camera must be able to capture an image of the retina in digital form that is suitable for the following processing.
- *Comparison* – the program in the device or in the computer extracts key features from the scanned image and compares these with the samples in database.
- *Representation* – every retinal pattern has to be represented in a way that can be quickly compared or saved into the database.

4.2.1 Optical Scanning System

The mechanical construction of an optical device is a complex issue. It is the most important part of the recognition device because the quality of an input image cannot

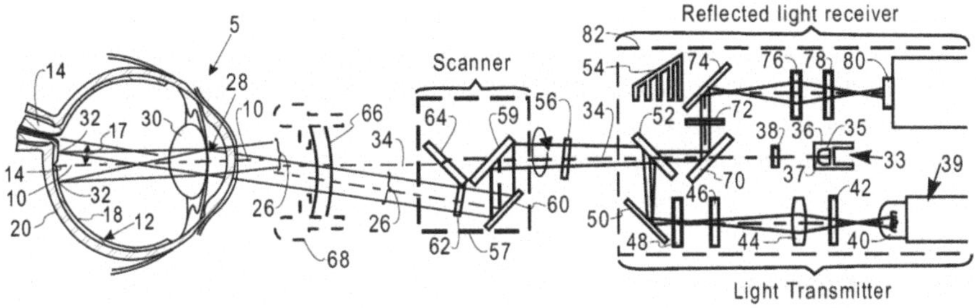

Fig. 3.20 The first version of the optical system used by EyeDentification System 7.5. (From patent US 4620318)

be increased by the following processing in any way, or it is minimally very difficult to enhance the quality. It is evident that the scanning device works on a principle of medical optical devices for eye examination. These so-called retinoscopes or fundus cameras are relatively complex devices, and that is reflected in their price.

Its principle is still similar to retinoscope where a light beam is focused on the retina and a reflected light is captured by a CCD camera. Retinoscope's light beam is set so that the eye lens focuses it as a point on the retinal area. The focused part reflects a part of the sent-back light beam to the eye lens, which modifies it again; the light beam leaves the eye under the same angle as when entering eye (reversible image). By this way, it is possible to get an image of the eye surface about 10° around the visual axis.

First products of the company EyeDentify, Inc., were using a relatively complex optical system with rotary mirrors for covering the scanned area on the retina. This system is described in a patent US 4620318 "Fovea-centered eye fundus scanner" from 1983. The device was performing a circular retinal scan, mostly because of the light reflection from the cornea, where points in the middle would be unusable if raster scanning was used. To balance scanning and visual axis, so-called UV-IR cut filters (hot mirrors reflect infrared illumination but transmit the visible light) and focus point, which the user focuses on, were used. A schematic drawing of the patent can be found in Fig. 3.20. The distance between the eye and optics was about 2–3 cm from the scanner. The system of balancing on an optical axis is a pivotal issue and is described more in detail in patent US 4923297 "Optical alignment system" from 1986.

The newer optical system from EyeDentify, Inc., is much simpler and also has the advantages of optical axis fixation with lower effort of the user needed, compared to the previous system. Its crucial part is the rotary scanning disc, which carries multifocal Fresnel lenses. This construction is described in patent US 5532771 "Eye fundus optical scanner system and method" from 1993.

To ensure that the scanned area is focused on the retina and that the user's eye lies on the scanning beam axis, fixation point/target, on which the user focuses his eye and which must stay in approximately same position for whole scanning, is used. That can be a series of optical networks with focal lengths of −7, −3, 0, and +3 diopters. It is assumed that the most users will be able to focus regardless of their optical defect. When the eye focuses on the target, the device is automatically balanced into the axis of the centered rotary disc on the eye fundus. If the user aligns two or more fixation patterns in one axis, infrared beam is centered on his pupil, and the information can be read.

4.2.2 Comparison

Whenever the user is looking into the optical system's scanning camera, his head can be rotated slightly from the original scanned position during enrollment. Rotation algorithm (phase corrector) is capable of rotating the data by few degrees. This process is done multiple times until the best possible match is reached (highest correlation).

Comparison of obtained samples is done in a few following steps:

- Using sampling the reference, eye record is transformed into the area with the same number of elements as the obtained area, which ensures the alignment (samples overlay).
- Both areas are normalized so that both have RMS (root-mean-square) value equal to 1 – intensities are normalized.
- Areas are correlated using a correlation equivalent to the time domain of the Fourier transformation.

The comparison quality is given by the correlation value where the time shift is equal to zero. It is in a range of +1 (absolute concordance) to -1 (absolute discordance). For a real application, experience has shown that the values around 0.7 can be considered as a concordance.

4.2.3 Representation

Representation of the retina is derived from an image composed out of annular areas (EyeDentification System 7.5 works on the principle of circular scanning). The size of the scanned area is chosen considering the worst possible scanning conditions (highly reduced pupil), but that is still enough for biometric identification. This means that for these purposes, it is not needed to obtain an image of a too large area and resolution.

In relation to the EyeDentify, Inc., device, two main representations of retinal pattern have appeared [27]:

- Original representation has 40 bytes. Those contain information about contrast, coded using real and imaginary coordinates of frequency spectrum generated by Fourier transformation.

- New representation has 48 bytes. It does not contain information about contrast in the time domain. The main advantage of time representation is its faster and more effective processing with lower demands on computing performance.

The template of the retinal pattern contains an area of 96 four-bit contrast numbers from 96 scans of concentric circles in the time domain, i.e., $96 \times 4 = 48$ bytes. Intensities of the time domain can carry a value in interval $<-8,7>$, while normalization on this resolution is used – adjustment to 4 bits of intensive distribution.

4.3 Characteristics of the Retina Recognition Technology

A selection of characteristics related to the suitability of retinal recognition is listed below.

4.3.1 Acceptability

Compared to the iris recognition in the case of the retina, acceptability is low. Many people are afraid of using this technology. They are convinced that a laser that could cause them an optical defect will be used. These concerns are however absolutely unnecessary because the laser is never used in this case. Another problem is the procedure of getting the retina image itself. It can take longer time depending on the user's cooperation and experience, which could bother some users.

With the retina, direct interaction of the user is also needed (to approach at a distance of few centimeters and focus on fixation points). At least with current methods, relatively big cooperation with the user is necessary. Therefore, the acceptability is really low.

4.3.2 Reliability

Concerning the retina recognition, its reliability is high. However, certain conditions during which it is not possible to obtain the retina picture of appropriate quality do exist. That is, mainly and particularly, unsuitable ambient lighting during which the user's pupil is too contracted. Other problems come with optical defects and eye dysfunctions.

Retina recognition is not very extensive, which may be the reason why not many objective tests of this method exist. In 1991, a multinational company Sandia National Laboratory [28] tested products of EyeDentify, Inc., on hundreds of volunteers. The result was zero FAR and FRR lower than 1%. However, at that time, the testing of biometric systems was in its infancy; therefore, we cannot be sure about the test's objectivity.

According to EyeDentify, Inc., the distribution frequency of each eye's pattern that was compared with any other was getting very close to the ideal Gaussian curve with expected mean value of 0.144 and standard deviation of 0.117 [27]. The corresponding probability in this distribution, expected value, and standard deviation of threshold rating 0.7 is about 1 to million [27].

The method of retina recognition is very prone to certain conditions, which must be kept during every scanning. Conditions which could increase false reject rate are, for example, incorrect distance between the scanner and the eye, unclean optics, edges of contact lenses, and glasses. Ambient lighting can also result in subconscious pupil contraction; for that reason, it is sometimes not possible to use the scanner outdoor during daylight.

4.3.3 Permanence

The biological structure of the retina hardly changes during the lifetime of an individual. However, the recognition can also be influenced by injury of various parts of the eye, e.g., the iris, lens, or other parts limiting outer access to the retinal surface. Retinal blood vessels pattern can also be affected by several diseases like diabetes, glaucoma, high blood pressure, or even heart disease.

4.4 Advantages and Disadvantages of Retinal Technology

From all popular biometrics, the recognition by the retina has the most restrictions. They are not insuperable; however, currently, there is no system that can remove these imperfections on a larger scale. Retinal recognition is also affected by high false reject rate because retinal scanning still requires a relatively high user cooperation which has a tangible impact on the quality of the retinal scan, causing a legitimate user to be rejected.

In comparison with other biometrics, retinal recognition offers relatively high universality, uniqueness, performance, and well resistance against frauds. However, retinal recognition systems are still extremely expensive and less user-friendly.

4.4.1 Advantages

- *Pattern permanence* – retinal blood vessels pattern hardly changes within the lifetime of an individual.
- *Number of unique features* – the rich blood vessels structure can contain up to 400 unique features. Identical twins can also have significantly different patterns.
- *Protection* – the retina is located inside the eye and, thus, is not exposed to threats from external environment. It is well protected and cannot be easily damaged such as fingerprints, hand geometry, etc.

- *Contactless scanning* – sanitary problems are eliminated.
- *Small size of template* – only 96 bytes has a current implementation which is very small in comparison to other biometrics. Thanks to this, it is very suitable for deployment in largely used databases and with a very short processing time.
- *Safety* – the retina is hidden inside the eyeball and is very difficult to acquire a retinal image without user cooperation and awareness. Even though the attacker would know retinal patterns, it is very difficult to imitate the optical system of the eye in order to counterfeit the sensor. After death, the retina degrades very quickly and thus cannot be used in the most cases for accurate postmortem identification.

4.4.2 Disadvantages

- *Fear of eye damage* – low level of infrared illumination used in this type of device is totally harmless for the eye; however, there exists a myth in the general public that these devices can damage the retina. It is required that users are familiarized with the system so that they can trust to it.
- *Outdoor and indoor usage* – a small pupil can increase false reject rate because the light has to come through the pupil twice (once toward the eye, then out of the eye), and the returning light beam can be significantly weakened if the user's pupil is too small.
- *User-friendliness* – the need to approach the eye very close to the sensor and focus the alignment point in the device may reduce the comfort of the device usage more than any other biometric methods. It is also related to sufficiently scanned image quality and using eyeglasses and contact lenses.
- *Strong astigmatism* – people with an optical defect (astigmatism) are not always capable of focusing their eye on a fixation target properly, and thus correct template cannot be generated.
- *High price* – it is expected that the price of the device, especially an optical apparatus for recognition by the retina, will always be higher than, for example, the price of a device for recognition of fingerprint or voice.

4.4.3 Commercial Applications and Devices

Retinal recognition has primarily been used in combination with access control to high-secured areas. This includes facilities such as nuclear development and plants, weapon development and production, government, military, secret organizations, etc. One of the best documented application deployments of retinal recognition was in the state of Illinois which used this technology to prevent welfare fraud by identification of welfare recipients [29]. Considering disadvantages, only a few companies from all over the world have introduced or been developing retinal recognition systems.

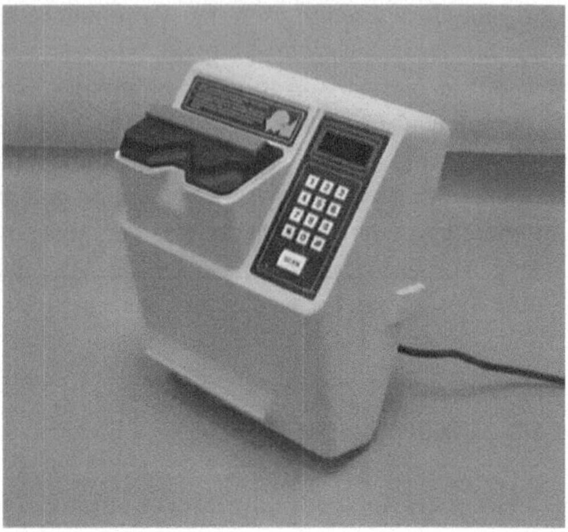

Fig. 3.21 EyeDentification System 7.5 [30]

• **EyeDentification System 7.5 (EyeDentify, Inc.)**
The pioneer in retinal recognition development is EyeDentify, Inc., company, which designed and produced the EyeDentification System 7.5 (Fig. 3.21) and its latest model ICAM 2001, which was introduced in 2001.

According to extant leaflet [29], EyeDentification System 7.5 has been equipped by three modes. Recognition operation mode is the retinal pattern of the user compared with all the eye signature templates stored in a local memory, and a person is allowed or denied depending on existing appropriate template in the database. Second, PIN verification mode: compares the user with only one template identified by a PIN number. The device also allows both eyes verification which reduces the chances of false accepts in PIN mode to one in a trillion. The third mode is enrollment and ensures a proper scan of the retina in order to generate a template and store into database. The user has to keep the eye approximately 15 millimeters above the lens focusing on a green visual alignment target. This process normally takes less than 1 min.

In the PIN verification mode, the recognition process takes approx. 1.5 s. In the recognition mode with 250 retinal templates stored in a local database, the recognition is accomplished in approx. 3 s. The device also provides the following features [29]:

- Stand-alone system or value added component is to increase the security of the currently deployed systems.
- Nonvolatile data storage with capacity of up to 1200 enrollees. The database can be backed up to an external memory.
- System management program control prevents unauthorized users from altering system configuration.
- Average throughput (since first interaction with the device to final acceptance/rejection decision) is from 4 to 7 s.

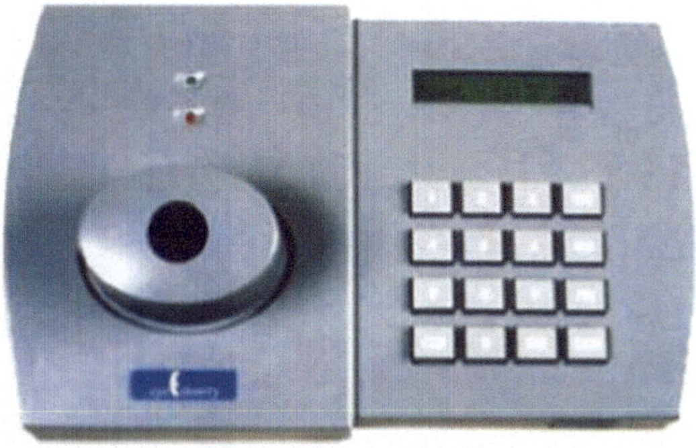

Fig. 3.22 ICAM 2001 [31]

- RS-232 interface for communication with an external system at a speed of 9600 bits/s or auxiliary port with a speed of 300–19,200 bits/s.
- Dimensions 30.5 × 38 × 20.3 cm.
- The price at product release time was 2500 USD.

• ICAM 2001 (EyeDentify, Inc.)

The last known scanning device produced by EyeDentify, Inc., the ICAM 2001 [31], was pulled out from the market due to its high price. Users found also the ICAM to be somewhat intrusive. The main difference between the older model was mainly in the size of template database (could handle up to 3000 retinal patterns) and also the device was much more compact with a dimension of 23.5 × 15.2 × 10 cm. False reject rate was specified by 12.4% (one try) and 0.4% (three trials).

The company TPI (Trans Pacific Int.) offered also a scanner similar to ICAM 2001. The product was called EyeKey; however, nowadays no information about it is known anymore. According to [32], the design was exactly the same – from this, it can be concluded that the TPI was presenting the ICAM 2001 as its own product with added value in increased user database to 300,000 users and with 15 s comparison time within all stored templates [31] (Fig. 3.22).

• Handheld Retinal Scanner (Retina Technologies, LLC.)

Another manufacturer is, for example, Retinal Technologies, since 2004 known as Retina Systems coming from Boston (US); however, any exact specifications of their system are not publicly available (Fig. 3.23).

One million USD of initial funding was received by Retinal Technologies for the development of an inexpensive way to deploy retinal scanning using a small handheld retinal camera that fits into the palm of the hand [33]. The price was set at just 50 USD only, which is much less compared to the solution of EyeDentify, Inc., company. It was claimed that the scanner uses a patented aspheric lens array capable of capturing a retinal image at distances up to three feet from the user's eye based on ophthalmoscope laser scanning. Templates are stored as codes with length up to 60 bytes. It was also claimed that glasses, contact lenses, and existing medical conditions do not interfere with the scanning of the retina. This technology was primarily intended for medical use in hospitals, but its appearance on the security

Fig. 3.23 Handheld retinal recognition device [33]

market was also expected. With the recent information found, Retinal Technologies was looking for another two million USD to market the technology [34]. No further information about this product can be found, and it is very possible that it was a marketing issue only with regard to low price and small dimension of the device totally different from the devices from EyeDentify, Inc.

5 Multimodal Biometric System

In today's world full of security issues, biometrics is an interesting approach. Ideally, the user interacts with a simple interface and in a matter of seconds, the biometric system scans selected biometric characteristic(s) and decides whether the user is allowed to pass or not.

However, such systems are not perfect, and there is always room for improvement. Recently, it has been discovered that a viable course of biometrics may be based on broader use of multi-biometric (multimodal biometric) systems in the future [35] which combine more than one biometric characteristic for evaluation (e.g., such as fingerprint and palm veins), unlike unimodal biometric systems, which use only a single source. The generalized model of a biometric system can be considered as a unimodal biometric system, i.e., such one which uses a single source of evidence for the identification of persons or verification of their identity. In contrast to this model stands a multi-biometric system. As the name suggests, a multi-biometric system uses more than one source of evidence (biometric characteristic) [35].

If designed correctly, such systems can be expected to be more accurate than their unimodal biometric counterparts [36]. Better security counts also among the most prominent benefits of multi-biometric systems. By using more sources of evidence, the security of such systems can be substantially improved.

In the case of a unimodal biometric system, specifically, the one which recognizes fingerprints, it might be easy to spoof the sensor with a spoof produced from latent fingerprint and cheat even very intricate liveness detection algorithms [37]. The deployment of a multi-biometric system can effectively prevent this risk by requiring another biometric modality. Multi-biometric systems can also help with situations in which unimodal biometric systems are considered to be discriminative. If an individual lacks a particular trait or if a trait is severely deformed so that the sensor cannot acquire it, then such individuals might be able to provide another biometric trait, and thus the system may allow them to enroll.

Another advantage lies in the fact that some biometric characteristics, e.g., voice, can be damaged by a noisy data signal. Multi-biometric systems can remedy this inconvenience by using a supplementary algorithm or a different modality.

Multi-biometric systems can operate faster in environments that necessitate a large database. Using more than one biometric trait as search criteria, a database that contains thousands of entries might be scanned more efficiently. For example, one trait would refine the list of potential candidates for an identity match, while another one could be then used to determine the identity from the reduced list [38].

On the other hand, multi-biometric systems are not without disadvantages. Usually, there are several methods of implementing such a system, and some perform poorer than others with certain biometrics, while others perform better. It is, therefore, important to contemplate the aims of the system and to design it accordingly.

A multi-biometric system usually brings forth the question of additional cost. Not only does the system have to accommodate additional resources such as a sensor or a chip for a surplus algorithm, but the cost of fusion of the acquired data has to be taken into account as well. Any new biometric trait required from users might also cause significant inconveniences. The question that arises from these facts is whether the costs incurred by the aforementioned are outweighed by the overall benefits of the system [35].

5.1 Biometric Fusion

An important aspect of a multi-biometric system is the fusion of gathered information. At a certain point during the recognition routine, it is necessary to merge the data into a single entity before proceeding further.

This in itself poses a significant challenge in the designing phase of a multi-biometric system development. As shown in Fig. 3.24, there are four separate operations that the system performs. At each of them, fusion can generally be introduced into the system.

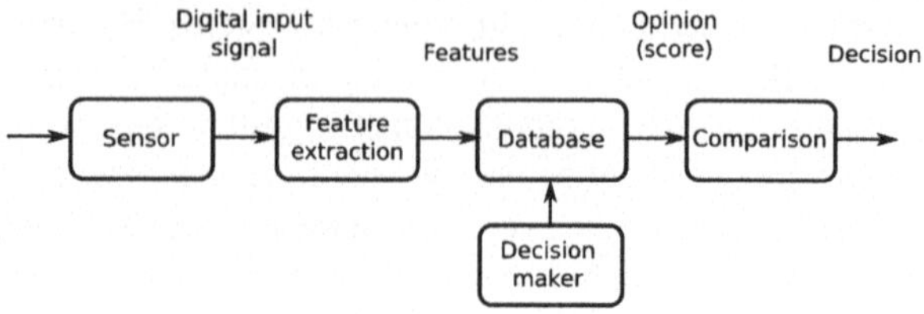

Fig. 3.24 Model of a common biometric system

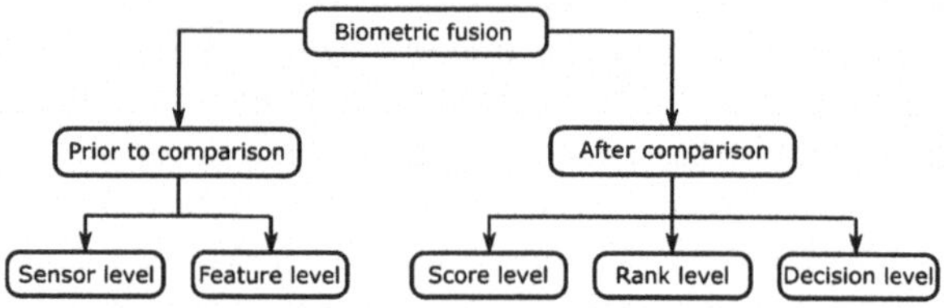

Fig. 3.25 Levels of biometric fusion [40]

It is worth noting that as the data advances through the system, their amount is compressed along the way. However, this does not necessarily imply that the sooner the fusion occurs, the better the results are [39].

While the data at the sensor level are arguably of a larger quantity than those at the feature level, the latter has usually been stripped of superfluous details and noise. On the other hand, it is possible that the feature extraction module may have produced specious results which could have been otherwise remedied at the sensor level.

The classification of biometric fusion is shown in Fig. 3.25.

Biometric fusion can be broadly divided into two sections – *fusion before comparison* and *after comparison*. The reason for this classification results from the fact that after comparison, the amount of information available to the system decreases by a significant margin which is commonly far greater than in the other cases [35].

5.1.1 Sensor-Level Fusion

The sensor-level fusion [41] involves joining multiple sources of raw evidence prior to extracting features. This can encompass text, images, videos, etc. At this level, the obtained data contain the most information available. In image processing, a particular method of fusion is employed, often referred to as mosaicking. In this process, a composite image is constructed from overlapping component images [35].

5.1.2 Feature-Level Fusion

In the feature-level fusion, sources of evidence are consolidated after features have been extracted from respective samples. Following this, fused feature data is then passed to a feature comparator module, and the system proceeds as if dealing with a single source of biometric evidence. Feature sets of distinct modalities or feature sets of identical modalities that have been extracted by different algorithms pose a challenge for numerous reasons [35]. It may be a problem to fuse two chosen modalities, if the basis on which they should be fused is not known. In these cases, it may be difficult to produce a fused set of features that would satisfy the demands on improvement over a unimodal biometric system. This might be exacerbated by the situation in which feature sets of different modalities are not compatible. One of them may vary in length, while the other one may be represented by a fixed-length set of features [42].

5.1.3 Rank-Level Fusion

It should be noted that the rank-level fusion is only applicable in those biometric systems that are set to identify a person, not to verify his/her identity [43]. It is still one of the more frequently applied methods of fusion. After processing the feature vector and acquiring the comparison score, the set of probable matching identities can be sorted in descending order, and thus, a ranked list of candidate identities can be created. The aim of this level of fusion is to merge the ranks produced by individual biometric modules in order to get a consolidated list of ranks for each identity.

5.1.4 Decision-Level Fusion

The decision-level fusion is particularly useful in situations where two or more finished biometric systems are available, and they need to be combined [41]. More often than not, the decision-level fusion is the only option in this case.

5.1.5 Score-Level Fusion

Score-level fusion is commonly used and preferred in multimodal biometric systems in general because matching scores contain sufficient information making genuine and impostor case distinguishable and are relatively easy to be obtained. Given a number of biometric systems, matching scores for a pre-specified number of users can be generated even with no knowledge of the underlying feature extraction and matching algorithms of each system.

5.2 Bimodal Eye Biometric System

Biometric systems in real-world applications are usually unimodal and thus can handle only one single biometric source of information (biometric characteristic). Unimodal systems are susceptible to a variety of issues such as intra-class variations, interclass similarities, noisy data, non-universality, and spoofing. By combining the *iris* and *retina* into one solution, it is possible to get very robust biometric recognition. The most important issue of a biometric system is the uniqueness of its biometric information included in the specific biometric characteristic, which influences the strength of such biometric system. The variability in biometric characteristics of the population can be described by biometric entropy. It is also related to a biometric fusion where they could be needed to quantify the biometric information for each biometric characteristic separately and the possible gain from their fusion.

In addition, the anatomic position allows capturing both images at once. An advantage of a bimodal system is also the variability of modes it can operate in. In the case of any requirement for the high secure system, both biometric characteristics can be processed on the same level of protection. In another case, the quality of scanning can be weighted, e.g., if one of these biometric characteristics is evaluated as less reliable (e.g., because of low-quality image acquisition), the second one is preferred if scanned properly. It also leads to a significant improvement of FMR (false match rate) and FNMR (false non-match rate) and higher reliability for the user – if there is no possibility to acquire one of the biometric characteristics for any reason, the system is still able to recognize him/her on the base of the second biometric characteristic.

The bimodal eye biometric system enables to fuse information at the feature level, matching score level and decision level. We identified several current papers addressing the problem of iris and retina fusion. In [44], the score-level fusion of the left and right irises and retinal features is presented. The weighted average of the scores was applied to all possible combinations of the two irises, and equal weights were assigned to each iris. The fused score is then obtained by a linear combination of these two scores. Since two different scores are obtained for the iris and retina, score normalization needs to be performed. Then, the final score is obtained from the scores of fused irises and retinas. Another approach in patent [45] simply attaches both iris and retina codes together. With the mentioned codes, it is possible to perform fusion by logical operations or join them into a bigger one. Several approaches can be used on the image level. For example, we can merge the images and evaluate them (e.g., by Daugman's algorithm). In this case, one of them can be used as a mask for another one, and their mutual relation such as angle or color is then computed.

Nevertheless, the first concept of the simultaneous iris and retina recognition in a single device was published by David B. Usher et al. [46] in 2008. However, the algorithms were not described in detail, and the preliminary results were focused mainly on image acquisition. The first device combining the iris and retina in one single device was introduced as a flagship of eye biometrics by Retica Systems Inc. [47] in 2006. The device is described as a combined hardware and software solution which allows capturing and fusing data from both the retina and the iris, creating the most secure and accurate biometrics in industry. Recognition device called Cyclops was dedicated for high level secured utilization such as border security, facilities, military and defense, nuclear plants, etc. Retica Systems Inc. had a plan of selling hardware or software solution or the software licenses for incorporation into existing biometric systems. The handheld version has been also supposedly under development. However, the whole project was probably a marketing issue only, because the device codenamed Cyclops had never been introduced and brought to the market [48].

In general, a proposal of ocular biometric device can be divided into four parts:

- *Combined optical system* constructed in order to acquire retina and iris images. In the case of the retina, it is very important to get a sharply focused image and suitable field of view with appropriate surface of the retina. This is related to pupil size and ambient light intensity which must be as low as possible preventing contractions of the pupil. On the other hand, when acquiring an iris image, it is very suitable to have small pupil in order to get the larger surface of the iris. The illuminating light forming bright Purkinje reflections [49] within the iris reflects from the exterior and interior surfaces of the lens and the cornea. These have to be suppressed or restricted out of the region of interest (e.g., restricted to pupil region). Unwanted reflections can be a big problem for any optical system in general and has to be kept at minimum including ambient light sources. Optical system also involves appropriate retina and iris illumination which must be with respect to the pupil size and allowed intensity, preventing the eye from damage.
- *Aligning system* to get the eye optical axis in one line with the optical axis of the device. This is very important when acquiring the retina because the beam of light entering the eyeball is limited by contraction of the pupil, and thus the required field of view is significantly decreasing with the mutual distance of both axes. This applies mainly for retina imaging. The eye and device axes positions must be such that the imaging axis is targeting the surrounding of the blind spot where the most amount of biometric features such as vessels bifurcations or crossings are located. Alignment can be done by two approaches or their combination – the user moves the head toward the device focusing a fixation light point built-in in the device, or the device is moved toward the user's eye depending on the feedback from a camera. Illumination during the process of alignment has also to be comfortable for the user with respect to pupil size as mentioned above. Near-infrared light is usually suitable. Alignment for the iris image acquisition is not so strict, given that the iris can be captured much more easily even from various distances and angles.
- *Image acquisition* – the first step preceding acquisition is an eye localization and identification of boundaries between the iris and the sclera and the iris and the pupil which are not usually concentric. Borders estimation is also related to an identification of areas of the iris that are covered by eyelashes, eyelids, and areas of bright reflections. Several methods have been described initially, based on Daugman's rubber sheet model (see Chap. 3.2.1) [50] or Wildes et al. [51] who also introduced a paper proposal using circular boundary model. Lately, Wildes also presented an algorithm using edge detectors and Hough transforms to segment the eye image, where also, two parabolas were used to find the eyelids. Paper [12] deals with enhanced Daugman's integro-differential method, optimizing its computation time and problem of locating the pupil center outside the

image. The same method is improved in [52] optimized for eye tracking. In a paper [53], it presents an approach for fast iris localization using contrast stretching and leading edge detection. Once the iris is localized and the exact position of the pupil is known, optical axes of the eye and the device must be aligned into one axis. Then it is possible to get an iris image and refocus the optical system to be able to take the retinal image. Focusing on the back of the eye has to be in near-infrared light spectrum to prevent the iris contractions. However, the focus in infrared range can vary against visible spectrum depending on the used wavelength. In other words, the object focused in infrared range is not focused in the visible spectrum and vice versa. It is a property of optical system, but both focuses (infrared and visible) are in mutual correlation shifted by a given constant. At the end of this localization and focusing process, it is finally possible to obtain proper iris and retinal image suitable for recognition itself. Retina and iris image acquisition requires various optical setups and cannot be performed at one time. After the iris image is obtained, it is important to align optical axes of the optics and eye to acquire the retinal image.

- *Image feature extraction and matching* – this is the last step of identification, and in the case of multimodal biometric systems, the extraction is usually followed by some kind of biometric fusion. Features can be joined at different levels of fusion prior to comparison (sensor and feature level) or after the comparison (score, rank, and decision level). Feature extraction and matching methods are described in more detail in previous chapters dealing with appropriate iris and retina biometric.

6 Retinal Features Extraction and Matching

Since the retinal vascular pattern is considered as the main source of biometric features, the first step is usually a segmentation of the vascular tree. The segmentation is followed by the feature extraction, which is based on the detection of bifurcations and crossings of the blood vessels.

Generally, the positions and/or mutual positions of these points, together with some additional features, are used for building a feature vector to cope with translation, rotation, and scale invariants. For example, the work [54] used orientations of blood vessels in specific bifurcation point together with four nearest bifurcations in their feature vector. The Mahalanobis distance gives recognition rate 98.87% for different datasets. In [55], there was defined the so-called principle of bifurcation orientation for each bifurcation, and together with their positions, the point pattern matching method was applied. They achieved total FRR = 4.15% and EER = 1.16% for a database containing 2063 images with small overlap (up to 25%) from 380 subjects diagnosed with diabetes.

One of the main disadvantages of this approach is the detection of bifurcations and crossings using morphological thinning. This step can lead to various results based on the thinning methods. Therefore, the major part of the current approaches tries to avoid this issue using the whole segmented vascular tree. Eigenvalue analysis of vascularity followed by multi-scale image registration to cope with the spatial transformation between the acquired and template image was introduced in [56]. They achieved 100% recognition rate for their own dataset of 284 retinal images. In [57] sampled binary vascular image along defined lines was used in order to extract binary signals for matching achieving over 95% success on about 400 retinal images. Barkhoda et al. [58] used a skeletonized image of vascularity for feature extraction in specific angular and radial partitions. Application of the fuzzy system with Manhattan distance measure leads to 99.75% accuracy using DRIVE dataset. On the other hand, in [59] there was used original intensity image for extraction intensity profiles along circles centered in the fovea, which must be detected. They achieved an averaged FAR per subject below 0.02 for 58 subjects.

6.1 Optic Disc Localization

Several papers deal with the optic disc localization. For example, [60] uses methods called principal component analysis and gradient vector flow snakes for optic disc borders recognition. This model is very computationally intensive but gives a high accuracy of the blind spot detection. Another approach mentioned in [61] is based on the assumption that an optic disc roughly takes up to 5% of the area of the brightest pixels. This method is very fast. Based on thresholding, it is very important to choose an appropriate threshold value which is computed from the average pixel intensities of background and foreground pixels. However, based on experiments on chosen retinal databases, this algorithm is not very precise.

Another work deals with the detection of the optic disc on an image with high levels of gray [62]. This approach works well if there are no pathologies in the image, which would be bright and very contrasting against the background. The principle of area threshold was used in [63] where disc outlines are detected using Hough transformation. That means that image gradient is calculated and as a disc, an area that corresponds the most to its shape is chosen. The problem of this attitude is that the optical disc does not always possess a circular or elliptical shape in the image. It can be overlapped by vessels that protrude it. The principle of Hough transformation was used also in [63]. Despite some improvements, problems were present while detecting the optical disc when the contrast of the images was too low or when the disc's shape was unconventional. Backward vessel tracing that comes from the optical disc was introduced in [80]. This method is one of the most successful for the localization of the optical disc. Its disadvantage is high resources consumption.

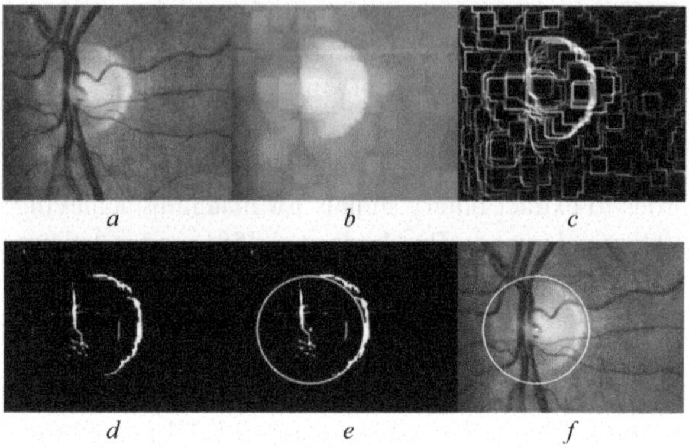

Fig. 3.26 Optic disc localization: (**a**) region of interest; (**b**) removed blood vessels; (**c**) Sobel edge detector; (**d**) eroded binary image; (**e**) Hough circle; (**f**) circle in the original image

6.1.1 Searching Pixels with Highest Average Intensity of Surroundings

This algorithm was inspired by the approach mentioned in [64]. This is based on low-pass filter application (a new pixel intensity value is decided by an average from the surroundings). The pixels of highest intensities are highlighted, and the brightest area is considered as the center of the region of interest (ROI). Subsequent operations are applied to this area only. The following procedures comprise the application of several filters on the ROI and circle detection using Hough circle transformation [65].

The morphological dilatation and Gaussian blur remove blood vessels from the ROI which may have a negative impact on edge detection. Then Sobel edge detector is used, and subsequently an image is converted to a binary form, using thresholding. Noise reduction is performed by morphological erosion. In the last step, Hough transformation is used for the optic disc circle detection. The result is the optic disc center and also its diameter. An example of the whole procedure is shown in Fig. 3.26.

6.1.2 Localization Based on Watershed Segmentation

This approach is based on watershed segmentation described in [66]. Various color specters were compared, red (RGB) channel comes out as the most effective for the detection of the optic disc outlines, where its outlines are the most continuous and also the most contrast to the background. Since that channel has a very small dynamic range and also since the optical disc belongs to the brightest objects in the image for the detection, it is better to use a brightness channel from the HSV model. For disc localization, the variation of gray level is used. As the optic disc is a bright visual object and vessels appear to be dark in the image, the gray level in papillary areas is higher than in other image areas. That is true only if no disease symptoms or other bright artifacts are present on the dark background. That can be addressed by using a shading corrector. In the image with the gray adjustment, the

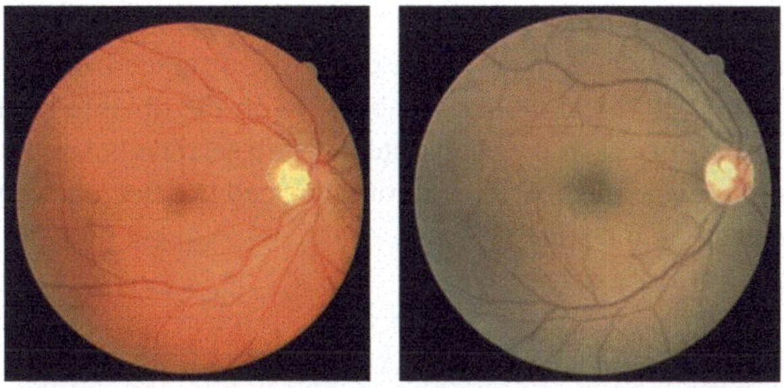

Fig. 3.27 Highlighted optic disc in fundus images

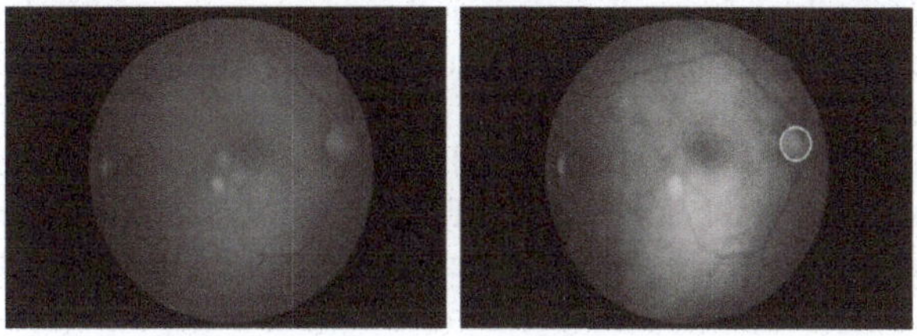

Fig. 3.28 Red channel of the retina image (left), filtered image with optic disc detected (right)

local variation for each pixel is calculated. The global maximum in this adjusted image is situated in the pupil or next to it, which makes it easier to work with a subset of pixels cut from the original picture which does not contain exudates which could disrupt the optical disc detection. Vessels which could distort the result are eliminated before the watershed segmentation [67] for finding the outlines of the optical disc is used. Then the watershed segmentation is used, and its result is the optic disc outline. The result can be seen in Fig. 3.27. The left image shows partial detection only; the highlighted region overlaps its border. The correctly localized optical disc and the borders of the optical disc corresponding to the borders of the highlighted region are shown in the right image.

6.1.3 CLAHE and Median Filter

The red color channel of the original image is used, because the optic disc is the most conspicuous there. Such image is processed by CLAHE [68] and median filter, then the Canny edge detection is performed, and Hough circle transformation [65] is used to detect the optic disc. An example is depicted in Fig. 3.28.

6.2 Fovea Localization

In literature, various approaches for macula detection exist. In [69], principle based on macula outlines detection was introduced with subsequent optimization using ant colony algorithm. The principle from [70], in which it is first necessary to localize the middle of the optic disc and vascular bed, was also introduced. Then the position of the macula is detected based on the distance from the middle of the optical disc. Thanks to thresholding combined with the obtained mask of the vascular bed, the darkest pixels in the candidate area are searched. Another approach from [71] uses a multilevel thresholding without the need to detect the candidate areas on the retina to detect the macula.

6.2.1 Anatomical Based Approach

Described fovea detection is based on anatomical characteristics of the eyeball, mentioned in [72]. It is based on the assumption that the angle of the optic disc center and the fovea joint is between $-6°$ and $+ 3°$ from the horizontal line. Their distance roughly corresponds to double of the optic disc's diameter. This defined sector is used, and the rest of the image is marked by the white color. This approach requires the optic disc diameter and position.

The whole procedure comprises a few steps. The ROI is chosen by the sector-shaped area given by the angle, center, and distance of the optic disc. The rest of the operations are similar to the optic disc detection described in Chap. 6.1.1. The low-pass filter is applied on the ROI, and the lowest intensity pixels are marked (the fovea is darker than the retinal background). Then the fovea is marked as a center of the largest detected area.

6.2.2 Multilevel Thresholding-Based Approach

The algorithm is characterized by low time consumption, and that is one of the reasons why it was also used. The suggested algorithm for the macula detection uses the principle of multilevel thresholding and localizing of ovals through all the image levels. The described approach was inspired by [71].

In the first step, the red channel is extracted from the RGB image because it contains less information than the other channels about vessels and veins in the retina, which can negatively influence the results of ellipse detection. The macula normally does not have a circular or oval shape, and its circumference is often irregular. For preprocessing, a blurring filter is used. Then the resulting image of the retina is segmented in a cycle, where each iteration means limit value increased by one during thresholding. The result of one iteration is shown in Fig. 3.29. This way, thresholding runs throughout all 256 levels, as the retina image in this stage is like a gray picture with 8-bit information for each pixel. By analysis of the resulting images, it has been discovered that threshold values from 100 to 230 have results that possess the most information about the macula position and thus are the most appropriate for the next step, during which ellipses are being searched in final particular images.

The outlines of ellipses are being searched throughout all thresholding levels. All found ellipses from all threshold levels are saved for further processing. Before that,

Fig. 3.29 Red channel of the retinal image after thresholding

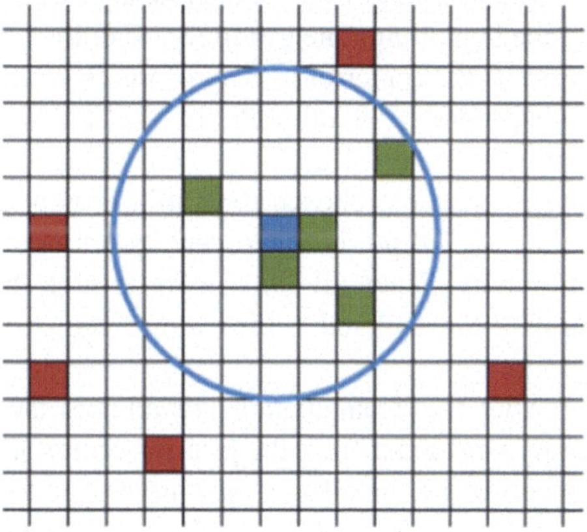

Fig. 3.30 Ellipses clustering

ellipses that do not fulfill the condition that their eccentricity must be larger than 0.5 and simultaneously smaller than 1.5 are removed. Eccentricity is calculated from a rectangle (defined by Eq. (3.5)), which bounds the ellipse with the width (w) and height (h) of this rectangle. The elimination of these inconvenient ellipses is important for removing of misleading areas because the shape of the macula is either a circle or an ellipse with a small eccentricity. On thresholding levels, the blurring filter is used again.

$$exc = \frac{\sqrt{w^2 - h^2}}{w} \tag{3.5}$$

This is done in order to highlight the round shape of areas on the image. Clustering of found ellipses by their center of gravity is depicted in Fig. 3.30. The blue pixel is the group's center of gravity; the green pixels are the ellipses' centers of gravity that belong to the group, and the red pixels are the centers of gravity of ellipses that do not belong to the group [71].

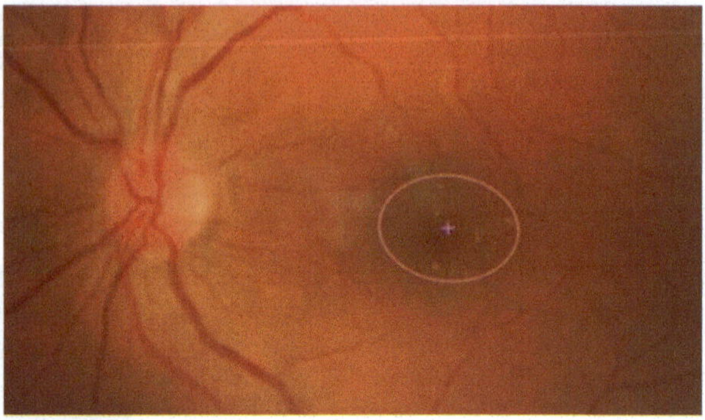

Fig. 3.31 Result of the macula detection in a fundus image

The last part of the macula detection algorithm consists of all the found ellipses. In the beginning, one group for all ellipses is created, and its center of gravity is the same as the center of gravity of the first ellipse. All ellipses are added to this group. Subsequently, all ellipses are shuffled in cycles. If the distance of the shuffled ellipse's center of gravity to the group's center of gravity to which it belongs is bigger than the chosen threshold, a new group is created. To this newly created group, the ellipse is added, and its center of gravity is selected as the group's center of gravity. When all ellipses go through this process, for each group, the center of gravity is calculated as the average center of gravity of all ellipses that belong to the group.

After this recalculation, all ellipses are evaluated again, and if their center of gravity is too far from the group's center of gravity and the found area of the macula circumscribes an ellipse, they are moved to another group, or a new group is created for them. That is repeated until no ellipse has to be moved between groups during the evaluation cycle. As the macular area, the biggest ellipse from the group is chosen. The fundus image with marked macula is shown in Fig. 3.31.

6.3 Blood Vessels Segmentation

The basic approaches for blood vessels detection may be divided into two groups – based on morphological operations and based on 2D filter applications realized by different methods such as matched filters [73], Gabor waves [74], etc. Segmentation in the first group is based on morphological operations and is less computationally intensive. For example, in [75] there is described a very fast segmentation using stop-hat operation.

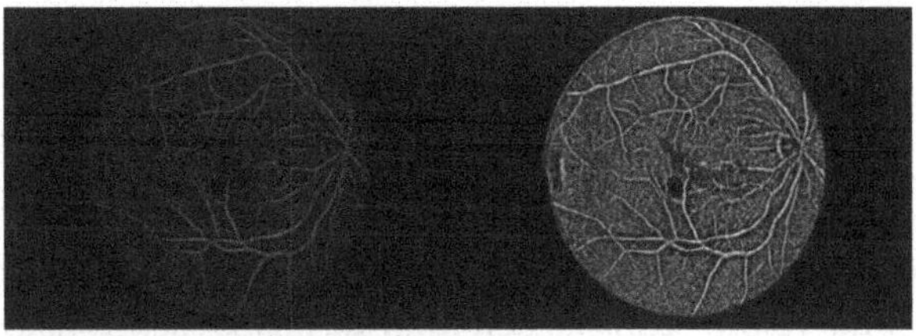

Fig. 3.32 Retinal image before (left) and after (right) thresholding operation

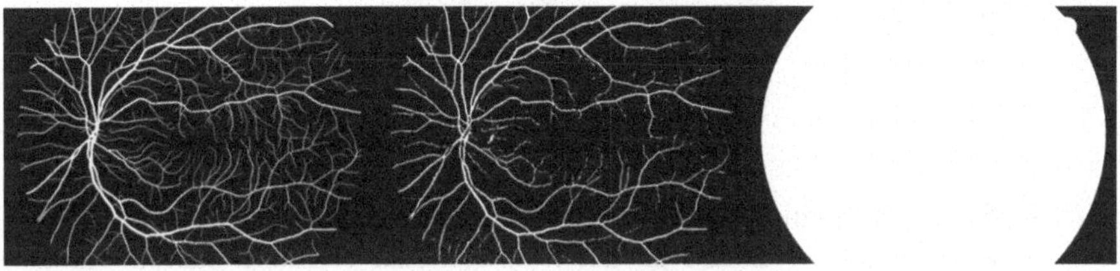

a *b* *c*

Fig. 3.33 Example of the described blood vessels segmentation: (**a**) manually segmented vessels; (**b**) matched filter algorithm used; (**c**) mask of the retinal region in original images

6.3.1 Thresholding

It is desirable to convert an image into a binary form for subsequent processing. This is made by thresholding. The threshold is computed for every input image with the assumption that blood vessels take up approximately 3–10% of the retina image (according to a specific type of imaging device). Another way is the use of adaptive thresholding. However, the main disadvantage of this method is the small white fragments, which are misclassified as vessels.

As shown in Fig. 3.32, not only the retinal veins are highlighted but also the noise which must be removed before bifurcations can be detected. This is achieved by filtering out blobs and by morphological dilatation of the image. This removes small holes and increases the vessels continuity.

6.3.2 Segmentation Using Matched Filter

The described approach is based on the matched filter for blood vessels detection. All the following operations are applied on the green channel of the given image because of higher contrast than in blue or red channel. The whole procedure is depicted in Fig. 3.33.

- *Automatic contrast adjustment* – despite the use of the green channel, blood vessels may have low contrast due to the poor quality of source images. In this case, it is necessary to adjust the contrast. Commonly used methods, such as histogram equalization, are not very suitable in the case of retinal images. The manual contrast adjustment has mostly the best results, but unfortunately, it cannot be applied in the case where the pictures have to be processed automatically. Inspired by [60], the method called *fast gray-level grouping* gives satisfying results for a given set of retinal images. The principle and implementation details are described in [61]. The main advantage of this method is the fact that the new histogram will have nearly a uniform distribution.

- *Matched filter* – the most important part of the blood vessels detection process is the vessels segmentation from the image background. A 2D filter response, which is defined by Eq. (3.6), is used.

$$\mathbf{K}(x, y) = -e^{\left(\frac{x^2}{2\sigma^2}\right)} \; for \; y \leq \left|\frac{L}{2}\right| \tag{3.6}$$

L stands for the minimal length of the vessel, where it does not change its orientation, andσ is the standard deviation in Gaussian (normal) distribution. The exact procedure of the filter generation is based on [76]. The obtained filter is 12 times rotated (each time for 15°), and all the 12 filters are applied on an image. Their responses are added together with their weight resulting in the final response.

6.3.3 Segmentation Using the Difference Between the Filtered Image and the Original

This method of segmentation has been found as the most suitable for the following fusion, thanks to its accuracy and computing speed.

- *Image enhancement* – the input image is enhanced by using the smoothing filters in order to reduce the noise and make the vasculature more visible. In order to obtain the most salient information from the image, the green channel is selected for further processing where the contrast is adjusted by using the contrast-limited adaptive histogram equalization algorithm [77]. This algorithm differs from the simple histogram equalization by calculating histograms for partitions of the image and is added in order to reduce the amplification of the noise inherent to adaptive histogram equalization.
- *Blood vessels detection* – the difference between the filtered image and the original, along with adaptive thresholding, is used to segment the blood vessels. The preceding step for segmenting the veins is the application of the median and blurring filters. This produces a relatively smooth image which is then compared with the non-filtered one. The differential image that results from this comparison is calculated according to the Eq. (3.7):

$$\text{diff}(x, y) = \frac{255}{\max}(\text{original}(x, y) - \text{filtered}(x, y)) \tag{3.7}$$

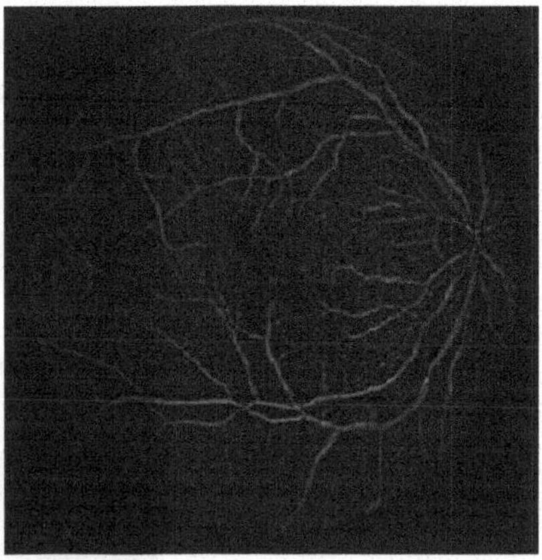

Fig. 3.34 Obtained differential image

where max is the maximum value of intensity difference of the pixels. Although the vascular structure is visible at this point, there is a significant level of noise, and the veins need to be segmented perfectly. The result of this method can be seen in Fig. 3.34.

6.3.4 Thinning

It is essential that thinned vessel must lie strictly in the middle of the original vessel. The simple but fast and well-working algorithm comes from [50]. The thinning is executed from four directions to ensure the position of thinned vessels in the middle of the original one. The algorithm can be described in accordance with [78] as follows:

7 While Points Are Deleted Do

(a) For all pixels $p(i, j)$ do:

If $2 \leq B(P_1) \leq 6$.
$A(P_1) = 1$
$P_2 \times P_4 \times P_6 = 0$ in odd iterations, $P_2 \times P_4 \times P_8 = 0$ in even iterations
$P_4 \times P_6 \times P_8 = 0$ in odd iterations, $P_2 \times P_6 \times P_8 = 0$ in even iterations

(i) Then delete pixel $p(i, j)$.

where $A(P_1)$ is the number of 0 to 1 transitions in a clockwise direction from P_9 back to itself and $B(P_1)$ is the number of non-zero neighbors of P_1. The result of thinning algorithm is depicted in Fig. 3.35.

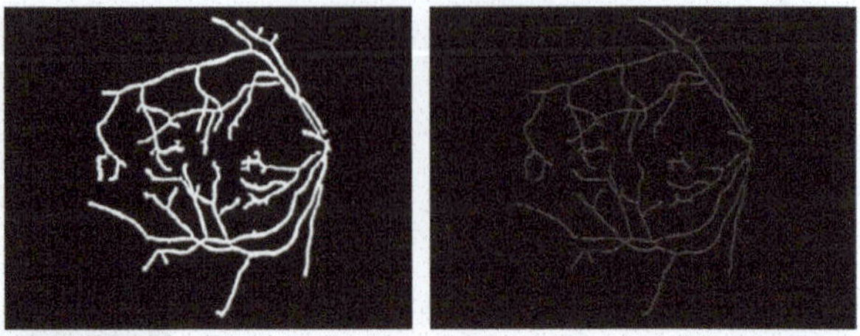

Fig. 3.35 Segmented blood vessels (left) after the thinning operation (right)

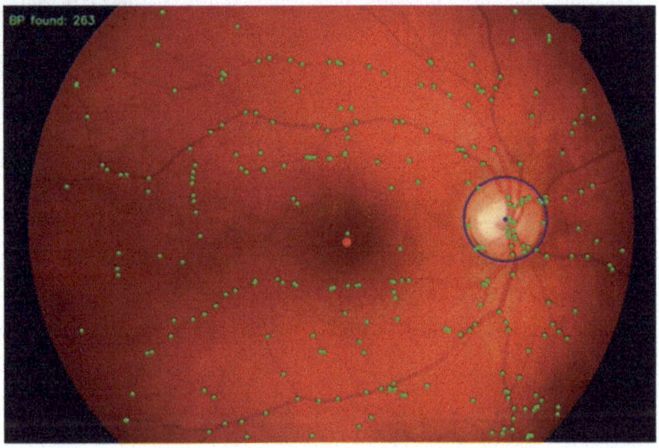

Fig. 3.36 Automatically detected bifurcations in an original image also with the optic disc and fovea

7.1 Bifurcation Localization

Bifurcations are obtained by evaluating every white pixel and its immediate neighborhood. If a bifurcation is to be marked, there must be at least three separate paths that diverge from a given pixel. To calculate this, the neighborhood is analyzed for the number of white pixels and their continuity. If three or more separate white areas are detected, the algorithm regards that pixel as a bifurcation and marks it and stores it in a list of points. If thinning yields an imperfect image with clustered bifurcations, the algorithm has to filter out such bifurcations. The only problem may be caused by the short pieces at the ends of the vessels created as a side effect of thinning. This problem is solved by the definition of a minimal length of the whole three vessels coming out from the point (Fig. 3.36).

7.1.1 Feature Comparison

The comparison module takes two feature vectors. The position of the optic disc is utilized to align bifurcations before image comparison. The adjusted vectors are compared, and the respective score is calculated in accordance with the level of similarity. The score is normalized so that it falls within the interval <0, 1> where a higher value indicates a better match.

First, the two vectors have to be aligned before comparison. This is achieved by taking the optic disc centers and translating the bifurcation points of one image. Since the rotation or angular displacement of images is minimal, only the translation is taken into account.

Next, the similarity score needs to be calculated. The algorithm is as follows:

1. For every bifurcation point b_1 in the smaller array of bifurcations B_1.
   (a) For every bifurcation non-matched point b_2 in the larger array of bifurcations B_2.
      (i) If Euclidean distance between b_1 and b_2 is shorter than the *threshold* and is currently the shortest, mark b_2 as selected.
   (b) If there is a match, mark selected b_2 as matched, and increase the number of matched bifurcations n.
2. Calculate the similarity score.

   Then the score is obtained accordingly to Eq. (3.8):

$$\text{score} = \frac{2n}{|B_1| + |B_2|} \tag{3.8}$$

8 Iris Features Extraction and Matching

There is a considerable range of articles focused on the iris recognition. One of the first automated iris recognition systems based on the utilization of Gabor wavelet filters was proposed by Daugman [79]. Several leading and most cited articles can be mentioned, just to outline the widely used approaches for the person's recognition based on the iris characteristic [9, 80, 81]. Usually wavelet-based [9, 79] or texture-based [80] methods for iris recognition are utilized. Other well-known approaches adopted discrete cosine transform (DCT) [81] to the iris pattern encoding. Somewhat recently, new methods for iris characteristic have been published in [82–84]. These approaches utilized the key point descriptors like SIFT (*scale-invariant feature transform*) [82] and SURF (*speeded up robust features*) [83] for the description of the local iris image areas. Also, inspired by earlier approaches, a new approach using wavelets and Gabor filters in combination with support vector machine and Hamming distance classifiers was proposed in [84].

The current iris recognition approaches usually achieve classification accuracy more than 99%. In spite of the current methods, they are very precise and reliable (for ideal images); still, some drawbacks concerning image quality and image acquisition do exist.

8.1 Pupil and Iris Localization

The pupil can be localized by applying an appropriate thresholding technique. First, however, a median filter is applied to the image. This step smoothes out the image and eliminates pixels with outlying values, which further helps in the segmentation procedure.

In order to determine the threshold value, the histogram of the image is calculated. In the lower half (the darker part) of the histogram, a pronounced peak can be found. This, together with the surrounding values, mainly denotes the pixels of the pupil. Therefore, the desired threshold has to be found around this peak. The chosen threshold is higher than the value of the peak in the histogram to ensure that the majority of the pixels of the pupil are included.

After thresholding is applied, the largest black area in the acquired image is bound to denote the pupil. Since it is elliptical in shape, detecting its center and radius can be determined simply by seeking its widest areas. The pupil itself is not entirely circular, but it can be substituted by a circle for the sake of simplicity and avoiding computational complexity.

While the pupil and its surroundings were distinguished from each other by a striking shift in the pixel intensity, the outline of the iris was not so distinct. Therefore, a different approach must be adopted. Although the shift in pixel intensity is not so pronounced, it is present nevertheless. To facilitate its detection, the contrast of the image needs to be adjusted. Together with the use of the median filter, this accentuates the current area of interest which is the outer edge of the iris.

While not being as sharply defined as in the case of the pupil, the outer edge of the iris can be detected by searching for places where the pixel intensity changes clearly over a certain distance. Within the input database, as mentioned above, this approximation gives satisfying results; however this method is not always applicable because of poorer quality of some images. To mitigate this issue, a fail-safe method using 1D Haar wavelets is employed. Although the iris is not entirely circular as well, it is still safe to substitute it by a circle. Additionally, the iris and the pupil are not concentric in general, but to make the algorithm faster and simpler, it has been assumed that they actually are.

Combined with the detected points where the edge of the iris is located, the radius of the iris can be calculated, and thus the extraction of the iris from the image is completed. The whole localization procedure is depicted in Fig. 3.37.

8.2 Features Extraction

The approach at this point varies, but in this algorithm, the unrolling of the iris precedes the segmentation of eyelids. For this unrolling, the Daugman's rubber sheet is used. At this point, the rubber sheet is reduced to a rectangular image with the width of 360 pixels.

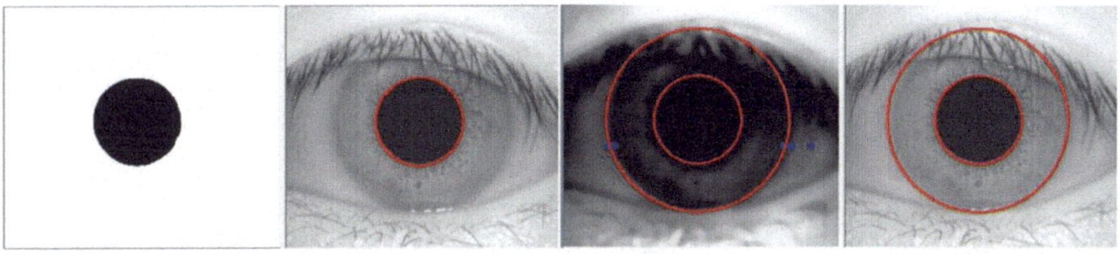

Fig. 3.37 Segmented pupil (left) and the iris. The blue points denote the detected shifts in pixel intensity

Fig. 3.38 Eyelid detection and mask generation. The yellow points denote the borders of examined regions

Eyelids and eyelashes are filtered out by the detector of pixel intensity change along the border of such rectangular image (the stripe). Given the fact that the rough location of the eyelids is predictable, the algorithm defines set boundaries within which the detection is performed. Once the horizontal borders are determined, the algorithm similarly detects the height of the eyelid. When this is done, two masks in the form of tetragons are generated, which have to be taken into account during the final phase of the feature extraction. Of course, this is true only if there are eyelids in the image (Fig. 3.38).

8.2.1 Features Extraction Using Gabor Filter

Gabor filter is represented by the following alternating function:

$$g_{\lambda,\theta,\varphi,\sigma,\gamma(x,y)} = e^{-\frac{(x\cos\theta+y\sin\theta)^2+\gamma^2(-x\sin\theta+y\cos\theta)}{2\sigma^2}} \cos\left(2\pi\frac{x\cos\theta+y\sin\theta}{\lambda}+\varphi\right) \quad (3.9)$$

with the values of wavelength $\lambda = 2$, orientation $\theta = 0°$, phase offset $\varphi = 0$, and aspect ratio $\gamma = 0$.

Before vector quantization, the issue of data reduction has to be addressed. As the height of the stripe can vary (depending on the radius of the iris), the answer to this problem also involves the solution of the issue of potentially varying dimensions of the stripe.

In order to resolve this, certain angular and radial slices of the rubber band are selected for quantization, so that the expected feature vector is of the desired size. During this part, it is necessary to take the eyelid mask into account and map it in accordance with the feature vector. The quantization itself is achieved by using cosine and sine filtering with the resulting vector size of 2048 bits. This represents a 128×8 pixel encoding of the resulting iris stripe, with one real and one imaginary

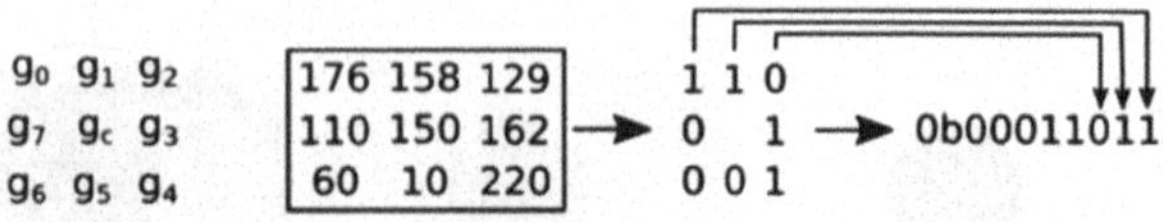

Fig. 3.39 LBP principle

Fig. 3.40 Example of LBP applied on iris image

bit for every pixel. At this point, the feature vector is complemented by a corresponding mask.

8.2.2 Features Extraction Using LBP

The local binary pattern is a method used for classification in computer vision. It is a powerful tool for texture classification first described in [85]. The method describes pixel in the image by a feature vector which is defined with regard to the surroundings [86]. In this case, a basic LBP has been used considering 8-neighborhood.

The LBP is computed by the following method. Each center pixel marked value g_c is compared with 8-neighborhood $g_0 - g_7$, where 0–7 is the index determining the feature position in the resulting vector of size 8 bits (1 byte). The vector value is given by comparison with the pixel in the center. Depending on if the value is smaller or higher, the value in resulting vector is 0 or 1. The base principle of the LBP is depicted in Fig. 3.39.

And from mathematical point of view, it can be expressed by Eq. (3.10):

$$LBP_{P,R} = \sum\nolimits_{p=0}^{P-1} s(g_p - g_c)2^P, s(x) = \begin{cases} 0, x > 0 \\ 1, x \leq 0 \end{cases} \tag{3.10}$$

where g_p is a value of the surrounding pixel with index p, g_c is a value of the center pixel, P is the surroundings size (in our case 8), and 2^P is a position in the resulting vector. The computed value is then stored on the original position of g_c pixel. Example of the resulting image is shown in Fig. 3.40.

8.3 *Features Comparison*

Unlike the retinal part, the iris feature comparison is relatively simple. Given two feature vectors, exclusive OR (XOR) operator is applied to the corresponding bits. In this algorithm, if the values are equal, the similarity score is incremented. The masks of respective vectors are used to filter out those pixels which do not include the iris. This reduces the number of pixels to be compared. Thus, the resulting score is normalized so that it fits within the interval between 0 and 1 as depicted in Eq. (3.11).

$$\text{score} = \frac{\text{score}}{2048 - \text{maskedBits}} \tag{3.11}$$

Because of potential differences in the angular position of the input image which were neglected in the feature extraction phase, the score is calculated for several slightly differing angles ranging approximately from $-16°$ to $16°$. The highest score is then selected as the final score and passed on into the decision-making process.

Another but actually very similar approach is to use Hamming distance of two feature vectors according to Eq. (3.12):

$$H = \frac{1}{L}\sum\nolimits_{j=1}^{L} A_j \oplus B_j \tag{3.12}$$

where L is the size of compared vectors.

9 Eye Livenesss Detection

9.1 Iris

During liveness testing of the iris, a few possibilities are feasible. The most common is the reaction of the iris to lighting changes when the pupil is stretching at lower and contracting at a more intensive light. This reflex is subconscious, and the reaction time is usually between 250 and 400 milliseconds. The pupil contracts and stretches a bit also, under permanent light conditions – this periodical effect is called hippus [19].

Another form of liveness detection can be performed by eye movement or winking according to scanner's voice commands.

Measurement of spectrographic attributes of tissues, fats and blood, are used by more modern devices. Blood reflects very well in the infrared illumination as well as pigment melanin in the iris. This effect is called coaxial back retinal reflection, called the "red eye effect" during photography, where the light is reflected back to a camera if strong light is used.

Purkinje reflexes from the retina and the lens surface can be also used for eye liveness detection. When the outer eye surface is illuminated by an appropriate light source, under certain conditions, an image reflected from the front and back retina surface can appear on the inner surfaces of the eye.

9.2 Retina

Retinal imaging is a relatively difficult process that cannot be easily imitated. To counterfeit that kind of scanner, it would be necessary to use very accurate model of eye optics with the same attributes as real eye, which is very difficult and almost impossible. Not many information about retina liveness detection exist; however, it is possible to use again medical information, e.g., inanimate retina has a different color. The light reflectivity of the retina or blood flow in blood vessels can be also detected.

Since the eye is a very sensitive organ, we cannot use any invasive method. Reciprocal liveness detection similarly to the iris can be used. However, such detection system could be counterfeit, if after successful liveness detection a real eye is changed for the fake one. For that reason, it is better to test the liveness using various methods. The first method is detecting the color of the macula. By this test, one can find out if the detected eye is living or dissected. Only after death that the macula becomes yellow; until that, it has a reddish color.

Another possibility is liveness detection based on eye movement. A similar principle is used in medicine during eye fundus examination. A medical doctor needs to see the whole retina and not only a part which can be seen from a direct view. Therefore, the device is equipped with a focus point which a patient looks at and toward which he moves the eye so that nearly the whole retina can be seen. This can also be utilized in liveness detection. The device is equipped with a similar focus point and in a few times is randomly moved. During each move, the retina is scanned, and the blind spot or macula position is compared. If the position is varying on each picture, the eye is evaluated as living.

10 Eye Diseases

Every part of our body can be affected by a disease during our lives, whether it is curable or incurable. By incurable disease, we will understand the kind of disability that cannot be eliminated surgically or anyhow else without consequence in the form of loss of biometrical information (e.g., amputation). Curable disease, on the other hand, is removable with minimal consequences (e.g., inflammation, laceration). The retina can be affected by both types of diseases, of course.

For a long time, the eye was an organ that was on the first look dark and untouchable. This started to change in the mid-nineteenth century, thanks to the discovery of ophthalmoscope. Using an ophthalmoscope, it was possible to see the big part of the inner eye. By examination of the eye's background, eye specialists can check the rear part of a patient's eye and determine its health condition. For this eye fundus examination, ophthalmoscope is used. Equipped with rotary lenses, it allows the zooming of individual parts of the retina up to 15 times.

Eye diseases and their potential detection in images play a very important role in biometrics. Unless recognition algorithms are optimized and adjusted, the users suffering from any disease impacting the biometric characteristic cannot fully use an acquisition device or cannot use the technology at all.

Any graphical operation in the retinal images affected by disease may have completely different impact on the result in comparison with a healthy retina. The number of features correctly detected in image depends on the size of area affected by the disease and on the algorithm ability to recognize features in inhomogeneous environments. Also medical visualization of retinal images is very important for ophthalmologists and can discover diseases related to (hence not only) the eye.

This chapter also describes the most spread diseases of the retina which is much more likely to get sick than the iris.

10.1 Equipment for Retinal Examination

10.1.1 Direct Ophthalmoscope

While using the ophthalmoscope, the patient's eye is examined from a distance of few centimeters through the pupil. Currently, several types of ophthalmoscope are known; however the principle is essentially the same: the eyes of both the examinee and the doctor are on one axis, and the retina is lighted by a light source that incidents on a semitransparent mirror or a mirror with a hole located in the observation axis at an angle of 45° [87]. The disadvantage of a direct ophthalmoscope is the relatively small examination area and the need of skill to operate and also the cooperation of the patient. The field of view (FOV) for this device is approximately 10 degrees. To increase the angle when observing the eye with a direct ophthalmoscope, different approaches can be used. These include placing the device as close to a patient's eye as possible or dilating the pupil. The principle of direct ophthalmoscope is shown in Fig. 3.41. The light beam falls on a small part of the pupil and does not overlap the observing ray, which minimizes the probability of disruptive reflections. The light beam is shown by yellow line; the observing ray is purple.

Complete techniques of ophthalmoscopy were published by a Czech scientist Jan Evangelista Purkinje in 1823. For the following 25 years, many people with miscellaneous specializations were working on creating an ophthalmoscope, but the first usable ophthalmoscope was introduced by Hermann von Helmholtz in 1851. He managed to do it based on the work of Brucke, who however had not been able to explain what the picture of the final rays that enter the observed eye and then come out of it was.

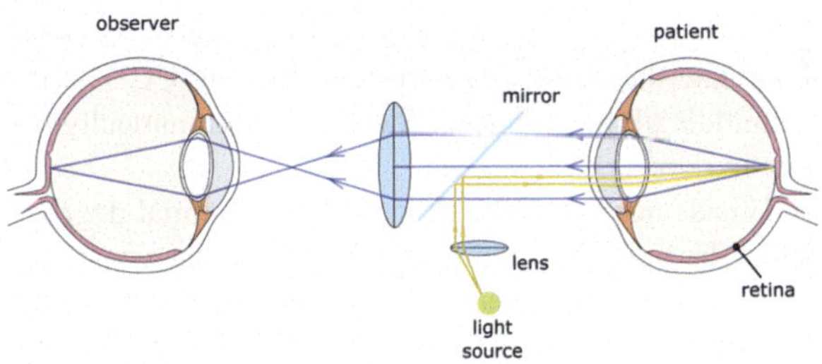

Fig. 3.41 The principle of direct ophthalmoscope [88]

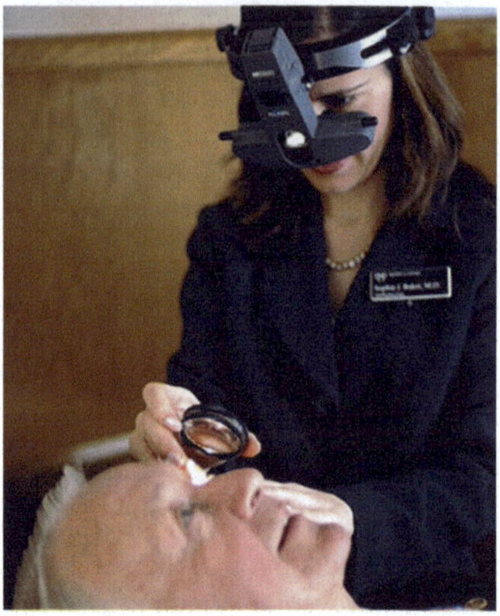

Fig. 3.42 Indirect ophthalmoscopy examination [89]

10.1.2 Binocular Indirect Ophthalmoscope

Most of the important red and yellow details in the retina such as vessels, hemorrhages, or exudates are visible against light red background of blood-filled eye choroid. Subtle and with the naked eye nearly invisible changes can be important symptoms of an ongoing disease. Even better results with direct ophthalmoscope can be achieved with binocular indirect ophthalmoscope [88] that provides a wide field of view, stereoscopic feeling, and also high-contrast resulting image. Logically, this results in other disadvantages – patient's pupil must be dilated, and the device is bigger, heavier, and more expensive. For patients, the most inconvenient is brighter radiation, sometimes even painfully penetrating (Fig. 3.42).

10.1.3 Fundus Camera

For more in-depth examination of the back part of the eye, fundus camera, which has currently and probably the greatest significance in retinal examination, is used. It allows creating a colored photography of nearly whole retinal surface. The optical principle of this tool is based on the so-called indirect ophthalmoscopy [90]. Fundus cameras are usually equipped with a source of white light, which they use to illuminate the retina and then scan it using CCD sensor. Some types can also find the middle of the retina and focus on it automatically using frequency analysis of the scanned image.

This device is usually described by maximal degree, under which it is possible to detect the light reflected from scanned ocular apparatus angle of view. The angle of 30°, which is considered as a standard angle of view, captures the image of an object 2–5 times bigger than it really is. Wide-angle fundus camera captures an image

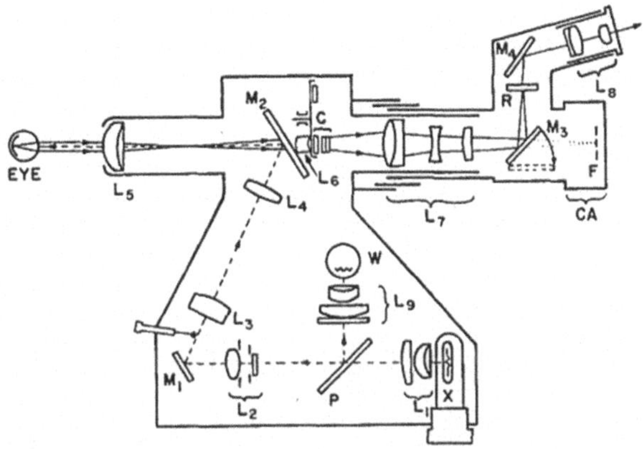

Fig. 3.43 The optical principle of the fundus camera [91]

under the angle of 45°–140°. Depending on the enlargement or reduction of the angle of view, the size of the object that is captured in an image is proportionately changed. Narrow-angle fundus camera uses an angle of view of 20° or less.

The light is generated either by a built-in lamp or using electronic lighting and then projection through a set of filters to a rounded mirror. This mirror reflects light to a group of lenses which focus the light. The mask on the uppermost lens shapes the light into a circle with gross edges. The light is then again reflected from the next round mirror with a central hole, stands out from the camera using lens objective, and continues on to a patient's eye through the cornea. Provided that both lighting systems and image are correctly aligned and focused, the resulting image of the retina stands out from the cornea again back to the device through an unilluminated part of the lens. The light continues through the mirror's central hole to a part of the device for astigmatic correction and to the lenses for dioptric compensation, and a final image is shown on the output lenses of the camera. The optical structure of fundus camera's optics is outlined in Fig. 3.43.

During scanning by the fundus camera, a patient sits in front of the camera's optics with his chin placed on a chin rest and his forehead in a device's forehead rest. The person that is taking the pictures will focus and align the optics to achieve the best possible result. The created photography is then used to determine the patient's diagnosis.

In Fig. 3.44, there is shown an example of the current state of the art of fundus cameras. Canon CR-1 is a non-mydriatic fundus camera with dramatically reduced brightness, improved overall comfort, and significantly shorter exams. One of the advantages of non-mydriatic fundus cameras is that the pupil of the eye does not have to be dilated or enlarged by the use of mydriatic eye drops; thus the examination is more comfortable. For digital imaging, standard DSLR camera is used to obtain a high-quality retinal image with resolution depending on the camera which is used. Canon CR-1 allows wide angle view of 45° and 2× digital magnification. Focusing is performed in two simple steps by aligning two halves of a split pupil image followed in and by the adjustment of split lines and working distance dots in

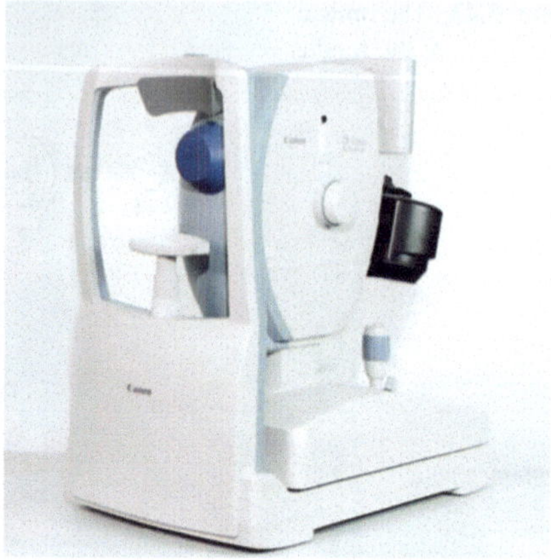

Fig. 3.44 Canon CR-1 non-mydriatic fundus camera

the retinal display. This ensures that the correct focus and working distance are achieved for the sharp image.

10.2 Macular Degeneration

Macular degeneration is a disease that is in 90% of cases formed with increasing age – then we also talk about age-related macular degeneration (ARMD), and it is the most common cause of blindness for patients over 65 years. It is estimated that more than eight million people in the United States have ARMD in some stadium. With increasing percentage of older people, its presence is still increasing, and it also increases with rising ability to handle other eye diseases. In the rest of the cases, the macular degeneration appears among children or young people in the form of Best disease or Stargardt disease [92]. These diseases are formed based on inheritance.

With macular degeneration, the retinal area that creates the middle of a field of view is damaged. As a consequence, a serious disorder of central field of view emerges. In its middle, a patient sees just gray shadow or even a black spot. The peripheral vision, however, stays unaffected. Macular degeneration can appear in two forms – dry (atrophic) and wet (exudative). Among the most common symptoms belongs blurry gray or black smudge in the center of the field of view (known as central scotoma). An affected person sees deformed straight lines, blurry font, or inappropriate shape of different objects. Color vision is also affected; the colors seem faded. Side vision stays sharp on one or both eyes [92]. An example of a retina affected by macular degeneration is depicted in Fig. 3.45.

Many epidemiologic studies use definition that describes ARMD as degenerative disease of individuals aged over 50, characterized by the presence of one of the following lesions [93]:

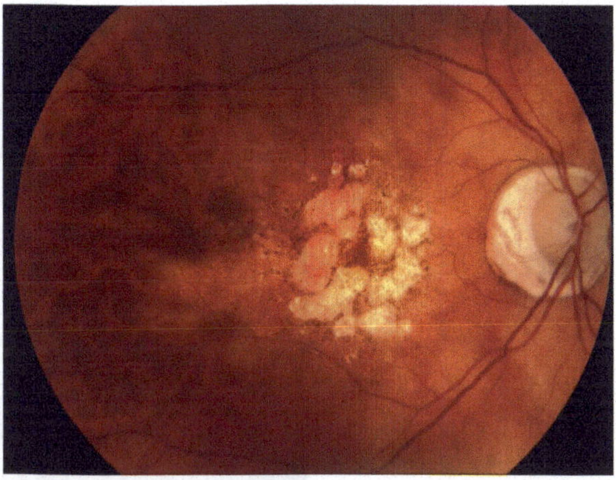

Fig. 3.45 Example of macular degeneration with druses

- *Soft (big, ≥ 63 μm) druses*. The presence of individual, soft, indistinct druses is considered as a bigger indicator of ARMD than the presence of soft clear druses. The presence of druses with size over 125 μm is also much more important than the presence of smaller druses.
- *Areas of hyperpigmentation* that are associated with druses, with the exception of hard druses surrounded by pigment.
- *Areas of depigmentation* associated with druses. These areas often appear as a shadow of druses and are most often sharply bounded.
- *Visual acuity* is not used for defining ARMD because advanced changes caused by ARMD can be present without changing central fixation.

10.2.1 Druses

This is easily visible as yellowish bearings lying deep in the retina. Druses are distinguished by size and shape, and sometimes they have crystalline look resulting from calcification [93].

As an eye where ARMD is not developed, it is considered that the eye on which no druses are observed or only a few small (smaller than 63 μm), druses with the absence of other ARMD symptoms. An eye in the initial stage of ARMD is one which contains a few (less than approx. 20) middle-sized druses (63–124 μm) or pigment abnormalities (increased pigmentation or depigmentation), without any other symptoms of ARMD. Advanced stage of ARMD is geographic atrophy that interferes into the center of the macula or choroidal neovascularization (CNV). Druses are described by the following characteristics [93]:

- *Types* – druses are generally separated into hard and soft with several subtypes. Soft druses are generally bigger and have a soft look. They have distinct thickness and predisposition to connecting; therefore, they show bigger variation in sizes and types. A cluster of soft druses has a squiggly appearance.

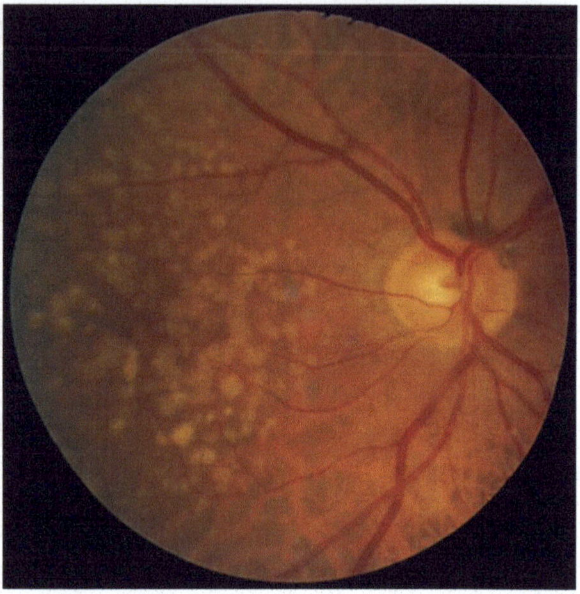

Fig. 3.46 Soft, yellow druses in the inner macula area

- *Degree of eye fundus affection* – this can be evaluated by a number of druses, area of affection, or druses' density which means whether they are separated, touching themselves, or if they cluster together.
- *Distribution* – the greatest importance is attributed to druses that are present in the inner macula, which is defined as an area of inner circle with a diameter of 3000 μm. In Fig. 3.46 can be seen an eye with many clearly visible, soft, yellow druses that are present mainly in the inner macula area.
- *Color* of druses is referred to as yellow, light, or white.

10.3 Diabetic Retinopathy

Diabetic retinopathy (DR) is a noninflammatory disease of eye retina. It is formed as a result of the overall damage of blood vessels during diabetes mellitus [94]. Classification of diabetic retinopathy is generally based on the seriousness of intraretinal microvascular changes and the presence or absence of retinal neovascularization. Retinopathy is classified as non-proliferative diabetic retinopathy (NPDR), if only intraretinal microvascular changes are present. This initial stadium then moves into the proliferative phase, in which new vessels are formed.

Wrongly compensated diabetes affects tiny vessels in the eyes, which are getting blocked, and as a result, the blood supply of the retina is reduced. Another form of retinal damage occurs when blood vessels are leaking and fluid is coming out causing retinal edema. Both insufficient blood supply and retinal edema destroy the ability to see. The eye tries to correct the situation by growing new blood vessels (neovascularization); however, that substandard and harmful breakup can cause eye extravasation (hemophthalmia) and tractional retinal detachment. Diabetic retinopathy has two forms: non-proliferative and proliferative [95].

Abnormalities of fundus during non-proliferative stage are microaneurysms and intraretinal abnormalities, which are the result of changes in retinal vessels transmittance and eventual blockage of retinal vessels. Vessels' blockage leads to bad blood supplying which appears as an increasing number of hemorrhages, vessel abnormalities, and intraretinal microvascular abnormalities. Retinal hyperperfusion is put in connection with the development of proliferative diabetic retinopathy.

Diabetic retinopathy is characterized by the presence of the following lesions [93]:

- *Microaneurysms of retinal capillaries* – Usually the first visible symptom of diabetic retinopathy. Microaneurysms are ophthalmologically defined as dark red dots with a diameter ranging between 15 and 60 μm. Most often, they are present in posterior areas. Despite that microaneurysms can appear even during other vessel diseases, they are considered to be a typical sign of NDPR. For individual microaneurysms, it is typical that they appear and then disappear with time. Microaneurysms alone without the presence of other symptoms of diabetic retinopathy don't have high clinical importance. In spite of that, the increase of their presence in the retina is connected with the development of retinopathy and increased probability that with their increase, other microvascular changes connected to diabetic retinopathy will appear, exists. In Fig. 3.47, there is shown a very severe diabetic retinopathy. It is characterized by excessive retinal bleeding and also by intraretinal microvascular abnormalities.
- *Hemorrhages* are also red dots that develop as a result of the weakening of microaneurysms' walls or tiny vessels and following the rupture of these walls. Hemorrhages exist in two forms. In the first one, hemorrhages have a character of

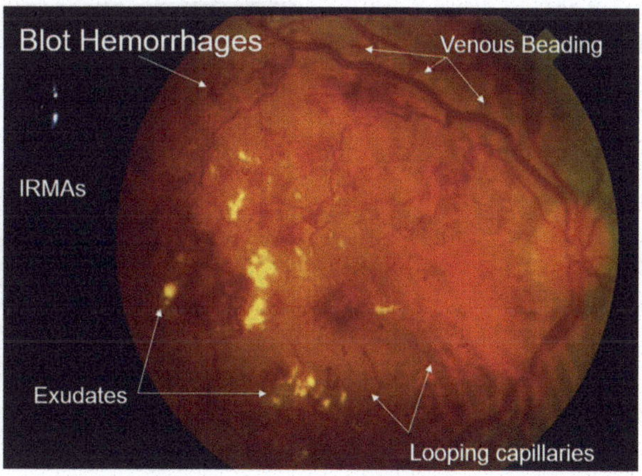

Fig. 3.47 Advanced stage of diabetic retinopathy [96]

dots. These pathologies look like little light red dots. The second type of hemorrhages looks blurry and is bigger.

- *Exudates* – With increased creation of microaneurysms, it is possible that excessive transmittance of retinal capillaries occurs. This leads to the formation of retinal edema, usually in the macular area. Macular edema is defined as the thickening of the retina caused by the gathering of liquid in the macula. Macular edema is often accompanied by hard exudates in the retina. These hard exudates are lipid sediments that are probably piled up under an influence of lipoprotein leak. Clinically, hard exudates are well-boarded, white-yellow intraretinal sediments, normally visible on the edges between edematous and non-edematous retina. The liquid that is creating the edema can come and leave with no visible consequences. Lipid sediments are on the contrary with the liquid connected with the retinal damage and permanent loss of vision, especially if they are located under the center of the macula.

10.4 Toxoplasmosis

Toxoplasmosis is a parasitic disease that ranks among zoonoses, which are diseases transmissible from animals to humans. They appear all over the world. In European countries, about 10–60% of inhabitants have created antibody toward toxoplasmosis, depending on their eating habits. In the Czech Republic, the seropositivity (the presence of antibodies in the blood) is about 20–40%. The disease usually appears only by increased temperature, flu conditions, headaches, fatigue, or swollen lymph nodes. Acute disease can sometimes transform into a chronical stage, often, the infection, however, occurs without any notice and is recognized only by finding specific anti-toxoplasma antibodies in the blood, which can in lower levels last for a whole life (latent phase of infection). A lot of toxoplasmosis' types exist – ganglionic, eye (Fig. 3.48), brain, gynecological, etc. Other forms of toxoplasmosis are not that common [98].

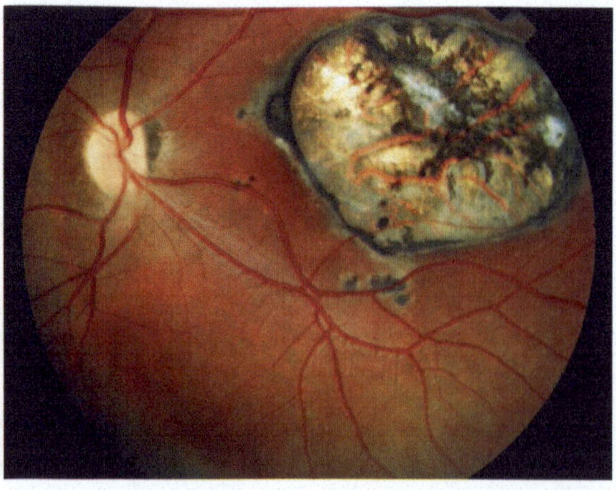

Fig. 3.48 An eye affected by toxoplasmosis [97]

11 Conclusion

In conclusion, we can say that a biometric technology has a big chance of success and is currently displacing conventional non-biometric solutions in many areas. Thanks to combining with cryptographic methods, biometric also penetrates the field of data protection. The field of biometric systems is very alive and still expanding, and innovative solutions that are perfectly usable in industrial applications are coming to the market every year. It can be assumed that even if the eye retina recognition is not currently very popular (there is no retina recognition device currently available in the market), it has undeniable advantages and will evolve and expand. Room for utilization improvement is primarily in high-secured areas such as nuclear plants, military purposes, or secret research laboratories.

Unimodal biometric systems have to contend with issues like non-universality, unacceptable error rates, intra-class variations, etc. Some of these limitations can be solved by the deployment of multimodal biometric systems that integrate the evidence presented by a multiple sources of information. In general, one of the most important issues of a biometric system is the uniqueness of its biometric information included in the specific biometric characteristic, which influences the strength of such system. The variability in biometric characteristics in the population can be described by biometric entropy. It is also related to a biometric fusion and is interesting to see an evaluation of the biometric information for each biometric characteristic separately and the possible gain from their fusion. Multimodal biometric systems elegantly solve several of the problems present in current unimodal systems. By the combination of multiple sources of information, multimodal systems increase population coverage, improve matching performance, and are more resistant against sensor counterfeit. Various fusion levels are possible in multimodal systems.

Acknowledgment This work was supported by the Ministry of Education, Youth and Sports from the National Programme of Sustainability (NPU II) project IT4Innovations excellence in science – LQ1602.

References

1. Wikimedia Commons, *Eyesection.svg*. [Online; accessed 5-July-2016]. URL: https://upload.wikimedia.org/wikipedia/commons/thumb/f/f5/Eyesection.svg/2000px-Eyesection.svg.png
2. D. Roberts, *Anatomy of the Eye. MD Support*, (2013), [Online; accessed 5-July-2016].URL: http://www.mdsupport.org/information/99-2
3. K. Franklin, P. Muir, T. Scott, L. Wilcocks, P. Yates, *Introduction to Biological Physics for the Health and Life Sciences* (Wiley Blackwell, 2010). ISBN 978-0470665930
4. Wikimedi Commons, *File:Retina.svg*. [Online; accessed 5-July-2016] https://commons.wikimedia.org/wiki/File:Retina.svg
5. Z. L. Stan (ed.), *Encyclopedia of Biometrics* (Springer, 2009). ISBN 978-0-387-73003-5

6. L. Yang, *Iris/Retina Biometrics*. CPSC 4600@UTC/CSE, [Online; accessed 5-July-2016]. URL: http://web2.utc.edu/~djy471/documents/b6.1.IRIS-Retina-utc.ppt
7. R.P. Moreno, A. Gonzaga, Features vector for personal identification based on Iris texture, in *Proceedings of the Irish Machine Vision and Image Processing Conference*, (Dublin, 2004)
8. P. Tower, The fundus oculi in monozygotic twins: report of six pairs of identical twins. A.M.A. Arch. Ophthalmol. **54**, 225–239 (1955)
9. J. Daugman, How iris recognition works, in *Proceedings of 2002 International Conference on Image Processing*, vol. 1, (2002)
10. M. Tistarelli, S.Z. Li, R. Chellappa, *Handbook of Remote Biometrics: For Surveillance and Security* (Springer, 2009). ISBN 978-1-447-12670-6
11. J. Daugman, How iris recognition works, in *IEEE Transactions on Circuits and Systems for Video Technology*, vol. 14, no. 1, (2004), [Online; accessed 5-July-2016]: URL: https://www.cl.cam.ac.uk/~jgd1000/csvt.pdf
12. R.B. Dubey, M. Abhimanyu, Iris localization using Daugman's intero-differential operator. Int. J. Comp. Appl., **93**, 6–12 (2014)
13. M. Adam, F. Rossant, F. Amiel, B. Mikovicova, T. Ea, Reliable eyelid localization for iris recognition, in *Conference: Advanced Concepts for Intelligent Vision Systems, ACIVS 2008*, (2008). https://doi.org/10.1007/978-3-540-88458-3_96
14. T. Johar, P. Kaushik, Iris segmentation and normalization using Daugman's rubber sheet model. Int. J. Sci. Tech. Adv. **1**(3) (2015). ISSN: 2454-1532
15. D.D. Zhang, *Automated Biometrics: Technologies and Systems* (Springer, 2013). ISBN 1461545196
16. A. Kumar, *Biometric Security: Iris Recognition*. [Online; accessed 10-August-2015] URL: http://www.slideshare.net/piyushmittalin/biometric-security-iris-recognition
17. J. Daugman, *Results from 200 Billion Iris Cross-Comparisons*. University of Cambridge, Technical report, (2005), ISSN 1476–2986. URL: https://www.cl.cam.ac.uk/techreports/UCAM-CL-TR-635.pdf
18. P. Khav, *Iris Recognition Technology for Improved Authentication*, (SANS Institute, SANS Security Essentials (GSEC) Practical Assignment, 2002), [Online; accessed 5-July-2016]. URL: https://www.sans.org/reading-room/whitepapers/authentication/iris-recognition-technology-improved-authentication-132
19. H. Bouma, L. Baghuis, Hippus of the pupil: periods of slow oscillations of unknown origin. Vis. Res., 11 (11), 1345–1351 (1971)
20. I.D. Iris, *El Salvador Sugar Mill Uses Iris Recognition for Time and Attendance*. [Online; accessed 5-July-2016]. URL: http://www.irisid.com/el-salvador-sugar-mill-uses-iris-recognition-for-time-and-attendance
21. Panasonic Corporation, *Iris Reader Access Control, BM-ET200*. [Online; accessed 5-July-2016]. URL: ftp://ftp.panasonic.com/pub/panasonic/cctv/BidSpecs/BM-ET200.rtf
22. ID Travel AG, *Biometric Systems for Secure and Rapid Access: IrisAccess 4000*. [Online; accessed 5-July-2016]. http://www.id-travel.ch/Downloads/FS_IrisAccess_en.pdf
23. IrisID, iCAM D1000, in *An Eye Fundus Scanner*. [Online; accessed 5-July-2016]. URL: http://www.irisid.com/productssolutions/hardwareproducts/icamd1000
24. Iritech, Inc., *IriShield™ Series*. [Online; accessed 5-July-2016]. URL: http://www.iritech.com/products/hardware/irishield%E2%84%A2-series#
25. J.P. Holmes, L.J. Wright, R.L. Maxwell, *A Performance Evaluation of Biometric Identification Devices* (Sandia National Laboratories, 1991). Technical Report SAND91-0276
26. C. Simon, I. Goldstein, A new scientific method of identification. N. Y. State J. Med. **35**(18), 901–906
27. R.B. Hill, *Biometrics: Retina Identification: Personal Identification in Networked Society* (Springer, 2006). ISBN 978-0-387-28539-9
28. J.P. Holme, L.J. Wright, R.L. Maswell, *A Performance Evaluation of Biometric Identification Devices*. Sandia report, SAND91–0276, (1991), [Online; accessed 5-July-2016]. URL: http://prod.sandia.gov/techlib/access-control.cgi/1991/910276.pdf

29. J. Farmer, *Stop All Federal Abuses Now! S.A.F.A.N.* Internet Newsletter, No. 264, (1997), [Online; accessed 5-July-2016]. URL: http://www.iahushua.com/WOI/illinois.htm
30. EyeDentify Inc., *The Ultimate in Positive Identification, EyeDentification system 7.5*. Leaflet, (1985), [Online; accessed 5-July-2016]. URL: http://simson.net/ref/biometrics/Biometrics/1985.Eyedentify.System7.5.pdf
31. Rayco Security Loss Prevention Systems, Inc., *Retina Verification, ICAM 2001, EyeDentify Retina Biometric Reader*. [Online; accessed 5-July-2016]. URL: http://www.raycosecurity.com/biometrics/EyeDentify.html
32. Trans Pacific (GNF) International, Inc., *EyeKey System*. [Online; accessed 5-July-2016]. URL: http://www.tpi-gnf.com/eky1.htm
33. Retinal Technologies, *A Handheld Scanner*. [Online; accessed 5-July-2016]. URL: http://biometrics.mainguet.org/types/eye_retinal.htm
34. C. Holmes, S. Walmsley, *Biometric Security in Today's Market: An Introduction to Fingerprint, Retinal, and Iris Scanning Technologies*. COP4910 – Frontiers in Information Technology, 2005, [Online; accessed 5-July-2016]. URL: http://pegasus.cc.ucf.edu/~cholmes/homepage/Biometrics.doc
35. A.K. Jain, A.A. Ross, K. Nandakumar, *Handbook of Multibiometrics* (Springer, New York, 2006). ISBN 978-038-7331-232
36. L. Hong, Y. Wan, A.K. Jain, Fingerprint image enhancement: algorithms and performance evaluation, in *IEEE Transactions on Pattern Analysis and Machine Intelligence*, (1998), pp. 777–789
37. T. Matsumoto, H. Matsumoto, K. Yamada, S. Hoshino, R.L. Renesse, Impact of artificial "gummy" fingers on fingerprint systems, in *Optical Security and Counterfeit Deterrence Techniques*, (2002), pp. 275–289. https://doi.org/10.1117/12.462719
38. Techbiometric, *Advantages of Multi-biometric Systems Over Unibiometric Systems*. [Online; accessed 5-July-2016]. URL: http://techbiometric.com/articles/advantages-of-multi-biometric-systems-over-unibiometric-systems
39. M. Faundez-Zanuy, L. O'Gorman, A.K. Jain, N.K. Ratha, Data fusion in biometrics, in *IEEE Aerospace and Electronic Systems Magazine*, (2005), pp. 34–38. https://doi.org/10.1109/MAES.2005.1396793
40. S. Paunovic, I. Jerinić, D. Starčević, Methods for biometric data connection in multimodal systems, in *Proceedings of the XIV International Symposium SYMORG 2014: New Business Models and Sustainable Competitiveness*, (2014), pp. 900–906 ISBN 978-8-676-80295-1
41. K. Nandakumar, *Multibiometric Systems: Fusion Strategies and Template Security* (ProQuest, 2008). ISBN: 978-0-549-61747-1
42. A.K. Jain, B. Chandrasekaran, N.K. Ratha, Dimensionality and sample size considerations in pattern recognition practice, in *Handbook of Statistics*, (1982), p. 835. https://doi.org/10.1016/S0169-7161(82)02042-2
43. N. Radha, A. Kavitha, Rank level fusion using fingerprint and iris biometrics, in *Indian Journal of Computer Science and Engineering*, (2012), pp. 917–923
44. L. Latha, S. Thangasamy, A robust person authentication system based on score level fusion of left and right irises and retinal features, in *Proceedings of the International Conference and Exhibition on Biometrics Technology*, vol. 2, (2010), pp. 111–120. https://doi.org/10.1016/j.procs.2010.11.014
45. D.F. Muller, G.L. Heacock, D.B. Usher, *Method and System for Generating a Combined Retina/Iris Pattern Biometric*, US Patent 7248720, (2007)
46. D. Usher, Y. Tosa, M. Friedman, Ocular biometrics: simultaneous capture and analysis of the retina and iris, in *Advances in Biometrics: Sensors, Algorithms and Systems*, (Springer, 2008), pp. 133–155 ISBN: 1846289203
47. Find Biometrics, *Retica Systems Inc. Announces the World's First Iris-Retina Biometric System*. 2006, [Online; accessed 5-July-2016]. URL: http://findbiometrics.com/retica-systems-inc-announces-the-worlds-first-iris-retina-biometric-system

48. PR Newswire, *Retica Systems Inc. Announces the World's First Iris-Retina Biometric System*. News 2006, [Online; accessed 5-July-2016]. URL: http://www.prnewswire.com/news-releases/retica-systems-inc-announces-the-worlds-first-iris-retina-biometric-system-55935287.html
49. A. Jóźwik, D. Siedlecki, M. Zając, Analysis of Purkinje images as an effective method for estimation of intraocular lens implant location in the eyeball. Optik Int. J. Light Electron Opt. **125**(20), 6021–6025 (2014). https://doi.org/10.1016/j.ijleo.2014.06.130
50. J.G. Daugman, *Biometric Personal Identification System Based on Iris Analysis*. US Patent 5291560, (1994)
51. R.P. Wildes, J.C. Asmuth, K.J. Hanna, S.C. Hsu, R.J. Kolczynski, J.R. Matey, S.E. McBride, *Automated, Non-invasive Iris Recognition System and Method*. US Patents 5572596 and 5751836, (1996)
52. M. Barbosa, A.C. James, Joint iris boundary detection and fit: a real-time method for accurate pupil tracking. Biomed. Opt. Express **5**(8), 2458–2470 (2014). https://doi.org/10.1364/BOE.5.002458
53. I.A. Saad, L.E. George, Robust and fast iris localization using contrast stretching and leading edge detection. Int. J. Emerg. Trends Technol. Comput. Sci. **3**, 61–67 (2014). ISSN 2278-6856
54. S. Qamber, Z. Waheed, M.U. Akram, Personal identification system based on vascular pattern of human retina, in *Cairo International Biomedical Engineering Conference, 2012*, vol. 2012, pp. 64–677 ISBN 978-1-4673-2800-5
55. H. Oinonen, H. Forsvik, P. Ruusuvuori, O. Yli-Harja, V. Voipio, H. Huttunen, Identity verification based on vessel matching from fundus images, in *Proceedings of IEEE International Conference on Image Processing*, (2010), pp. 4089–4092 ISBN 978-1-4244-7993-1, ISSN 1522-4880
56. C. Mariño, M.G. Penedo, M. Penas, M.J. Carreira, F. Gonzalez, Personal authentication using digital retinal images. Springer Pattern Anal. Appl **9**(1), 21–33 (2006). ISSN 1433-7541
57. C. Köse, C. İkibaş, A personal identification system using retinal vasculature in retinal fundus images. Expert Syst. Appl. **38**(11), 13670–13681 (2011)
58. W. Barkhoda, F. Akhlaqian, M. Amiri, M. Nouroozzadeh, Retina identification based on the pattern of blood vessels using fuzzy logic, in *EURASIP Journal of Advances in Signal Processing*, (2011), pp. 113–121 ISSN: 1687-6180
59. H. Borgen, P. Bours, Wolthusen S. D., Visible-spectrum biometric retina recognition, in *Proceedings of International Conference on Intelligent Information Hiding and Multimedia Signal Processing*, (2008), pp. 1056–1062 ISBN 978-0-7695-3278-3
60. G.V. Saradhi, S. Balasubramanian, V. Chandrasekaran, Performance enhancement of optic disc boundary detection using active contours via improved homogenization of optic disc region, in *International Conference on Information and Automation*, (ICIA, 2006), pp. 264–269 ISSN 2151-1802
61. P.C. Siddalingaswamy, G.K. Prabhu, Automated detection of anatomical structures in retinal images. Int. Conf. Comput. Intell. Multimed. Appl. **3**(10), 164–168 (2007). ISBN 0-7695-3050-8
62. S. Tamura, Y. Okamoto, K. Yanashima, Zero-crossing interval correction in tracing eye-fundus blood vessels. Pattern Recogn. **3**, 227–233 (1988). https://doi.org/10.1016/0031-3203(88)90057-x
63. A. Pinz, S. Bernogge, P. Datlinger, et al., Mapping the human retina, in *IEEE Transactions on Medical Imaging, No. 4*, (1998), pp. 606–619. https://doi.org/10.1109/42.730405
64. D.W.K. Wong, J. Liu, N.M. Tan, et al., Automatic detection of the macula in retinal fundus images using seeded mode tracking approach, in *IEEE Engineering in Medicine and Biology Society, Institute of Electrical & Electronics Engineers (IEEE)*, (2012). https://doi.org/10.1109/embc.2012.6347103
65. J. Parker, *Algorithms for Image Processing and Computer Vision* (Wiley Computer Publishing, New York, 1997). ISBN 04-711-4056-2

66. C. Sinthanayothin, J.F. Boyce, H.L. Cook, et al., Automated localization of the optic disc, fovea, and retinal blood vessels from digital color fundus images. Br. J. Ophthalmol. **83**(8), 902–910 (1999). 10.1136/bjo.83.8.902
67. S. Umbaugh, *Digital Image Processing and Analysis: Human and Computer Vision Applications with CVIPtools* (CRC Press., ISBN 978-1-4398-0205-2, Boca Raton, FL, 2011)
68. T. Johar, P. Kaushik, Iris segmentation and normalization using Daugman's rubber sheet model. Int. J. Sci. Tech. Adv. **1**(3) (2015). ISSN: 2454-1532
69. G. Kavitha, S. Ramakrishnan, Identification and analysis of macula in retinal images using Ant Colony Optimization based hybrid method, in *2009 World Congress on Nature & Biologically Inspired Computing (NaBIC), Institute of Electrical & Electronics Engineers (IEEE)*, (2009). https://doi.org/10.1109/nabic.2009.5393783
70. M. Mubbashar, A. Usman, M.U. Akram, Automated system for macula detection in digital retinal images, in *IEEE International Conference on Information and Communication Technologies*, (2011). https://doi.org/10.1109/icict.2011.5983555
71. D.W.K. Wong, J. Liu, N.M. Tan, et al., Automatic detection of the macula in retinal fundus images using seeded mode tracking approach, in *IEEE Engineering in Medicine and Biology Society, Institute of Electrical & Electronics Engineers (IEEE)*, (2012). https://doi.org/10.1109/embc.2012.6347103
72. P. Verlinde, G. Chollet, Comparing decision fusion paradigms using k-NN based classifiers, decision trees and logistic regression in a multi-modal identity verification application, in *Second International Conference on Audio and Video-Based Biometric Person Authentication*, (2003)
73. S. Chaudhuri, S. Chatterjee, N. Katz, M. Nelson, M. Goldbaum, Detection of blood vessels in retinal images using two-dimensional matched filters, in *IEEE Transactions on Medical Imaging, No. 3*, (1989), pp. 263–269 ISSN 0278-0062
74. B. Jähne, *Digital Image Processing Concepts, Algorithms, and Scientific Applications* (Springer, Berlin Heidelberg, 1997). ISBN 978-3-662-03479-8
75. A. Aquino, M.E. Gegúndez, D. Marín, Automated optic disc detection in retinal images of patients with diabetic retinopathy and risk of macular edema, in *World Academy of Science, Engineering and Technology, No. 60*, (2009), pp. 87–92
76. J. Parker, *Algorithms for Image Processing and Computer Vision* (John Wiley & Sons, 1996). isbn:0471140562
77. P.S. Heckbert, *Graphics Gems IV*, Graphic Gems Series (AP Professional, 1994). ISBN 0-12-336155-9
78. N. Dede, *Implementation of Thinning Algorithm in OpenCV*. OpenCV Code, [Online; accessed 5-July-2016]. URL: http://opencv-code.com/quick-tips/implementation-of-thinning-algorithm-in-opencv
79. J.G. Daugman, High confidence visual recognition of persons by a test of statistical independence. IEEE Trans. Pattern Anal. Mach. Intell. **15**(11), 1148–1161 (1993). ISSN 0162-8828
80. L. Ma, T. Tan, Y. Wang, D. Zhang, Efficient iris recognition by characterizing key local variations. IEEE Trans. Image Process. **13**(6), 739–750 (2004)
81. D.M. Monro, S. Rakshit, D. Zhang, DCT-based iris recognition. IEEE Trans. Pattern Anal. Mach. Intell. **29**(4), 586–595 (2007). ISSN 0162-8828
82. S. Sun, S. Yang, L. Zhao, Non-cooperative bovine iris recognition via SIFT. Neurocomputing **120**, 310–317 (2013). https://doi.org/10.1016/j.neucom.2012.08.068
83. H. Mehrotra, P.K. Sa, B. Majhi, Fast segmentation and adaptive SURF descriptor for iris recognition. Math. Comput. Modell., Elsevier **58**(1–2, 132), –146 (2013). ISSN: 0895-7177
84. H. Rai, A. Yadav, Iris recognition using combined support vector machine and hamming distance approach. Exp. Syst. Appl. **41**(2), 588–593 (2014)
85. T. Ojala, T. Pietikäinen, D. Harwood, Performance evaluation of texture measures with classification based on Kullback discrimination of distributions, in *Proceedings of the 12th IAPR International Conference on Pattern Recognition*, vol. 1, (1994), pp. 582–585 ISBN 0-8186-6265-4

86. M.Y. Shams, M.Z. Rashad, O. Nomir, M.Z. El-Awady, Iris recognition based on LBP and combined LVQ classifier. Int. J. Comput. Sci. Inf. Technol. **3**(5) (2011). https://doi.org/10.5121/ijcsit.2011.3506
87. J. Macek, *Klasifikace a rozpoznávání patologických nálezů v obrazech sítnice oka (Classification and Recognition of Pathologic Foundings in Eye Retina Images)*. Master's thesis, Faculty of Information Technology, Brno University of Technology, (2015)
88. M. Yanoff, *Ophthalmology*, 3rd edn. (Mosby Elsevier, 2009). ISBN 978-0-323-04332-8
89. Mayo Clinic, *Indirect Ophthalmoscopy*. [Online; accessed 5-July-2016]. URL: http://www.mayoclinic.org/tests-procedures/eye-exam/multimedia/indirect-ophthalmoscopy/img-20006175
90. S.E. Sherman, *History of Ophthalmology: The History of the Ophthalmoscope*, vol 2 (Springer, 1989), pp. 221–228 ISBN 978-0-7923-0273-5
91. R.L. Wiggins, K.D. Vaughan, G.B. Friedmann, Holography using a fundus camera. Appl. Opt. (1), 179–181 (1972). https://doi.org/10.1364/AO.11.000179
92. J. Orellana, A.H. Friedman, *Clinico-Pathological Atlas of Congenital Fundus Disorders: Best's Disease* (Springer, 1993), pp. 147–150., ISBN 978-1-4613-9322-1
93. A.P. Schachat, P. Wilkinson Ch, D.R. Hinton, P. Wilkinson, *Retina*, 4th edn. (Mosby Elsevier, 2005). ISBN 978-0-323-04323-6
94. L. Poretsky (ed.), *Principles of Diabetes Mellitus*, 2nd edn. (Springer, 2010). ISBN 978-0-387-09840-1
95. W. Gloria, *Diabetic Retinopathy: The Essentials*. LWW; 1 Har/Psc edition, (2010.), ISBN 1605476625
96. American Optometric Association, Diabetic eye disease, in *Diabetic Eye Disease*, (2009), [Online; accessed 5-July-2016], URL: http://www.slideshare.net/MedicineAndHealth14/diabetic-eye-disease
97. W. Lihteh, *Ophthalmologic Manifestations of Toxoplasmosis*. [Online; accessed 5-July-2016]. URL: http://emedicine.medscape.com/article/2044905-overview
98. J.D. Camet, H. Talabani, E. Delair, F. Leslé, H. Yera, A.P. Brézin, *Toxoplasmosis – Recent Advances: Risk Factors, Pathogenesis and Diagnosis of Ocular Toxoplasmosis* (Intech, 2012). ISBN 978-953-51-0746-0
99. C. Tisse, L. Martin, L. Torres, M. Robert, Person identification technique using human iris recognition, in *Proceedings of ICVI 2002*, (2002), pp. 294–299
100. H.E. Lahn, *Iridology: The Diagnosis from the Eye* (Kessinger Publishing, 2010). ISBN 978-1162622729
101. M.U. Akram, S. Khalid, S.A. Khan, Identification and classification of microaneurysms for early detection of diabetic retinopathy. Pattern Recogn. **46**(1), 107–116 (2013). https://doi.org/10.1016/j.patcog.2012.07.002.s
102. S. Qamber, Z. Waheed, M.U. Akram, Personal identification system based on vascular pattern of human retina, in *Cairo International Biomedical Engineering Conference, 2012*, (2012), pp. 64–677 ISBN 978-1-4673-2800-5

Chapter 4
3D Hand Geometry Recognition

Michal Dvořák and Martin Drahanský

1 Introduction

How can a person be reliably identified or verified and how secure is this method? The area of biometrics offers a solution to this issue by claiming that to determine one's identity, only some of his/her physical traits are needed. Many security systems of today identify people using various biometrics, most often fingerprint, facial markers, or retina scan. These systems, however, are not applicable in every situation either due to environmental, price, or other reasons, and another method may be required. Hand geometry represents an alternative approach to human identification and verification that avoids issues that make more conventional approaches inconvenient, such as placement in areas where the face needs to be covered or where fingerprints can become obscured (non-clean environment).

The fundamental idea of hand-based geometry identification is built around the assumption that everyone possesses a hand and its shape is unique. In other words, if a reference shape of user's hand is on record, this person's identity can be established by comparing his immediate hand shape with the one on record. It was this assumption that gave birth to the first systems designed to do just so.

The first problem that needs to be solved is to identify the characteristics that can be both efficiently collected without compromising the requirement for uniqueness. 2D hand geometry then represents a viable abstraction method where the parameters are such as silhouette or length, width, and thickness of fingers [1].

In order to expand into a 3D system, a measurement of depth has to be introduced; this then in turn allows parameters such as curvatures of individual fingers and wrist as well as plastic deformations on the back or palm of the hand to be measured.

M. Dvořák (✉) · M. Drahanský
Brno University of Technology, Centre of Excellence IT4Innovations, Brno, Czech Republic
e-mail: idvorakmi@fit.vutbr.cz; drahan@fit.vutbr.cz

M. S. Obaidat et al. (eds.), *Biometric-Based Physical and Cybersecurity Systems*,
https://doi.org/10.1007/978-3-319-98734-7_4

2 2D Hand Geometry Acquisition

To better understand the issues presented by hand biometrics, let's first take a brief look on principle of 2D acquisition.

The method itself is tantalizingly simple, since all that is needed is a silhouette of a hand. For this task, all that is required is a CCD or CMOS camera with sufficient resolution, and whereas for fingerprint identification, this resolution may be high and thus require a more expensive sensor, e.g., 500 DPI; the hand geometry systems will operate with much lower-resolution cameras in commercial systems usually 100–200 DPI but viable with even less than 100 DPI [2].

During 2D acquisition a static image of a back of a hand is usually captured, which is followed by the extraction of required features. Depending on system, additional images may be acquired from different angles in order to gather additional features. This usually takes form of capturing the image from the top and side in order to gain all previously mentioned features [1]. As can be seen on Fig. 4.1, image from this direction allows for efficient gathering of majority of required features.

The system may be further modified to improve the acquisition process and simplify the feature extraction stage, such as including reflective background to enhance edges or add cameras to have more sources of features and pins which ensure that the hand is in a predictable position; the principle however remains the same. Figure 4.2 shows the possible set of features to be extracted. All of currently available solutions are based on this principle.

2.1 Existing Commercial Solutions

As far as authors are concerned, there are to this day only two commercial systems being currently developed, both under the Schlage. These are devices of HandPunch® and HandKey® series, respectively.

At the time of writing this chapter, the newest model of the series was HandKey II [3], HandPunch GT-400 [4], and HandPunch 4000 [5]. HandKey II, as can be seen

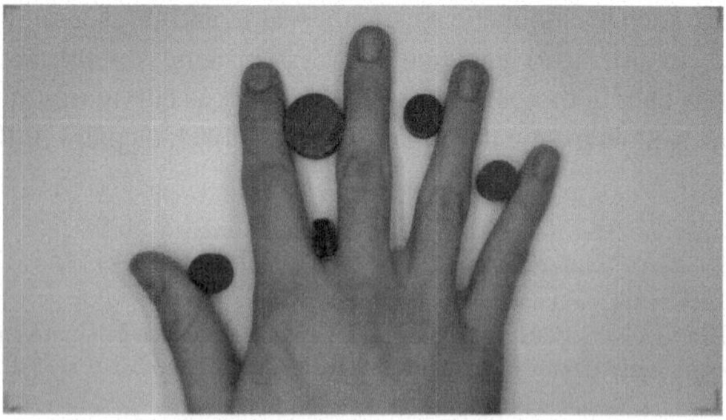

Fig. 4.1 Gathered image of a hand for 2D hand geometry extraction of features

on Fig. 4.2, is foremost an access device, designed to primarily serve as a security access mechanism which is the reason why it has features such as duress code. The base user memory of the device is 512 user fields which is expandable. Per available information, the hand geometry features are after extraction compressed to 9 byte large template.

The HandPunch series is per description to primarily serve as a time and attendance terminal. HandPunch 4000, like the HandKey II, employs 9 byte large template size; the default user memory is 530, with offered expandability to 3498. The HandPunch GT series differs in design but appears to be similar in function. The template size is different, as it is represented by 20 bytes instead of 9 bytes. The device also has a larger default user field size of 1000 and no offer of expansion. On Fig. 4.3 the appearance of HandPunch device can be seen.

Fig. 4.2 Image of HandKey II biometrics hand geometry reader [3]

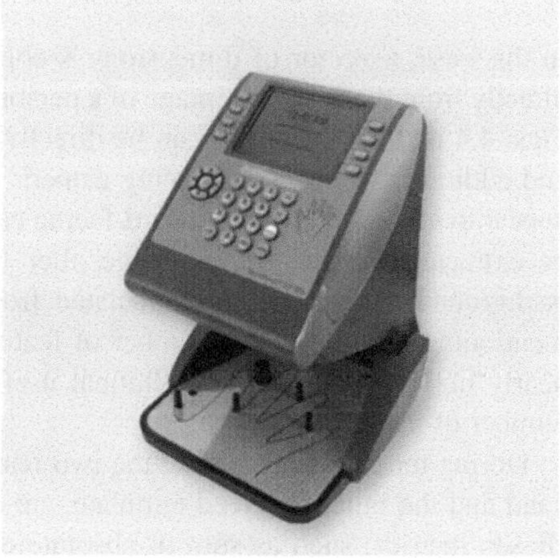

Fig. 4.3 Image of HandPunch GT-400 biometrics hand geometry reader [4]

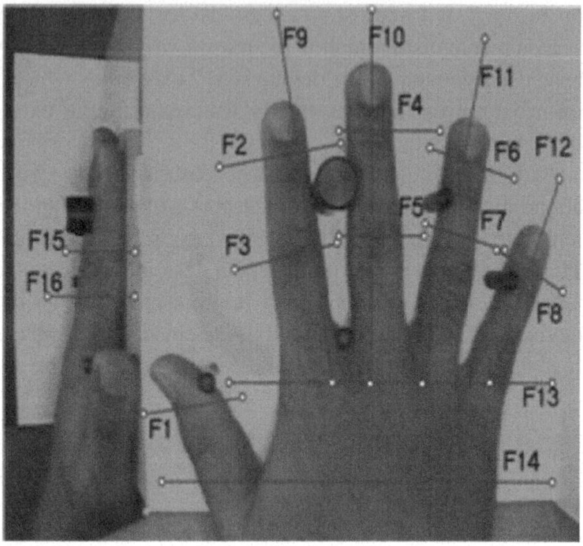

Fig. 4.4 Viable hand 2D hand geometry features [2]

2.2 *Identification Methods*

To identify a user based on their hand geometry, methodology needs to be devised to extract unique set of features, by which comparison of one set can be determined to belong to the same person. Since the likelihood of the source images being completely identical is minute, due to small placement and orientation differences, as well as small changes in hand shape over time, the methods need to accommodate a tolerance in the measurements. In general, methods based on direct measurements and hand shape alignment are used today.

2.2.1 Methods Based on Direct Measurement

In this case, a vector of dimensions is constructed based on measurements gained directly from the acquired image of a person's hand. As can be seen, for example, on Figs. 4.4 and 4.14 features can be directly inferred from downward-facing camera and additional 2 from side-facing camera. All the commercially available systems appear to utilize solely the upward-facing camera. The features themselves can either be extracted from a binary image after the segmentation of the hand from the background or they can be calculated from the defined pixel ranges. It is worth mentioning that while the number of features that can be extracted from image is nearly limitless, very little additional useful information is gained with increased number of features (Fig. 4.5).

During the matching phase, the two feature vectors, one of the newly measured hand and the other of saved template, are compared against each other based on a chosen metrics, such as sum of absolute difference (1), weighted sum of absolute

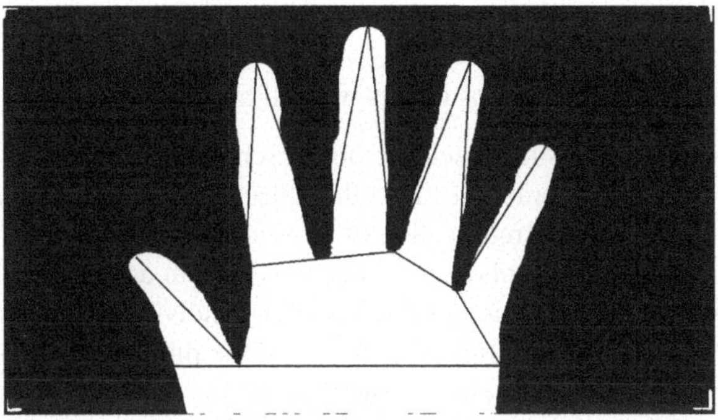

Fig. 4.5 Other possible set of extractable features

difference (2), Euclidean distance (3), or weighted Euclidean distance (4) [1]. The measured features are defined as $x_i = [x_1, x_2, \ldots, x_z]$, template features as $y_i = [y_1, y_2, \ldots, y_z]$, and σ_i as a weight of the given feature.

$$\sum_{i=1}^{z} |x_i - y_i| \tag{4.1}$$

$$\sum_{i=1}^{z} \frac{|x_i - y_i|}{\sigma_i} \tag{4.2}$$

$$\sqrt{\sum_{i=1}^{z} (x_i - y_i)^2} \tag{4.3}$$

$$\sqrt{\sum_{i=1}^{z} \frac{(x_i - y_i)^2}{\sigma_i^2}} \tag{4.4}$$

Both advantage and disadvantage of this method are the requirement for pins. For the feature extraction, same set of pixels is used every time, since the range of movement of the hand is restricted by these pins. The pins are accomplished by guiding the hand into a predefined position. The advantage is that this method is fast and/or requires very low computational power as there is no need to algorithmically determine the position of each feature. The disadvantages are the pins themselves; the necessity of them prevents the method to become touchless and disqualifies the usage of dynamic acquisition, as it requires proper position before the image is acquired. And while the commercial devices have an antibacterial layer, it still can be a potential health risk.

The shortcomings of this method can be addressed by methods based on silhouette matching.

2.2.2 Methods Based on Silhouette Alignment

In this method, fixed position is not required, even though it can be applied even to a pin-guided system. Instead of measuring fixed distances, , a measure of difference between a new outline/silhouette of a hand and the outline stored as a template, is calculated.

The general steps of this method are as follows:

- Preprocessing based on the chosen system.
- Outline extraction, since the object should be easily separable from background, only a preferred method of thresholding is required to acquire a binary image and usage of morphologic operations to gain an outline of the measured hand.
- Finger extraction, the outlines of individual fingers need to be extracted. To identify them, tips and valleys of the fingers are detected. This step is especially necessary in pinless systems.
- Align the corresponding finger pairs with template. To do this, any algorithms that measure distance of two metric subspaces can be used, for example, modified Hausdorff distance [6] where the orientation and position of one outline are gradually changed to achieve closest alignment with the reference.
- Calculate the distance of the pair. MAE (*mean alignment error*) is then calculated, an average distance difference between corresponding points of the outlines.
- If MAE is less than the decided threshold, it is declared that the measured hand is of the same person as the one saved in template.

On Fig. 4.6 the fingers of measured hand aligned on top of the template saved in database can be seen. MAE would now be calculated, and based on thresholding it would be decided whether the analyzed hand matches the template. (In this case it clearly does not.)

While both these methods serve the purpose they are designed for, they do exhibit several limitations.

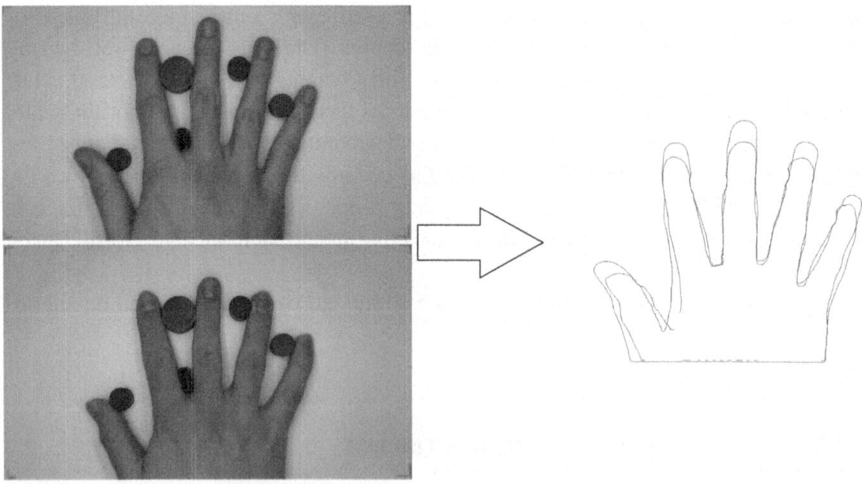

Fig. 4.6 – Hand image used as a source and second hand image (left), outline overlap after finger positioning (right)

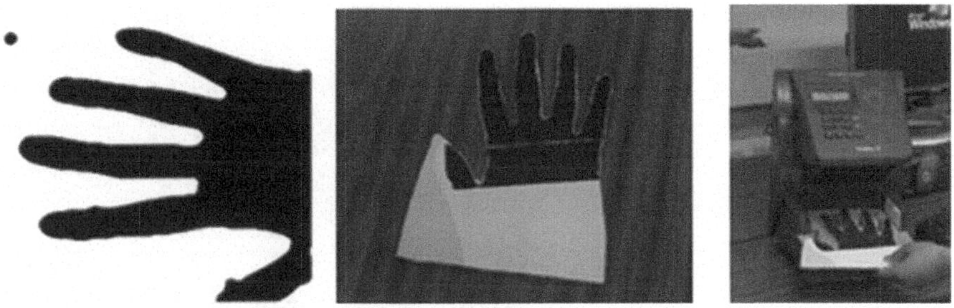

Fig. 4.7 Paper silhouette spoof used to bypass commercial device [8]

2.3 Limitations of Currently Employed 2D Hand Geometry-Based Biometrics

The simplicity of acquisition and low demands on the resolution of the sensor present are, however, several shortcomings; while some are still being addressed, others are insurmountable due to the limitations of 2D acquisition .

Despite the improvement of existing commercial hand recognition tools such as HandKey II [3], due to the fact that only relatively small amount of information is available to begin with, either the database size or success rate of identification has to be sacrificed, as proved by the continued research into uniqueness of one's hand and its measurements as viable identification parameter of large population [7]. The transition from 2D to 3D hand recognition adds features and feature vectors, thus increasing an entropy and offering a solution to the issue.

Spoofing presents another challenge as was demonstrated in [8] where a simple paper silhouette was used as a spoof that could be used to introduce a successful impostor. To detect a spoof, the system has to be able to acquire data that is difficult to replicate when developing a spoof, either by including detection of large amount information or by detecting a property of human hand that is inherently unavailable in an artificial spoof, such as detecting whether currently analyzed object is in fact part of the living human body (Fig. 4.7).

The approaches to verifying liveliness of hand are multiple; among the most common are the approaches based on detecting body heat for confirmation, which however increases complexity and costs of the system, especially of system that is to be touchless. The resistivity/conductivity of the human skin can be used as well; it too however imposes the requirement for the system to be touch based. From the optical approaches, which would meet the criteria for making the system touchless, is the most common one, the utilization of vein detection. This can be achieved either by a surface detection utilizing a near-infrared light (NIR) or subsequent detection, i.e., Fig. 4.8.

Alternatively, system can capture a dispersed light passing through the hand as in Fig. 4.9. With this approach, usually only fingers are used as the whole hand presents too much tissue for the NIR and the absorption of NIR by hemoglobin is no longer

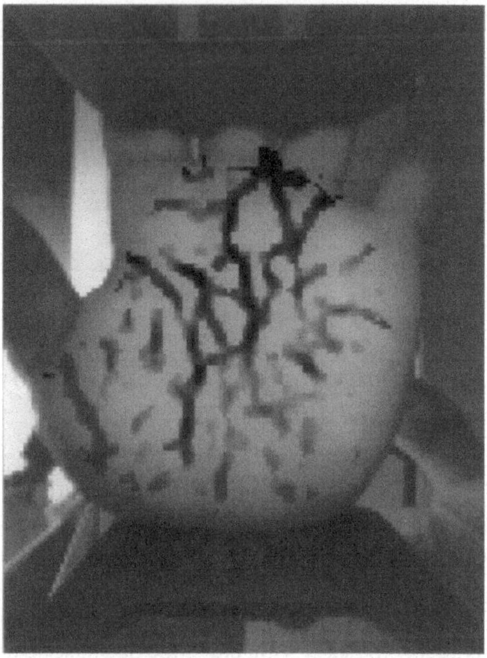

Fig. 4.8 Surface vein detection using NIR

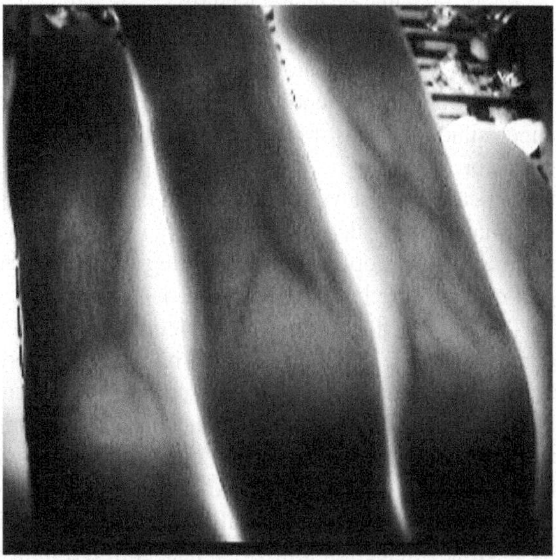

Fig. 4.9 Vein detection using NIR light source behind object

easily detectable without special light source. Either of these approaches is easily implementable into the hand biometric system, as the NIR can be used as a light source in CCD- or CMOS-based systems. The vein detection can then be used either as only a liveliness verification or preferably as an additional source of unique and identifiable features.

The multidimensional approach can assist by increasing the amount of information needed for construction of spoof. Necessary dimensions gathering from uncooperative target become a nontrivial task, as a majority of 3D scanning approaches requires scanned object in a specific position or the accuracy will decrease. As far as authors are concerned to date, no one has presented a method of acquiring a 3D scan of a hand from an uncooperative user with accuracy necessary to create a viable spoof. As will be explored further in subsequent parts of this chapter, the experiments have been performed in this area and support the assumption of increased entropy.

The last limitation that needs to be mentioned here is the difficulty in creating a pinless and contactless solution. As far as the authors are concerned, there are no existing pinless and by extension contactless commercial devices available now. While the issue of pinless hand recognition has been approached even in pure 2D hand recognition systems, it has been demonstrated that utilization of depth sensing devices is a viable method of approaching this problem [9] as well as the need for the physical contact with the sensing device.

3 3D Acquisition

The principle of 3D acquisition in biometrics has been successfully tested. The price and high computational demands limited these tests to academic research in the past. However, as the price of viable 3D acquisition devices keeps decreasing and the computational capabilities of even mobile hardware increasing, the 3D acquisition has become a viable alternative to 2D hand geometry systems.

On Fig. 4.10 a 3D image captured using low-cost Creative VF0800 camera [10] can be seen. On the image, it can be seen that, along with required 3D features, majority of 2D features can be inferred as well, which allow extraction of the 3D and

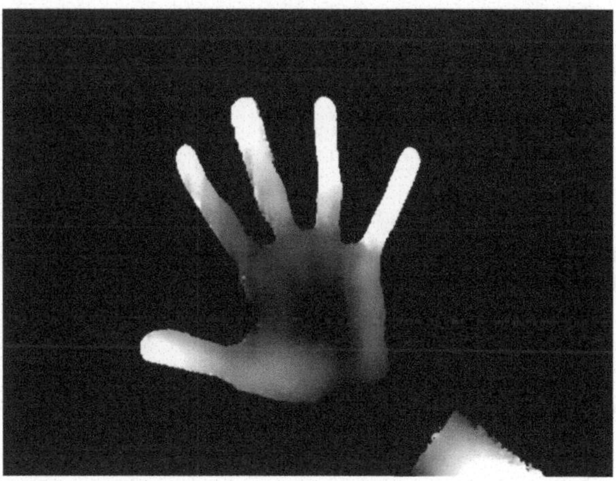

Fig. 4.10 3D image of hand using low-cost 3D camera

2D features with no additional hardware. As the depth map also includes information necessary to determine absolute dimensions of scanned object, it can also be used as a basis for a contactless system.

While any 3D mapping method may be used, in 3D hand geometry, it can be observed that majority of academic publications use a system based on principle of active triangulation, be it the industrial 3D laser digitizers [11, 12] or systems based on light pattern projection [13].

3.1 *Using 3D Scanner*

Utilization of professional laser scanner in biometrics can be seen in [11] – in this case a Minolta Vivid 910 has been used for creation of large database (3540 items) of right hand scans, where the device can work with accuracy in three axes up to X, 0.22 mm; Y, 0.16 mm; and Z, 0.10 mm to the Z reference plane [14].

The method presented in [11] localizes the position of four fingers on intensity map and based on this information extracts corresponding data from range map. The cross sections of each finger are extracted at chosen distances along the finger's length. This paper then proceeds to calculate features based on curvature and vector of normal of the finger segment. Figure 4.11 shows a depth data of cross section, calculating curvature and normal features.

In the paper, the 2D and 3D data is experimentally matched, and EER (equal error rate) and AUC (area under ROC curve) are calculated (Table 4.1).

As expected, the combined 2D and 3D geometry data provides the best performance.

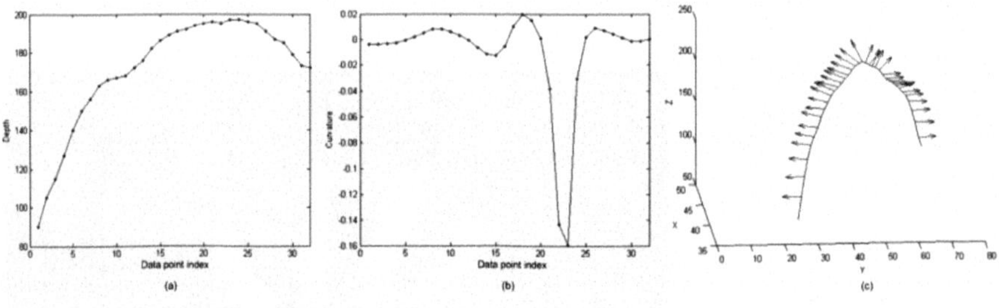

Fig. 4.11 (**a**) Cross-sectional finger segment and (**b**) its computed curvature features; (**c**) normal features computed for a finger segment [12]

Table 4.1 EER and AUC of 2D, 3D, and a combined matcher [12]

Matcher	EER [%]	AUC
3D hand geometry	3.5	0.9655
2D hand geometry	6.3	0.9722
(2D + 3D) hand geometry	2.3	0.9888

3.2 *Using Structured Light*

The use of structured light is based on the projection of visible or infrared pattern of defined properties onto a surface of an object, in our case a human hand. On the surface of a plastic object, the pattern appears deformed; by recording the direction and magnitude of this deformation, and comparing it to expected position, the depth of the pixel can be calculated. This approach has been used in several publications, where various sources of the pattern have been tested.

3.2.1 IR Pattern Projection

This approach became viable thanks to the spread of affordable 3D cameras such Intel RealSense or Microsoft Kinect. The cameras in this case provide both an RGB image and a depth map, with corresponding coordinates.

In [12] it was demonstrated that by using Intel RealSense camera, 2D and 3D features may be collected. After the preprocessing, during which markers such as fingertips, finger valleys, and wrist lines are identified, a vector including 41-dimensional vector of 2D features and 137-dimensional vector of 3D features can be extracted. Features are:

- Finger length (2D)
- Finger valley distance (2D)
- Finger width (2D)
- Wrist to valley distance (2D)
- Finger axis surface distance (3D)
- Finger width (3D)

Figures 4.12 and 4.13 show the respective features on the depth images and intensity images.

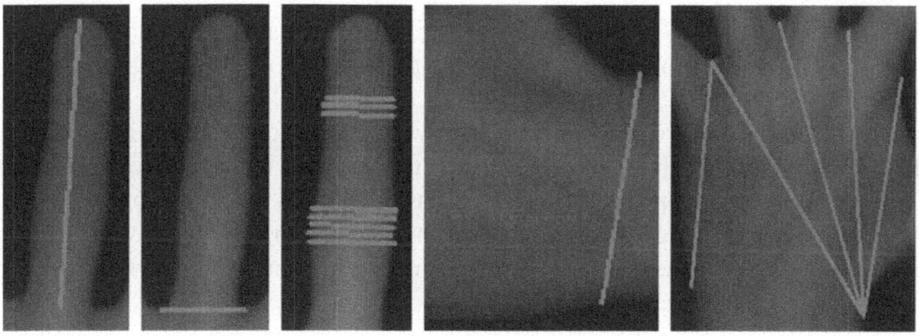

Fig. 4.12 Extracted 2D features [13]

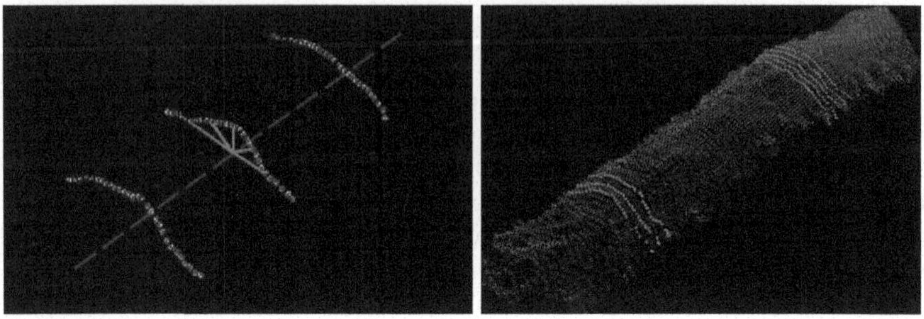

Fig. 4.13 Extracted 3D features [13]

Table 4.2 Overall performance of Bronstein and Drahansky system using LMNN metric learning approach [12]

		FRR [%]	
Features	ERR [%]	@FAR = 0.5%	@FAR = 1%
2D	2.76	6.50	1.63
3D	2.92	10.70	6.42
2D + 3D	1.61	4.81	2.23

Table 4.3 Qualitative comparison of hand biometric-based matchers [12]

Method	Features	Templates	Database	FAR [%]	FRR [%]	EER [%]
Jain and Duta	2D	1–14	53	2.00	3.50	N/A
Jain et al.	2D	1 (avg)	50	2.00	15.00	N/A
Woodard and Flynn	2D + 3D	1 (avg)	177	5.50	5.50	5.50
Malassiotis et al.	2D + 3D	4	73	3.60	3.60	3.60
Kumar et al.	2D + 3D	5	100	5.30	8.20	N/A
Kanhangad et al.	2D + 3D	5	177	2.60	2.60	2.60
Bronstein and Drahansky	2D + 3D	1 (avg)	88	1.61	1.61	1.61

The feature vectors are in this case matched using the *Mahalanobis distance* [15], utilizing *large margin nearest neighbors* (LMNN) [16] the weights for individual features.

As the 2D and 3D feature vectors are calculated separately, a comparison of individual vector performance as well as the performance of their combination can be examined (Table 4.2).

From these data, it can be inferred that while 3D features alone with low-cost hardware, in combination with 2D features, the resultant method proves superior. The improvement is especially apparent in case of images where one of the source vectors suffers high level of noise. Where second stream serves to improve to overall recognition rate, this method proved to be comparable even with the state-of-the-art approaches (Table 4.3).

3.2.2 Lasers and Diffraction Grating

A shortcoming of generic commercial light pattern-based approach is its lack of detail and operation on short range, and even though the results were still satisfactory, the system can be improved upon by designing the light projection for the application. Both the wavelength and the actual pattern can then be chosen, based on a priory information about the measured object.

In [17] the array of 532 nm lasers is being used with simple two MS LifeCam HD 3000 camcorders used for acquisition. Optics on laser turns the dot projection into line projection, and diffraction grating then allows to turn the single line into an array of parallel lines. On Fig. 4.14, it can be observed how the resultant line pattern appears on hand and how it can be separated from the background.

As can be seen, on Fig. 4.15 the concept has been proven, and the model of 3D hand may be reconstructed despite the low cost of entire setup.

So far, this method has only been presented as a viable scanning method and no database has been produced. As is, the method performs a preprocessing necessary for 2D hand geometry identification and thus can be seen as a viable approach to

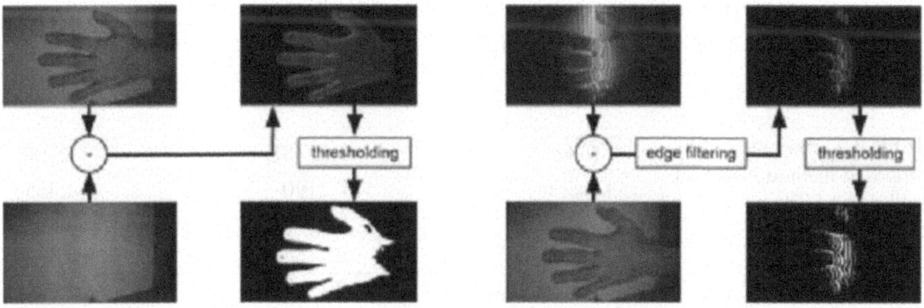

Fig. 4.14 The process of extracting the deformation markers, created by diffracting line projection [17]

Fig. 4.15 Resultant surface reconstruction from the line projection [17]

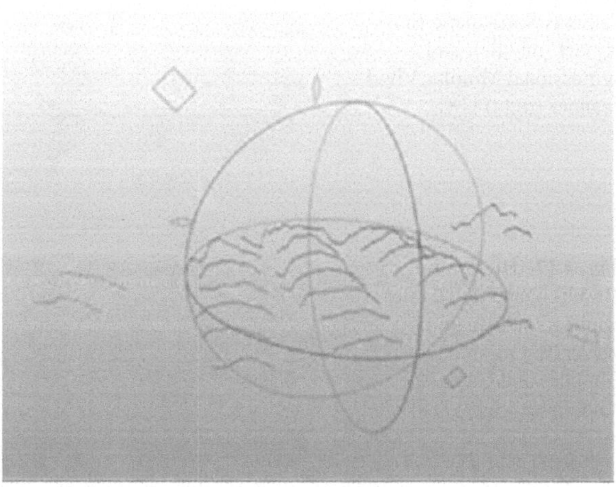

increasing the entropy of the hand biometrics. The accuracy of method has been demonstrated by scanning an object of known dimensions.

The cube with side of length 25.5 mm has been scanned, and the root mean squared and normalized root mean squared have been calculated (Table 4.4).

3.3 *Time-of-Flight (TOF) Scanning*

TOF cameras due to the principle of their operation are more suited for scanning large objects. However with emergence of low-cost systems such as SoftKinetic DS325 and Kinect v2 RGB-D camera, even this form of image capture has become available.

While the research in hand biometry using the TOF has been limited solely to gesture detection, there has been an investigation on potential biometric applications, especially with regard to face recognition [18]. The use of TOF cameras has been tested in order to capture a depth map of user's facial features from various technologies, as can be seen on Fig. 4.16.

Various approaches to filtering 3D meshes have been investigated, such as feature preserving mesh denoising, and show that despite the high noise level, the information can be extracted with accuracy high enough for rudimentary biometric identification (Fig. 4.17).

Table 4.4 Accuracy of the proposed method based on calibration [17]

Model	NRMSE	RMSE
Laser 0	0.2190	1.4558
Laser 1	0.1993	0.7977
Laser 2	0.1764	0.4317
Merged	0.1764	1.0474

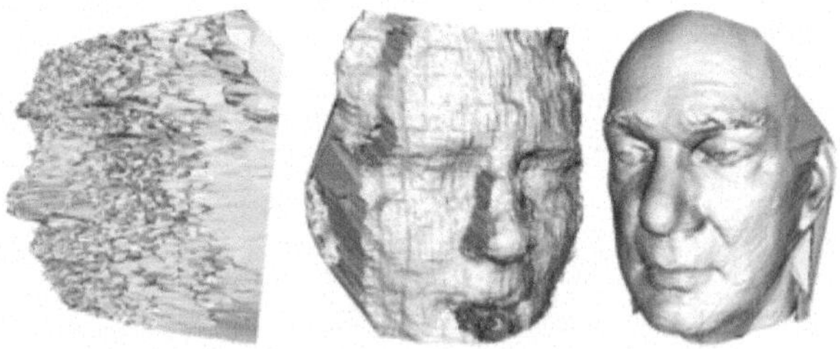

Fig. 4.16 Example 3D acquisition using TOF cameras SoftKinetic (left), Kinect (middle), and professional Minolta Vivid scanner (right) [18]

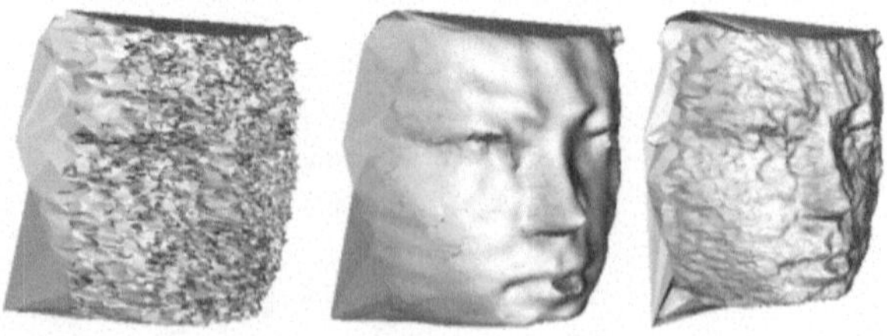

Fig. 4.17 High-noise DS325 image (left), after application of feature preserving mesh denoising (middle) and Gaussian smoothing (right) [18]

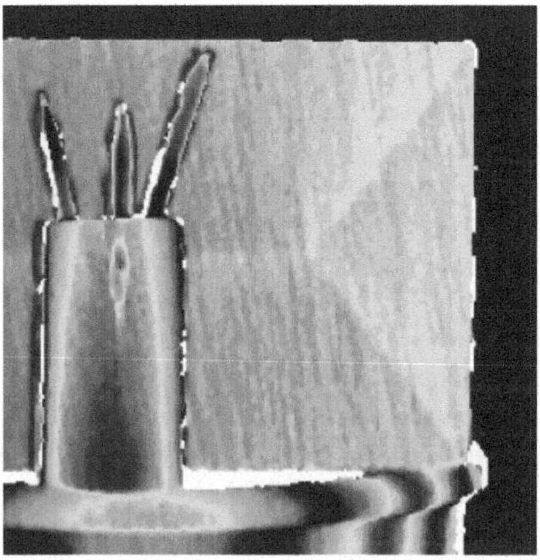

Fig. 4.18 Example of 3D image created using stereovision [20]

3.4 Stereovision Approach

Stereo vision utilizes multiple cameras fixed in known position. Due to a distance between them, an observed object will appear displaced on the two images acquired from two cameras. The magnitude of this displacement can be used to calculate a position of the object in three axes, and if resolution is high enough, the shape of an object can be determined, as the deformations too appear displaced on two images.

Currently there are multiple produces offering a 3D stereo image acquisition with declared resolution high enough for biometric usage, for example, ENSENSO N10-304-18 with resolution in Z axis up to 0.109 mm [19]. The most prevalent issue of stereovision technology appears when scanning homogenous object; however, as the hand due to papillary lines is not homogenous, this issue in our case does not apply. In theory any calibrated two camera setup can, given enough information, generate a 3D model using specialized software (Fig. 4.18).

To bypass homogeneity, random noise projection is sometimes used. By projecting the noise onto the object, the parallax calculation can be calculated for every visible projected pixel.

4 Utilization of Line Scanners in Biometric Acquisition

Direction in image acquisition that is now being explored by the authors is utilization of line scanners for 2D and 3D multimodal hand biometrics. The advantages present themselves in the form of acquisition of a dynamic object (hand). This would lead to increasing a throughput at a gate utilizing this system, while maintaining image quality necessary for multimodal feature extraction, namely, fingerprints and palm prints.

4.1 Camera System

The proposed goal of this system is to be able to perform an image acquisition such that 2D hand geometry can be extracted, with resolution and accuracy high enough to allow for subsequent fingerprint and palm print recovery with a hand moving up to 0.5 ms^{-1}. The system also must be expendable to allow for 3D hand geometry features to be extracted.

To meet these demands, a line scanner would need to be capable of output resolution of 500 DPI on the cross section of at least 200 mm, requiring the physical resolution of at least 4000 px. To meet the vertical resolution requirement, with the same output resolution requirement, the minimum required framerate is 10 kHz.

The Basler racer raL6144-16gm [21] has been chosen, as the 6144 ph resolution is sufficient and as 17 kHz frame rate would allow for hand moment of up to 850 mms^{-1}. Secondary reason for choosing this model is the sensor size of 43 m, allowing for utilization of full-frame optics.

The optics have been chosen based on the requirements for maximum distance of object from sensor and especially with depth of focus in mind, in order to accommodate the touchless requirement for the system. AF Nikkor 50 mm f/1.8D has been determined to meet these requirements.

4.2 Proof of Concept

Figure 4.19a) shows an image acquired via line scanner fixed on a moving platform moving at a speed of 0.5 m/s. Figure 4.19b) then focuses on a detail of a little finger, demonstrating that output resolution is high enough for a fingerprint extraction and therefore high enough for 2D hand geometry extraction. Due to high noise of the background, ideal segmentation is not trivial, and for demonstration on Fig. 4.20, a mask has been used to remove majority of generated noise.

To demonstrate the possibility of fingerprint extraction, we have used the commercial tool VeriFinger [22]. As can be seen on Fig. 4.21, as expected, the setup being realized as proof of concept only, the lighting proved to be non-ideal and produced glare resulted in a void appearing in the segmented picture of the fingerprint. At the same time however it can be noticed that there is a large number of positively identified fingerprint markers, serving as proof that the system as is is sufficient for the defined task.

Palm print feature extraction can be performed in a similar manner, as the lighting in the POC was not ideal for this sort of data processing and voids can be observed due to glares; overall, however, it can be confirmed that this setup is viable for palm print extraction as well, thus allowing a statement that the POC has met the requirement placed on it. Figure 4.22 presents an image ready for palm print feature extraction.

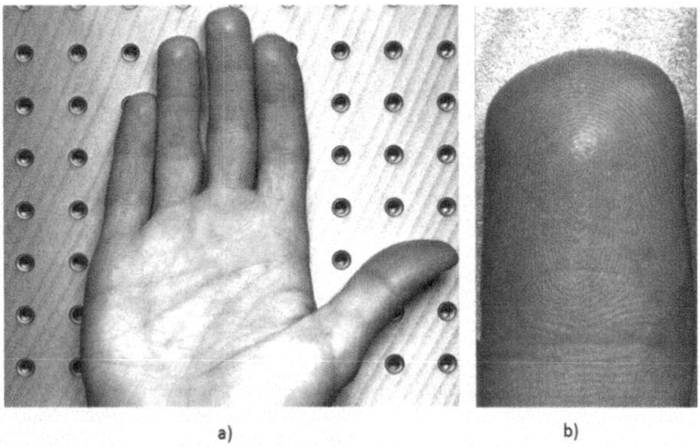

Fig. 4.19 (**a**) Palm print using line scanner mounted on moving platform; (**b**) detail of ring finger

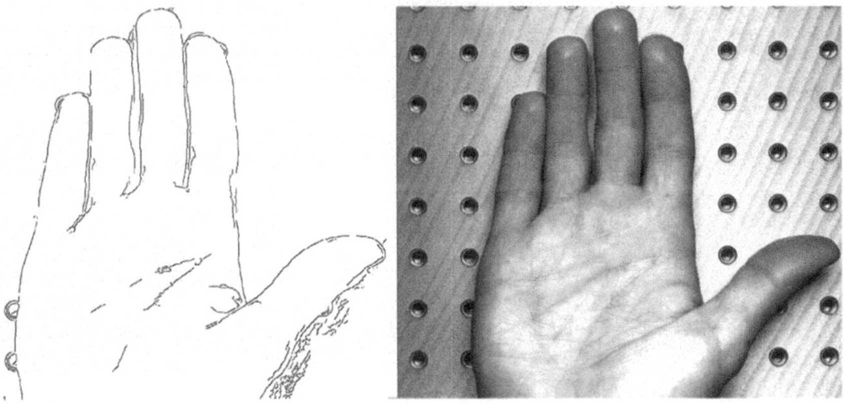

Fig. 4.20 Mack filtering and edge detection used to extract outline (left) from an image of hand acquired via line scanner (right)

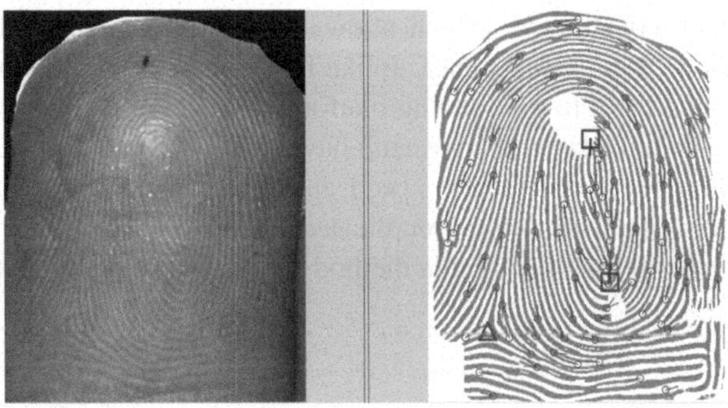

Fig. 4.21 Fingerprint image using line scanner (right), extracted fingerprint using VeriFinger SDK (right)

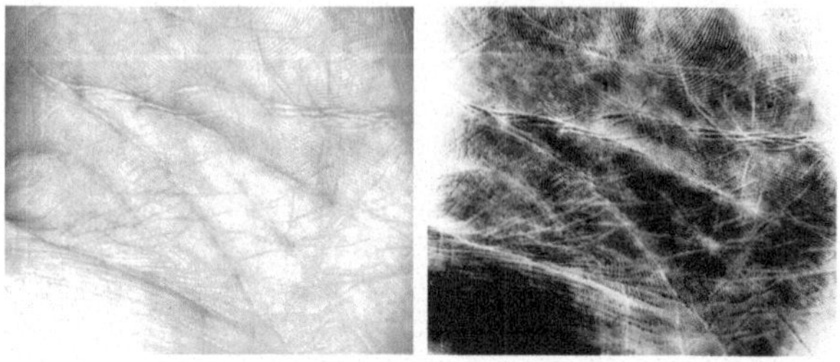

Fig. 4.22 The POC palm print extraction from the image, source image (left) and preprocessed palm print (right)

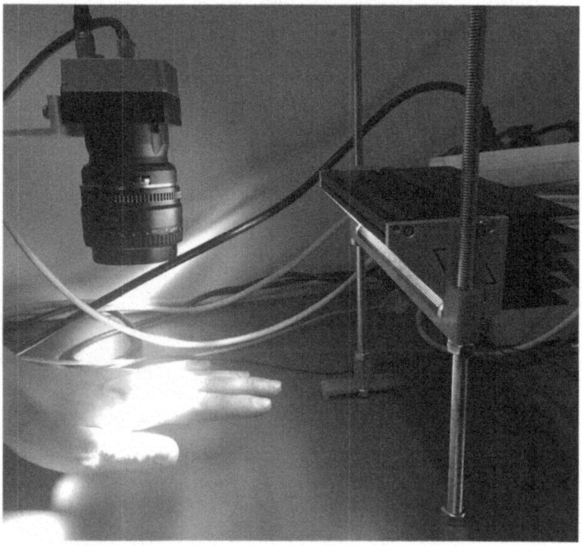

Fig. 4.23 Setup of system with free-hand movement

4.3 Reconstruction of Hand Image in Touchless System

Reconstruction of hand model using a data from a non-calibrated movement is currently being explored. Based on the proof of concept, a camera system has been constructed, which allows a free movement of the hand under the field of view of the line scanner, unlike POC where the hand was moving at constant speed while fixed to the moving platform. The setup is presented on Fig. 4.23.

Initial tests with this setup have been performed, and as can be seen on Fig. 4.24, the uneven movement speed and direction lead to deformations that make hand geometry extraction complicated; however, the overall details presented serve as proof of viability of this method for biometric feature extraction.

4.4 Utilization of 3D Line Scanner of 3D Feature Extraction

As has been briefly mentioned in the earlier part of the chapter, stereovision is one of the possible approaches in acquiring the 3D features of hand geometry. 3D PIXIA

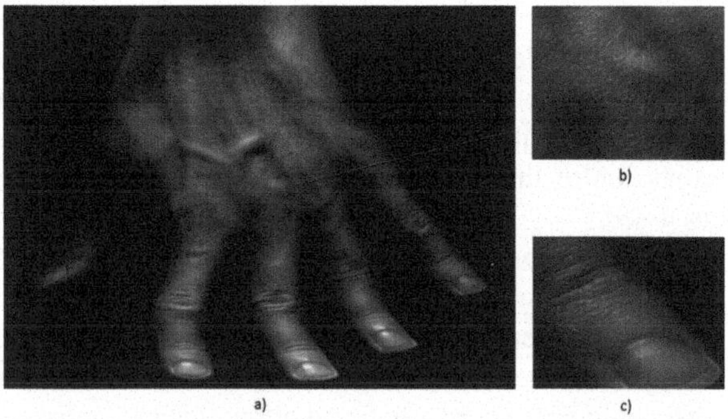

Fig. 4.24 (**a**) Handprint using a fixed line scanner scanning a moving hand; (**b**) detail of middle knuckle; (**c**) detail of little finger

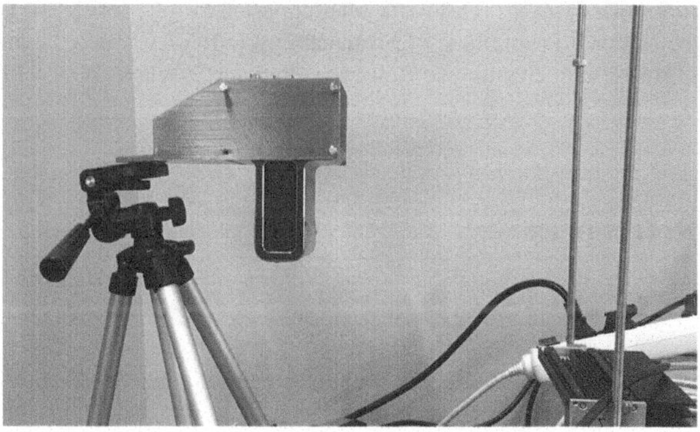

Fig. 4.25 Preview of 3DPIXA setup for 3D line scan acquisition

[23] has been identified as a camera, whose specification parameters meet the requirements necessary for accurate 3D feature acquisition (Fig. 4.25).

As of now, only theoretical and preliminary preparations have been made.

5 Conclusion

In this chapter, an overview of current state of development in 2D and 3D hand-based biometrics has been presented. Existing methods of 2D acquisition and identification have been described as well as systems that are now available commercially. The shortcomings of 2D-based biometrics especially in relation to their susceptibility to spoofing have been explored, and possible approaches to solving these shortcomings with existing methods that allow for more secure system were established.

3D-based geometry has been introduced as a viable direction in development, and advantages of this method presented over 2D geometry have been outlined. Current development in this area has been presented, as well as other viable methods that warrant a future research.

Novel method of biometric image acquisition has been presented, utilizing line scanner. The design and test of proof-of-concept device have been shown. The capability of multimodal acquisition has been demonstrated by performing a test extraction of palm print, fingerprint, and outline of a hand captured by this system. Expansion of this system that would allow for extraction of 3D features has been discussed.

3D hand geometry biometrics builds upon the 2D hand geometry a good method for biometrics. 2D method has proven to be useful in applications where other methods of identification are inconvenient. It has been unable to become truly widespread, due to limited template size. With the advance of 3D approach, this wish could become rectified and allow this technology to spread.

Acknowledgment This work was supported by the Ministry of Education, Youth and Sports from the National Programme of Sustainability (NPU II) project IT4Innovations excellence in science, LQ1602, and "Secure and Reliable Computer Systems" IGA – FIT-S-17-4014.

References

1. M. Drahanský, F. Orság, et al., *Biometrie* (Computer Press, Brno, 2011). ISBN 978-80-254-8979-6
2. A.K. Jain, A. Ross, S. Pankanti, A prototype hand geometry-based verification system, Proc. AVBPA (1999), pp. 166–171,
3. HandKey II, *Allegion* [online]. Dublin, Ireland (2014) [cit. 2017-04-25]. Available at: http://us.allegion.com/Products/biometrics/handkey2/Pages/default.aspx
4. HandPunch GT-400, *Allegion* [online]. Dublin, Ireland (2014) [cit. 2017-04-25]. Available at: http://us.allegion.com/Products/time_attendance/g_series/gt400/Pages/default.aspx
5. HandPunch 4000, *Allegion* [online]. Dublin, Ireland (2014) [cit. 2017–04-25]. Available at: http://us.allegion.com/Products/time_attendance/f_series/hp_4000/Pages/default.aspx
6. M.-P. Dubuisson, A.K. Jain, A modified Hausdorff distance for object matching. Pattern Recognition, 1994. Vol. 1-Conference A: Computer Vision & Image Processing., in *Proceedings of the 12th IAPR International Conference on*, Vol. 1. IEEE, (1994)
7. M.A. Ferrer, J. Fabregas, M. Faundez, J.B. Alonso, C. Travieso, Hand geometry identification system performance, in *43rd Annual 2009 International Carnahan Conference on Security Technology*, Zurich, (2009), pp. 167–171
8. H. Chen, H. Valizadegan, C. Jackson, S. Soltysiak, A.K. Jain, Fake hands: spoofing hand geometry systems, in *Biometric Consortium 2005*, (Washington, DC, 2005)
9. S.-I. Joo, S.-H. Weon, H.-I. Choi, Real-time depth-based hand detection and tracking. Sci. World J. **2014**, 284827, 17 pages (2014). doi:https://doi.org/10.1155/2014/284827
10. Intel RealSense™ Camera [online] (2015), [cit. 2017–04-17]. Available at: https://software.intel.com/en-us/RealSense/SR300Camera
11. V. Kanhangad, A. Kumar, D. Zhang, Combining 2D and 3D hand geometry features for biometric verification, in *2009 I.E. Computer Society Conference on Computer Vision and Pattern Recognition Workshops*, (Miami, FL 2009), pp. 39–44
12. V. Kanhangad, A. Kumar, D. Zhang, A unified framework for contactless hand verification. IEEE Trans. Inf. Foren. Sec., 6 (3), 1014–1027 (2011)). https://doi.org/10.1109/TIFS.2011.2121062
13. J. Svoboda, M. Bronstein, M. Drahanský, Contactless biometric hand geometry recognition using a low-cost 3D camera, in *Proceedings 2015 International Conference on Biometrics*, (IEEE Biometric Council, Phuket, 2015), pp. 452–457 ISBN 978-1-4799-7824-3

14. Vivid 910: Non-contact 3D digitizer [online]. In (2002), s. 6 [cit. 2017-04-17]. Available at: http://www.upc.edu/sct/documents_equipament/d_288_id-715.pdf
15. P.C. Mahalanobis, On the generalised distance in statistics (PDF). Proc. Nat. Inst. Sci. India **2** (1), 49–55 (1936). Retrieved 29 Sept 2016
16. K.Q. Weinberger, J.C. Blitzer, L.K. Saul, Distance metric learning for large margin nearest neighbor classification. Adv. Neural Inf. Proces. Syst. **18**, 1473–1480 (2006)
17. J. Svoboda, O. Klubal, M. Drahanský, Biometric recognition of people by 3D hand geometry. in *The International Conference on Digital Technologies 2013*, Zilina, (2013), pp. 137–141
18. Š. Mráček, et al., 3D face recognition on low-cost depth sensors. Biometrics Special Interest Group (BIOSIG), in *2014 International Conference of the. IEEE* (2014)
19. Ensenso N10. *Imaging development systems* [online]. Obersulm, Germany [cit. 2017-04-25]. Available at: https://en.ids-imaging.com/ensenso-n10.html
20. Nové modely 3D kamer Ensenso. Analyza obrazu [online]. (2015) [cit. 2017-04-25]. Available at: http://www.analyza-obrazu.cz/aktuality/nove-modely-3d-kamer-ensenso/
21. RaL6144-16gm - Basler racer. Basler: The power of sight [online]. Germany [cit. 2017-04-25]. Available at: https://www.baslerweb.com/en/products/cameras/line-scan-cameras/racer/ral6144-16gm/
22. VeriFinger SDK. Neurotechnology [online]. [cit. 2017-04-25]. Available at: http://www.neurotechnology.com/verifinger.html
23. 3D Line Scan Camera 3DPIXA compact 30µm. Chromasens [online]. [cit. 2017-04-30]. Available at: https://www.chromasens.de/en/product/3d-line-scan-camera-3dpixa-compact-30um

Chapter 5
Fundamentals and Advances in 3D Face Recognition

Soodamani Ramalingam, Aruna Shenoy, and Nguyen Trong Viet

1 Introduction

The demand for establishing identity has existed for thousands of years in human history. Examples can be found across the fields from military to economics: when population increases significantly, governments need effective methods to manage identification systems. That challenge has become even larger with millions of individuals passing through nations' borders every day. As a result, identification systems are required to perform its task not only in a local context but on a global scale as well. The push for more research in this field is also contributed by the growth of personal handheld devices containing personal and confidential information, and therefore any unauthenticated access must be prevented.

Personal characteristics such as fingerprint and signature have been used for long for identification. They form the basics of biometrics, in which the chemical, physical and behavioural attributes are studied, transformed to measurable metrics and applied to distinguish one individual from the crowd.

Biological traits must satisfy a certain number of factors [1] to be accepted as suitable for evaluating:

- Universality: All individuals of the cohort must possess this trait.
- Uniqueness: Each individual must have distinguished characteristics.
- Permanence: Trait of each individual remains more or less the same over time.
- Collectability: Samples of trait can be collected in sufficient amount from target group.

S. Ramalingam (✉) · N. T. Viet
School of Engineering and Technology, University of Hertfordshire, Hatfield, UK
e-mail: s.ramalingam@herts.ac.uk

A. Shenoy
National Physical Laboratory, Teddington, UK
e-mail: Aruna.Shenoy@npl.co.uk

M. S. Obaidat et al. (eds.), *Biometric-Based Physical and Cybersecurity Systems*,
https://doi.org/10.1007/978-3-319-98734-7_5

- Performance: Potential time, cost and accuracy to perform the identification.
- Acceptability: How willingly people will cooperate to submit their trait for testing.
- Forgery resistance: Defence against the fraudulent methods where examined trait is imitated.

Generally, there are two main types of biometrics: physiological and behavioural. Physiological characteristics refer to the data from individual body with typical traits being fingerprint, face, iris and DNA, while behavioural biometrics means the modality is the learned habit of target such as keystroke, signature or voice. The physiological methods are known to provide higher accuracy rate and permanence but fall behind in acceptability. For example, DNA gives extremely high accuracy and anti-fraud attributes but is not an option when it comes to practical application on mass due to its cost and high time-consuming [2]. Another more common technique is fingerprints: it has high-accuracy, low-cost device but unfortunately consumes colossal amount of calculating power to identify one person [3].

Face recognition emerges as a promising way to fulfil requirements since all individuals possess recognisable facial characteristics. Furthermore, facial recognition has been known suitable for applying on large-scale population. Consider the environment inside an airport; checking fingerprints of every traveller results in long queues at custom gate and several officers' efforts. Meanwhile, it would take less than a dozen cameras for capturing, identifying and then pulling out any wanted target. It makes it even suitable as faces are socially acceptable biometric and may not require a considerable co-operative effort from any individual unlike for obtaining a fingerprint. Faces are there everywhere.

From the time of the first manually sampled system in 1963 [4] to the age of smartphones, face recognition systems have strived forward with huge leaps and still gaining speed in terms of performance, cost and complexity. While the first systems were very susceptible to even the small change in pose, their successors today already reached the accuracy more than 90% [5] and implemented into commercial systems [6].

The rest of the chapter is organised as follows: in Sect. 1, we consider fundamental aspects of a generic face recognition system including definitions and terminologies related to 2D face recognition (2DFR) system. In Sect. 2, we introduce the principles of 3D face recognition (3DFR) systems, their key differences from 2D face recognition systems and their building blocks. We take a look at the applications of 3DFR including border control as well as its applicability as a mobile biometric. In Sect. 3, we consider two recent case studies from a design and development perspective. This includes a stereo vision and a Microsoft KINECT-based 3DFR systems from previous and current work by the authors. Section 4 explores benchmarking mechanisms and standards applicable for face biometric technologies. Section 5 touches upon ethics and privacy issues related to commercial systems and a conclusion in Sect. 6.

1.1 Biometrics and Border Control

Biometric technology has been adopted by various countries in diverse scenario applications to meet the requirements of an efficient identification system. Airports and border control have benefitted the most from biometrics. They reduce the operational cost and are convenient to the travellers without compromising security or throughput. Of the various modalities of biometrics such as fingerprint, finger vein, palm vein, iris and face recognition, just to mention a few, due to the ease of presentation, face recognition has been a preference by the International Civil Aviation Organisation (ICAO). In the past, the face recognition systems have used 2D images. In the recent past, there has been tremendous advances in technology both hardware and software that has prompted the use of 3DFR in the various use cases where previously 2DFR has been predominantly used. In this section we discuss how 3DFR has developed in the last decade and its applicability, particularly in border control.

1.1.1 Border Control Across the Globe

Countries with constant conflicts across their borders such as India-Pakistan and Israel-Palestine and countries with a lot of population migration due to war and suffering are currently engaged in implementing biometrics to enable efficient border control and combat terrorist activities by the use of biometrics. For instance, India has the biggest ever recorded biometric ID system in the world and employs Nippon Electric Company (NEC) multimodal biometrics called Aadhaar which is the new Unique Identification Authority of India (UIDAI). It includes automated fingerprint identification and automatic face recognition and as of April 2017 has registered more than a billion Indian under this programme.

Indonesia's SITA iBorders will supervise the immigration at the country's 27 air- and seaports with the use of 300 workstations. The SITA's kiosks with advanced border management systems include Australia, Bahrain, Canada, Kuwait, Mongolia, New Zealand, South Africa, Spain and the USA [7]. SITA's security system includes in its core a highly scalable software engine called the BioThenticate. It implements the core functions such as enrolment, identification, credentialing and verification of individual identities. The BioThenticate enrol unit captures biometric data from multiple sources such as machine-readable documents and can be deployed as stand-alone workstations. Customised biometric credentials are created by BioThenticate. The verifying unit does a live comparison with the data stored on a biometric credential or the engine database. This involves a secure template generation and identification matching of either one-to-many or one-to-one followed by one-to-one matching. The BioThenticate server architecture is hardware-independent [7].

The UK government started issuing electronic passports to most of its citizens since 2006. The additional security on the passport includes a facial biometric of the holder on the chip. The e-passport readers at various airports across the country enable swift and efficient facial recognition verification [8].

Although 2D systems are commonly deployed, 3D face recognition is clenching the biometric markets today. 3DFR overcomes the disadvantage of changing facial expressions by applying isometrics [9, 10]. There are a number of advantages in using 3D in comparison to the 2D face recognition on its own as they can offer the benefits of being expression [9, 11–13], pose [12–14], age and illumination invariant [15]. However, recent research has suggested that the multimodal face recognition techniques by using 2D and 3D methods increase the accuracy and performance [13, 16–18]. For instance, in the USA, Animetrics, Inc., which is a face recognition biometric service provider, has been working on 3D face recognition for border control. The patented work by Animetrics Inc., includes generation of 3D models from 2D face images of varying orientation and scale, generation of databases from the 3D models for training and testing and also normalisation using 3D models in the face recognition systems [19].

1.2 Market Survey of 3DFR Systems

The field of 3D face recognition (3DFR) is quite new but advancing quite rapidly. At the algorithmic level, the techniques vary depending on the modes of model representation (or registration) [20–22], feature extraction [23] and matching [24–26]. A good set of survey papers [27, 28] provide varied systems on generic 3DFR. These cover a range of techniques starting from imaging, representation, matching and both greyscale and colour images. Feature extraction has recently gained a lot of prominence as it dictates the performance of a recognition system. In this section, we consider a review of current techniques that is related to 3D facial feature extraction.

3D face recognition (3DFR) systems and 3D fusion with 2D are in high demand due to their ability to overcome the shortfalls of the 2D only systems. Currently, commercial face recognition systems that use 3D technology overcome issues with illumination and pose that impact the overall performance of the system. 3DFR systems solve this problem by providing information on texture and additionally facial depth. This however comes at a higher cost and slower speed. Multimodal (2D + 3D) fusion are also in the development and use at the moment. The sensors used in 2D/3D recognition systems have adopted IR imaging techniques to overcome limitations by illumination and pose.

1.2.1 Research-Based 3DFR Systems

Recent inventions in camera and imaging systems provide the means to measure and reconstruct 3D faces. Using three-dimensional coordinates eliminates the effects of lighting condition and skin colours due to the fact that all the data is in form of coordinates, containing much more information than a flat image. The additional depth dimension allows researchers to analyse and study about the curvatures [29] and estimate samples' pose [30]. Automatic face segmentation on depth maps is carried out in [31] where K-means clustering technique that segments the foreground from the background is adopted. Verification is carried out on FRGC v1 and v2 datasets with 99.9% accuracy of neutral faces and 96.5% overall. The popular idea of using local shape descriptors extends from 2D to 3D [32] making sparse representations of face models feasible [33–35]. 3D capturing devices usually return data as point clouds, based on which a face's mesh is built and aligned to the desired angle [36] for further analyses.

Face recognition systems benefit from multimodal feature (MMF) sets, and their performance can outway that of individual modalities [37]. Multimodal systems utilise multiple information sources enabling increased performance, reliability and filling in missing information. MMFs play a key role in fusing information towards decision-making in a face recognition system. In situations where several modalities may be identified such as multiple sensor configurations or combinations of feature sets, the problem becomes that of selecting the right modality for the application. The cumulative match curve (CMC) which is a set of performance plots typically used in biometric systems may exhibit similar responses of the modalities under the same environmental conditions, or the number of parameters to deal with is large making the feature selection a difficult task. In such cases, subjective judgements that do not have a 100% certainty or due to lack of data or incomplete information lead to decision-making under uncertainty [38]. We consider some of these multimodal systems in the rest of the section.

1.2.1.1 2D + 3D Face Recognition Systems

Even though research in 3DFR claim having solved problems of pose invariance as compared to 2D, most research work in 3D continues to focus on pose invariance [35, 39]. It is well acknowledged that face recognition systems underperform as a single modality. The success of multimodal systems and in particular 2D + 3D face recognition algorithms is becoming a popular but simpler approach to improving recognition accuracies [39–41]. In [25], Wang et al. utilise 3D + 2D image features and fuse final recognition using PCA which showed improved performance in comparison to single modal systems. Such systems typically require manual selection of fiducial points for pose normalisation [42, 43]. In addition, matching requires perfect image alignments and filling missing points through interpolation.

The work by Gordon [44] uses disparity maps to model faces and employs curvature estimations on range data along with depth information for face recognition. The paper reports high accuracy (70–100%) and viewpoint invariance. Lee and Milos [45] segment range images into convex regions based on the signs of mean and Gaussian curvatures leading to an extended Gaussian image (EGI).

1.2.1.2 Multidisciplinary Systems

The marriage of image processing and computer graphics provides robust performance under noisy conditions by use of morphable models [43, 46]. An emerging area is that of geodesic distance measurement [47, 48], which is the shortest distance between two points and is a good candidate for feature extraction. Geodesic distances provide a basis for mapping 3D space into a 2D image. These approaches assume that human face is isometric, which indicates the preservation of geodesic distance in various expressions. Moments are used as features and treated as a face signature in [49].

1.2.1.3 Multi-Spectrum Systems

Microsoft KINECT [50]-based 3D data capture for face recognition has recently gained popularity. Typically, such systems use both colour and depth data for locating different fiducial points in 3D. In [51] gender classification of 101 participants constituting 39 males and 45 females in the age group 18–60 is carried out. Variations in poses, illumination, expression and disguise have been attempted in [52] using 4784 samples of 52 subjects. The construction of avatars from average morphable models from depth data is experimented in [19] by energy fitting algorithms in 3D. The average shape from a set of samples lends itself well to building avatar models and tested on a small database of 20 subjects. In [53], a complex 3D transformation between the 3D model feature vertices and the corresponding points on 3D texture map is carried out. A high success rate of 92.3% of samples was synthesised successfully and achieved the peak signal-to-noise ratio (PSNR) of 38.7 dB compared with 35.4 dB of a previous study.

1.2.1.4 Multiple Image-Based Systems

A video database captured with BU-3DFE scanner has been used for face synthesis built from texture requiring feature labelling and complex 3D transformation [31]. An Eigen-based recognition technique from multiple images has been reported with high recognition rates. The FRGC v2 hybrid dataset combining 2D + 3D from a Minolta Vivid camera has been used in [54] for face recognition. Again, this technique requires a complex procedure for pose correction, spherical face representation and location of fiducial points before applying ICP. A very high recognition rate of over 99% for neutral poses and 95–98% for non-neutral poses at FAR = 0.001 has been reported on 4950 images. More recently, a multi-criteria decision-making (MCDM) for multimodal fusion of features in a 3D face recognition system is proposed in [55].

1.2.2 Commercial 3DFR Systems

The Artec ID [56] claims that its product named Broadway 3D is the first device in the world to match human performance in less than a second. Walking, or even running, simply passing by the device with the glance is enough to enrol the individual and recognise the identity of the individual for verification. It is capable of overcoming the issues humans face with the ability to differentiate nuanced geometry with an accuracy of up to fractions of a millimetre (such as in the case of identical twins).

Japanese Ayonix [57] is an interactive real-time face detection and recognition application for identifying individuals supporting policing and security staff or marketing and operations departments as an integrated application. This enables services in retail business to identify individuals and provide "VIP" treatment to its frequent shoppers and enable staff to meet their needs and requirements efficiently. This enables them to better understand their services to customers and promote customer behaviour/customer satisfaction.

Morpho systems have developed a 3D face reader [58] that has been successfully used in a wide range of applications including airport and seaport sensitive areas, government facilities, mines, power and petrochemical plants, banking and financial institutions, hospitals and laboratories, stadiums and entertainment facilities, corporate buildings, data centres and prisons.

3D facial recognition systems have substantial potential, even if it isn't fully realised yet; its level of accuracy isn't quite high enough to be used in very large high-traffic areas. But it is good enough to already be implemented into a number of other sectors where 2D systems have been trialled too such as commercial sectors, marketing, healthcare and hospitality.

1.2.3 Face Recognition on Mobile Devices

The upcoming new area for biometrics is in the mobile industry. Online identity verification using face and fingerprint are becoming a reality with more and more applications geared towards mobile online authentications using biometric as a second factor along with a PIN or password. However, factors such as lack of accuracy, lack of security of the personal shared data, cyber threats and the high implementation cost are likely to hinder growth in this area.

As more and more applications demand remote authentication of a face biometric, thermal cameras are also being used. This enables liveness detection checks too because of the heat map generated. This not only collects information such as such of the head; it will not be massively impacted in its performance with variations in facial hair, make-up or eyeglasses.

Identical twins have features which are identical which make it difficult for any 2D face recognition system to distinguish them. To improve the performance, they have now resorted to multimodal recognition by fusion of other biometrics such as fingerprint and iris in addition to face. Previous studies have shown that features which are well distinguishable between identical twins are not always the same features that are distinguishable for any two non-twin faces [42]. These problems have been now addressed by 3D face recognition systems. The current state-of-the-art 3D systems seem to provide a superior performance in comparison to 2D face recognition systems in the task of distinguishing features [39]. Despite all these improvements, face expression variations still are a challenge for both 2D and 3D face recognition systems.

More and more mobile devices use biometric authentication, and it is predicted that all majority of mobile devices will by default have an embedded biometric capability with affordable and attractive prices. This will provide opportunities for users to be familiar with biometric technology and be able to ubiquitously use them for several authentication applications such as banking, payments, online login, enterprise mobility, government services, etc. The cost of such mobile devices is drastically down making it attractive for the customers [59]. In addition, more and more people are willing to make mobile transactions and are comfortable using biometric as a second authentication factor [60]. However, an issue with mobile biometrics is the security. Mobile devices need a reliable method of protecting the device from unwanted users particularly while it has been activated to be used with a biometric [61].

1.3 Factors Influencing Choice of Biometric Systems

Although all 2DFR systems make use of these three generic functions of image capture, preprocessing and feature extraction followed by matching, the application use cases (such as access control, surveillance, border control, etc.) focus on many aspects such as:

1. *Performance*: The performance evaluation of any biometric system would provide qualitative ability of the system. This includes metrics such as total time taken for enrolment, time taken for a successful verification, how successful is the system in correctly authenticating an individual, how many impostors are authenticated falsely, how many people were unable to be enrolled on the system in the first place, etc. Depending on the type of scenario or the real-life application area, that performance metric becomes more relevant [62]. For example, the border security checks with face recognition systems for verifications look at how efficiently a system can achieve verification with a shortest time. In an area with large queues of people, systems should be capable of faster successful turnaround without compromising security. However, when face recognition system is used as an application to verify an individual at an office premises, the speed at which this action is done is not that significant. The various parameters used to describe the performance of a system will be discussed in Sect. 1.3.1.

2. *Usability*: This refers to the ease of using a system successfully [63]. It is defined as "the extent to which a product can be used by specified users to achieve specified goals with effectiveness, efficiency and satisfaction in a specified context of use" (ISO 13407:1999). Biometric systems seem to have different functionalities and features depending on the modality, providing a different user experience. Any system will be successful only if it is well-received by the public – if it is easy to use, easy to remember and efficient. This is not only applicable to the end-users but also to the administrators and system analysts irrespective of the age and gender of the individual. Usability of biometric systems in itself is a major topic as any system should be usable by all people irrespective of height, weight, age, gender, experience, knowledge and disabilities. Usability studies also focus on the environment in which the system is operated. The application areas must consider temperature, noise, and lighting/illumination, position of the device, traffic, low or high temperature and humidity with/without instructions.
3. *Transaction time*: A successful verification with shortest transaction is expected in ideal conditions. Hence, research is driven towards developing faster and efficient systems. The transaction time includes time from the presentation of the biometric sample to the decision declaration [62]. Currently 3D FR systems are struggling to beat the 2DFR system transaction time, although they have much higher performance rates. Hence, deployment of these systems in a high-traffic area is still a challenge.
4. *Spoofing*: A biometric system verifies the biometric sample submitted by the individual with the template provided at enrolment. Biometric spoofing refers to tricking the biometric system with an artefact or an act of impersonation [64]. It is also referred to as presentation attack. The main risk of using biometrics in high-secure areas is the ability of spoofing the system with artificial artefacts. It is hence necessary to make sure that the system is not just capable of providing sufficient performance, but it is also resistant to presentation attacks.

1.3.1 Performance Metrics of FR Systems

The performance factors [62] are significant depending on the deployed applications [65, 66]. The performance of any system must be very high. It should meet the thresholds that are set up to satisfy the purpose. However, the threshold of these performance metrics [59] can vary depending on where they are used. The metrics mainly used to describe the biometric performance of a 2DFR system would be [66]:

1. *False accept rate (FAR)*: The probability that the system incorrectly accepts the biometric sample as match to an identity when it belongs to another subject. This is also called as false match rate (FMR).
2. *False reject rate (FRR)*: The probability that the system incorrectly rejects the biometric sample as an improper match when it belongs to the claimed identity. This is also called as false non-match rate (FNMR).
3. *Equal error rate (EER)*: The rate at which the FAR is equal to FRR.

The total time taken for enrolment and for subsequent verifications are crucial for applications involving people flow management. However, there must be a reliant fall-back option in place to handle problems in application areas. Systems used in high-secure areas must have high performance and anti-spoofing capabilities too. Face recognition systems need to meet the varied requirements of different applications including scale in terms of end-users, the security level required and therefore threshold levels for operating the systems [67]. These requirements will need to answer the following questions [62]:

1. How many people will be using it and what sort of population will use it (age, gender)?
2. What is the throughput? How many individuals should the system verify each time?
3. Is a single modality enough? Is it for a high security access control or for entry building of critical infrastructure where two-level authentication is required?
4. What should the threshold be? The rejection or the acceptance of a user is decided by the score generated while matching. If the score is above, it is accepted. If it is below the threshold, it is rejected. Hence, the threshold level decided the how strict the rules for entry are. This purely depends on the security level of the area.
5. How expensive is the system deployment? How many systems are required and what level of support is required?

After a thorough study of these factors, the system in general (or face recognition) will be installed. Although all eventualities are taken care of with installation of any sort of technology, however, with biometric systems such as face recognition systems, a proper fall-back system should be in place. Face recognition system is widely used and accepted by people and hence used in large high-traffic areas such as airports. When these systems fail, it could cause a massive backlog and queues within a very short duration of time. Use of other face images such as ID cards could be an option in such an eventuality.

A balance between FAR and FRR decides on the applicability of a face recognition system. As can be seen in Fig. 5.1, high security applications cannot risk a high FAR and therefore would compromise on FMR. Likewise, Forensic applications would need to consider every possibility and therefore would compromise on FRR than FAR. Civilian applications that allow easy access would have both low FAR and FRR. The FAR and FRR intersect at a certain point. The value of the FAR and the FRR at this point, which is of course the same for both, is called the equal error rate (EER).

1.4 Challenges in 3D Face Recognition

The challenges of deploying 3D face recognition [69] systems in a variety of use-case scenarios are primarily cost, but there are other aspects that are equally important such as:

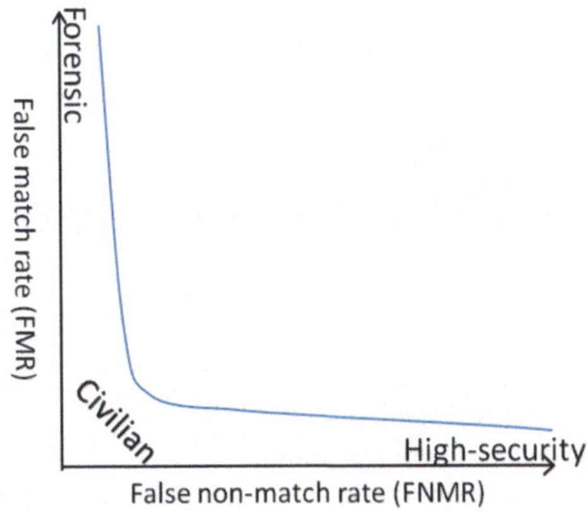

Fig. 5.1 FAR vs FRR and choice of application [68]

Large depth of field is preferred – Depth of field (DOF) is also called the focus range and is defined as the distance between the nearest point and the farthest point on the face image. To get a sharp image, a large DOF is preferred. When the DOF becomes smaller, the focus is on a tiny area of the image maybe suitable with cameras used in scene imagery. However, it is not suitable for 3D face recognition images where the entire face area is necessary to suitable degree for processing and successful verifications.

The acquisition time should be like the 2D face systems. Current 3D face systems are slower than the 2D face recognition systems. The throughput time is an important aspect for decision of application area for a 3D face recognition system. To be able to use them in high-traffic areas such as the application areas where 2D are currently used, they should be fast enough to be able to process several individuals in a minute.

Accuracy – Either for sole use for 3D only systems or 3D in fusion with 2D, the accuracy rate of recognition should be much higher for application of these expensive systems.

Currently, there are very few databases that are developed for 3D to be used as test databases to compare performances of 3D face recognition systems making it even harder for objective comparisons of systems.

2 3D Face Recognition: Design and Development

In this section, we consider the building blocks of both 2DFR and 3DFR systems. Both configurations are similar in most respects except that of the initial stage of image capture where the imaging sensors differ and as a result the information contained in the captured images. Other than that, any algorithm that is applicable for preprocessing, feature extraction and matching is equally applicable to both modalities.

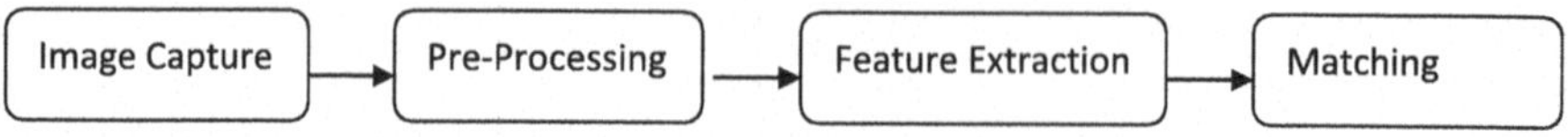

Fig. 5.2 Phases of facial recognition

2.1 2D Face Recognition (2DFR)

A 2DFR system consists of the following stages of processing as shown in Fig. 5.2:

1. Face detection
2. Preprocessing and normalisation
3. Facial feature extraction
4. Face recognition – identification or verification

Face detection: When an image is submitted to the FR system, it starts with detecting the presence of a face in the image. If it detects face, there it will locate the position of the face (s) in the image. Successful detection of a face depends on the quality of the image, pose angle, illuminations and occlusion of the face by either facial hair or cover ups.

Preprocessing: The input images are typically preprocessed to remove any noise and normalise them to obtain a uniform set of image database. This includes normalisation with respect to scale, resolution, illumination and rotation.

Feature extraction: Once a face is detected, the system extracts a set of features N of the face. The feature vectors will form the templates in the reference database. Feature sets may include the shape of the lips, eyes, nose, cheekbone eyebrows, hairline etc. They are unique to the person and help humans identify an individual. Features can be either internal or external. Internal features are clearly visible on the frontal pose of the face, whereas external features include the hair and ears. The facial features of the face are extracted in the form of a feature vector and are a sufficient representation of the face. The distance between the eyes, the length of the nose and the distance between the two corners of the lips are all facial features and can be treated as facial vectors unique to that individual and enable discrimination between two different individuals.

Matching and recognition: The final stage in any FR system is recognition. Matching depends on how the system is to be used at the time of test. Recognition of a face can be in the form of identification or verification.

1. *Identification* (1: many matching) compares the test face with several other faces in the database, and the result is the identity that best matches the test face.
2. *Verification* (1:1 matching) compares the test face image with another face image from the database to validate the identity.

The process of matching is itself carried out using the set of features extracted (in N-dimensional feature space) by calculating the distance between known feature points from training images and unknown feature points representing the testing images.

2.1.1 Local and Global Feature-Based Recognition

Traditional systems obtain (flat) 2D facial images from conventional cameras and then apply customized algorithms for deriving desired features from the samples. Such algorithms mostly belong to one of two approaches depending on whether the extracted information from the image is a set of features from fiducial points on the face (local) or the whole face (global) [70].

Local feature-based matching: A facial image is divided into blocks and features extracted for these individual blocks. The resulting features are concatenated to form a feature vector. Examples of such local feature descriptors include local binary patterns, Gabor jets and others. For recognition classifiers such as KNN (K nearest neighbour), Euclidean distance, artificial neural network (ANN) and others are used.

Global feature-based matching: With global techniques, features such as distances between fiducial points such as the eyes, nose, mouth, ears, etc. are measured and recorded for comparing during the recognition process. This technique was commonly implemented in the first generations of automatic facial identify machines [71]; its main drawback is the significant drop in accuracy in case the targets' head pose changes, and therefore an improvement method needs to be derived. In 1991, Matthew A. Turk and Alex P. Pentland et al. [72] introduced mapping images onto the planes where features are represented by vectors, called "Eigen faces", which used the face as a whole. Using principal component analysis (PCA), faces are transformed into sets of Eigen vectors and covariance matrix which describe the attribute of the whole face surface, not the absolute metrics between separated features; therefore to some extent, it is less prone to errors relating to the head's pose. For two decades since the paper of Turk and Pentland, a substantial number of algorithms incorporating PCA have appeared, usually combined with other novel methods to improve the performance [73–76]. Other features include SIFT (scale invariant feature transform) and SURF (speeded-up robust features).

Currently the existing 2D face recognition systems are not robust due to their sensitivity to illumination, facial expressions and inability to cope with wide variations in pose. Consider the case of a busy airport; it is impossible to get homogenous lighting conditions, casted shadow and pose variations among all passengers not to mention the exposure of their faces to the cameras, all of these drastically affected the systems' performance [77]. Researchers have constantly attempted to overcome these issues [78–80] and gained some improvements but still the issues exist. Over the last decade, improvements in the speed and cost of 3D face recognition systems along with their improved performance have increased the use of 3D systems in varied use-case scenarios. The 3D depth information is independent of the pose and illumination, making it suitable as a viable biometric identification system in practical and vigorous application areas.

2.2 3D Face Recognition (3DFR) Fundamentals

Research in 3D face recognition systems is becoming increasingly popular due to the development of more affordable 3D image acquisition systems and the availability of 3D face databases. Such systems have made quite a progress in solving problems of localization, pose and illumination variances. However, these problems continue to exist. With security applications such as border crossing, it is difficult to acquire idealistic images without being constrained and intrusive at capture points.

Pioneering studies for 3D trend began in the 1990s; images were obtained from special system designed for getting information about the depth of the subject, and it proved to be a promising candidate for next step in recognition development [81] due to the notable immunity for illumination as well as facial angle. However, getting more data for each sample means that 3D face recognition requires intensive computational power and spacious storage which held back the growth of this trend until recent years – the era of smartphones and affordable memory. Microprocessors have become much more capable than two decades ago and continue to grow stronger each year, and handheld phones are no longer mere devices for making calls only but fully operational personal computers in all means [82]. The quality of cameras, captured images and their resolution has improved vastly over the recent times. These factors contributed to 3D face recognition as an alternative to biometric identification. The methods used to get the 3D facial models can be summarised into three main categories: synthesising from 2D images, range image and 3D scanning.

2.2.1 3D Facial Models

Synthesizing 3D images from 2D usually means getting one or several 2D images of the target, then morph and project them onto prepared 3D head models. It is like covering a head sculpture by a canvas with a portrait of the subject. This approach helps to shrink down the size of storage since only 2D images are captured, and it solves the problem of facial posing to some extent and provides good recognition rate [83]. Unfortunately, those systems consume high computational power for morphing images and can still be affected by illumination [83].

Range image or commonly known as depth map represents the distance of object to a certain point of view (e.g. cameras). It can be illustrated as a greyscale image, but a pixel's intensity gives the distance and not colour information. Special devices are needed for capturing the depth map. Depth maps can be derived from either passive sensor such as stereo camera or measuring distance active sensor. Stereo cameras are made up of two cameras with a small distance between them just like human eyes; therefore images obtained from them are slightly different. This difference is then

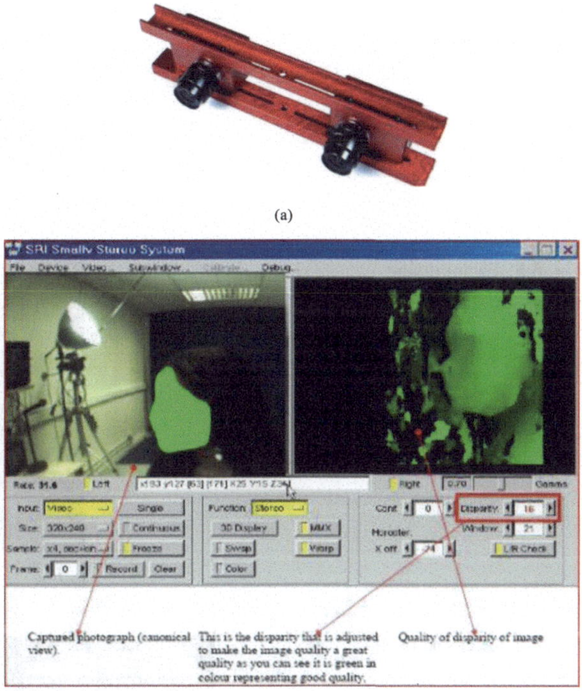

(a)

(b)

Fig. 5.3 Commercial stereo vision system from Videre design [84]. (**a**) Commercial stereo camera. (**b**) User interface and disparity estimation

used to create a perception of depth imitation similar to what happens in the human brain for binocular views (Fig. 5.3).

Active cameras have a different approach: they commonly have one projector for emitting infrared and another lens to get back the reflected rays from objects. Calculating the angles of the returning ray, it is possible to measure the distance from camera to target. This type of depth camera is now used in gaming industry for interactive game. They have high frame rate and medium precision therefore able to keep track and get useful 3D facial images in real time [85, 86] with recognition rate achieved more than 80% [87]. Representatives of these types of affordable devices can be purchased in game stores such as Microsoft Kinect and Intel RealSense 3D Camera (Figs. 5.4 and 5.5).

3D scan is commonly known as the most precise mean to generate 3D model of the objects Fig. 5.6. For example, Minolta laser camera captures 3D models with the precision up to 50 μm [90]. The most remarkable advantage of 3D scans is their independence from view point and lighting condition. In other words, if objects are located one behind the one, it is still observable in 3D scan while shadowed in depth image. This situation would be treated as occlusion in 2D imaging. Thus, 3D imaging systems can display facial features correctly and contain surface curvatures that are useful features. The disadvantage of these systems is that they are bulky and bear excessive cost making it unsuitable for casual use.

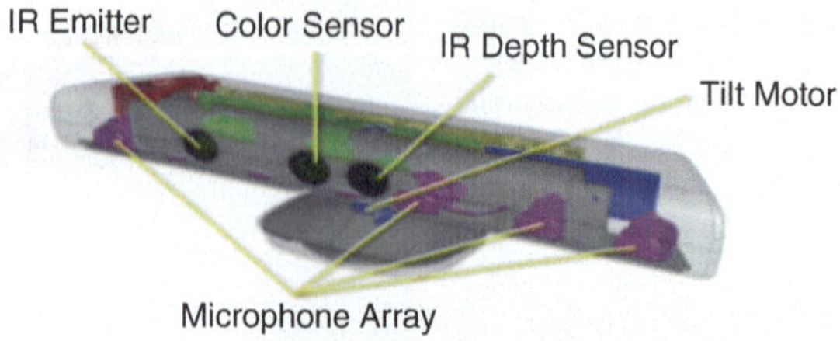

Fig. 5.4 Microsoft KINECT camera [88]

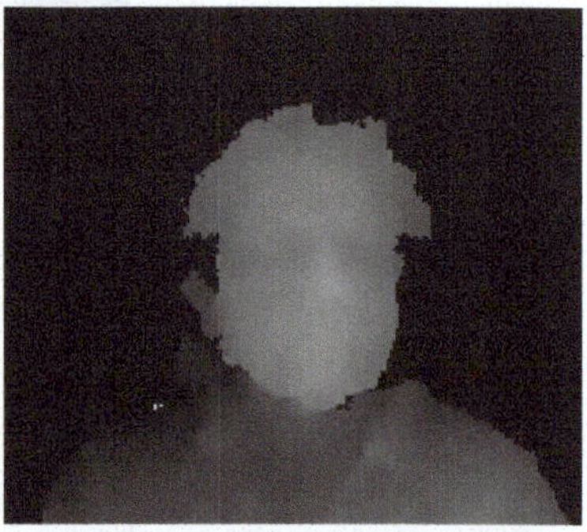

Fig. 5.5 Facial image captured with Kinect [89]

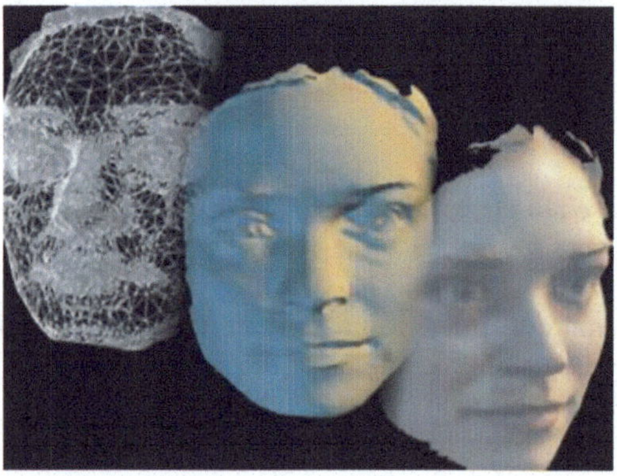

Fig. 5.6 Example of 3D scans [32]

2.2.2 3D Face Recognition: Building Blocks

Facial recognition systems usually comprise of four phases as shown in Fig. 5.2: capture input image from camera, preprocess the images to normalise and reduce erroneous data, extract desired features and finally apply the facial recognition algorithm. This process is common for both 2D and 3D face recognition systems.

A source image from cameras contains error in most of the cases; they probably come from the devices' limited precision, material of the object or occlusions. Captured (raw) images usually have spikes (sudden change of data value in compared to surrounding environment) or missing data in certain areas of the faces such as the eyebrows or hair. Among those factors, occlusions appear to be a considerable obstacle for recognition algorithms. In [91], authors infer from experiments that large occlusions like sunglasses or hats reduce recognition performance by 10–35%.

2.2.3 3D Face Recognition: Preprocessing

Depending on the quality of images obtained from 3D vision systems, the level of preprocessing will vary. Here, we consider some typical preprocessing for images from the stereo vision system in [21].

De-noising: A variety of image processing methods has been developed for surface smoothing. Replacing pixel data with average values of its neighbours is the main idea of median filter. With some modification in the algorithms, it demonstrates itself a sufficient method for 3D images [92]. Seemingly more complex filters also implemented like Laplacian ones, in which the 3D vertexes are relocated, result in getting rid of noises efficiently [93], or Gaussian smoothing [94] to keep the data of curves. Another useful technique is interpolation that is used for filling in the holes, and cubic variant works well despite the high demand for computing power [95]. See Fig. 5.7.

Morphological operations: Morphological operations are typically used in digital images to study the shape of the images. They are implemented using a structuring element (SE) at all possible locations in the image and comparing with the neighbourhood pixels. The centre pixel of SE is the pixel to be iterated on. The operations of erosion and dilation are principal morphological operations applied on the centre pixel of SE. *Erosion* determines the minimum in the neighbourhood of the SE and is assigned to its centre pixel. *Dilation* determines the maximum in this neighbourhood and assigns it to the centre pixel. Erosion and dilation have opposite effects. Erosion removes pixels in the image, and dilation grows layers on the image.

Dilation helps to reduce the noise in an input 3D image by iteratively growing the pixels in the neighbourhood according to the structuring element Fig. 5.8. However, this results in losing its shape. To retain the shape, erosion follows dilation. The results of iterative erosion are shown in Fig. 5.9.

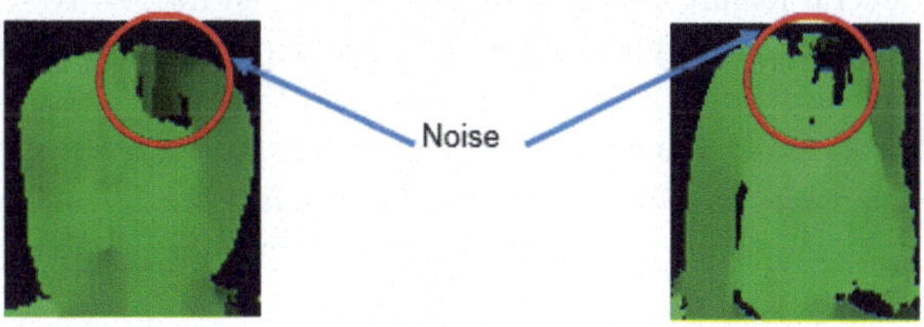

Fig. 5.7 Noise in disparity maps

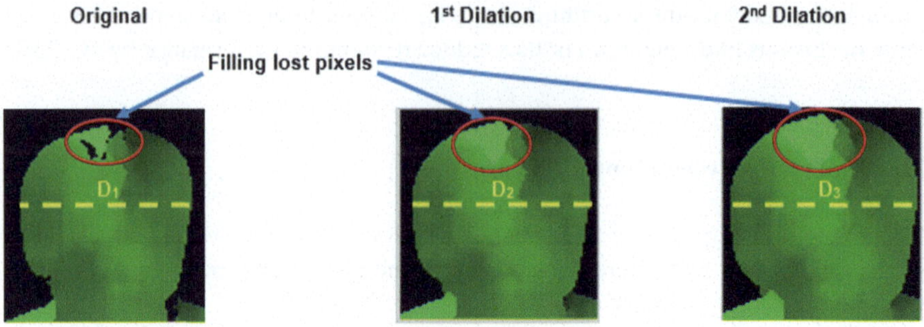

Fig. 5.8 Dilation results on disparity map

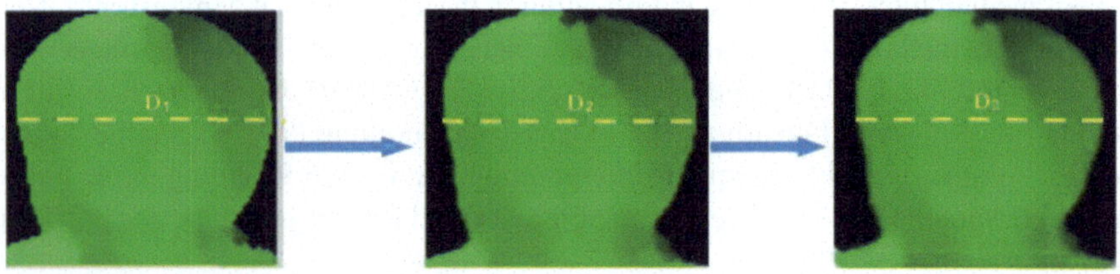

Fig. 5.9 Erosion results on disparity map

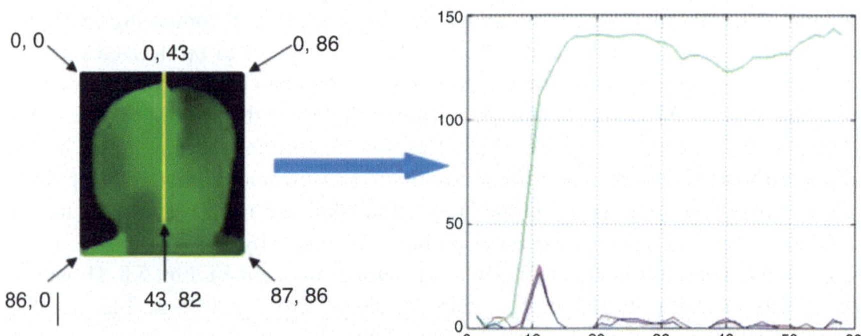

Fig. 5.10 Intensity profile generation for the shown line segment

2.2.4 3D Face Recognition: Feature Extraction

Several feature extraction techniques exist for 3DFR systems which are covered in Sect. 1.2.1. In this section, we consider 3D profile generation as an example. The intensity profile of an image is the set of intensity values taken from regularly spaced points along a line segment or multiline path in an image as shown in Fig. 5.10. The graph appears as a signature and hence is termed by the authors as profile signatures. Here, they represent the RGB values. Since the image is monochrome (G component only), the signature has a more prominent green colour value compared to blue and red colour signatures.

A 3D face image is sampled at equidistant points along the *X* or *Y* axis, and corresponding profile signatures may be generated to form a set of feature vectors. In addition, a second level of features may be generated on these signatures to provide a sophisticated feature set. Such a mechanism is detailed out in the case study in Sect. 3.1.

3 3D Face Recognition: Case Studies

In this section, two case studies of 3D face recognition systems are considered that differ fundamentally in the type of systems used to capture the input images. The first of the case studies deals with disparity maps from a stereo vision system as well as range data from the FRVT dataset. The second system considers a MS KINECT-based data capture. A generic feature extraction algorithm that suits these image sets is considered.

3.1 Case Study 1: Stereo Vision and Range Image Maps-Based 3D Face Recognition

We consider two case studies of 3DFR systems. Case study 1 considers two databases, namely, an in-house student DB and the FRGC database. While the image capture and preprocessing stages vary, the same feature extraction and matching techniques are applied to both databases.

A. *Image Capture*

1. *Student database-*DB_1*:* A student database captured from a stereo vision systems [96, 97] consisting of 100 students as subjects with 10 canonical views per subject (fixed sample sizes) under a controlled illumination environment. Small variations in pose were allowed. The canonical views span 180° and therefore an approximate 18° separation between two consecutive samples. See Fig. 5.11.
2. *FRGC database-*DB_2: FRGC V1.0 data consists of 275 subjects with varying sample sizes leading to a total of 943 images captured with a Minolta Vivid camera [98]. Database sets provide range image (2.5D) with x, y and z coordinates of each vertex, and they are transformed into depth image so that the distances are displayed proportionately by greyscale intensities. A median filter is applied to smoothen any noise, and a K-means clustering algorithm divides the image into foreground and background regions. Through Sobel edge detection, homogeneous regions are formed, and assuming the face area is elliptical in shape, facial regions are successfully located.

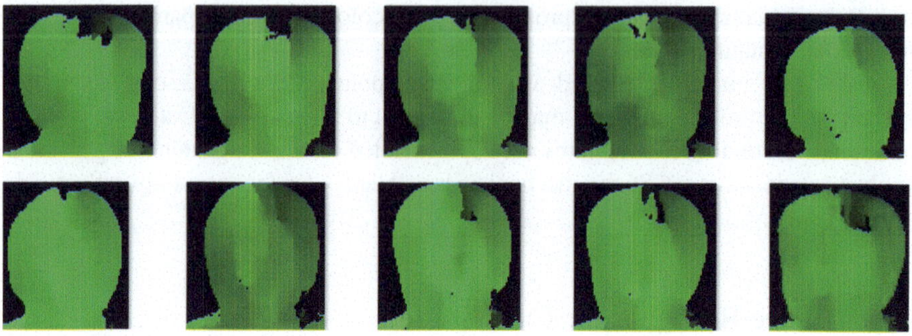

Fig. 5.11 Canonical views of a subject

B. *Image Normalisation*
The student DB was acquired in an illumination-controlled environment and hence did not require further normalisation. The FRGC database required illumination normalisation using the standard histogram equalisation technique available in MATLAB. Both the student and FRVT databases were manually cropped and resized to an image size of 128 × 128 pixels. Thus, the DBs were normalised with respect to scaling and illumination. The rest of the steps are common for both databases. See Fig. 5.12.

C. *3D Profile Signatures*
With the DB images, signatures were derived at the intersections of facial surface with evenly spaced vertical planes. The signatures act as profile curves at sample points along the *Y*-axis (90°) of the image. For convenience, a fixed set of signatures is derived for each image. Similarly, other directional signatures are also derived. These are unary features. Concatenated feature sets are derived by combining signatures at two (binary features) or more (*n*-ary features) intersecting angles. The 3D signatures appear as a compressed image due the effect of sampling in 3D. Sampling takes place at points of intersection of a stack of planar surfaces oriented in a particular angle with the images.

D. *Model Representation*
Models are built to form a feature database suitable for matching. Two approaches are followed:

(a) *An average model* constructed by averaging the normalised canonical views as in the student database.
(b) *Individual model* where sample images are retained as multiple models of face images. The individual models are useful when there are insufficient samples for the subjects as in the case of the FRVT dataset where the number of samples/subject is one for some part of the database. The within-class distance is larger in the former case compared to the latter as it is a fuzzy representation encompassing the average information from all the samples of a subject. Therefore, with the average model representation, it is not expected to produce a 100% degree of match score between the query and

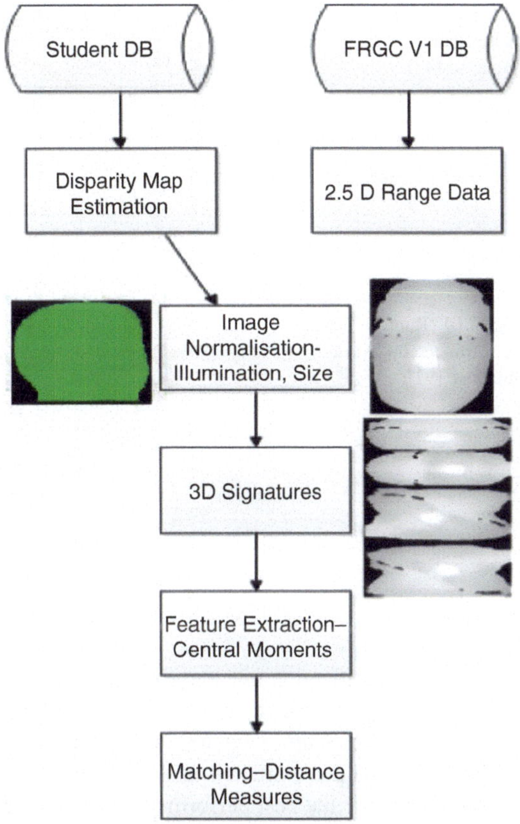

Fig. 5.12 3DFR system architecture

the target images. However, this does not imply that it is a poor representation as it allows an implicit modelling of imprecision within the dataset.

E. *Classification and Query Processing*
Face recognition is treated as a classification problem where a query image is mapped to one of many classes. The Fisher's linear discriminant analysis (FLDA) is a technique that is quite popular among the vision community [99]. Such classifiers perform LDA training via scatter matrix analysis. For this case study, the FLDA is applied on the set of central moments extracted on the original feature vectors.

The Euclidean distance measure is used in the feature space to perform the recognition task [97]. A query image from the *probe feature sets* is matched against the database. The result is a ranked set of images based on the above distance measures between the query and the models.

F. *Classification and Query Processing*
The performance metrics are explained in Appendix A. Performance evaluation of the 3DFR system is carried out using (i) CMC (cumulative match curve) of

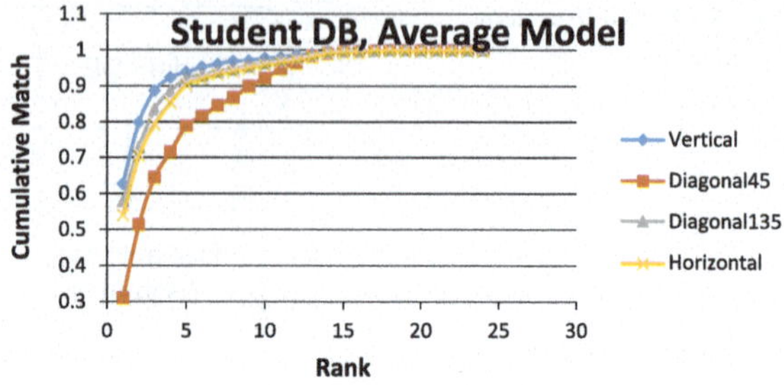

Fig. 5.13 Rank vs score performance on average database [97]

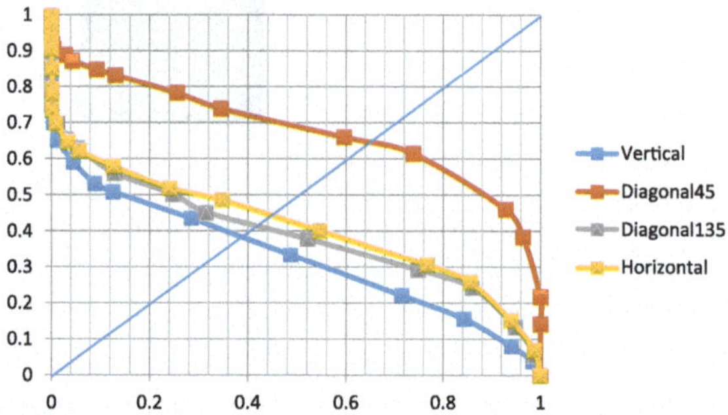

Fig. 5.14 Rank vs score performance on average database [97]

rank vs identification rate and (ii) equal error rate (EER) plots of false acceptance rate (FAR) vs false rejection rate (FRR). The charts provide a relative performance of each feature when tested against an average model of the database. The profile signature extracted at 45° is the worst performer, whereas the vertical signatures are the best performer. The same inference is reflected based on EER (Figs. 5.13 and 5.14).

3.2 *Case Study 2: KINECT-Based 3D Face Recognition*

In [100], the authors describe a Microsoft KINECT-based 3D face synthesis mechanism for generating a face image database. It is an affordable device with the ability to capture RGB and depth image at the same time. Kinect operates like a scanner by emitting infrared beams and then analyse reflected signals for measuring distance [88]. KINECT is well supported by Microsoft with drivers compatible with Windows. Several examples are available for C and C# developers on Microsoft's official website.

3.2.1 KINECT-Based Disparity DB_3

Microsoft KINECT-based 3D data capture for face recognition has recently gained popularity. Main challenge for any imaging system is processing the raw data from the capturing devices, which can contain loss of information due to the nonideal conditions of lighting, distance and face angles, a constant problem for researchers to deal with. Other issues like the hair and the participant's clothes' shadow also interfere with face understanding leading to reduced performance of the system. This work attempts to address these issues through low-level image processing. This section deals with such a preprocessing mechanism for generating a set of 3D image database. The key stages of the system development focus on the following.

3.2.2 Image Capture and Preprocessing

The commercial Microsoft KINECT camera used in [100] is as shown in Fig. 5.4. The depth data for a sample image derived in KINECT is as shown in Fig. 5.5. Here the closest point to the camera has an intensity value of 0 and the furthest point 255 and therefore requires inverting the image. Secondly, the horopter range is too far leading to background scene interference. The foreground object is therefore segmented from the background scene using thresholding, results of which are shown in Fig. 5.15.

The preprocessing techniques considered are (i) segmentation of the foreground object from the background, (ii) localization of the face region, (iii) face cropping and (iv) noise removal. The input image is negated to interpret depth in a manner that brighter pixels represent shorter depth and closer to the camera. Note this difference between Figs. 5.5 and 5.16. In order that we have a normalised image and to enable an easy capture of the face image, the user interface props a window on the PC guiding the user to adjust the face position to lie within the window. This allows easy cropping of the region of interest (ROI), i.e. the face as shown in Fig. 5.17. Figure 5.17 shows the resulting cropped face image. This image needs further preprocessing to extract just the face region numbered "3" and segment out other regions labelled 1 and 2 corresponding to the background and neck. These regions are segmented by calculating their mean intensity values and thresholding on that

Fig. 5.15 Image captured with KINECT sensor [89]

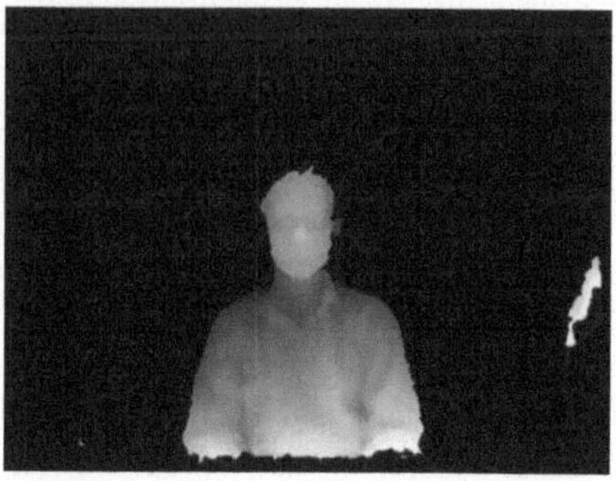

Fig. 5.16 Foreground image segmented and inversed

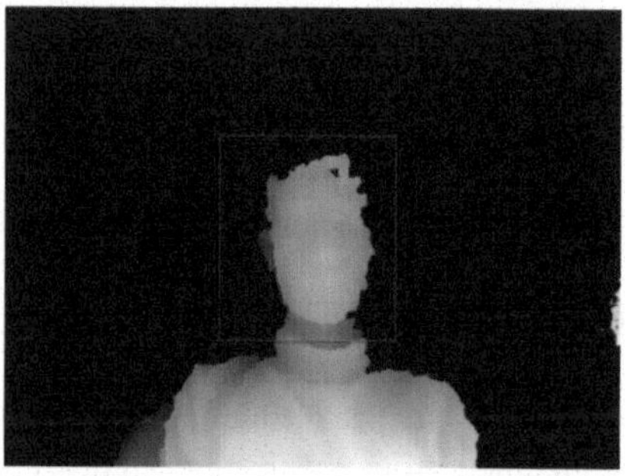

Fig. 5.17 Facilitating localisation of the ROI

basis. Typically, the face region is larger than the rest, and this knowledge is used to crop just the region of interest. The resulting image is shown in Fig. 5.19.

3.2.3 Noise Removal in Depth Images

Typically, there are two types of noise in the depth data, namely, holes and spikes. Holes occur as black spots within the ROI and indicate lack of information due to sensor noise. This requires a filling strategy that does not affect the rest of the image [100]. Spikes occur due to specular regions within the face, namely, the eye, nose tip and teeth [101]. Eye lens contribute to a positive spike, specular reflection from the eyes causes a negative spike, and teeth and nose form small spikes. Glossy facial make-up or oily skin can also contribute to spikes. Both types of noise are removed by iterative noise removal procedures. The spikes leave holes in the image that will need to be filled in using morphological operations. Further, textures such as hair leave holes in the image; again similar filling operations are followed. Figures 5.18, 5.19, 5.20, 5.21, 5.22, and 5.23 show these steps [89].

3.2.4 Face Modelling

Once the noisy images are cleaned up, the next step is that of generating a model. This requires locating certain fiducial points on the face. In [100], the eyes, jaw and

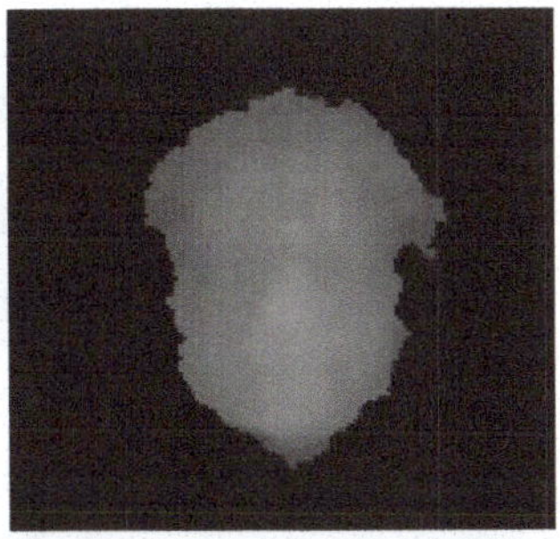

Fig. 5.18 Cropped image

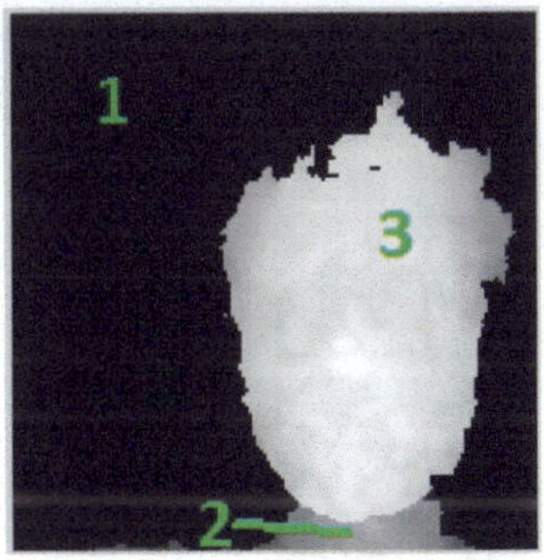

Fig. 5.19 Segment labelling

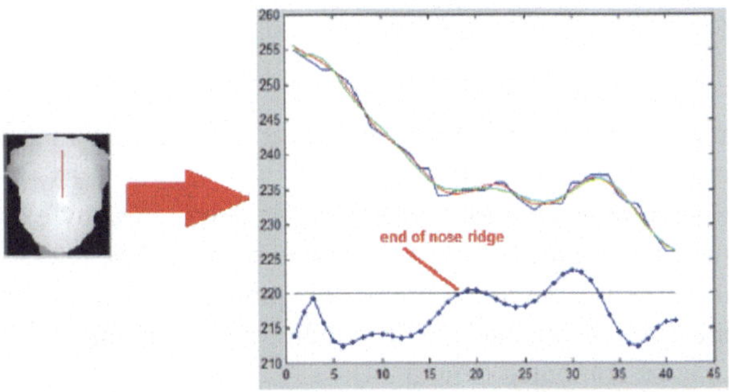

Fig. 5.20 Nose ridge determination through gradient descent (top graph) [91]

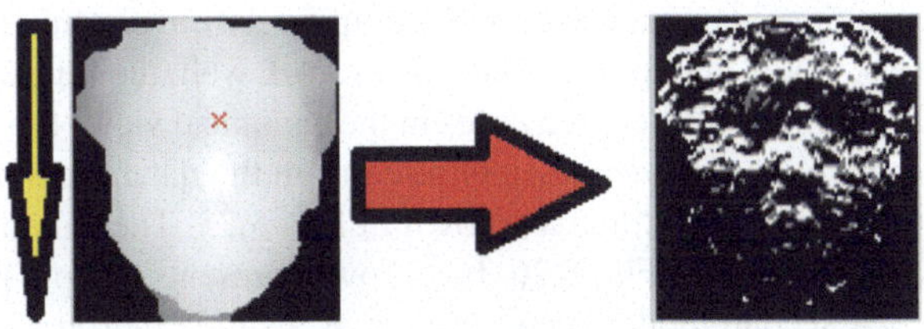

Fig. 5.21 Downward gradients

nose are located. If we consider a frontal image, by traversing from top to bottom at the centre of the image, the temple can be determined. The first change in gradient from a negative to positive value determines the nose ridge. See results in Fig. 5.20.

By using both upward and downward gradients as shown in Figs. 5.21 and 5.22, the eye regions are located as shown in Fig. 5.23. With the upward gradient, only

Fig. 5.22 Upward gradients

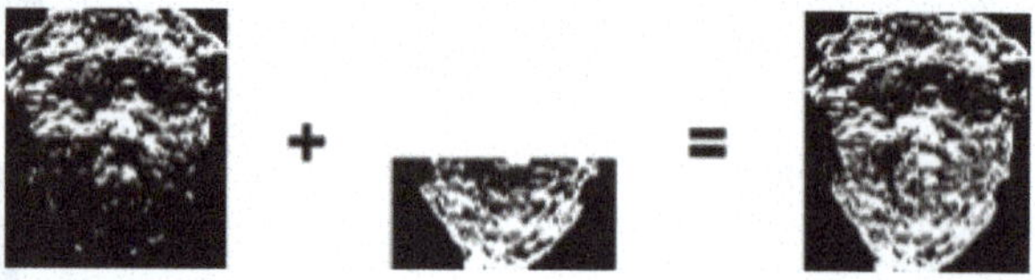

Fig. 5.23 Sum of downward and upward gradients to locate the eyes

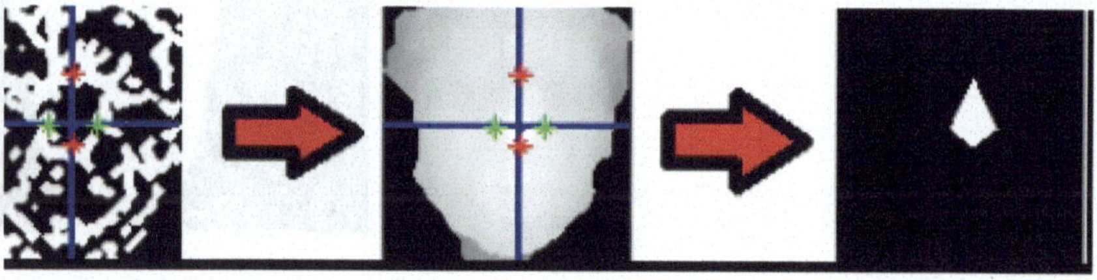

Fig. 5.24 Nose mask derivation [87, 100]

sections between the jaw and nose is retained which is summed with downward gradient as shown in:

The nose tip region may now be determined by examining the principal curvatures around the temple point and nose ridge. The intersection of this line with that of the eyes is used to determine this point as shown in Fig. 5.12. Principal curvatures are useful in locating the nose region as a protrusion. Thus, a mask of the nose region may be computed (Fig. 5.24).

Now that the fiducial regions are located, an image database is formalised. The resulting images are as shown in Fig.5.13 from canonical images of 15 subjects. It is to be noted that for the synthesis to work well, the fiducial points must be noticeable in the image. Where the poses in the canonical views are only partially visible, errors occur as can be visually identified from the database images. Thus, local regions in the face are identified as shown in Fig. 5.25. The canonical views of three subjects are as shown in Fig. 5.26. Each row represents sample poses for each individual.

A second model is constructed as a single morphed model from the samples. A simplistic approach is to average the sample images. It is important that the 1–1 correspondence between samples is maintained for the models to work well. An example of this process is shown in Fig. 5.27. It is noted that the average models occupy an approximate area but retain the shape of the face. Such an approximation allows imprecise matching to occur.

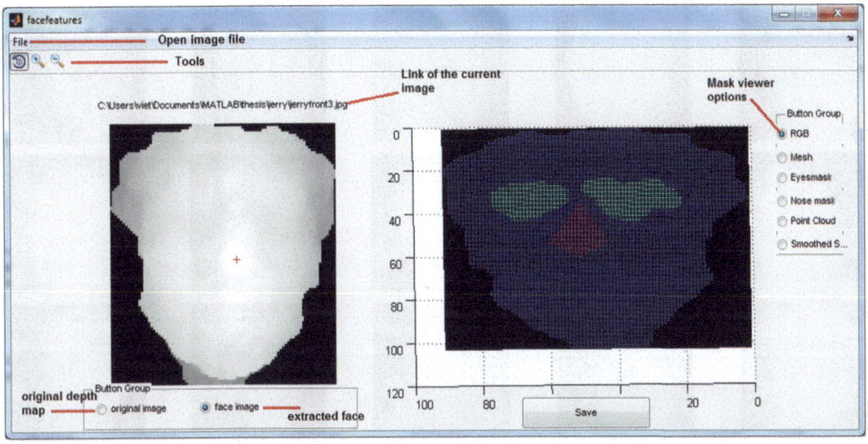

Fig. 5.25 KINECT-based face image synthesis: Fiducial points on face located

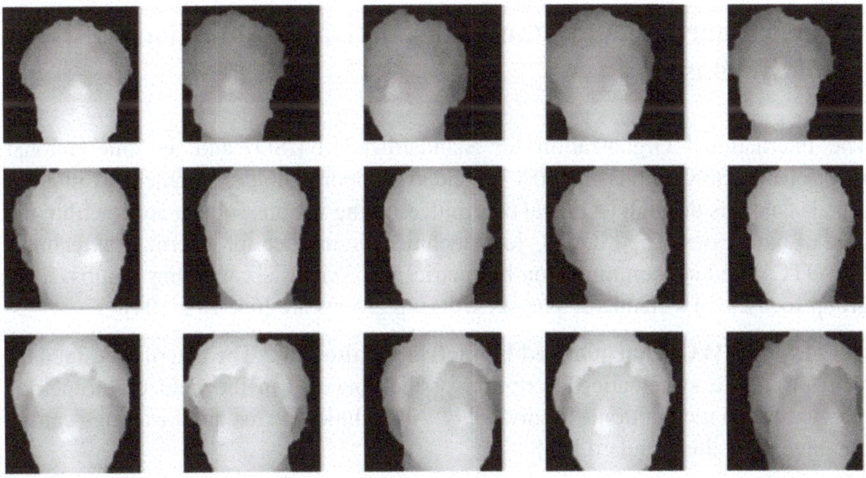

Fig. 5.26 Database with individual models: Canonical views of KINECT-based depth images for three subjects with five samples/subject

3.3 KINECT-Based 3D Face Query Matching

Given that we now have a procedure for generating morphed faces from a KINECT sensor and preprocessing that generates a smoothed database of images, we can use the feature extraction and matching techniques in Sect. 3. For brevity, the process is not repeated for this database here.

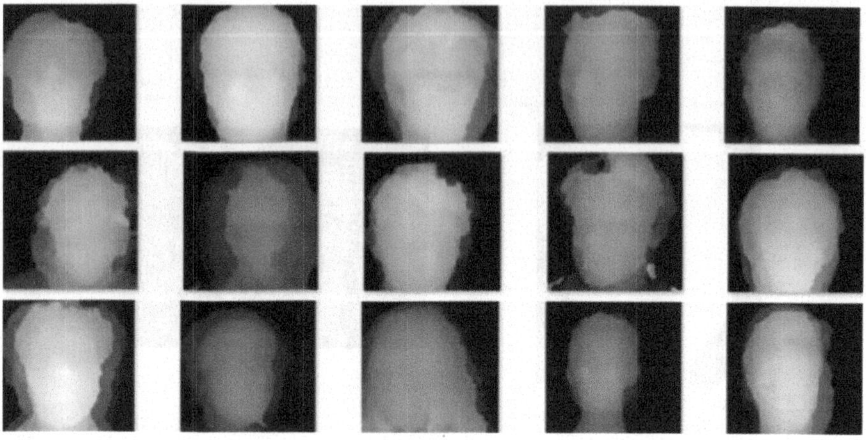

Fig. 5.27 Database with average models with one sample/subject

4 Benchmarking and Standardisation and Evaluation of Systems

The International Organization for Standardization (ISO) and the International Electrotechnical Commission (IEC) standard supports the development of international standards through technical committees in the various subject areas within the field of biometrics. The ISO/IEC joint technical committee for information technology (JTC1) and subcommittee on biometrics (SC37) has six working groups. Each group focuses on a standard. The six working groups are [64, 102–104]:

JTC 1/SC 37/WG 1 harmonized biometric vocabulary: The aim of this standard is to provide a systematic description of the concepts in the field of biometrics relevant to the recognition of individuals. All technical terms and meanings can be obtained from this standard [66].

JTC 1/SC 37/WG 2 biometric technical interfaces: This document focuses on the standardisation of the interfaces between various biometric component and subsystems. It also includes data transfer and security aspects in the process within [63, 105, 106].

JTC 1/SC 37/WG 3 biometric data interchange formats: It involves standardising the content, meaning and representation of biometric data formats of particular biometric technology or technologies [107, 108].

JTC 1/SC 37/WG 4 biometric functional architecture and related profiles: It addresses the standardisation of biometric functional architecture and related profiles. These profiles refer to the biometric-related standards relevant to how the configurable parameters in the system should be set, to achieve interoperability [108, 109].

JTC 1/SC 37/WG 5 biometric testing and reporting: It involves testing and reporting methodologies, requirements, functional and technical performance metrics that cover biometric technologies, systems and components [108, 110–112]. It addresses the cross-jurisdictional and societal aspects in the application of international biometrics standards. It focuses on the aspects of design and implementation of biometric technologies and its usability, health and safety, legal requirements and other cross-jurisdictional and societal considerations relevant to personal information.

JTC 1/SC 37/WG 6 cross-jurisdictional and societal aspects of biometrics: It addresses the cross-jurisdictional and societal aspects in the application of international biometrics standards. It focuses on the aspects of design and implementation of biometric technologies and its usability, health and safety, legal requirements and other cross-jurisdictional and societal considerations relevant to personal information [113].

Use of standards while developing and implementing biometrics enables interoperability, repeatability and interchangeability when needed. At a time when the technology is moving more and more into a cloud-based system, the data exchange and interoperability of data become easier and consistent with the adoption of standards. This enables biometric system manufacturers to follow the guidelines in development and to reduce technical risks when working in a larger geographic scope.

Standards in progress within biometrics do not just cover the vocabulary, data formats, sensors, image quality testing requirements and methodology but also liveness detection to safeguard system against spoofing attacks. Although ISO mainly develop standards, there are some technical reports in other areas such as the cross-jurisdictional and societal aspects of biometrics.

With emerging technologies and newer methodologies, standards must be adaptable to certain extent. Hence, there are periodic reviews and updates. If there are extensive changes, the working group projects are reinitiated. Standards are well used once developed. For example, programmes such as personal identity verification (PIV) in the USA [114] since 2004 and the recent world's largest programme UIDAI (unique identity authority of India] have popularly used them.

The National Institute of Technology in the USA performs routine biometric evaluations for manufacturers and conduct challenges in various biometric modalities to identify the best performing biometric algorithm. Similarly, the National Physical Laboratory in the UK provides testing and evaluation of biometric products for both the government and manufacturers. The Home Office Centre for Applied Science and Technology also uses NPL services for peer review of its own evaluations [115].

HECTOS, a project funded under the EC FP7 security research programme, brings together various leading organisations from across Europe to study how a harmonised approach to evaluation and certification schemes of physical security products could be developed. One of the case studies in this project is biometrics. It looks at various aspects of testing and evaluation of biometrics [64, 116, 117].

5 Ethics and Privacy

Though the idea of using biometrics for security and border control seems very interesting with promising results, there are many ethical issues surrounding these. As discussed earlier biometrics could include different parameters such as fingerprint, iris, DNA, voice and face images. Not everyone is happy to have these parameters scrutinised especially the travellers. The usability for the point of view of the monitoring officer and from the point of view of the travellers/subjects must be taken into consideration. The government in the UK scrapped the idea of the use of biometrics in the design of national ID card scheme which was intended to generate the national identity register. The reasons for scrapping this scheme go beyond saving the economy, as more and more people in the UK feel that this is an infringement of personal freedoms and civil liberties [10]. The environment and climate can also cause changes in the humidity and texture of the palm/fingers and may result in false negative causing an embarrassment to the subject at the check points. These emotional issues cannot be undermined, and it is necessary to strike a balance between the use of these advanced techniques and ethical issues surrounding them.

Also, the extensive use of video surveillance in general along with advances in face recognition has raised ethical and privacy issues of the people identifiable in the recorded data. Although CCTV recording have been in use for decades, the image quality was just good enough for the system to identify an individual only if they were on a database and the images had to be of a very good quality even for a somewhat likely match. Due to high-definition cameras and surveillance systems that produce more sharp, clear and detailed images, the highly computational algorithms are now capable of finding matches easier of not just the person of interest but also of people in that frame. With these developments and advances in technology, more and more issues of privacy of the public in such recordings are becoming a major issue especially in the law and enforcement area. Using CCTV images and the mobile tracking could enable track people without their permission. These raise several issues. Hence, there is a lot of interest in face de-identification to protect the identity of the individual. This is fully supported by law as protecting an individual's privacy both ethically and socially is a legal requirement under EU data protection directive [118]. Standards covering the ethical and privacy aspects at manufacturing and implementing level of the biometric product are under development [111, 119].

Methods such as pixilation, masking, blurring the face in the frame and blacking out the face area in images have been tried in the past to protect the privacy of the individuals of no interest. However, with advances in face recognition algorithms, these processes can be reversed to obtain the missing contents/data in the image. Robust de-identification techniques such as the k-anonymity-based methods have been used successfully [68]. The main goal of de-identification algorithms enables organisations recording data and further sharing it with law and enforcement authorities by complying with the EU directive. All k-anonymity-based methods de-identify a face image by replacing it with the average of k faces from a gallery and hence achieve privacy protection by guaranteeing a recognition rate lower than 1/k. Issues such as ghost artefacts/misalignment of faces tend to appear particularly in the eye and mouth areas due to the vast variety in eye position and mouth shape/location exhibited by different expressions [120]. Discussing various other methods is beyond the scope of this chapter but provides an insight how technology advances are happening to protect privacy. This is currently a hot topic of research as de-identification can regenerate the images only when appropriate and thus retain the privacy of the individual.

6 Conclusions

In the first part of the chapter, we introduce the concept of face biometrics leading on to 2D face recognition (2DFR) principles. The concepts of 2DFR are extended to 3D face recognition (3DFR). We look at the research and market surveys of face recognition as a biometric and specifically for border control applications. Specific challenges in 3DFR systems are explored. Some of the key biometric performance metrics are covered.

In the second part, two case studies that use some of the techniques discussed in the first part are dealt with in detail. Both case studies are 3DFR systems using a stereo vision and MS KINECT sensors. Their preprocessing techniques differ as the sensors are different. But the rest of the processes starting from feature extraction are unified for these sensors' databases.

In the last section, we consider standards and performance metrics from a standards perspective. Further, ethical and privacy matters are key factors defining acceptance by end-users and the public and hence considered in this chapter.

Appendix A: Performance Metrics

In this appendix, we consider the notations and terminologies commonly used to evaluate biometric systems [121].

- *Gallery and probe sets*: For purpose of performance evaluation, the feature set F is divided into partitions of gallery G that forms the database of templates of enrolled subjects and probe P that forms the set of query samples. Depending on the specific performance metric to be determined, the elements of the gallery and probe sets, $g \in G$ and $\in P$, respectively, will vary. For example, the probe set could be a subset of the gallery during the training phase of a face recognition system and mutually exclusive during the testing phase.
- *Identification:* Identification in a biometric system is the process of determining the identification of an individual from the database. The identification process

matches a probe as a query against the gallery and returns similarity scores, $\forall g \in G$. The scores are usually normalised in the range [0,1].

- *Verification* is the process of confirming that a claimed identity is correct by comparing the probe with one or more enrolled templates.
- *Open-set and close-set identification:* Identification is close-set if a person is assumed to be previously enrolled and open-set otherwise (as in the case of a watch list whose identity is not known previously).
- *False acceptance rate (FAR):* an empirical estimate of the probability that an impostor has been falsely verified to bear a correct identification.
- *False rejection rate (FRR):* an empirical estimate of the probability that a person with true identification has been falsely rejected by the system.
- *Equal error rate (EER):* The rate at which FAR = FMR.
- *Identity function:* A function $id(g)$ that returns the identity as an integer indexing the database templates and given by $id : \mathcal{X} \longrightarrow \mathcal{U}$ where $\mathcal{U}$ is a set of unique identities. Let U_{g} denote these set of identities in G and U_{p} the identities in P. As mentioned before, for some testing conditions of training and testing phases, $U_{\mathrm{g}} \cap U_{\mathrm{p}} = \varnothing$.
- *Identification rate:* Closed-set performance evaluation requires the sorting of similarity scores during a matching process of the probe against the gallery which are now in a natural increasing order of ranking. The identification rate $I(k)$ is defined as the fraction of probes at rank k or below:

$$I(k) = \frac{|\{b|\mathrm{rank}(b) \leq k, \quad \forall_{b \in B}\}|}{|U_{\mathrm{p}}|},$$

where $|U_{\mathrm{p}}|$ is the size of the probe set.

- *Cumulative match curve (CMC):* The CMC chart is a plot of k vs $I(k)$. It is a non-decreasing function. The example in [121] is quoted here. If there are 100 probes and a system has 50 outputs with 50 rank 1 outcomes, 40 rank 2 outcomes, 5 rank 3 outcomes, 3 rank 4 outcomes and 2 rank 5 outcomes, then the number of elements with rank k or less is {50, 90, 95, 98, 100} for ranks $k = \{1, 2, 3, 4, 5\}$, respectively. Hence, the identification rate is 50% for rank 1 performance, 90% for rank 2 performance and so on. As k increases, the identification rate increases and eventually attains 100%.

References

1. D. Scheuermann, S. Schwiderski-Grosche, B. Struif, *Usability of Biometrics in Relation to Electronic Signatures* (GMD-Forschungszentrum Informationstechnik, Sankt Augustin, 2000)
2. P.S. Teh, A.B.J. Teoh, S. Yue, A survey of keystroke dynamics biometrics. Sci. World J. **2013**, 24 (2013)
3. A.K. Jain, P. Flynn, A.A. Ross, *Handbook of Biometrics* (Springer-Verlag, New York, 2007)

4. M. Satone, G. Kharate, Feature selection using genetic algorithm for face recognition based on PCA, wavelet and SVM. Int. J. Electr. Eng. Inf. **6**, 39–52 (2014)
5. A.F. Abate, M. Nappi, D. Riccio, G. Sabatino, 2D and 3D face recognition: A survey. Pattern Recogn. Lett. **28**, 1885–1906 (2007)
6. P.J. Phillips, P. Grother, R. Micheals, D.M. Blackburn, E. Tabassi, M. Bone, Face recognition vendor test 2002, Presented at the Proceedings of the IEEE International Workshop on Analysis and Modeling of Faces and Gestures, 2003
7. SITA, Biometrics Centred Border Management: Helping governments reliably and securely verify travellers' identities, SITA2010
8. UK Her Majesty Post office, Guidance on Biometric Passport and Passport Reader, 2016
9. A.M. Bronstein, M.M. Bronstein, R. Kimmel, Expression-invariant 3D face recognition, in *Audio- and Video-Based Biometric Person Authentication: 4th International Conference, AVBPA 2003 Guildford, UK, June 9–11, 2003 Proceedings*, ed. by J. Kittler, M.S. Nixon (Springer, Berlin Heidelberg, 2003), pp. 62–70
10. S. Mansfield-Devine, Comment on biometrics. Biometric Technol. Today, **July–August**, 12 (2010)
11. V.D. Kaushik, A. Budhwar, A. Dubey, R. Agrawal, S. Gupta, V.K. Pathak, et al., An Efficient 3D Face Recognition Algorithm, in *2009 Third International Conference on New Technologies, Mobility and Security*, 2009, pp. 1–5
12. H. Yuxiao, J. Dalong, Y. Shuicheng, Z. Lei, Z. Hongjiang, Automatic 3D reconstruction for face recognition, in *Sixth IEEE International Conference on Automatic Face and Gesture Recognition, 2004. Proceedings*, 2004, pp. 843–848
13. V. Blanz, T. Vetter, Face recognition based on fitting a 3D morphable model. IEEE Trans. Pattern Anal. Mach. Intell. **25**, 1063–1074 (2003)
14. B. Brecht. Facing the future with 3D facial recognition technology. Biometric Technol. Today. Jan 2009, 8–9 (2009)
15. A. Suman, Automated face recognition, applications within law enforcement, Market Technol. Rev., Oct 2006 (2006)
16. C. Beumier, M. Acheroy, Face verification from 3D and grey level clues. Pattern Recogn. Lett. **22**, 1321–1329 (2001)
17. X. Chenghua, W. Yunhong, T. Tieniu, Q. Long, Automatic 3D face recognition combining global geometric features with local shape variation information, in *Sixth IEEE International Conference on Automatic Face and Gesture Recognition, 2004. Proceedings*, 2004, pp. 308–313
18. Y. Wang, C.-S. Chua, Y.-K. Ho, Facial feature detection and face recognition from 2D and 3D images. Pattern Recogn. Lett. **23**, 1191–1202 (2002)
19. Z. Michael, M. Michael, G. Gunther, S. Marc, S. Jochen, Automatic reconstruction of personalized avatars from 3D face scans. Comput. Animat. Virtual Worlds **22**, 195–202 (2011)
20. A. Ansari, M. Abdel-Mottaleb, M.H. Mahoor, Disparity-based modelling for 3D face recognition, in *ICIP*, 2006, pp. 657–660
21. V. Blanz, K. Scherbaum, H.P. Seidel, Fitting a morphable model to 3D scans of faces, in *IEEE ICCV*, 2007, pp. 1–8
22. X. Lu, A. Jain, Deformation modeling for robust 3D face matching. IEEE Trans. Pattern Anal. Mach. Intell. **30**, 1346–1357 (Aug 2008)
23. Y.A. Li, Y.J. Shen, G.D. Zhang, T. Yuan, X.J. Xiao, H.L. Xu, An efficient 3D face recognition method using geometric features, in *2010 Second International Workshop on Intelligent Systems and Applications*, 2010, pp. 1–4
24. Y. Pan, B. Dai, Q. Peng, Fast and robust 3D face matching approach, in *Image Analysis and Signal Processing*, 2010, pp. 195–198
25. Y. Wang, J. Liu, X. Tang, Robust 3D face recognition by local shape difference boosting. IEEE PAMI **32**, 1858–1870 (2010)

26. N. Uchida, T. Shibahara, T. Aoki, H. Nakajima, K. Kobayashi, Face recognition using passive stereo vision, in *IEEE International Conference on Image Processing*, 2005, pp. 950–953
27. P. Sharma, M. Goyani, 3D face recognition techniques - a review. Int. J. Eng. Res. Appl. IJERA **2**, 787–798 (2012)
28. S. Huq, B. Abidi, S. G. Kong, M. Abidi, A survey on 3D modeling of human faces for face recognition, in *3D Imaging for Safety and Security*, vol. 35, ed. by A. Koschan, M. Pollefeys, M.A. Abidi (Springer, Dordrecht, 2007), pp. 25–67
29. N.U. Powar, J.D. Foytik, V. K. Asari, H. Vajaria, Facial expression analysis using 2D and 3D features, in *Proceedings of the 2011 I.E. National Aerospace and Electronics Conference (NAECON)*, 2011, pp. 73–78
30. Y. Sheng, A.H. Sadka, A.M. Kondoz, Automatic single view-based 3-D face synthesis for unsupervised multimedia applications. IEEE Transactions on Circuits and Systems for Video Technology **18**, 961–974 (2008)
31. M.P. Segundo, L. Silva, O.R.P. Bellon, C.C. Queirolo, Automatic face segmentation and facial landmark detection in range images. IEEE Trans. Syst. Man Cybernet. Part B Cybernet. **40**, 1319–1330 (2010)
32. T.S.N. Uchida, T. Aoki, H. Nakajima, K. Kobayashi, Face recognition using passive stereo vision, in *IEEE International Conference on Image Processing*, 2005, pp. 950–953
33. D. Huang, K. Ouji, M. Ardabilian, Y. Wang, L. Chen, 3D face recognition based on local shape patterns and sparse representation classifier, in *Advances in Multimedia Modeling, Lecture Notes in Computer Science*, vol. 6523, ed. by K.T. Lee, W.H. Tsai, H.Y. Liao, T. Chen, J.W. Hsieh, C.C. Tseng, (Springer, Berlin/Heidelberg, 2011), pp. 206–216
34. H. Tang, Y. Sun, B. Yin, Y. Ge, 3D face recognition based on sparse representation. J. Supercomput. **58**, 84–95 (2011)
35. K. Tae-Kyun, J. Kittler, Design and fusion of pose invariant face-identification experts. IEEE Transactions on Circuits and Systems for Video Technology **16**, 1096–1106 (2006)
36. V. Bevilacqua, F. Adriani, G. Mastronardi, 3D head normalization with face geometry analysis, genetic algorithms and PCA. J. Circuits Syst. Comput. **18**, 1425–1439 (2005)
37. H. Zhou, A. Mian, L. Wei, D. Creighton, M. Hossny, S. Nahavandi, Recent advances on singlemodal and multimodal face recognition: A survey. IEEE Trans. Hum.-Mach. Syst. **44**, 701–716 (2014)
38. D.-L. Xu, J.-B. Yang, Y.-M. Wang, The evidential reasoning approach for multi-attribute decision analysis under interval uncertainty. Eur. J. Oper. Res. **174**, 1914–1943 (2006)
39. F. Hajati, A.A. Raie, Y. Gao, Pose-invariant multimodal (2D+3D) face recognition using geodesic distance map. J. Am. Sci. **7**(10), 583–590 (2011)
40. P. Xiong, L. Huang, C. Liu, Real-time 3d face recognition with the integration of depth and intensity images, in *Eighth International Conference on Image Analysis and Recognition - Volume Part II, ser. ICIAR'11*, Berlin, Heidelberg, 2011, pp. 222–232
41. K.W. Bowyer, K. Chang, P. Flynn, A survey of approaches and challenges in 3D and multi-modal 3D + 2D face recognition. Comput. Vis. Image Underst. **101**, 1–15 (2006)
42. Z. Sun, A.A. Paulino, J. Feng, Z. Chai, T. Tan, A.K. Jain, A study of multibiometric traits of identical twins. Proc. SPIE Biometric Technol. Hum. Identif. VII **7667**, 76670T-1–76670T-12 (2010)
43. V. Nirgude, A. Gulve, S. Waghmare, Face recognition system using principal component analysis & linear discriminant analysis method simultaneously with 3d morphable model. UACEE Int. J. Artif. Intell. Neural Netw., 40–44 (2011)
44. G.G. Gordon, Face recognition based on depth and curvature features, in *Computer Vision and Pattern Recognition, 1992. Proceedings CVPR'92., 1992 I.E. Computer Society Conference on*, 1992, pp. 808–810
45. J.C. Lee, E. Milios, Matching range images of human faces, in *[1990] Proceedings Third International Conference on Computer Vision*, 1990, pp. 722–726

46. F.B. ter Haar, R.C. Veltkamp, 3D face model fitting for recognition, in *Lecture Notes in Computer Science, Part IV*, ed. by D. Forsyth, P. Torr, A. Zisserman, 5305th edn., (Springer-Verlag, Berlin Heidelberg, 2008)
47. F. Hajati, A.A. Raie, Y. Gao, 2.5D face recognition using patch geodesic moments. Pattern Recogn. **45**, 969–982 (2012)
48. S. Berretti, A.D. Bimbo, P. Pala, 3D face recognition using iso-geodesic stripes. IEEE Pattern Anal. Mach. Vis. **32**, 2162–2177 (2010)
49. M. Bronstein, R. Kimmel, A. Spira, 3D face recognition without facial surface reconstruction, in *European Conference on Computer Vision*, 2004
50. W. Liu, A.S. Mian, A. Krishna, B.Y.L. Li, Using Kinect for face recognition under varying poses, expressions, illumination and disguise, Presented at the Proceedings of the 2013 I.E. Workshop on Applications of Computer Vision (WACV), 2013
51. H. van den Yannick, *Gender Classification with Visual and Depth Images* (Tilburg University, 2012)
52. B.Y.L. Li, A.S. Mian, W. Liu, A. Krishna, Using Kinect for face recognition under varying poses, expressions, illumination and disguise, in *2013 I.E. Workshop on Applications of Computer Vision (WACV)*, 2013, pp. 186–192
53. Y. Sheng, A.H. Sadka, A.M. Kondoz, Automatic 3D face synthesis using single 2D video frame. Electron. Lett. **40**, 1173–1175 (2004)
54. A. Mian, M. Bennamoun, R. Owens, An efficient multimodal 2D-3D hybrid approach to automatic face recognition. IEEE Trans. Pattern Anal. Mach. Intell. **29**, 1927–1943 (2007)
55. S. Ramalingam, Fuzzy interval-valued multi criteria based decision making for ranking features in multi-modal 3D face recognition. Fuzzy Set Syst. **337**, 25–51 (2018)
56. Artec Broadway 3D, 3D face recognition walk-through device. (3 Aug 2017). Available: https://www.artecid.com/products/artec-broadway-3d
57. Ayonix, *Ayonix Public Security* (Ayonix, Tokyo)
58. Morpho (ed.), *Identification - Morpho 3D Face Reader: Fast, Convenient, Secure Facial Recognition* (Morpho, France, 2017)
59. A. Perala, Cheaper biometric smartphones flooding global market: acuity (Jan 23, 2017). Available: http://findbiometrics.com/mobile-biometrics-primer-403020/?omhide=true
60. N.L. Clarke, S.M. Furnell, Authentication of users on mobile telephones – A survey of attitudes and practices. Comput. Secur. **24**, 519–527 (2005)
61. H.A. Shabeer, P. Suganthi, Mobile phones security using biometrics, in *International Conference on Computational Intelligence and Multimedia Applications (ICCIMA 2007)*, 2007, pp. 270–274
62. ISO, Standards on information technology, access control scenario and grading scheme committee, in *Biometric Performance Testing and Reporting Part 5*, vol. ISO/IEC 19795-5:2011, (International Standards Organisation, Geneva, 2016), p. 36
63. ISO, International standards for HCI and usability, in *ISO 13407: Human-Centred Design Processes for Interactive Systems*, (International Standards Organisation, Geneva), 1999
64. V.N. Nirgude, V.N. Nirgude, H. Mahapatra, S.A. Shivarkar, Face recognition system using principal component analysis & linear discriminant analysis method simultaneously with 3D morphable model and neural network BPNN method. Glob. J. Adv. Eng. Technol. Sci. **4** (2017)
65. S. Marcel, BEAT – biometrics evaluation and testing. Biometric Technol. Today **2013**, 5–7 (2013)
66. ISO, ISO/IEC 2382-37:2012 Information technology -- vocabulary -- part 37: Biometrics, ed, 2012
67. Common criteria for information technology security evaluation-part 1: introduction and general model, vol. Version 2.3, ed, 2005
68. E.M. Newton, L. Sweeney, B. Malin, Preserving privacy by de-identifying face images. IEEE Trans. Knowl. Data Eng. **17**, 232–243 (2005)

69. K.W. Bowyer, K. Chang, P. Flynn, A survey of approaches and challenges in 3D and multi-modal 3D+2D face recognition. Comput. Vis. Image Underst. **101**, 1–15 (2006)
70. R. Dass, R. Rani, D. Kumar, Face recognition techniques: a review. Int. J. Eng. Res. Dev. **4**, 70–78 (2012)
71. W.W. Bledsoe, *A Facial Recognition Project Report* (Panoramic Research, Palo Alto, 2016)
72. T. Mathew, P. Alex, Eigenfaces for recognition. J. Cogn. Neurosci. **3**, 71–86 (1991)
73. C. Ki-Chung, K. Seok Cheol, K. Sang Ryong, Face recognition using principal component analysis of Gabor filter responses, in *Recognition, Analysis, and Tracking of Faces and Gestures in Real-Time Systems, 1999. Proceedings. International Workshop on*, 1999, pp. 53–57
74. J.M. Kim, M.A. Kang, A study of face recognition using the PCA and error back-propagation, in *2010 Second International Conference on Intelligent Human-Machine Systems and Cybernetics*, 2010, pp. 241–244
75. Y. Jian, D. Zhang, A.F. Frangi, Y. Jing-yu, Two-dimensional PCA: a new approach to appearance-based face representation and recognition. IEEE Trans. Pattern Anal. Mach. Intell. **26**, 131–137 (2004)
76. M.Z. Alom, A. Khan, R. Biswas, M. Khan, Night mode face recognition using adaptively weighted sub-pattern PCA, in *2012 15th International Conference on Computer and Information Technology (ICCIT)*, 2012, pp. 119–125
77. W.L. Braje, D. Kersten, M.J. Tarr, N.F. Troje, Illumination effects in face recognition. Psychobiology **26**, 371–380 (1998)
78. A. Wagner, J. Wright, A. Ganesh, Z. Zhou, H. Mobahi, Y. Ma, Toward a practical face recognition system: robust alignment and illumination by sparse representation. IEEE Trans. Pattern Anal. Mach. Intell. **34**, 372–386 (2012)
79. Y.M. Lu, B.Y. Liao, J.S. Pan, A face recognition algorithm decreasing the effect of illumination, in *2008 International Conference on Intelligent Information Hiding and Multimedia Signal Processing*, 2008, pp. 378–381
80. L. Wu, P. Zhou, X. Xu, An illumination invariant face recognition scheme to combining normalized structural descriptor with single scale retinex, in *Biometric Recognition: 8th Chinese Conference, CCBR 2013, Jinan, China, November 16–17, 2013. Proceedings*, ed. by Z. Sun, S. Shan, G. Yang, J. Zhou, Y. Wang, Y. Yin, (Springer International Publishing, Cham, 2013), pp. 34–42
81. J.Y. Cartoux, J.T. Lapreste, M. Richetin, Face authentification or recognition by profile extraction from range images, in *[1989] Proceedings. Workshop on Interpretation of 3D Scenes*, 1989, pp. 194–199
82. T. Bajarin, Why your smartphone will be your next PC. TIME (25 Feb 2013). Available: http://techland.time.com/2013/02/25/why-your-smartphone-will-be-your-next-pc/
83. B. Weyrauch, B. Heisele, J. Huang, V. Blanz, Component-Based Face Recognition with 3D Morphable Models, in *2004 Conference on Computer Vision and Pattern Recognition Workshop*, 2004, pp. 85–85
84. K. Konolige, Projected texture stereo, in *2010 I.E. International Conference on Robotics and Automation*, 2010, pp. 148–155
85. Z. Zhang, Microsoft kinect sensor and its effect. IEEE MultiMedia **19**, 4–10 (2012)
86. R. Berri, D. Wolf, F. Osório, Telepresence robot with image-based face tracking and 3D perception with human gesture interface using kinect sensor, in *2014 Joint Conference on Robotics: SBR-LARS Robotics Symposium and Robocontrol*, 2014, pp. 205–210
87. F. Gossen, T. Margaria, Comprehensible People Recognition Using the Kinect's Face and Skeleton Model, in *2016 I.E. International Conference on Automation, Quality and Testing, Robotics (AQTR)*, 2016, pp. 1–6
88. Microsoft. Kinect for Windows. (4 Aug 2013, 2017). Available: https://developer.microsoft.com/en-us/windows/kinect/develop
89. S. Ramalingam, N.T. Viet, 3D face synthesis with KINECT, Presented at the Proceedings of the 2013 I.E. International Conference on Systems, Man, and Cybernetics, 2013

90. Non-contact 3D digitizer. (2004). Available: http://www.dirdim.com/pdfs/DDI_Konica_Minolta_Vivid_9i.pdf
91. Z. Zhou, A. Wagner, H. Mobahi, J. Wright, Y. Ma, Face recognition with contiguous occlusion using markov random fields, in *2009 I.E. 12th International Conference on Computer Vision*, 2009, pp. 1050–1057
92. H. Yagou, Y. Ohtake, A. Belyaev, Mesh smoothing via mean and median filtering applied to face normals, in *Geometric Modeling and Processing. Theory and Applications. GMP 2002. Proceedings*, 2002, pp. 124–131
93. J. Vollmer, R. Mencl, H. Müller, Improved Laplacian smoothing of noisy surface meshes. Comput. Graphics Forum **18**, 131–138 (1999)
94. B. Gökberk, A. Ali Salah, L. Akarun, R. Etheve, D. Riccio, J.-L. Dugelay, 3D face recognition, in *Guide to Biometric Reference Systems and Performance Evaluation*, ed. by D. Petrovska-Delacrétaz, B. Dorizzi, G. Chollet, (Springer, London, 2009), pp. 263–295
95. A. Mian, M. Bennamoun, R. Owens, Automatic 3D face detection, normalization and recognition, in *Proceedings 2006 Third International Symposium on 3D Data Processing, Visualization and Transmission 3DPVT 2006*, IEEE, 2006, pp. 735–742
96. S. Ramalingam, R. Venkateswarlu, Stereo face recognition using discriminant eigenvectors, in *WSES International Conference on Speech, Signal and Image Processing 2001 (SSIP 2001)*, Malta, 2001, pp. 2621–2626
97. S. Ramalingam, 3D face recognition: feature extraction based on directional signatures from range data and disparity maps, in *2013 I.E. International Conference on Systems, Man, and Cybernetics*, 2013, pp. 4397–4402
98. P.J. Phillips, W.T. Scruggs, A.J.O. Toole, P.J. Flynn, K.W. Bowyer, C.L. Schott, et al., FRVT 2006 and ICE 2006 large-scale experimental results. IEEE Trans. Pattern Anal. Mach. Intell. **32**, 831–846 (2010)
99. P.N. Belhumeur, J.P. Hespanha, D.J. Kriegman, Eigenfaces vs. fisherfaces: recognition using class specific linear projection. IEEE Trans. Pattern Anal. Mach. Intell. **19**, 711–720 (1997)
100. S. Ramalingam, N.T. Viet, 3D face synthesis with KINECT, in *2013 I.E. International Conference on Systems, Man, and Cybernetics*, 2013, pp. 4433–4438
101. A. Mian, N.E. Pears, 3D face recognition, in *3D Imaging, Analysis and Applications*, (Springer, London, 2012), pp. 311–366
102. G. Cannon, A. Yamada, P. Statham, Biometric security standards, in *Encyclopedia of Biometrics*, ed. by S.Z. Li, A.K. Jain, (Springer, Boston, 2009), pp. 1–9
103. Biometric Institute., http://www.biometricsinstitute.org/
104. Planet biometrics. Available: http://www.planetbiometrics.com/
105. S. Elliott, *JTC 1 SC 37 – Biometrics International Standards* (Biometrics Standards, Performance, and Assurance Laboratory, Purdue University, US), 2002
106. ISO, ISO/IEC/SC 37 WG2 Biometric technical interfaces. https://www.iso.org/committee/313770.html, 2002
107. JTC 1/SC 37/WG 3 Biometric data interchange formats. https://www.iso.org/committee/313770.html, 2002
108. ISO, Biometric technologies and security, in *International Biometric Standards Development Activities*, vol. ISO/IEC JTC 1/SC 37, (National Institute of Standards and Technology (NIST), Gaithersburg)
109. JTC 1/SC 37/WG 4 Biometric functional architecture and related profiles. https://www.iso.org/committee/313770.html, 2002
110. JTC 1/SC 37/WG 5 Biometric testing and reporting. https://www.iso.org/committee/313770.html, 2002
111. ISO, Biometric performance testing and reporting, in Information Technology. https://www.iso.org/committee/313770.html, 2002
112. ISO/IEC 19795 Series of International Standards: Information technology — Biometric performance testing and reporting, 2007–2012. https://www.iso.org/committee/313770.html, 2002

113. JTC 1/SC 37/WG 6 Cross-Jurisdictional and Societal Aspects of Biometrics. https://www.iso.org/committee/313770.html, 2002
114. N.B. Nill, Test procedures for verifying image quality requirements for personal identity verification (PIV) single finger capture devices, Centre for Integrated Intelligence Systems, Massachusetts, Report W15P7T-05-C-F600, Dec 2006
115. *Current and Future Uses of Biometric Data and Technologies* (Common Select Committee, Science and Technology Committee, London, UK, 2014)
116. HECTOS Deliverable D4.1, Selected product types, use cases and validation criteria for biometric case studies, Harmonized Evaluation, Certification and Testing of Security products (HECTOS). (2015). Available: http://hectos-fp7.eu/dissemination.html
117. HECTOS Deliverable D4.2, Working set of performance requirements and associated evaluation methodologies for the selected biometric case studies, Harmonized Evaluation, Certification and Testing of Security Products (HECTOS). (2016). Available: http://hectos-fp7.eu/dissemination.html
118. Directive 95/46/EC of the European Parliament and of the Council of 24 October 1995 on the protection of individuals with regard to the processing of personal data and on the free movement of such data. Off. J. L **281**, 31–50 (1995)
119. ISO/IEC JTC 1/SC 27/WG 5, Standard on Identity management and privacy technologies. https://www.iso.org/committee/313770.html, 2002
120. A. Shenoy, L. Meng, F. Rezwan, A. Ariyaeeinia, Retaining expression on de-identified faces, in *International Biometric Performance Conference (IBPC)*, NIST, Gaithersburg, 2012
121. S. Marcel et al., Description of metrics for the evaluation of biometric performance, in *BEAT Biometrics Evaluation and Testing* (2012)

Part II
Emerging Technologies

Chapter 6
Advances in Key Stroke Dynamics-Based Security Schemes

Mohammad S. Obaidat, P. Venkata Krishna, V. Saritha, and Shubham Agarwal

1 Introduction

The security system of any organization needs to be strong and robust; otherwise there is a high chance that adversaries exploit the system by exposing very important information, which is unlicensed, altering confidential information, rejecting authorized admittance to the system, etc. Today, most of the purchase, ticket booking, and information sharing are taking place online where security is highly required; this is referred to as e-security [1–6]. Regardless of whether the framework is on the web or not, security is essential, and subsequently, it is constantly prescribed to utilize some of the strategies like standards for utilizing passwords, encryption of passwords, utilizing intense antivirus software, not interfacing the framework with imperative information to the network, and better mailer-daemon frameworks to channel obscure sources or obscure expansions, among others. There have been significant advances in the field using various technologies for authentication. This includes the use of alphanumeric strings, biometrics, numeric pins, and others. Over the years these various implementations failed to serve the purpose for which they were

M. S. Obaidat
Department of Computer and Information Science, Fordham University, Bronx, NY, USA
e-mail: m.s.obaidat@ieee.org

P. Venkata Krishna (✉)
Department of Computer Science, Sri Padmavati Mahila Visvavidyalayam, Tirupati, India
e-mail: dr.krishna@ieee.org

V. Saritha
Department of Computer Science and Engineering, Sri Padmavati Mahila Visvavidyalayam, Tirupati, India

S. Agarwal
Amity School of Engineering and Technology, Amity University, Noida, UP, India
e-mail: shubham.agarwal@owasp.org

M. S. Obaidat et al. (eds.), *Biometric-Based Physical and Cybersecurity Systems*,
https://doi.org/10.1007/978-3-319-98734-7_6

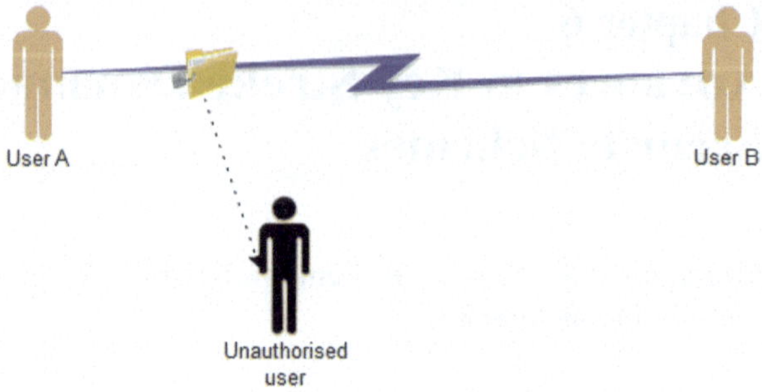

Fig. 6.1 Unauthorized access

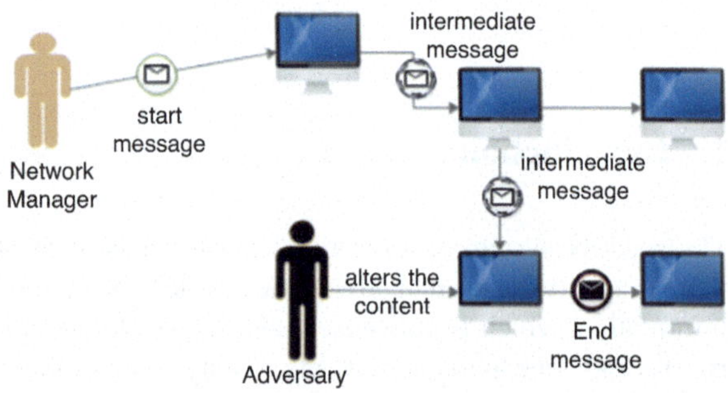

Fig. 6.2 Adversary modified the information during the transmission from source to destination

implemented as history witnessed the breach of every technology proving that nothing is failsafe.

It is observed from the surveys made by the "Federal Bureau of Investigation (FBI, Los Angeles)" and the "American Computer Security Institute (CSI)" that the cost of recovery from attacks varies from thousands to millions of dollars. So, more investments are made to provide high security in the network system, which fortifies the safety of the information and also to train the employees to be aware of the security systems. The rate at which the investments are increasing day-by-day proves that it is highly required to develop a cost-effective strategy to provide security to the systems [1–3]. Recently on 11th of May 2017, there is a high-level hacking that took place which shattered many countries.

This chapter focuses the discussion on the use of an efficient technology that proved to be effective in providing securing admittance to systems.

Some of possible cases of security cracks are:

(a) Access to unprotected confidential information by unauthorized users as shown in Fig. 6.1.
(b) Modification of confidential data by the adversary node as shown in Fig. 6.2.
(c) When the customer gives instruction to the stock broker, the intruder sends the similar instruction multiple times which causes loss to the customer; see Fig. 6.3.

Security attacks, security mechanisms, and security services are the three main features of any information security system. Security mechanisms are utilized to battle against security assaults and threats to enhance the security of the framework.

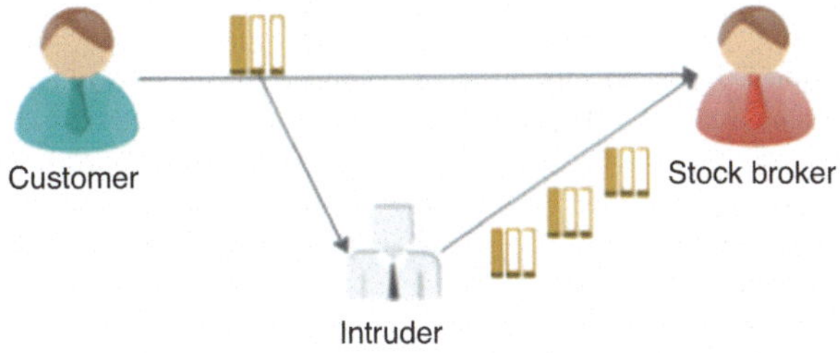

Fig. 6.3 Intruder sending the instructions repeatedly causing loss to the customer

Security attacks may be passive or active attacks [4, 5]. Information is not modified in the case of passive attacks, but instead it gains the access to the information illegally as shown in Fig. 6.1. It is difficult to detect such types of attacks. The most sensational data breach attack was executed over Yahoo, in the year 2016, where the account information of around 500 million users was leaked.

The active attack is illustrated in Fig. 6.2. It can be observed that the intruder is modifying the information before it reaches the destination. Active attacks are further classified as mimic, echo, alteration, and denial of service. In mimic, an unauthorized node behaves as an authorized one to acquire additional rights. Echo is demonstrated in Fig. 6.3. Messages are transmitted incessantly which leads to congestion and halts the system to be operated in ordinary mode.

The three key features that any security structure needs to possess are (a) truthfulness, which enables only the legal users to alter the resources of the association; (b) accessibility, which avoids denial of service attacks; and (c) secrecy, which enables the admittance of the resources only to the legal users of the organization [1–6].

1.1 Various Styles for User Authentication (Table 6.1)

Keystroke dynamics on the other hand provide a method so that only the person who initially registered for the account can open it even if his password is made public. Hence, it is observed that adding keystroke dynamics in authentication services would exponentially increase the security of the system.

At present, besides human intelligence, nothing can be used to beat the social engineering attack. In this chapter, a biometric-based approach toward strengthening the security of computer systems and networks is presented. In particular, this is known as keystroke dynamics, which is used to extend security of authentication of the legitimate user based on typing skills.

The typing style of a person on a computer keyboard is determined based on the time at which the key is pressed and the time at which it is released. This information is the basis for keystroke dynamics. This information is initially recorded and processed using efficient neural networks or pattern recognition procedure to obtain

Table 6.1 Various styles of user authentication

Knowledge-based authentication – grounded on the facts of the private data of the individual		
Type	Example	Source of the example
Static – It is grounded on predecided set of shared confidences	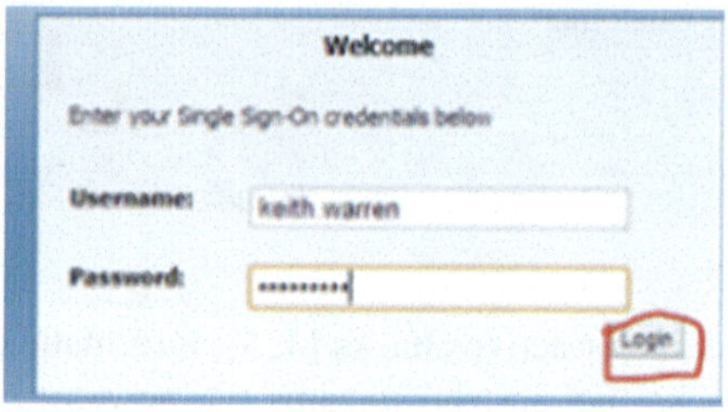	https://identityandaccessmgmt.wordpress.com/2014/01/26/oaam-11g-r2-enable-knowledge-based-authentication/
Dynamic – It is based on the queries produced from an extensive base of personal data	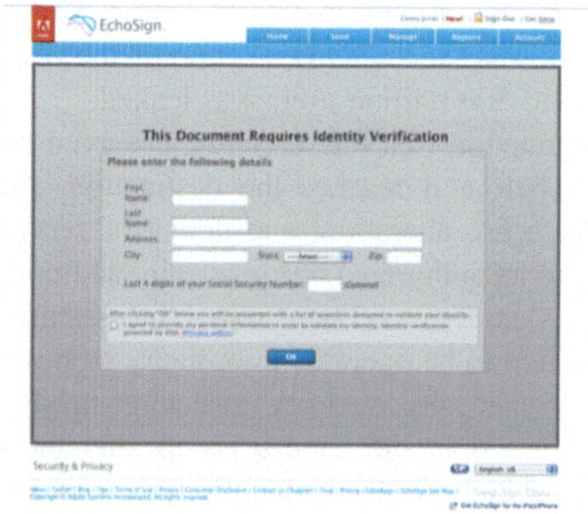	http://echojason.typepad.com/echosign_tips_and_tricks/2012/03/signer-authentication-20-knowledge-based-authentication-and-enhanced-web-identity-authentication-1.html
Object-based authentication – using any objects		
For example, introducing a person by another in case of opening an account in the bank	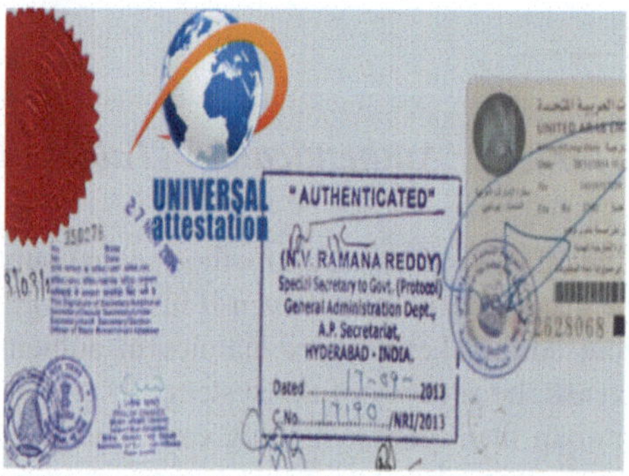	https://www.google.co.in/imgres?imgurl=http%3A%2F%2Funiversalsolutions.biz%2Fwp-content%2Fuploads%2F2015%2F03%2Fhrd-mea-embassy--

(continued)

Table 6.1 (continued)

Biometrics-based authentication – It depends on the exclusive natural features of an individual		
Physiological–Physical		
Finger prints		https://www.pinterest.com/pin/478929741604267074/
Face recognition	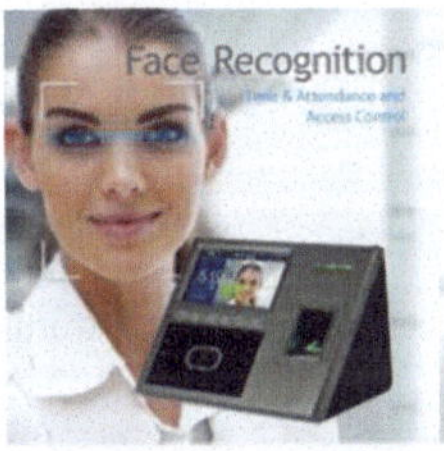	http://credenzeinfra.com/face-recognition-system.php
Retina	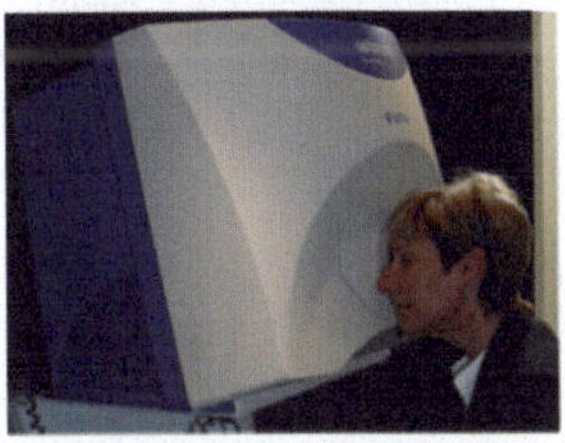	http://itblogs.in/biometrics/retina-recognition-technique/
Iris	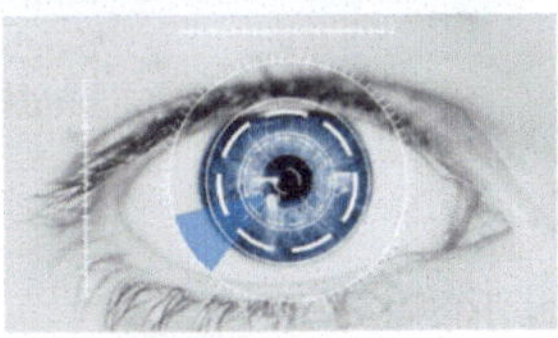	https://www.dasec.h-da.de/teaching/dasec-scientific-talk/2015-12-03-on-biometrics/
Hand geometry		http://blog.synerion.com/biometric-time-clocks-what-are-they-what-can-they-do

(continued)

Table 6.1 (continued)

DNA	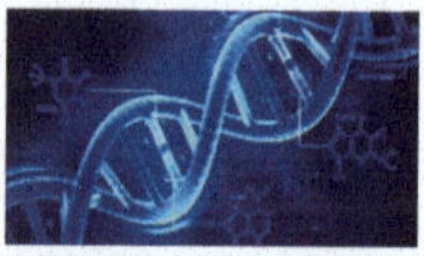	https://www.pinterest.com/explore/dna/
Behavioral–behavior		
Keystroke dynamics		http://biomedia4n6.uniroma3.it/research/biometrics.html
Voice recognition	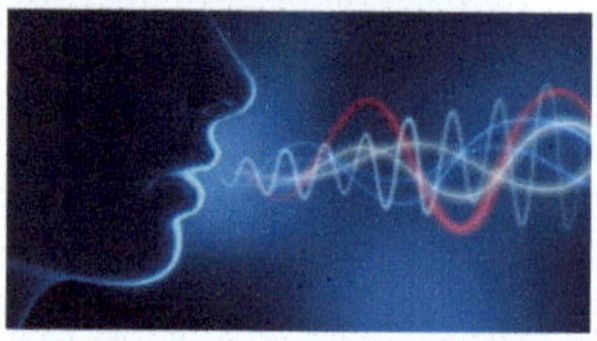	http://www.information-age.com/think-you-speak-voice-recognition-replacing-password-123461752/
Gait	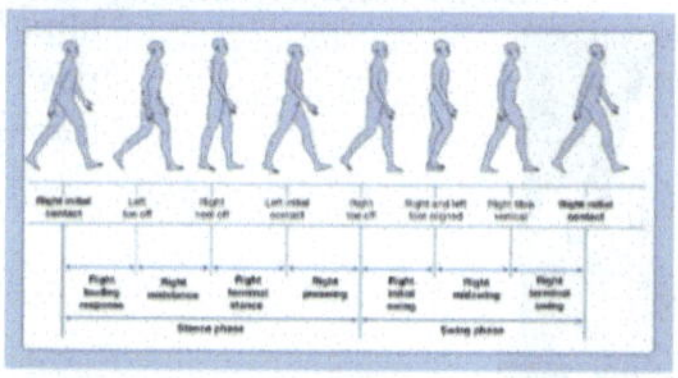	https://clinicalgate.com/gait-analysis-technology-and-clinical-applications/
Hand-written signature		http://biomedia4n6.uniroma3.it/research/biometrics.html#signature
Mouse dynamics	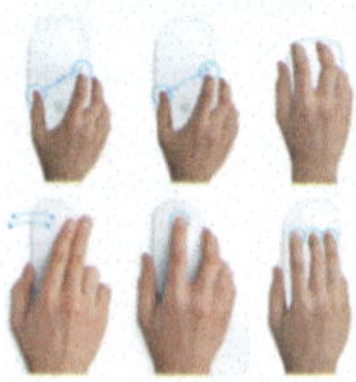	http://mac.appstorm.net/tag/trackpad/

a prime outline, and later when the person needs to be authenticated, the new information is compared with the key outline.

2 State of Art

In [7], the authors studied the consequences of added touchscreen features to the authentication and identification act using dataset of 42 users.

The verification and identification of the flaw hosts is made using keystroke dynamics [8]. The assessment is taken place to check the robustness in contrast to fraud attacks. In particular, a verification system called TUBA is introduced, which can be operated remotely for checking user's typing style. The power of TUBA is assessed using a thorough exploratory assessment including two simulated bots, which are connected serially [8].

Keystroke dynamics scheme is used in cloud validation framework [9]. This framework is appropriate to static and dynamic text keyed on customary or advanced keyboards like touchscreen. The proposed framework uses diverse component extraction strategies in preprocessing phase to limit the amount of element space. In addition, distinctive combination guidelines are assessed to integrate the various component extraction techniques so that a group of the utmost important elements are selected. Due to enormous number of user tests, a clustering technique is connected to the user profile formats to decrease the authentication time. The framework is connected to three various standard datasets utilizing three unique classifiers [9]. To explore the achieved client illustrations and validate the same at terminals, various classifiers are produced by the investigators. It is assumed that a single Gaussian distribution is utilized to deliver the digraph plots in keystroke data by the Gaussian density approximator, which is used in the basic machine learning procedures. The authors in [10] tried to minimize the hypothesis by permitting multiple distributions to produce digraphs with the help of Gaussian mixture model (GMM). The testing is done using the information gathered in precise conditions [10]. The perceptive difficulties of a specified job and the statistic components of the user who is typing are considered to analyze whether the typing style depends on both components or only on one of them. The authors in [11] use new fusion structures in the view of dialect creation and keystroke dynamics, which records accurately the changing aspects of the phonetic decisions of the user.

Generally, the keystroke dynamics is used with the typing styles of the user on the traditional keyboard. The authors in [12] implemented keystroke dynamics for the touchscreen devices. The user is authenticated depending on the verification algorithm. Both the desirable and undesirable test cases of the user are used for the instigation by means of two-class classifier. As gathering of undesirable test cases of the user is not practically feasible at all times, then the authentication is carried out using single-class classifier and desirable test cases to identify and extricate the unauthorized persons. The authors in [12] developed and tested the framework which can authenticate based on only desirable test cases or both desirable and undesirable test cases.

Commonly, typing is done using both the hands, and keystroke dynamics is also mostly tested based on these patterns only. But, One-handed Keystroke Biometric Identification Competition (OhKBIC) [13] is conducted in the 8th International

Association for Pattern Recognition (IAPR) International Conference on Biometrics (ICB). The exertion of this competition is presented in [13]. The sentence typed by 64 members is gathered as an inimitable dataset. Test cases are taken to pretend regular keying style and the challenging person keying with a single hand. Contestants are intended to outline the models, which can categorize a dataset that has both regular keyed patterns and challenging person keyed test cases. The competition is to obtain maximum correctness of classification, and the obtained outcome of each contestant is submitted. The procedures which gave maximum correctness are projected in [13].

The former research proved that the score regulation and techniques exclusively applicable to the particular application helps in enhancing the performance of the biometric recognition schemes, which are based on the vocal sound and the sign. Hence, the authors in [14] tried to implement these techniques on keystroke dynamic verification systems. There are two objectives that are projected by the authors in [14]. Using various thresholding methodologies to test their impact on keystroke dynamics verification schemes for live functioning circumstances is the first objective. Utilization of score normalization methodology to enhance the performance of keystroke dynamics is the second objective. Totally 114 samples are used for testing purpose. The authors proved that the performance is enhanced by 20% [14]. The authors in [15] tested among three various hand positions during keying. Specifically, touch features are considered for assessment purpose. Mostly, features based on time are used, but it is proved that the features based on postures will perform better [15]. When both features are integrated and considered, then the authentication accuracy is increased, and it is shown using equal error rates, which are reduced by 36.8%. Various frequently used assessments are enumerated by training and testing the information related to a single session or across sessions, possessor information or the other, and static or dynamic positions of the hand during keying in order to prove that the applicability is improved. No restrictions in the hand positions increase the usability also. The authors also designed a probable outline to evaluate the system during indefinite hand positions during keying [15].

The text, which is lengthy and keyed using various devices, is considered to meet the objective of extending the keystroke dynamics-based user authentication (KDA). It is analyzed whether the performance of the authentication accuracy depends on the device being used, the size of the text, and the procedure for the authentication being followed. The evaluation is carried out using three types of keyboards – PC keyboard, soft keyboard, and touch keyboard, and it is concluded that the PC keyboard gives more accuracy when compared to the other two. Furthermore, the size of the text can enhance the performance in terms of accuracy [16].

3 Workflow of Keystroke Dynamics

The main measures of keystroke dynamics to authenticate a person are the speed at which the typing is done and the pressure with which the keys are pressed. The raw metrics for these measures are the dwell time and flight time. The time period for which the key is kept pressed is referred to as dwell time and the time duration amid the key-down and key-up of two consecutive keys is referred to as flight time.

The basic workflow of keystroke dynamics is shown in Fig. 6.4. Initially raw data is collected from the user and is analyzed. Data collection method for every technique described in the chapter remains similar. When a phrase is typed, the dwell time and flight time are calculated and recorded. Different types of algorithms are used for feature extractions. After feature extraction, the classification techniques are applied. The user is authenticated after various comparisons. Keystroke dynamics does not need any additional hardware for evaluation. In fact, same models are extrapolated to more sophisticated technologies such as touch dynamics.

The average typing speed may vary when compared to the typing speed of a word or two. The pace at which particular words are typed may differ for some other

Fig. 6.4 Workflow of keystroke dynamics

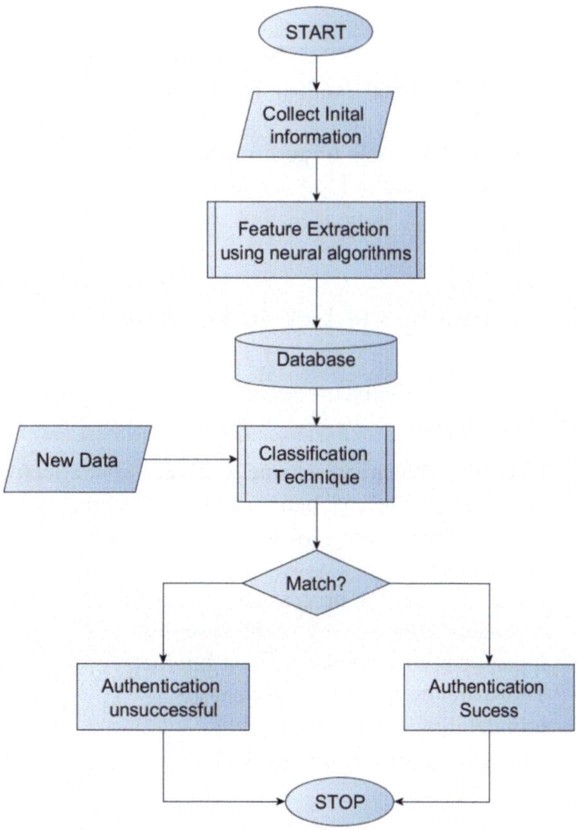

words even though it is same person. For example, the pace at which English language words are typed will be different when typing any other language words like Spanish, French, Arabic, Chinese, etc. The movement style of the mouse and the rate at which the mouse is being used or moved can also be considered in keystroke dynamics.

Besides the flight time and dwell time, many other metrics are there for user verification. Some of them are the speed with which the user is typing, seek time, investigating typical errors, the rate at which errors occur, the process of correcting the errors, and the pressure during the keystroke. The keystrokes can be analyzed using average rate at which the keys are pressed, the comparison of the time between the press of two keys, digraph (delay in time between two consecutive keystrokes), and trigraph (delay in time among three consecutive keystrokes).

The effectiveness of the system is evaluated using false rejection rate (FRR), false acceptance rate (FAR), and equal error rate (EER). EER is used to acquire overall correctness. EER is also known as crossover error rate (CER):

$$\text{FRR} = \frac{\text{Number of legal users disallowed}}{\text{Total number of legal users admitting the system}}$$

$$\text{FAR} = \frac{\text{Number of illegal users allowed}}{\text{Total number of illegal users tried to admit into system}}$$

A detailed analysis of performance of different techniques can be found in the research conducted in [1–6, 17–20].

4 Advantages of Keystroke Dynamics

As a behavioral biometric scheme, keystroke dynamics scheme has the following main advantages:

- Keystroke dynamics scheme is a very cost-effective procedure, as it does not require any additional hardware. It requires only the keyboard for it process. This is in addition to only software for keyboard control and for testing and authenticating process.
- It is easy to set up; process and training is not required to use.
- It enhances the security of the systems.
- It can be used with and without Internet.
- Transparency and noninvasiveness.
- Inimitability.
- Unremitting observation and validation.

5 Disadvantages

Although keystroke dynamics has many advantages, it has some drawbacks. The main ones are:

- Lower accuracy than physiological biometric schemes.
- User's vulnerability to exhaustion.
- Change in typing style from time to time.
- The typing speed may vary when the user is injured.
- There are different types of keyboards; hence there is a chance that the typing patterns vary from device to device for the same user.

6 Applications

It is used for providing security to log-ins, determining hackers, user identification, and anticipation of scams. Moreover, the continuous monitoring mechanisms are used to observe the user using the system.

Today, security is mainly required at any place like home, industrial plants, and institutions' campuses, among others. Hence, various kinds of software and hardware like BioTracker, AdmitOne Security, and BioCheck are coming into the market for providing security. Some of these products do continuous monitoring to confirm the authentication of the user, and some are one time checking.

Keystroke dynamics are sometimes used in combination with the other type of authentication method. For example, keystroke dynamics can be combined with the traditional password method, or it can be combined with voice recognition. This way of multi-level/multimodal authentication increases the level of security being provided for the system. The data related to keystroke dynamics of the users like the time at which user has logged in and logged out and the websites being used are stored, and analysis will be performed in order to identify or determine whether the user is properly utilizing the resources of the organization or sharing it illegally. Examples include sharing of software licenses and software as a service (SAAS) applications.

7 Keystroke Dynamics Appropriateness

Various conditions are identified to verify the appropriateness of keystroke dynamics.

- Anybody who knows how to operate a keyboard can use this software anywhere.
- False acceptance rate (FAR) and false rejection rate (FRR) are not sufficient to validate a user.

- The typing speed or other metrics, which are used in keystroke dynamics, vary from time to time.
- When a system is logged in and the user is changed, then it can be identified using keystroke dynamics.
- The utilization of this software might be an abuse against native laws. Hence, legal advice is required.
- Maintenance of data secrecy is needed.

8 Keystroke Dynamics Features

There are two types of features in keystroke dynamics. They are conventional and non-conventional features. The features which are based on the time are referred to as conventional features. The features which can be obtained when the input text is long and are not based on time are referred to as non-conventional features.

8.1 Conventional Features

Dwell time – the time for which the key is pressed is referred to as dwell time. It is also known as hold time.

If P_1 represents the press time of the key1 and R_1 represents the release time of the key1, then dwell time, $T_{\text{dwell}} = P_1 - R_1$.

Flight time – There are four types.

Press-Press (FT_{PP}) – the elapsed time between the press of the first and the second key, also referred to as digraph:

$$FT_{PP} = P_2 - P_1$$

Press-release (FT_{PR}) – the elapsed time between the press and release of two successive keys:

$$FT_{PR} = R_2 - P_1$$

Release-press (FT_{RP}) – the elapsed time between the release and press of two successive keys, also referred to as seek time:

$$FT_{RP} = P_2 - R_1$$

Release-release (FT_{RR}) – the elapsed time between the release of two successive keys:

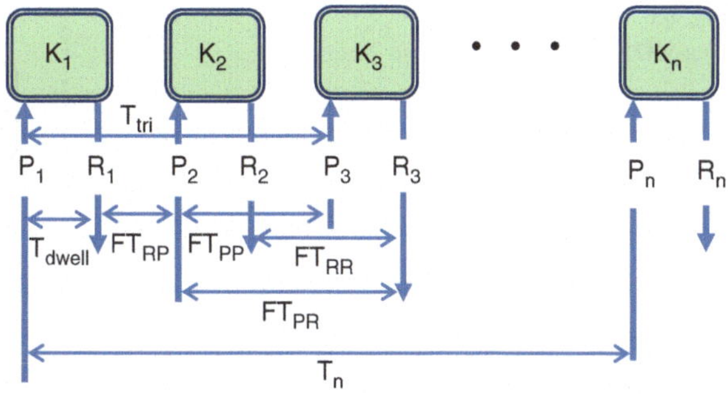

Fig. 6.5 Illustration of conventional features

$$\mathrm{FT_{RR}} = R_2 - R_1$$

Trigraph –the time duration amid the press of the first and the third key:

$$T_{\mathrm{tri}} = P_3 - P_1$$

n-graph – the time duration amid the press of the first and the *n*th key:

$$T_n = P_n - P_1$$

All the conventional features are illustrated in Fig. 6.5.

8.2 *Non-conventional Features [21]*

The time taken to extract these features is usually longer than the time taken to extract conventional features. These features are used to analyze the typing pattern of the users. There are two types of non-conventional features. They are semi-typing and editing features.

Word-per-minute – this measure gives the number of words that are typed in a minute. If it is measured for a minute, then WPM is directly obtained. If the measured time is more than a minute, then the total number of words typed is divided with the measured time to obtain WPM.

Negative up-down (Neg-UD) – when the key is pressed before releasing the previous key, overlapping occurs. This scenario occurs mostly when the user types very fast. Neg-UD is illustrated in Fig. 6.6.

Neg-UD is computed as the fraction of the number of Neg-UD overlapping to the number of pairs typed.

Negative up-up (Neg-UU) occurs when the later key is released before the former key is released. Neg-UU occurs if there exists Neg-UD, but it is to be noted that Neg-UU need not occur if Neg-UD occurs. Neg-UU is computed as the proportion

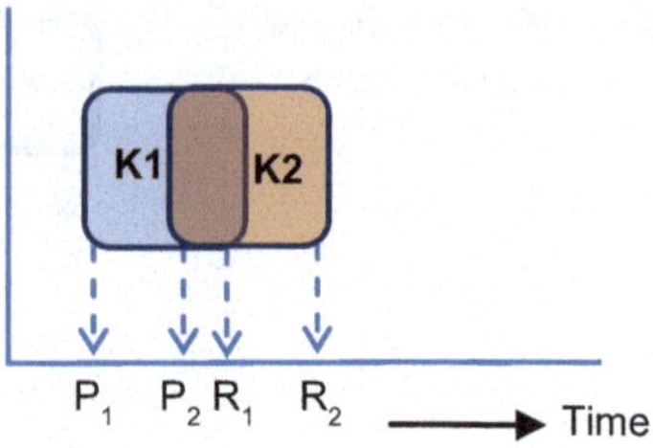

Fig. 6.6 Illustration of Neg-UD

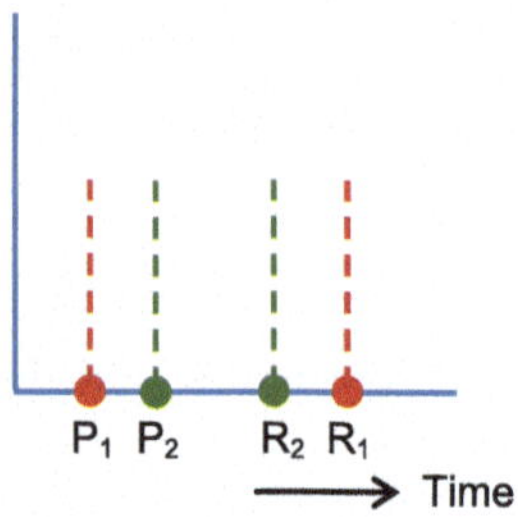

Fig. 6.7 Illustration of Neg-UU

of the number of Neg-UU overlapping to the number of pairs typed. Neg-UU is illustrated in Fig. 6.7.

8.2.1 Editing Features

These features are completely independent of time. They analyze the way the user is editing the text, for example, regularity of making errors, style of correcting the errors, etc.

Error rate – It is the proportion of flaws the user has made and corrected. It is the fraction of the times the delete and backspace button are keyed when compared to the total keys pressed.

Caps lock usage – It is the ratio between the number of times the capslock button is used to make the letter(s) into capital form and the total number of capital letters.

Shift key usage – One of the usages of the shift key is to type capital letters. Here, variations of the way the shift key is used need to be considered, e.g., using right/left shift key and releasing the letter key before/after the shift key.

9 Feature Subset Selection

When the data is very huge, it is very difficult to perform the operations. Hence, a subset of the data is extracted before classification is performed. In this process, it is required to reduce the repetitive or inappropriate information. The feature selection helps in minimizing the intricacy and in making it easy to infer. The appropriate selection of the data subset enhances the accurateness. Feature subset selection is an

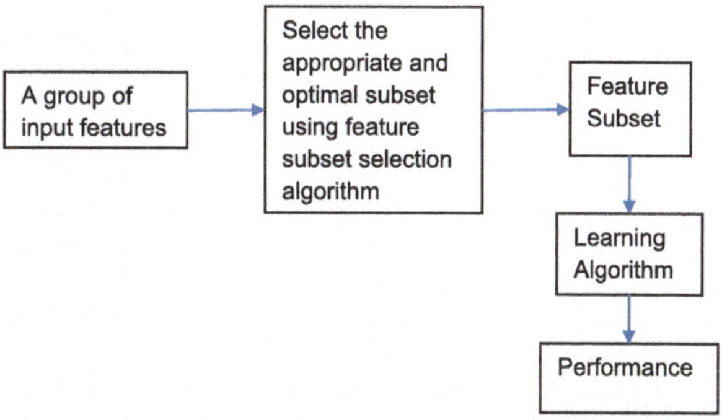

Fig. 6.8 Filter process

optimization problem as the subset extracted needs to perform well by reducing or increasing specific parameters. Hence, any optimization algorithms like ant colony optimization, genetic algorithms, and particle swarm optimization, or simulated annealing, can be used to perform feature subset selection. Machine learning techniques like neural networks or learning automata can be used in order to perform feature subset selection and inevitably choose a suitable subset of features.

There are three types of feature subset selection. They are filter methods, wrapper methods, and embedded methods [22].

9.1 *Filter Method*

This is commonly used as a phase before processing and the process is completely independent of the machine learning procedures. Rather, elements are chosen on the premise of their scores in different measurable tests for their connection with the result variable. Fig. 6.8 shows the filter process.

9.2 *Wrapper Method*

This process is expensive when compared to filter process. The steps involved in wrapper method to obtain best feature subset are summarized below:

- Initially, haphazardness is added to the given dataset by making rearranged duplicates of all components.
- Next, a component significant measure is utilized.
- At each cycle, it verifies whether a genuine element has a higher significance or not and always expels highlights, which are considered exceptionally insignificant.
- Lastly, the process is stopped either when all components get affirmed or rejected, or it achieves a predefined breaking point of runs.

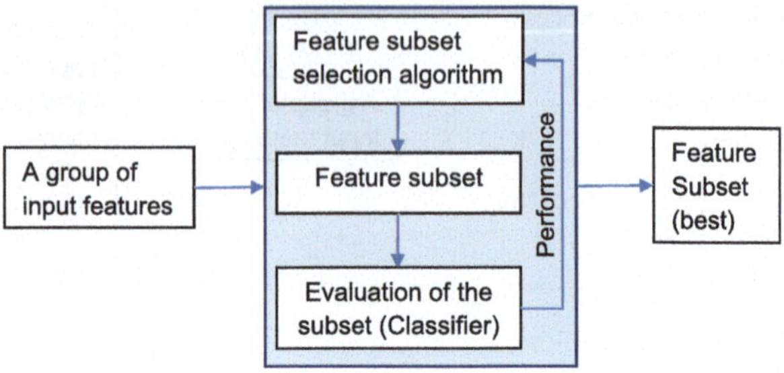

Fig. 6.9 Wrapper method

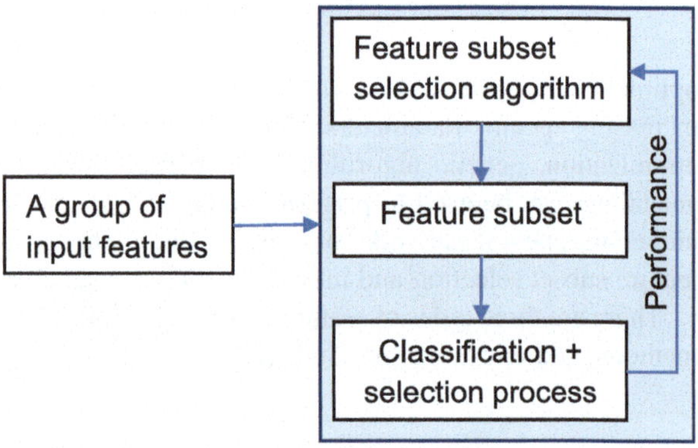

Fig. 6.10 Embedded method

Few examples of wrapper methods are forward feature selection, backward feature elimination, recursive feature elimination, etc. The wrapper method is shown in Fig. 6.9.

9.3 Embedded Method [23]

Here, selection of feature subset is a fragment of the model creation procedure. Learning algorithm accomplishes both feature selection and classification, whereas only classification is done in the wrapper method. Examples on embedded methods include regularized trees, memetic algorithm, and random multinomial logit. The process of embedded method is shown in Fig. 6.10.

10 Classification Methods

- Statistical methods
  - Distance-based classification [24, 25]

    - Euclidean – The distance between two points p and q is the line segment with end points as p and q. When Euclidean distance is used as a measure in keystroke dynamics, p and q are the test vector and the mean vector, respectively [26].
    - Mahalanobis distance – This is the distance amid the point p and the distribution D. It gives the measure, which indicates how many standard deviations far is the point p from the distribution D [27, 28].
    - Manhattan distance – When two points, p and q, are considered, the distance between them is the measure of the points calculated along the axes at 90° [28].
    - Minkowski distance – This is a generalization of Euclidean distance and the Manhattan distance, which is measured in normed vector space [24].

  - Mean – The average of the keystroke measures is considered for authentication [29].
  - Median – The middle point of the test vector is utilized in the authentication process [30].
  - Standard deviation – The standard deviation of the test vector is calculated and is used for verification purpose [31].

- Pattern Recognition and Machine Learning Methods

  - Bayes classifier – It is a system in light of Bayes' Theorem with a doubt of opportunity among pointers. In clear terms, a Naive Bayes classifier assumes that the proximity of a particular component in a class is unimportant to the closeness of some different components [32].
  - Fisher's linear discriminant (FLD) – It is a technique utilized as a part of measurements to locate a straight amalgamation of components that portray or isolate at least two classes of items. The subsequent blend might be utilized as a direct classifier, or generally for reducing the dimensions formerly advanced classification [33].
  - Support vector machine (SVM) – It is a classifier, which is capable of making fine dissimilarities properly by introducing unraveling hyperplane [21].
  - K-means algorithm – It assumes section n recognitions into k sets in which each observation has a place in the cluster with the nearest mean, satisfying as a model of the set. This results in partitioning of the data space into Voronoi cells [34].
  - Auto-regressive model – It is a depiction of a kind of arbitrary procedure. Intrinsically, it is utilized to depict certain time-fluctuating procedures in nature, financial matters, and so on. According to the AR model, the result variable is dependent only on its past numeri and stochastic variable. Hence, the model is represented in differential equation form [35, 36].
  - Decision tree – It is represented in the form of a tree, and this tree helps in making decision in terms of efficiency and cost, among others, as the branches of the tree depict the probable significances, results, economics, and efficacy [21].

  - Breiman and Cutlers random forests algorithm – It is a collaborative learning technique by producing expansive classification trees and accumulating them while training is carried out. It can choose variable on its own, and consequently it is vigorous against disturbances [37, 38].

- Neural networks
- Neural networks can be classified into the following main paradigms [1–6]:

  - Supervised

    - Back propagation – This is a classical example on supervised neural networks. It is a technique to compute the incline of the forfeiture concerning the weights in the artificial neural system. It is generally utilized as a fragment of procedures, which enhance the execution of the system [39].

  - Unsupervised

    - Hopfield neural network – This is a classical example on unsupervised neural networks. It is an artificial neural system, which is repetitive. Meeting the local minimum is ensured. However, there is rare chance of converging to a wrong outline instead of the stockpiled outline [40].

  - Perceptron algorithm – This is an automated machine engineered to speak to or reform the capacity of the mind to perceive and segregate [41].
  - Sum of products (SOP) – The output of two units and the corresponding weight value are multiplied to produce the input of another unit. All the inputs of semi-linear units are added to obtain the input of SOP architecture [4].
  - Auto associative neural networks – They get pre-feedback in order to deliver an estimation of the identity mapping among the inputs and the outputs of the network. Backpropagation or any other learning method can be utilized to enhance the system. The important component of an auto-associative network is a tailback of the sizes among information and yield [42].
  - Deterministic RAM network (DARN) – It is another variation of artificial neural networks. Lookup tables are maintained in order to store neuronal functional information, which helps in the authentication process. It is a weightless neural network as there will be no weights among the nodes [43].

- Hybrid techniques – More than one of the above techniques can be combined to design a hybrid technique.
- Other approaches

  - Fuzzy c-means clustering – Here, one value may belong to one or more sets. This is more recurrently used in pattern recognition applications [44].
  - Global alignment algorithm – Earlier information is not needed. It is exceptionally effective, and it can be utilized in the similar way as the online systems [45].
  - Time interval histogram – Single memoryless nonlinear mapping of time interims can essentially enhance the behavior of the system [46].

- Reinforced password using fuzzy logic – The features of the keystrokes are estimated using fuzzy logic, and five different rules of fuzzy logic are used for the estimation, and the center of gravity method is utilized [47].
- Hardened password – Hardened password is produced by integrating the password with the difference of time between the strokes of consecutive keys and the period for which the keys are pressed [48].

Various other techniques of classification have been used in the literature and are presented in [18].

11 Benchmarking Datasets

The basis of the keystroke dynamics somewhere lies around processing data; hence necessity of high-performance datasets also arises.

Availability of good datasets often reduces the search time as the time that might be consumed in the process of data collection can be brought down tremendously. There are five significant datasets available for the special purpose of keystroke analysis. These are:

- *GREYC KEYSTROKE* [49]: Here, there are 133 different users who participated in formation of this dataset. The collection of data is done on two different keyboards with all the users typing the same text "greyc laboratory."
- *Web-GREYC* [50]: There is no restriction of text on this version; author claims data comprises of 118 users whose log-in credentials and passwords were collected for the period of a year.
- *BH Keystroke Dynamics Database* [51]: This is an innovated new database, which has over 2057 samples of over 117 subjects. This database is divided into two different subsets A and B, which are taken in a cybercafe environment and online systems, respectively.
- *BioChaves* [52]: Here, there exist 4 datasets A, B, C, and D which have 10, 8, 14, and 15 users, respectively: 47 in total. In set A and B, four fixed phrases were typed by the user which are "chocolate," "zebra," "banana," and "taxi," while the classes C and D have fixed Spanish phrase "computador calcula."
- *CMU* [53]: Here, in total we have 8 different sessions with 51 users typed in the phrase ".tie5Ronal."
- *CMU-2* [54]: This is a free text-based dataset for collecting keystroke data. This involves data collection from around 20 users.
- *Pressure sensitive* [55]: This is a dataset, which provides pressure-sensitive data. It is gathered by 104 diverse users with the static texts as inputs "pr7qlz," "jeffrey allen," and "drizzle."

12 Conclusion

To conclude, biometric-based keystroke dynamics scheme provides adequate features to be used to identify the users in different scenarios. We provided in this chapter a review of the basics and major works in keystroke dynamics as a behavioral biometric-based authentication scheme to secure access to computer and network systems.

As discussed in the previous sections, applications of keystroke dynamics are huge and are only limited by imagination. Further research is required for increasing the stand-alone reliability of the keystroke systems so as to deal with typographical errors. Neither the work presented in this chapter nor that of any other researcher has dealt with the reliability of typographical errors. An argument can be non-conventional keystroke features, but they fail to show the increase in reliability as expected. Also, it is concluded that artificial network paradigms remain more successful than classical pattern matching technique for classification in this context because of their ability to learn.

References

1. M.S. Obaidat, A methodology for improving computer access security. Comput. Secur. **12**(7), 657–662 (1993)
2. M.S. Obaidat, D.T. Macchiarolo, An online neural network system for computer access security. *IEEE Trans. Ind. Electron.* **40**(2), 235–242 (1993). https://doi.org/10.1109/41.222645
3. M.S. Obaidat, D.T. Macchairolo, A multilayer neural network system for computer access security. IEEE Trans. Syst. Man Cybern. **24**(5), 806–813 (1994)
4. M.S. Obaidat, B. Sadoun, Verification of computer users using keystroke dynamics. IEEE Trans. Syst. Man Cybern. **27**(2), 261–269 (1997)
5. M.S. Obaidat, B. Sadoun, Keystroke dynamics based authentication, in *Biometrics: Personal Identification in Networked Society*, ed. by A. Jain, R. Bolle, S. Pankanti (Eds), (Kluwer, Boston, 1999), pp. 213–230
6. M.S. Obaidat, N. Boudriga, *Security of e-Systems and Computer Networks* (Cambridge University Press, Cambridge, 2007)
7. M. Antal, L.Z. Szabó, I. László, Keystroke dynamics on android platform. Procedia Technol. **19**, 820–826 (2015)
8. D. Stefan, D. Yao, Keystroke-dynamics authentication against synthetic forgeries, in *Sixth International Conference on Collaborative Computing: Networking, Applications and Worksharing (CollaborateCom* 2010*)*, Chicago, IL, pp. 1–8, 2010
9. A. Abo-alian, N.L. Badr, M.F. Tolba, Keystroke dynamics-based user authentication service for cloud computing. Concurrency Comput. Pract. Exp. **28**(9), 2567–2585 (2016)
10. H. Çeker, S. Upadhyaya, Enhanced recognition of keystroke dynamics using Gaussian mixture models, in *MILCOM* 2015 - *2015 I.E. Military Communications Conference*, Tampa, FL, pp. 1305–1310, 2015. doi: https://doi.org/10.1109/MILCOM.2015.7357625
11. D.G. Brizan, A. Goodkind, P. Koch, K. Balagani, V.V. Phoha, A. Rosenberg, Utilizing linguistically enhanced keystroke dynamics to predict typist cognition and demographics. Int. J. Hum. Comput. Stud. **82**, 57–68 (2015)
12. M. Antal, L.Z. Szabó, An evaluation of one-class and two-class classification algorithms for keystroke dynamics authentication on mobile devices, in 2015 *20th International Conference*

on Control Systems and Computer Science, Bucharest, pp. 343–350, 2015. doi: https://doi.org/10.1109/CSCS.2015.16
13. J.V. Monaco et al., One-handed keystroke biometric identification competition, *2015 International Conference on Biometrics (ICB)*, Phuket, pp. 58–64, 2015. doi: https://doi.org/10.1109/ICB.2015.7139076
14. A. Morales, E. Luna-Garcia, J. Fierrez, J. Ortega-Garcia, Score normalization for keystroke dynamics biometrics, *2015 International Carnahan Conference on Security Technology (ICCST)*, Taipei, pp. 223–228, 2015. doi: https://doi.org/10.1109/CCST.2015.7389686
15. D. Buschek, A. De Luca, F. Alt, Improving accuracy, applicability and usability of keystroke biometrics on mobile touchscreen devices, *Proceedings of the 33rd Annual ACM Conference on Human Factors in Computing Systems*. ACM, pp. 1393–1402, April 2015. doi: https://doi.org/10.1145/2702123.2702252
16. P. Kang, S. Cho, Keystroke dynamics-based user authentication using long and free text strings from various input devices. Inf. Sci. **308**, 72–93 (2015). https://doi.org/10.1016/j.ins.2014.08.070
17. I. Traore, I. Woungang, M.S. Obaidat, Y. Nakkabi, Online risk-based authentication using behavioral biometrics. Multimed. Tools Appl. J. **71**(2), 575–605 (2014)
18. M. Karnan, M. Akila, N. Krishnaraj, Biometric personal authentication using keystroke dynamics: a review. Appl. Soft Comput. **11**(2), 1565–1573 (2011)
19. I. Traore, I. Woungang, B. Khalilian, M.S. Obaidat, A. Ahmed, Dynamic sample size detection in learning command line sequence for continuous authentication. IEEE Trans. Syst. Man Cybern. B **42**(5), 1343–1356 (2012)
20. B. Sayed, I. Traore, I. Woungang, M.S. Obaidat, Biometric authentication using mouse gesture dynamics. IEEE Syst. J. **7**(2), 262–274, Greece (2013)
21. A. Alsultan, K. Warwick, H. Wei, Non-conventional keystroke dynamics for user authentication. Pattern Recognit. Lett. **89**, 53–59 (2017)
22. D. Shanmugapriya, P. Ganapathi, A wrapper-based classification approach for personal identification through keystroke dynamics using soft computing techniques, in *Identity Theft: Breakthroughs in Research and Practice*, (IGI Global, Hershey, 2017), pp. 267–290. Web. 3 Jul 2017. doi:. https://doi.org/10.4018/978-1-5225-0808-3.ch013
23. https://www.analyticsvidhya.com/blog/2016/12/introduction-to-feature-selection-methods-with-an-example-or-how-to-select-the-right-variables/
24. Y. Zhong, Y. Deng, A survey on keystroke dynamics biometrics: approaches, advances, and evaluations, in *Recent Advances in User Authentication Using Keystroke Dynamics Biometrics*, (Science Gate Publishing, 2015), pp. 1–22
25. M.T.J. Modi, H.G. Upadhaya, M. Thakor, Password less authentication using keystroke dynamics a survey, in *International Journal of Innovative Research in Computer and Communication Engineering, IJIRCCE*, pp. 7060–7064, 2014
26. I.H. Shimaa, M.S. Mazen, H. Hala, User authentication with adaptive keystroke dynamics. Int. J. Comput. Sci. Issue (IJCSI) **10**(4), pp. 126–134 (2013)
27. S. Cho, C. Han, D.H. Han, H. Kim, Web-based keystroke dynamics identity verification using neural network. J. Organ. Comput. Electron. Commer. **10**(4), 295–307 (2000)
28. S. Hocquet, J. Ramel, H. Cardot, User classification for keystroke dynamics, in *Advances in Biometrics, International Conference*, ICB, pp. 531–539, 2007
29. S. Modi, S.J. Elliott, Keystroke dynamics verification using a spontaneously generated password, in *Proceedings of the 40th Annual IEEE International Carnahan Conference on Security Technology (ICCST'06)*, pp. 116–121, Oct 2006
30. K. Revett, S. T. deMagalhaes, and H. M. D. Santos, "Enhancing login security through the use of keystroke input dynamics," in *Advances in Biometrics, Proceedings*, vol. 3832, pp. 661–667, Springer, Berlin, Germany, 2006
31. S.T. de Magalhaes, K. Revett, H.M.D. Santos, Password secured sites—stepping forward with keystroke dynamics, in *Proceedings of the International Conference on Next Generation Web Services Practices (NWeSP'05)*, pp. 293–298, August 2005

32. M.M. Hoobi, Keystroke dynamics authentication based on Naïve Bayes classifier. Iraqi J. Sci. **56**(2A), 1176–1184 (2015)
33. Z. Zainuddin, A.S. Laswi, Implementation of the LDA algorithm for online validation Based on face recognition. *J. Phys. Conf. Ser.* **801**(1), pp. 1–7 (2017). http://iopscience.iop.org/article/10.1088/1742-6596/801/1/012047/pdf
34. R. Shikder, S. Rahaman, F. Afroze, ABM Alim Al Islam, Keystroke/mouse usage based emotion detection and user identification, in *2017 International Conference on Networking, Systems and Security (NSysS)*, Dhaka, pp. 96–104, 2017. doi: https://doi.org/10.1109/NSysS.2017.7885808
35. Z. Changshui, S. Yanhua, AR model for key stroker verification. IEEE Int. Conf. Syst. Man Cybernet. **4**, 2887–2890 (2000)
36. W. Eltahir, M. Salami, A. Ismail, W. Lai, Dynamic keystroke analysis using AR model. IEEE Int. Conf. Ind. Technol. **3**, 1555–1560 (2004)
37. L. Breiman, Random forests. Mach. Learn. **45**, 5–32 (2001)
38. N. Bartlow, B. Cukic, Evaluating the reliability of credential hardening through keystroke dynamics, in *2006 17th International Symposium on Software Reliability Engineering*, Raleigh, NC, pp. 117–126, 2006. doi: https://doi.org/10.1109/ISSRE.2006.25
39. A. Salem, D. Zaidan, A. Swidan, R. Saifan, Analysis of strong password using keystroke dynamics authentication in touch screen devices, in *2016 Cybersecurity and Cyberforensics Conference (CCC)*, Amman, pp. 15–21, 2016. doi: https://doi.org/10.1109/CCC.2016.11
40. S. Wang, H. Wang, Password authentication using Hopfield neural networks. IEEE Trans. Syst. Man Cybern. Part C Appl. Rev. **38**(2), 265–268 (2008)
41. A. Rezaei, S. Mirzakuchaki, A recognition approach using multilayer perceptron and keyboard dynamics patterns, *2013 First Iranian Conference on Pattern Recognition and Image Analysis (PRIA)*, Birjand, pp. 1–5, 2013. doi: https://doi.org/10.1109/PRIA.2013.6528445
42. P.H. Pisani, A.C. Lorena, Negative selection with high-dimensional support for keystroke dynamics, in *2012 Brazilian Symposium on Neural Networks*, Curitiba, pp. 19–24, 2012. doi: https://doi.org/10.1109/SBRN.2012.15
43. S. Yong, W.K. Lai, G. Goghill, Weightless neural networks for typing biometrics authentication, in *Knowledge-Based Intelligent Information and Engineering Systems, KES 2004. Lecture Notes in Computer Science*, ed. by M.G. Negoita, R.J. Howlett, L.C. Jain, vol. 3214, (Springer, Berlin, Heidelberg, 2004). doi: https://doi.org/10.1007/978-3-540-30133-2_37
44. S. Mandujano, R. Soto, Deterring password sharing: user authentication via fuzzy c-means clustering applied to keystroke biometric data. in *Proceedings of the Fifth Mexican International Conference in Computer Science*, pp. 181–187, 2004
45. K. Revett, A bioinformatics based approach to behavioural biometrics, in *2007 Frontiers in the Convergence of Bioscience and Information Technologies*, Jeju City, pp. 665–670, 2007. doi: https://doi.org/10.1109/FBIT.2007.143
46. R. Jugurta, M. Filho, E.O. Freire, On the equalization of keystroke timing histograms. Pattern Recognit. Lett. **27**(13), 1440–1446 (2006)
47. W.G. de Ru, J.H.P. Eloff, Enhanced password authentication through fuzzy logic. IEEE Exp. Intell. Syst. Appl. **12**(6), 38–45 (1997)
48. F. Monrose, M.K. Reiter, S. Wetzel, Password hardening based on keystroke dynamics, in *Proceedings of the Sixth ACM Conference on Computer and Communications Security*, Kent Ridge Digital Labs, Singapore, pp. 73–82, 1999. ISBN: l-58113-148-8
49. R. Giot, M. El-Abed, C. Rosenberger, GREYC keystroke: A benchmark for keystroke dynamics biometric systems, in *2009 I.E. Third International Conference on Biometrics: Theory, Applications, and Systems*, Washington, DC, pp. 1–6, 2009. doi: https://doi.org/10.1109/BTAS.2009.5339051
50. R. Giot, M. El-Abed, C. Rosenberger, Web-based benchmark for keystroke dynamics biometric systems: a statistical analysis, in *2012 Eighth International Conference on Intelligent Information Hiding and Multimedia Signal Processing*, Piraeus, pp. 11–15, 2012. doi: https://doi.org/10.1109/IIH-MSP.2012.10

51. Y. Li, B. Zhang, Y. Cao, S. Zhao, Y. Gao, J. Liu, Study on the BeiHang keystroke dynamics database, in *2011 International Joint Conference on Biometrics (IJCB)*, Washington, DC, pp. 1–5, 2011. doi: https://doi.org/10.1109/IJCB.2011.6117485
52. J. Montalvao, C.A.S. Almeida, E.O. Freire, Equalization of keystroke timing histograms for improved identification performance, in *2006 International Telecommunications Symposium*, Fortaleza, Ceara, pp. 560–565, 2006. doi: https://doi.org/10.1109/ITS.2006.4433337
53. K. Killourhy, R. Maxion, Why did my detector do that?!, in *International Workshop on Recent Advances in Intrusion Detection, Proceedings of 13th International Symposium, RAID 2010*, Ottawa, ON, Canada, 15–17 Sept 2010
54. K.S. Killourhy, R.A. Maxion, Free vs. transcribed text for keystroke-dynamics evaluations, in *Proceedings of the 2012 Workshop on Learning from Authoritative Security Experiment Results*. ACM, pp. 1–8, 2012. ISBN: 978-1-4503-1195-3, doi: https://doi.org/10.1145/2379616.2379617
55. J.D. Allen, An analysis of pressure-based keystroke dynamics algorithms, Dissertation. Southern Methodist University, ProQuest Dissertations Publishing, 2010. 1477849

Chapter 7
Behavioral Biometrics Based on Human-Computer Interaction Devices

Chi Lin and Mohammad S. Obaidat

1 Introduction and Background

Wi-Fi devices have been utilized to offer network connections to various kinds of mobile and wireless devices with high speed, low latency, and good secure characteristics. It has already resulted in the prevalence of Wi-Fi devices and ubiquitous coverage of Wi-Fi networks, providing the chances to extend Wi-Fi's capabilities beyond communication, especially in sensing physical environment, physical conditions, and so on. When such ubiquitous signal is propagating through the air, any environmental challenges, such as small-scale or large-scale variations, affect the received wireless signal, which is commonly known as shadowing and small-scale fading. With such measurable changes in the received signals, activities in the physical environment, which potentially make slight changes, could be detected or even inferred.

In wireless communication field, the channel state information (CSI for short) is, in fact, the channel used to represent the characteristic of the communication link. It describes the attenuation factor of the signal on each communication path, which is the value of the elements of channel gain matrix H, such as signal scattering, environmental fading, and power decay of distance. In addition, CSI enables the communication system to adapt to the current channel condition, guaranteeing the high reliability and high speed in multi-antenna system, which is the basis for providing high-quality communication. CSI value is usually used to describe how

C. Lin (✉)
School of Software, Dalian University of Technology, Dalian, China

Key Laboratory for Ubiquitous Network and Service Software of Liaoning Province, Dalian, China
e-mail: c.lin@dlut.edu.cn

M. S. Obaidat
Department of Computer and Information Science, Fordham University, Bronx, NY, USA

M. S. Obaidat et al. (eds.), *Biometric-Based Physical and Cybersecurity Systems*,
https://doi.org/10.1007/978-3-319-98734-7_7

a signal propagates from the transmitter to the receiver through the open environment and to represent the combined effect of transmitting characteristics, for example, scattering, fading, and power decay with distance. CSI makes it possible to adapt transmissions to current channel conditions, which is crucial for achieving reliable communication with high data rates in multi-antenna systems.

In general, the receiver evaluates CSI and gives feedback to the sender, which will need the reverse evaluation in TDD system. Thus, CSI can be divided into CSIR and CSIT. The CSI value actually characterizes the channel frequency response (CFR) for each subcarrier between each transmit-receive (TX-RX) antenna pair. Let MT denote the number of transmit antennas, MR denote the number of receive antennas, and Sc denote the number of OFDM subcarriers. Moreover, let $X(f, t)$ and $Y(f, t)$ be the frequency domain representation of the MT-dimensional transmitted signal and the MR-dimensional received signal, respectively. The two signals are correlated by the expression

$$Y(f,t) = H(f,t) \times X(f,t) + N,$$

where $H(f, t)$ is the complex channel frequency response of the carrier frequency f measured at time t, and N is the MR-dimensional noise.

Human identification recognition has become increasingly important in pervasive computing and human-computer interactions (HCI) which have broad application prospects. In recent decades, researchers have attempted to develop methods for identifying human with behavioral biometric.

Comparing to traditional techniques, which utilize expensive devices and have fundamental limitations of requiring line of sight and compromising human privacy, Wi-Fi signal-based human identification recognition systems have the low-deployment-cost property due to the popularity of Wi-Fi signals.

Identity recognition is one of the most popular Wi-Fi recognition techniques in recent years. With the continuous development of science and technology, virtual reality, augmented reality, wearable computing equipment, and other technologies, the traditional human-computer interaction technology gradually cannot meet people's requirements. In order to better achieve the human computer interaction, and accurately identify people's identity, Wi-Fi-based identification has become popular by researchers in the field. Compared with the traditional image and pattern recognition techniques, and other static information processing technology, dynamic information recognition is no doubt more accurate and more comprehensive.

In the study of human action behavior recognition, most of the schemes adopted are based on the identification of human body behavior [2], sensor identification [3], 3D modeling recognition [4], and so on. These principles can be divided further into the following three categories: computer vision-based identification, sensor-based identification, and wireless signal-based identification.

Vision-based identification of the environment has many restrictive conditions. It needs to be sufficient in the case of light. In addition, the vision-based identification usually requires specific devices such as camera for recording videos. Moreover, processing such data or information usually requires large computational and storage overhead. The processing time is long, which cannot identify a certain person in a timely manner. Besides that, sensors and RFID tags have been used for identity recognition. It requires that the humans carry the sensor device or RFID tags with them. The wireless signal-based identification uses specific equipment, such as the oscillator, to receive the signal. At present, radio frequency-based identification can achieve higher recognition accuracy with the support of professional equipment, but it is not popular.

Although the abovementioned schemes perform well in human body identification, and some of them can even meet the requirements of the commercial applications, they have requirements on the measurement conditions. Wi-Fi-based human identification with its low price and good performance is becoming popular worldwide. A large number of applications based on Wi-Fi signals have been devised. Compared to other methods, Wi-Fi signal-based recognition's biggest advantage is the small dependence on the device. Only a wireless router is needed, with Linux system equipment.

The core of the various types of Wi-Fi signal-based recognition systems such as human identification, gesture recognition, motion recognition, and other applications is processing and analysis. Commonly used methods are RSS and CSI. Wi-Fi signal processing principle based on RSS is basically processing the Wi-Fi fingerprint. The RSS signal is firstly converted into RSSI signal, a kind of processed RSS signal, which has removed the influence of the RSS signal interference component.

The system collects the RSSI values from different AP in different physical locations. The physical coordinates and the value of each RSSI signal form a fingerprint, which will be stored into the database, and then the system will start performing the recognition. The recognition system collects the RSSI value from the available AP, which will form a set of associated signal observations. Then it uses the nearest neighbor algorithm to match the data in the database to select the most matching estimated position. That is the distance information for each AP. In general, the RSSI-based schemes are based on triangulation or per-stored Wi-Fi fingerprints, to estimate the position.

CSI-based Wi-Fi signal processing principle is based on monitoring the channel state changes to obtain the required information [7]. In the real world, the Wi-Fi signal is affected by the obstacle and the electromagnetic wave, and the propagation of the Wi-Fi signal usually follows a multipath [11] with distortions such as delay, attenuation, and frequency diffusion in different paths. Due to the time-varying characteristics of the channel, the multipath effects of the channel will change with the probability of statistical knowledge, which conforms to the Gaussian distribution. When some of the paths on the channel changes are detected, the information of the certain region can be obtained within a certain range.

There is a big difference between RSS and CSI. Firstly, RSS refers to the data link layer, and CSI stands for the signals collected from the physical layer. CSI has more details than Wi-Fi information. RSS is mainly applied to indoor positioning and other occasions with its low accuracy, and CSI is applied to action recognition, gesture recognition, and other relative cases and settings with fine-grained high accuracy.

2 Related Works

Recently human activity recognition method on the basis of Wi-Fi signal, such as WiSee [8], E-eyes [9], Keystroke [10], etc., has been proposed based on the observation that different human activities introduce different interferences for multipath distortions in the received Wi-Fi signals [13, 14]. The main advantage compared to the cameras, sensors, or RFID-based methods is that they take advantage of the ubiquity of Wi-Fi signals. They do not require the user to carry or wear additional equipment because they rely on the received Wi-Fi signal reflected by the person.

Generally speaking, human activity recognition can be clustered into two categories: device-based and device-free. Device-based methods require the subjects to equip with special devices (e.g., smart watch/phone, RFID tags, audio recorder, etc.) to assist the recognition system to collect data, while device-free methods can greatly reduce the effect brought to the subjects when implementing the recognition process. Vision-based and radar-based methods can be regarded as device-free methods. However, vision-based methods suffer from the high computation complexity and privacy concerns, while radar-based methods may cause concerns with respect to adverse health results. Recently, newly emerged methods utilizing the variation of wireless signals have been proposed, which can avoid those issues. These methods leverage wireless signals to implement human activity recognition. They can be classified into the following categories: (1) received signal strength indicator (RSSI)-based approaches, (2) channel state information (CSI)-based approaches, and (3) specialized hardware-based approaches.

2.1 *RSSI-Based Approaches*

In a traditional IEEE 802.11 system [14, 15], RSSI is the relative received signal strength in a wireless environment. It is an indication of the power level being received by the receive radio after the antenna and possible cable loss. Therefore, higher RSSI value indicates a stronger signal [16, 17].

RSS-based human recognition systems receive the signal strength changes caused by human and transform them to received signal strength (RSS). Due to the low resolution, the RSS values provided by the commercial device can only make recognition coarsely with low accuracy. Existing RSS-based human recognition systems can be utilized for activity recognition with the recognition rates of over 80% for crawling, lying down, standing up and walking [1, 2], and indoor localization [3].

However, due to its intrinsic property of extracting energy information from wireless signals, its measurement is coarse-grained, which cannot be utilized for fine-grained human activity recognition.

2.2 CSI-Based Approaches

Device-free motion recognition techniques are widely used to recognize human activities with the variations of wireless signals. For these techniques, the common challenge is how to accurately understand and precisely model the changes of wireless signals. Some researchers proposed to use received signal strength (RSS) as an indicator to infer the changes of wireless signals. However, the disadvantage of RSS values is that they can only provide coarse-grained information about channel variations, which cannot gather information about small-scale fading and multipath effects caused by micro-movements.

CSI values are available in COTS Wi-Fi network interface cards (NICs) such as Intel 5300 [4] and Atheros 9390 [5]. Recently CSI values have been used for activity recognition and localization. WiKey [6] can recognize keystrokes in a continuously typed sentence with an accuracy of 93.5%. WiFall is proposed to detect falling of single human in an indoor environment [7]. Zhou et al. proposed using CSI value to detect the human presence in an environment [8]. Frog Eye proposed by Zou et al. can count the number of people in a crowd [9]. E-eyes recognizes a set of nine daily activities such as washing dishes and taking a shower [10]. Our scheme is CSI-based. We use CSI values to identify human identities in an indoor environment [18–20].

2.3 Hardware-Based Approaches

Fine-grained radio signal measurements can be collected by special hardware or software-defined radio (SDR). Device-free human activity recognition has also been studied using special hardware and software-defined radio. Using the micro-Doppler information, radars can measure the movement speeds of different parts of human body [7]. A special hardware with USRPs is used by WiSee, to extract small Doppler shifts from OFDM Wi-Fi transmissions to recognize human gestures [8]. WiTrack uses specially designed frequency-modulated carrier wave signal to track human movements behind a wall with a resolution of approximately 20 cm [9]. In general, all those schemes require specialized devices in realizing the recognizing process. As an ubiquitous signal, Wi-Fi signal can be found or received everywhere; however, little attention has been paid on how to make full use of such signals. Here, we intend to utilize Wi-Fi signal for identity recognition, in what follows, related theories and methods are demonstrated in detail [21–24].

3 Theories for Recognition Methods

Here, machine-learning theories are used to recognize human identification. The principal component analysis (PCA) is used to reduce the dimension of the CSI values to reduce the storage overload. Moreover, the support vector machine (SVM) is employed to help in the identification. Next, we list and discuss the usage of these techniques in detail.

3.1 Support Vector Machine (SVM)

Support vector machine, also called support vector network, is a kind of supervised study algorithm related to data analysis in classification and regression analysis. Given a group of training cases, each case divided into one specific class of the two classes, SVM training algorithm creates the model that distributes a new case to one class out of the two specific classes, making it a non-probabilistic binary linear classifier. In SVM model, the cases are represented as discrete points, which enable these of different classes to be separated by a distinct gap. Then, all the new cases are mapped to the same space and decided which class they belong to by observing which side of the gap they fall on [25, 26].

In the proposed scheme, SVM is used for classifying CSI signals for determining the identity of a person. When a set of examples are input, SVM will be utilized to calculate the probability of each CSI signals, and the one with the highest probability will be set as the recognition result.

3.2 Principal Component Analysis (PCA)

Principal component analysis (PCA) is a kind of dataset simplification and analysis technology. PCA is often used to reduce the dimensionality of the matrix and maintain the characteristic that has the biggest contribution to the derivation, which is accomplished by ignoring the high-rank principal component but maintaining the low-rank principal component. In this way, the low-rank component can always maintain the most important part of the data, which at the same time strongly relies on the accuracy of the data.

PCA is the simplest of multivariate analysis based on eigenvectors. In general, its operation can be considered to reveal the internal structure of the data in a way that best explains the variance of the data. If the multivariate dataset is visualized as a set of coordinates (one axis per variable) in the high-dimensional data space, the PCA can provide the user with a lower-dimension picture that is projected from the object's view or "shadow" the richest view. This is done by using only the first few principal components, so that the dimension of the transformed data is reduced.

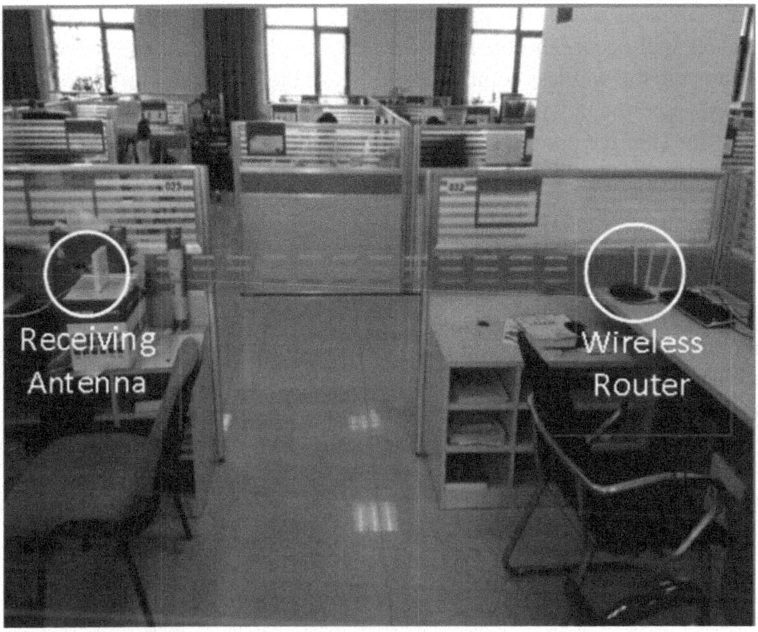

Fig. 7.1 Experiment setup

PCA is used for collecting the characteristic of the CSI values. Moreover, since the dimensionality of the data processed is so high, we use PCA to reduce the dimension, which simultaneously reduces the computational and storage overhead for data processing and storing.

4 Test-Bed Experiment Installations and Configurations

In our lab experiment, two commercial off-the-shelf (COTS) Wi-Fi devices are used (see Fig. 7.1).

A TP-Link TL WR886N router is used for consciously sending Wi-Fi signals, and a laptop implemented with Intel5300 network card with three antennas is used for receiving signals. The system works with Linux 802.11n CSI tool running on Ubuntu 12.04. After gathering data from the Wi-Fi signals, related machine-learning algorithms developed by Python 3.5 and scikit-learn are utilized.

In order to intuitively obtain the change of CSI signal, a real-time visualization platform to capture the CSI signals is implemented. This platform was running in the windows operating system with the visual studio 2016 and Matlab 2016b. It displays new figures and a vector about CSI signal each time it receives the CSI signals for intuitively showing the variances of the CSI values.

When a person walks through our experimental environment, variances and disturbances will occur, and the received CSI signals will change dramatically due to the characteristic wireless communications. All the CSI data information collected through the experiment will be stored and saved in the CSI profile pool. These data are useful for later data processing and identity recognition.

5 Experiments

To build an easily deployable and low-cost solution for fine-grained human identification recognition, we devise a system that senses and recognizes the identities of people through Wi-Fi with machine learning, namely, Wide. Here, we discuss the preliminaries, goals, system overview, and the core components of our devised system.

5.1 Design Goals

In order to identify human identities through analyzing detailed CSI from a single commodity Wi-Fi device (i.e., a router), our system's design, implementation, and requirement of our system involve a number of challenges [27–29]:

- Ubiquitous recognition. The system should be easy to deploy on existing commercial Wi-Fi without the need to introduce any dedicated hardware devices, and there is no need for users to wear any additional sensors. In addition, the identification process should be unobtrusive without the need for additional gestures or movement to identify. To highlight its prevalence, it should only use existing Wi-Fi traffic or beacons at the deployed AP without using dedicated user-generated traffic [30].
- Robust to open environment dynamics. The interferences from open environments, such as movement, waving, and other behavioral movements, can dynamically change the collected CSI time series values. Therefore, the system should be able to provide accurate identification by mitigating such disturbances and errors [31].
- Adaptable to personal diversity. Our system should be able to be used by multiple users without user-specific calibration. Thus, due to lack of consistency, it should be resilient to cope with individual diversity. It should be able to precisely recognize the identity of a certain person no matter what he wears or how fast he moves [20].
- Fast identification with small storage overhead. Our system should be able to identify people with high precision in a very short time. On one hand, related background is needed to be stored in the system so as to provide fundamentals for

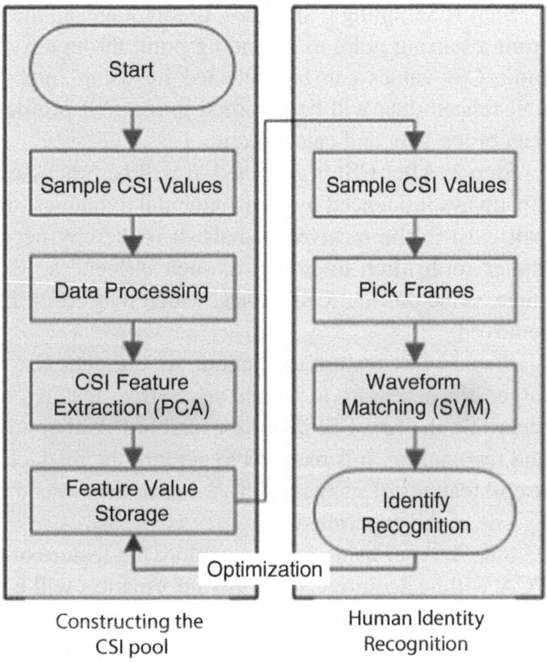

Fig. 7.2 Flowchart of *Wide*

identity recognition. On the other hand, a large storage overhead may cast heavy burdens on calculations.

5.1.1 System Overview and Design

To realize the ubiquity and simplicity of the system design, in our scheme, only a router for initiating Wi-Fi signals and a laptop for receiving such signals and calculating CSIs are required.

As is shown in Fig. 7.2, Wide is composed of two parts: (a) constructing CSI pool and (b) human identity recognition. In addition, in order to continuously optimize the accuracy of Wide, we apply a feedback technology. Whenever it succeeds in identifying people, the corresponding CSI features will be stored in the CSI pool, and as a result of this feedback technology, performance will be gradually enhanced.

When the Wi-Fi signal is sent from the router, the laptop will obtain the corresponding CSI value, which includes the combined effect of scattering, fading, and power decay with distance. In Wide, the CSI value is used to represent the feature subcarriers in OFDM. It plays an important role in the identification process.

As shown in Fig. 7.2, a detailed flowchart of Wide is described next.

In the first stage, we need to collect enough information so as to construct a CSI profile pool. The details are as follows.

Step 1: Sampling CSI values. In this stage, all the volunteers are required to move from a starting point to an ending point through a deterministic route. At the same time, CSI values can be collected by using Intel 5300 network adapter. All the CSI-related data will be recorded into a CSI profile, which will be used for future data processing and calculations.

Step 2: After CSI is collected, it will be processed and saved. As the CSI signals are always influenced by environmental dynamics; therefore, noise and disturbances will exist in the received signals. It is thereby necessary to use the filter, such as Butterworth filter, to get rid of such influences. Moreover, in the data processing stage, some unusual CSI values, which may result from communication failure, are removed.

Step 3: CSI feature extraction. As the data size of recording one CSI signal is large, it is necessary to reduce and extract features of such signals. Moreover, only recording the raw CSI data may lead to trivial contributions; for further classifying and recognition, it is required to explore the instincts of CSI signals and mine some useful features of such data. PCA is used to extract the features of the CSI signals one by one in the experiment.

Step 4: After using PCA to explore the feature of the collected CSI, the results of PCA will be acquired. Then this information will be stored as a profile of a certain person. Such information labeled by the identity of its owner will be saved in the database. After collecting all the information of all the users (i.e., the volunteers), the CSI pool is constructed. All the features of the CSI information throughout the experiments will be stored as the background information for future recognizing.

In the second stage, the recognition experiment begins; see the right part of Fig. 7.2. Volunteers moves from A to B sequentially. The CSI values are sensed from the network adapter, and the results can be displayed by user's laptop.

In the recognition process, we have four steps.

Step 1: Sampling CSI values. This step is identical to the first step in constructing the CSI pool. We sample the CSI values of the volunteers through our test-bed devices.

Step 2: Pick frames. Due to the high frequency of sending packets, it is not necessary to use all the collected CSI values for recognition process. We hereby pick, some CSI values, for processing. Once a CSI frame is quite different from two-sequential frames forward and backward to some extent, it will be removed as it may have resulted from the environmental dynamics or noise. Finally, we pick enough CSI frames and conduct data processing and PCA to obtain the features.

Step 3: After the features are extracted from the sampled CSI values, we begin our classification process. Here, we use the SVM to determine the identity of each volunteer by analyzing the CSI values generated in the abovementioned process.

Step 4: Identification and recognition. In this process, the result will be displayed in the screen by using the SVM + PCA technique. Each time, Wide outputs a result, which has the maximum probability in recognizing. Once a person is successfully recognized, we will label such results and extract the CSI values. Then we record and update the CSI feature values recorded in the sampling process. Such a feedback

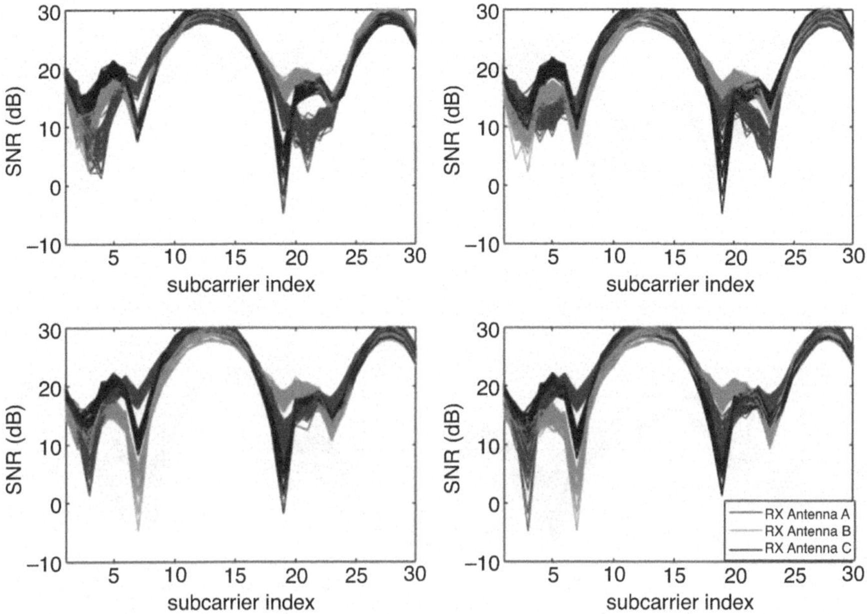

Fig. 7.3 CSI signals in open environment with no interferences

process will continuously enhance the recognition accuracy of the proposed scheme, and eventually, we obtained around 99% accuracy.

5.1.2 Recognizing Static People

Dynamic person identification is not an easy task. In the process of movement, people will produce a large number of variables and interference, which will pose a great impact to data processing. To simplify the complex program and directly verify that the Wi-Fi signal can be used to identify human identities, we designed a static experiment with the same principle as the dynamic person identification experiment.

First of all, we made a clear experiment environment with no one doing any activity. We collect the CSI signals in the empty environment via our visualized platform. Parts of Fig. 7.3 depict this. The figures are almost the same with a little interference caused by unknown movements. It proves that the CSI signals are stable signals in a certain environment, which means that we can use the nearest neighbor algorithm or other algorithms like it to match the data stored in the datasets to select the most matching estimated answers.

Then in order to verify that we can identify the human identity with the Wi-Fi signals, we collected CSI information of the experimenters as is shown in Fig. 7.4. We made the experimenters stand in a certain place by themselves (in this experiment, we stand at the middle of the line connected the antenna and router). There is a

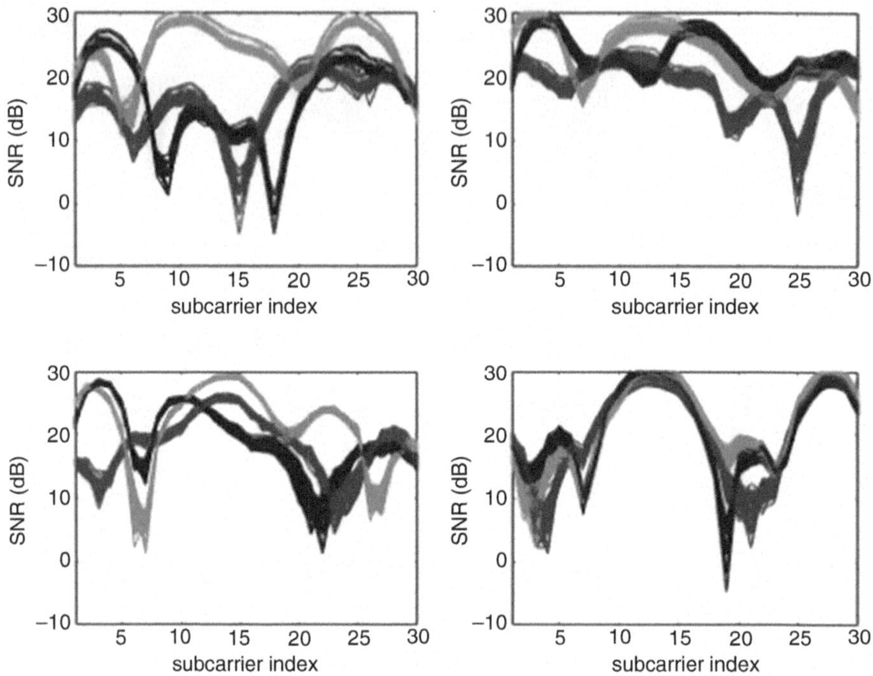

Fig. 7.4 CSI signals of different volunteers

huge difference between each of them. In our view, the CSI signals can be used effectively to identify the human identities.

5.1.3 Dynamic Identification Recognition

To identify human identities when people are moving, we design and develop our system Wide. Wide consist of two main parts: (a) constructing CSI pool and (b) human identity recognition. In order to build Wide, we should construct a CSI pool first for recording CSI information as the background, which is extremely useful for later identification and recognition.

5.1.4 Prepare the CSI Pool

The CSI signals will change when someone interferes with it, and it will stay stable if there are no new interferences. In our scinario, volunteers are required to move from the starting point A to the ending point B in a normal way as they walk, then they stop at some specific point which helps us collect more detailed and accurate CSI information.

We use the database to store the received signals and then use preprocessing method such as filtering and node reduction to acquire more accurate CSI data. All data collected by our database are regarded as a CSI pool. This information is used as the background knowledge for future identity recognition.

When constructing the CSI pool, we intend to collect various kinds of data of a certain person. For example, we collect the CSI signals in our pool when a volunteer is moving at different speed and wearing different clothes.

5.1.5 Human Identity Recognition

After we build the CSI pool, we begin our identification process. To prove the resilience of Wide, we require the volunteers to pass through the start point A to the end point B at any speed without knowing Wide. Then we put the CSI information into Wide to analyze the data. Wide will display the results of this identity through calculations. We then analyze the accuracy of our scheme to show its superiority.

5.2 A Case Study: Recognizing Identities for Volunteers

In order to demonstrate the performance of our program, a test-bed experiment is conducted. In our experiments, Wide consists of two commercial off-the-shelf (COTS) Wi-Fi devices, as is shown in Fig. 7.1. We use the TP-Link TL WR886N router to send Wi-Fi signals consciously and use a laptop with an Intel 5300 card with three antennas to receive signals. Wide is based on the Linux 802.11n CSI tool running on Ubuntu 12.04. After collecting data from the Wi-Fi signal, the related machine-learning algorithm developed by Python 3.5 and scikit-learn is implemented. In Wide, each transmitter-receiver (TX-RX) antenna pair of the transmitter and receiver has 30 subcarriers. Let MTx and MRx denote the number of transmitting and receiving antennas. Thus, there are 30 * MTx * MRx CSI streams in the time series of the CSI values.

Wide has a CSI pool for characterizing human identity profile as a background, which should be previously built/collected through our off-the-shelf devices. Therefore, we use the platform for timely visualizing the variances of CSI. Whenever our laptop receives a signal from the router, it will display a new CSI figure on the screen. Firstly, as shown in Fig. 7.7b, when no one moves, we find that the CSI is always stable, with only slight interference being detected, which is affected by other people (see Fig. 7.7). Then, when the volunteers moved forward (see Fig. 7.5), we started building the CSI pool. In order to identify people in a timely manner, we need to record in advance related background CSI information for each person. So first, we need to collect and record useful CSIs. In our experiments, in order to completely capture all the detailed CSI, the volunteers need to move in a deterministic route with different stride frequencies. In this process, when the volunteers move from the starting point A to the end point B, we collect the detailed CSI throughout the

Fig. 7.5 Recognizing volunteers

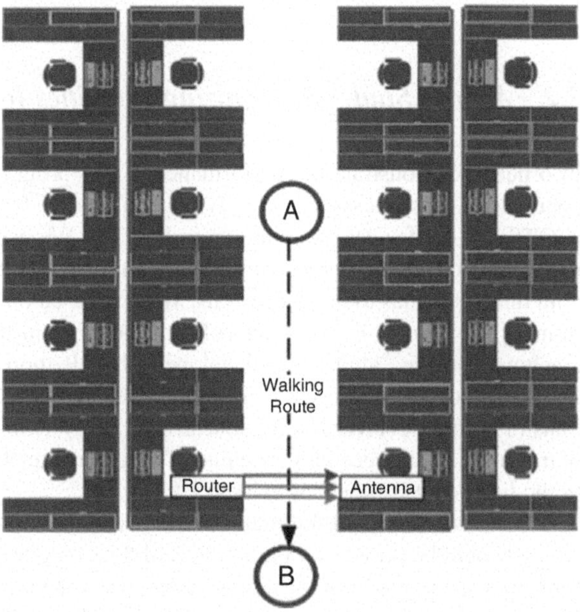

Fig. 7.6 Route of walking

process (see Fig. 7.7a). We set the frequency of sending Wi-Fi signals to 0.02 s to capture any changes during the moving process. Then we choose each person's feature point as the basis for recognition.

Then we start our identification process. In order to prove that Wide is resilient to individual diversity, ten volunteers can pass through the start to the end at any speed, without knowing Wide (see Fig. 7.6). Each time the volunteer arrives at the end, Wide outputs its recognition results on the screen. Then we learn the experimental results of the test-bed experiment.

Obviously, such a recognition process is taken in an open environment, in which disturbances and noise exist everywhere and all the time. Therefore, to get an accurate result, we need to process our data through filtering or migrating the impact of noise.

5.3 Recognizing Accuracy Evaluation

Firstly, we build the CSI pool in advance. Ten volunteers move at different speeds from the starting point to the end. The router sends Wi-Fi signals every 0.02 s, and the laptop displays the accumulated sensed CSI on our screen. We aggregate and select 7000 datasets as training datasets as the CSI pools and input them into SVM for training and classification (i.e., identify people). And then we begin the experiment of identification, the volunteers walking from the starting point to the end. We use Wide to sample, select, and record useful CSIs. The whole experiment is carried out in an open environment where other people may often move, bypass, speak, and do other actions as usual. We mainly analyze the recognition results in the face of interference in an open environment.

We found that, on average, when there was interference, i.e., other people moving, talking, and so on, the identification accuracy in the open environment is 98.7%. Then we tested Wide's performance when there was no interference (see Fig. 7.7b) with an accuracy of 100%. After that, we tried to further improve the accuracy of Wide, through statistical methods to eliminate environmental disturbances. This has led to increased accuracy of 99.7%.

5.4 Dimensionality Reduction

In Wide, although there is a high accuracy, it usually takes a long time to identify people; this is not suitable for practical applications. Because the direct use of 30 * MTx * MRx CSI streams will result in high computational cost and storage overhead for recording information. Therefore, we need to compress such data. As shown in Fig. 7.7, all subcarriers show the relevant changes in their time series. Thus, a promising approach is to use principal component analysis (PCA) to reduce the dimension of the CSI values. We are likely to accelerate Wide performance in terms of dimensionality reduction. Previously, Wide recorded each frame as a 1 * 90 vector, which resulted in a large computational overhead in the calculation. To reduce this overhead, we use PCA to reduce data dimension to reduce computing and storage overhead. Then, we verify the performance of our recognition accuracy when selecting different dimensions of data and input into Wide. The results for different dimensionalities are shown in Table 7.1.

In data processing process, we choose different dimensions through the PCA for the data. Then we examine the performance of the PCA by analyzing the percentage of the principal components. In Table 7.1, we note that most of the principal

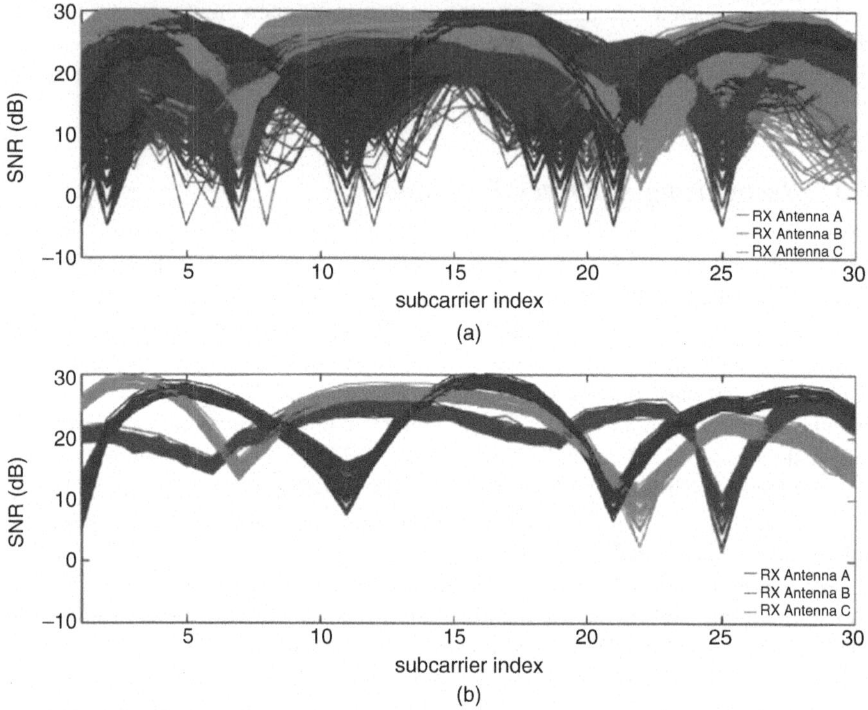

Fig. 7.7 Collected CSIs when (**a**) disturbance exists and (**b**) no disturbance exists

Table 7.1 Recognition results for different dimensionalities

Dimensionality after PCA	Percentage of principal component	Accuracy (%)	Training time (s)	Storage overhead (kB)	Recognizing time (ms)
90	1.0	99.7	81	25257	1.028
80	0.99977	99.6	117	2245′	0.993
50	0.99718	99.4	137	14032	0.889
30	0.99671	99.2	119	8419	0.774
20	0.98151	99.2	101	5613	0.785
10	0.96534	98.9	83	2807	0.858

components are involved after the implementation of PCA. Then, we test the recognition accuracy under different dimensions. We note that when we only use ten dimensions, the recognition accuracy can still reach 98.9%, indicating excellent recognition ability. As the CSI pool acts as the core of Wide, we analyze the performance of the training time and storage cost. After that, when we deal with the extracted features of CSI values recorded in the CSI pool, we measure the training time. We found that time configuring parameters required in Wide would take up to 137 s. With regard to storage overhead, we observed that storage costs were significantly reduced due to the reduced dimensionality of PCA. When the dimension is 10, it only needs 2807 KB to store data in the CSI pool, which is 10 times smaller than the 90 dimension. Finally, we measure the identification time of Wide to prove its timely identification behavior. As the last column lists, on average, Wide is able to complete the identification process within 1 ms, which is quite small compared to the volunteer's movement time.

Therefore, we can conclude that Wide can identify human identities with high precision while using a very short time.

5.5 *Limitations and Challenges*

Our test-bed experimental results show that Wi-Fi signals can be used to identify human identities with high accuracy in an open environment. However, restrictions for Wide still exist, which cannot be ignored. After we extended the distance between the router and the antenna, we found that the accuracy was significantly reduced, because when they were close to each other, the collected signal was much stronger than the other disturbance. In addition, when more volunteers participate in our experiments, the recognition speed is slow, and the accuracy is gradually reduced. Therefore, in the future particular efforts will be made to improve the performance of Wide. Wide triggers a new paradigm for identifying moving people by analyzing Wi-Fi signals in an open environment. Besides that, there are still many challenges to take full advantage of the Wi-Fi signal, which we focus on and discuss in the next.

5.5.1 Counting Human Numbers

Due to the characteristics of multipath transmission, any slight change in the environment may result in a change in the CSI value. This feature is particularly useful for counting without the use of special equipment. Wi-Fi signals can be potentially used to evaluate variations such as counting the number of users in open places, halls, classrooms, movie theaters, etc. Changes in CSI values at the entrance or exit can directly reflect changes in the number of people. For example, the total number will be minus one each time a change signal is detected in the exit. Thus, how to correlate a changing CSI time series with entry and existing actions is challenging and has a huge potential prospect in applications.

5.5.2 Privacy Leakage

In our test-bed experiment, the Wi-Fi signal can be used to identify human identities, which may potentially be exploited by an attacker to recognize the identity or appearance of the target victim. We can imagine that through Wide, when the

Wi-Fi signal is available without the victim's consciousness, the opponent can track the user. Whenever the victim arrives, the corresponding CSI value will be changed, and his unique CSI feature will be sampled. Thus, by using Wide, he will be localized and tracked. This trajectory tracking attack/identity will greatly compromise privacy, especially the user's indoor location. In addition, this privacy leakage attack is not easy to detect, because the victim may even be unaware of the prevalence of Wi-Fi signal characteristics and potential adversarial router.

5.5.3 Accurate and Fine-Grained Localization

The spatial characteristics of wireless signals are the basis for distinguishing and determining the location of wireless indoor positioning, so the use of Wi-Fi signals to develop accurate and fine-grained positioning technology will become more attractive. Due to the prevalence of Wi-Fi signals, it should be noted how to map different CSI changes by analyzing different Wi-Fi signals with different locations and how to distinguish and locate the exact location.

5.5.4 Multiple Identity/Gesture Recognition

Currently, the Wi-Fi signals are only used to accurately identify individual identities or gestures in closed or relatively silent areas. The features of the CSI value for multipath distortion are not fully explored, and any movement will cause a signal change. Obviously, it is not enough to identify only one identity or one posture, and it is more attractive and challenging to detect multiple identities or positions in practice. Thus, in future work, we will construct a model to take into consideration time changes associated with CSI values.

6 Conclusions and Future Research

In this introductory chapter, we describe a new approach to recognize the identity of a person through analyzing Wi-Fi signals and its potential application prospects. The chapter starts with the definition of Wi-Fi signals and CSI (channel state information) and their potential applications. Particular emphasis is placed on the characteristics of the CSI, which indicate that CSI can be used for recognizing the identity of people. Then a CSI-based human identity recognition scheme, Wide, and its test-bed experiment are demonstrated. By collecting, analyzing, and processing Wi-Fi signals to recognize identities of people, it is revealed that the proposed scheme can recognize people with promising accuracy in a short time.

Our system based on the CSI has a high accuracy, but the use of the network card and other equipment has certain restrictions. In the future work, we must focus on three aspects: low cost, ease of use, and high precision. For the time being, the CSI-based identification system has great prospects for development.

As part of our future research, we will concentrate on these following aspects:

1. Touching Screen Applications

Touch screen is a very good way of human-computer interaction, but there are still some shortcomings on today's touch screen, such as high cost, fragility, etc. Moreover, the touch screen must go with the physical contact, which leads to inconvenience and causes damage to the screen easily. In addition, most of the screens now are still non-touch screen; it will cost a lot if we convert all of them into touch screens.

As mentioned above in this chapter, WiKey [6] can recognize keystrokes in a continuously typed sentence with a high accuracy. Wi-Fi signals could be used on indoor localization. WiSee [8] can recognize human gestures. It can be predicted that we can design a system which can recognize the gestures by the human who touches the screen and react to the human, just like the touch screen. Moreover, the system can react with no physical contact to avoid the screen damage by the rude behaviors.

2. Device-Free Indoor Sensing Systems

People have to stop in front of the door to show their identity by key, ID card, or something else, which will spend extra time and make the movement discontinuous. There is also a risk that the things they use to identify their identities may be stolen by others.

Our test-bed experiment and other experiments have proved that the Wi-Fi signals have a great ability to identify human identity, which could be used at the door sensor to help identify human identity in front of the door.

3. Motion Sensing

Common motion sensing game players such as the Wii, the Kinect, and so on all have some weakness which needs to be solved. They may require something you should hold or have a low accuracy, which impede the progress. With the development of science and technology, virtual reality games gradually become the hot topic that require new approaches to recognize human activities and human identities.

The Wi-Fi signal-based human recognition and localization with a high accuracy could just fit the problem. As mentioned earlier, WiSee [8] can recognize the gestures by the human. It is achievable to capture motion via Wi-Fi signal. Thus, the combination of the virtual reality and the Wi-Fi signal-based human recognition will bring great changes to the game industry.

4. Mobile Wi-Fi Sensing

Latest research focused on the application including recognizing and localizing when the access point or router is statically deployed. Little attention has been paid for recognizing application for a mobile access point. For example, a person carrying a phone configuring as a mobile access point, how can we make full use of such CSIs for localization or even recognition. In our future work, we will concentrate on how to use the mobile Wi-Fi access point as the signal source and utilize the varying CSI signals for developing more interesting applications. Moreover, in the future works, we intend to intensely explore the characteristics and develop more interesting applications.

Reference

1. S. Sigg, S. Shi, F. Buesching, Y. Ji, L Wolf, Leveraging Rf-channel fluctuation for activity recognition: Active and passive systems, continuous and RSSI-based signal features, in *Proceedings of International Conference on Advances in Mobile Computing & Multimedia*, (ACM, 2013), page 43
2. S. Sigg, M. Scholz, S. Shi, Y. Ji, M. Beigl, Rf-sensing of activities from non-cooperative subjects in device-free recognition systems using ambient and local signals. IEEE Trans. Mob. Comput. **13**(4), 907–920 (2014)
3. Z. Yang, Z. Zhou, Y. Liu, From RSSI to CSI: Indoor localization via channel response. Acm Comput. Surv. **46**(2), 25 (2013)
4. D. Halperin, W. Hu, A. Sheth, D. Wetherall, Tool release: gathering 802.11n traces with channel state information. ACM SIGCOMM CCR **41**(1), 53 (2011)
5. S. Sen, J. Lee, K.-H. Kim, P. Congdon, Avoiding multipath to revive in building WiFi localization. in *Proceeding of ACM MobiSys*, (2013), pp. 249–262
6. K. Ali, A.X. Liu, W. Wang, M. Shahzad, *Keystroke recognition using WiFi signals*, in Proceedings of ACM MobiCom, (2015), pp. 90–102
7. P. Van Dorp, F. Groen, Feature-based human motion parameter estimation with radar. IET Radar Sonar. Nav. **2**(2), 135–145 (2008)
8. Q. Pu, S. Gupta, S. Gollakota, Shwetak Patel, Whole-home gesture recognition using wireless signals, in *Proceedings of the 19th annual international conference on Mobile computing & networking*, (ACM, 2013), pp. 27–38
9. F. Adib, Z. Kabelac, D. Katabi, R. Miller. 3d tracking via body radio reflections, in *Usenix NSDI*, vol. 14, (2013)
10. Y. Wang, J. Liu, Y. Chen, M. Gruteser, J. Yang, H. Liu. E-eyes: In-home device-free activity identification using fine-grained WiFi signatures, in *Proceedings of ACM MobiCom*, (2014)
11. K. Ali, A.X. Liu, W. Wang, M. Shahzad, *Keystroke recognition using WiFi signals*, (ACM MobiCom, 2015), pp. 90–102
12. Y. Wang, K. Wu, L.M. Ni, Wifall: device-free fall detection by wireless networks[J]. IEEE Trans. Mob. Comput. **16**(2), 581–594 (2017)
13. G. Wang, Y. Zou, Z. Zhou, et al., We can hear you with wi-fi[J]. IEEE Trans. Mob. Comput. **15** (11), 2907–2920 (2016)
14. D. Halperin, W. Hu, A. Sheth, et al., Tool release: gathering 802.11 n traces with channel state information[J]. ACM SIGCOMM Comput. Commun. Rev. **41**(1), 53–53 (2011)
15. K. Wu, J. Xiao, Y. Yi, et al., CSI-based indoor localization[J]. IEEE Trans. Parallel Distrib. Syst. **24**(7), 1300–1309 (2013)
16. K. Wu, J. Xiao, Y. Yi, et al., Fila: fine-grained indoor localization[C], in *INFOCOM, 2012 Proceedings IEEE*, (IEEE, 2012), pp. 2210–2218
17. X. Liu, J. Cao, S. Tang, et al., Wi-Sleep: contactless sleep monitoring via WiFi signals[C], in *Real-Time Systems Symposium (RTSS), 2014 IEEE*, (IEEE, 2014), pp. 346–355
18. X. Wang, L. Gao, S. Mao, et al., DeepFi: deep learning for indoor fingerprinting using channel state information[C], in *Wireless Communications and Networking Conference (WCNC), 2015 IEEE*, (IEEE, 2015), pp. 1666–1671

19. W. Wang, A.X. Liu, M. Shahzad, et al., Understanding and modeling of wifi signal based human activity recognition[C], in *Proceedings of the 21st Annual International Conference on Mobile Computing and Networking*, (ACM, 2015), pp. 65–76
20. S. He, S.H.G. Chan, Wi-Fi fingerprint-based indoor positioning: recent advances and comparisons [J]. IEEE Commun. Surveys Tuts. **18**(1), 466–490 (2016)
21. J. Han, C. Qian, X. Wang, et al., Twins: device-free object tracking using passive tags[J]. IEEE/ACM Trans. Networking **24**(3), 1605–1617 (2016)
22. K. Ali, A.X. Liu, W. Wang, et al., Keystroke recognition using wifi signals[C], in *Proceedings of the 21st Annual International Conference on Mobile Computing and Networking*, (ACM, 2015), pp. 90–102
23. Y. Wen, X. Tian, X. Wang, et al., Fundamental limits of RSS fingerprinting based indoor localization[C], in *Computer Communications (INFOCOM), 2015 I.E. Conference on. IEEE*, (2015), pp. 2479–2487
24. Z. Zhou, Z. Yang, C. Wu, et al., Lifi: line-of-sight identification with wifi[C], in *INFOCOM, 2014 Proceedings IEEE*, (IEEE, 2014), pp. 2688–2696
25. B. Wei, W. Hu, M. Yang, et al., Radio-based device-free activity recognition with radio frequency interference[C], in *Proceedings of the 14th International Conference on Information Processing in Sensor Networks*, (ACM, 2015), pp. 154–165
26. Z.P. Jiang, W. Xi, X. Li, et al., Communicating is crowdsourcing: Wi-Fi indoor localization with CSI-based speed estimation[J]. J. Comput. Sci. Technol. **29**(4), 589–604 (2014)
27. X. Wang, L. Gao, S. Mao, et al., CSI-based fingerprinting for indoor localization: a deep learning approach[J]. IEEE Trans. Veh. Technol. **66**(1), 763–776 (2017)
28. C. Wu, Z. Yang, Z. Zhou, et al., Non-invasive detection of moving and stationary human with WiFi[J]. IEEE J. Sel. Areas Commun. **33**(11), 2329–2342 (2015)
29. Xu H, Yang Z, Zhou Z, et al. Enhancing wifi-based localization with visual clues[C]. in *Proceedings of the 2015 ACM International Joint Conference on Pervasive and Ubiquitous Computing*. (ACM 2015), pp. 963–974
30. N.U. Hassan, A. Naeem, M.A. Pasha, et al., Indoor positioning using visible led lights: a survey [J]. ACM Comput. Surv. (CSUR) **48**(2), 20 (2015)
31. A. Makki, A. Siddig, M. Saad, et al., Survey of WiFi positioning using time-based techniques [J]. Comput. Netw. **88**, 218–233 (2015)

Chapter 8
Continuous Authentication Using Writing Style

Marcelo Luiz Brocardo, Issa Traore, and Isaac Woungang

1 Introduction

A great deal of literature has been published regarding weaknesses that are inherent in traditional alphanumeric passwords. By design, passwords can be broken using dictionary or brute-force attacks. They can be forgotten, stolen, or shared. As an alternative, it has been recommended to use strong password schemes based on biometrics or generated using tokens or a combination of several of these schemes in the form of a multifactor authentication scheme. It has been shown, however, that no matter how strong the initial authentication controls are, there is still some residual risk of hackers being able to bypass such controls. For instance, the botnets have been used in many hacking instances to circumvent strong user authentication using a second factor such as a dedicated one-time password token and succeeded in stealing hundreds of thousands of dollars from online bank accounts belonging to small businesses.

The main reason why authentication circumvention is a serious issue is because of the static nature of the current approach to user authentication; the verification of the user credentials happens only once and typically at the beginning of the session. This indicates that if some malicious individuals can bypass this initial step, then they will be able to use the system resources freely without having to worry about being detected.

M. L. Brocardo (✉)
Department of Business and Administrative – ESAG, Santa Catarina State University – Udesc, Florianópolis, Brazil
e-mail: marcelo.brocardo@udesc.br

I. Traore
Department of Electrical and Computer Engineering, University of Victoria – UVIC, Victoria, BC, Canada
e-mail: itraore@ece.uvic.ca

I. Woungang
Department of Computer Science, Ryerson University, Toronto, ON, Canada
e-mail: iwoungan@scs.ryerson.ca

M. S. Obaidat et al. (eds.), *Biometric-Based Physical and Cybersecurity Systems*,
https://doi.org/10.1007/978-3-319-98734-7_8

One possible solution is to reinforce the initial (static) authentication with a CA process, which involves checking the user's biometric profile periodically throughout a computing session [1–3]. This ensures that the user always remains the same from the initial login time (where he/she claims specific identity) to the logout.

CA may be carried out actively or passively. Active CA requires the user to provide her authentication credentials when requested, while passive CA is carried out by collecting, and checking transparently, authentication data without the user being aware of such activity. Active CA can be implemented using traditional authenticators; however, it could irritate the users by asking them to provide explicit authentication credentials several times during a session. On the other hand, while passive CA does not suffer this limitation, it faces some restrictions regarding the kind of modality that can be used for authentication purpose.

Three emerging passive modalities include keystroke dynamics, mouse dynamics, and stylometric analysis. Keystroke dynamics is a behavioral biometrics that aims to identify humans based on the analysis of their typing rhythms on a keyboard [4]. Mouse dynamics is a behavioral biometrics that allows for identifying individual users based on their mouse usage while interacting with a computer [5]. Stylometric analysis consists of recognizing individuals based on their writing styles [6, 26, 27]. All these modalities can be collected passively using standard computing devices (such as mouse, keyboard) throughout a session without any knowledge of the user.

While a significant amount of research has been accomplished on the above modalities, their accuracies are still below those of well-established traditional biometrics such as fingerprint or iris scanning.

In this chapter, we discuss about open research challenges for which adequate solutions must be found in order to provide a firm basis for these technologies and even position them as a viable alternative to password-based authentication. While there are several publications on CA using mouse dynamics and keystroke dynamics, the use of stylometry for such purpose is still at an early stage. Many of the challenges inherent in this field are amplified when dealing with writing style data. We introduce an approach which tackles effectively many of these issues.

The rest of this chapter is organized as follows: Section 2 discusses the research challenges faced by CA using stylometry. Section 3 gives an overview of the state of the art of stylometric analysis. Section 4 outlines our general approach. Section 5 describes the feature space used in our work. Section 6 describes the experimental evaluation of the proposed approach. Section 3 concludes this chapter and discusses some future work.

2 Challenges in Using Stylometry for Continuous Authentication

While there is a rich body of research on stylometry, significant research challenges must be addressed when using such modality for CA. We discuss some of these challenges in this section.

2.1 Data Quality Issues

One of the challenges in analyzing behavioral data such as stylometry extracted from computer-mediated communications is the unstructured nature of the data. For example, in some cases, the sparse information conveyed makes it difficult to extract meaningful discriminating signals. Additionally, such information is characterized by a strong variability due to the inherent instability of the human behavioral and cognitive characteristics as well as changes in environmental conditions (e.g., hardware) when collecting such data. Thus, adequate data analysis techniques must be developed to ensure effective and efficient user recognition.

2.2 Free Actions Analysis

A specific challenge related to CA is the need to ensure that the authentication process will be conducted passively without any active involvement of the user. This requires collecting and analyzing what we refer to as free actions. Free actions are sample human-computer interaction (HCI) data such as free text, which are freely generated by a user without following a predefined template. In contrast, fixed actions refer to sample HCI data generated using predefined templates. While fixed actions are adequate for static authentication at login time, free actions must be used for CA to ensure transparency. However, free actions analysis is more challenging than fixed actions analysis because in the former case, authentication must be carried out using unseen examples at training time. For instance, in free actions analysis, while training may involve sample actions related to word processing and e-mail applications, authentication may be carried on samples that are collected while the user is playing a game or doing some programming task.

2.3 Performance Trade-Off

Accurate decision-making is the primary performance factor for any biometric system. For biometric systems, accuracy is typically measured in terms of false rejection rate (FRR) and false acceptance rate (FAR). A false rejection (FR) occurs when the system rejects a legitimate user, while a false acceptance (FA) occurs when the system accepts an impostor as a legitimate user. The equal error rate (EER) is another common metric used to assess the performance of a biometric system. EER corresponds to the operating point where FAR = FRR.

An important challenge faced by biometric authentication is the fact that a minimum amount of data sample must be captured for accurate decision-making. This means that in addition to accuracy, another key performance indicator to take into account is the authentication delay or time to authenticate (TTA). In general, the greater the TTA, the more accurate the system is. Lower TTA is preferable; however, this means there are smaller windows of vulnerability for the system. Thus, when using stylometry for CA, there is a need for developing analysis techniques that allow for an adequate trade-off between accuracy and TTA.

2.4 *Concept Drift*

Another key challenge faced by biometric authentication is the user behavior, or cognition may evolve over time, which means a change in the user profile. This is referred to as the problem of concept drift. Investigating the concept drift is essential for accurate user authentication.

2.5 *Security and Privacy Risks*

Stylometric data carry private user information which could be misused if accessed by intruders. Adequate techniques must be developed to mitigate privacy leaks.

CA based on stylometry can also be the target of different security threats including forgery, account takeover, signature tampering, insider attack, and defective implementation. An insider attack can be carried by a legitimate user being monitored by attempting to shut down the authenticator. Although this could be mitigated by limiting the privileges of regular users, the system will still be vulnerable in case of account takeover by an intruder who succeeds in escalating their privileges. Software defects may also crash the authenticator, creating a window of vulnerability where the protections provided will be unavailable. A thorough testing of the authenticator can help mitigating such threat.

The authenticator is not immune to forgery. It can be the target of automated attacks (referred to as generative attacks), where high-quality forgeries can be generated automatically using a small set of genuine samples [7]. An adversary having access to writing samples of a user may be able to effectively reproduce many of the existing stylometric features. It is essential to integrate specific mechanisms in the recognition system that would mitigate forgery attacks.

3 Related Work

Authorship analysis using stylometry has so far been studied in the literature primarily for the purpose of forensic analysis. Writing style is an unconscious habit, and the patterns of vocabulary and grammar could be a reliable indicator of the authorship. Stylometric studies typically target three different problems including authorship attribution or identification, authorship verification, and authorship profiling or characterization.

Authorship attribution consists of determining the most likely author of a target document among a list of known individuals. Earliest successes in attempting to quantify the writing style were the resolution of disputed authorship of Shakespeare's plays by Mendenhall [8] in 1887 and the Federalist Papers by Mosteller and Wallace in 1964 [9]. Recent studies on authorship identification investigated ways to identify patterns of terrorist communications [10], the author of a particular e-mail for computer forensic purposes [11–13], as well as how to collect digital evidence for investigations [14] or solve a disputed literary, historical [9] or musical authorship [15–17].

Work on authorship characterization has targeted primarily gender attribution [18–20] and the classification of the author education level [21].

Authorship verification consists of checking whether a target document was written or not by a specific author. There are few papers on authorship verification outside the framework of plagiarism detection [6], most of which focus on general text documents.

Koppel and others introduced a technique named "unmasking" where they quantify the dissimilarity between the sample document produced by the suspect and that of the other users (i.e., impostors) [6]. They used support vector machine (SVM) with linear kernel and addressed the authorship verification as a one-class classification problem. They used a dataset composed of ten authors, where 21 English books were split in blocks of 500 words. The overall accuracy was 95.7% when analyzing the feature set composed by the 250 most frequent words.

Iqbal et al. experimented not only with variants of SVM, including SVM with sequential minimum optimization (SMO) and SVM with RBF kernel, but also linear regression, Adaboost.M1, Bayesian network, and discriminative multinomial Naïve Bayes (DMNB) classifiers [12]. The proposed feature set included lexical, syntactic, idiosyncratic (spelling and grammatical mistakes), and content-specific features. Experimental evaluation of the proposed approach on the Enron e-mail corpus yielded EER ranging from 17.1% to 22.4%.

Canales et al. combined stylometry and keystroke dynamic analysis for the purpose of authenticating online test takers and used k-NN algorithm for classification [22]. The experimental evaluation of their proposed scheme involved 40 students with sample document size ranging between 1710 and 70,300 characters, yielding FRR = 20.25%, FAR = 4.18% and FRR = 93.46%, and FRR = 4.84% as performance when using keystroke and stylometry, respectively. The combination of both types of features yielded EER of 30%. The feature set included character-based, word-based, and syntactic features. They concluded that the feature set must be extended and certain type of punctuations may not necessarily represent the style of students when taking online exams.

Chen and Hao proposed to measure the similarity from e-mail messages by mining frequent patterns [23]. A frequent pattern is defined as the combination of the most frequent features that occur in e-mails from a target user. The basic feature set included lexical, syntactic, content-specific, and structural features. They used

PCA, k-NN, and SVM as classifiers and evaluated the proposed approach using a subset of the Enron dataset involving 40 authors. The experimental evaluation of their proposed scheme yielded 84% and 89% classification accuracy rates for 10 and 15 short e-mails, respectively.

More recently, an authorship verification competition was organized during the PAN – uncovering plagiarism, authorship, and social software misuse – evaluation lab at CLEF 2013 [24]. In total, 18 teams competed in two categories: intrinsic verification (as one-class problem) and extrinsic verification (as two-class problem). The evaluation dataset was composed of a set of d documents per author for training and a single document per author for testing. The dataset contains textbooks and newspapers in English, Greek, and Spanish, where d could be 30 for English, 30 for Greek, and 25 for Spanish language. Most of the teams used the simple character n-gram and word-based features, as well as a shallow architecture for classification. Although the competition was a good initiative to spread the concept of authorship verification, there was no significant improvement in accuracy since the best team achieved a rate of correctly classified documents of only 75% [25], and the corpus was composed of long text samples. Furthermore, no performance figure was provided about the classification error rate.

4 General Approach

The principle of CA is to monitor the user behavior during the session while discriminating between normal and suspicious user behavior. In case of suspicious behavior, the user session is closed (or locked), or an alert is generated. As shown in Fig. 8.1 (Adapted from [28]), the flag to prompt another authentication is based on the time or amount of data (i.e., the delay between consecutive reauthentication).

Our goal in this work is to apply authorship analysis technique for continuous user authentication. The proposed framework relies on authorship verification, which is the centerpiece of any authentication system. Our authorship verification methodology is structured around the steps and tasks of a typical pattern recognition process as shown in Fig. 8.2. While traditional documents are very well structured and large in size, providing several stylometric features, short online documents (such as e-mails and tweets) typically consist of a few paragraphs written quickly and often with syntactic and grammatical errors. In the proposed approach, all the sample texts used to build a given author profile are grouped into a single document. This single document is decomposed into consecutive blocks of short texts over which (continuous) authentication decisions happen.

Existing datasets consist of a set of candidate users and a set of text samples from these users. The basic assumption for the dataset is that it must contain sufficient information to discriminate different users. In order to have the same canonical form, we apply preprocessing canonicizer filters to standardize the text.

In addition, it is important to combine all texts from the same author creating a long text and then divide the combined text into smaller blocks of text. Each block of

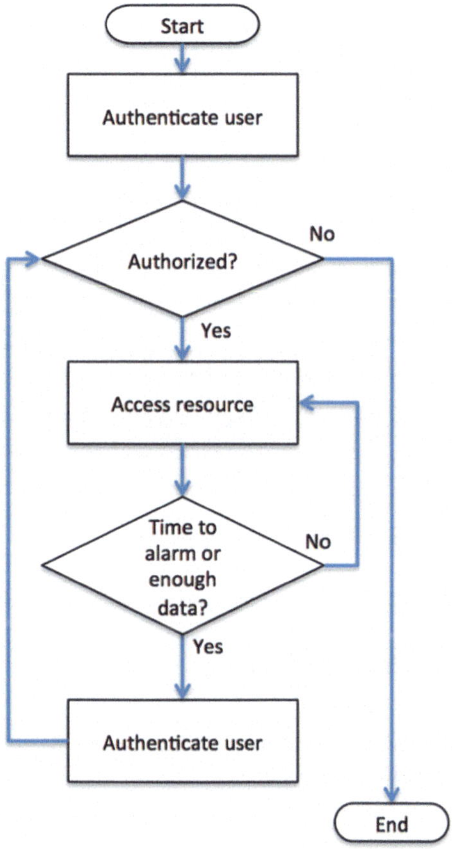

Fig. 8.1 Generic architecture of continuous authentication system

text is treated as a sample over which authentication decision occurs. This approach allows simulating repeated authentication windows, which is the foundation of CA.

Predictive features (*n-best*) are extracted from each block of text creating training and testing instances. The classification model consists of a collection of profiles generated separately for individual users. The proposed system operates in two modes: enrolment and verification. Based on the sample training data, the enrolment process computes the behavioral profile of the user.

The verification process compares unseen block of texts (testing data) against the model or profile associated with an individual (i.e., one-to-one identity matching) and then categorizes the block of text as genuine or impostor. In addition, the proposed system addresses the authorship verification as a two-class classification problem. The first class is composed of (positive) samples from the author, whereas the second class (negative) is composed of samples from other authors.

Different machine learning models can be used for the classification. As outlined later, we investigated in this work different classifiers, including both shallow and deep learners.

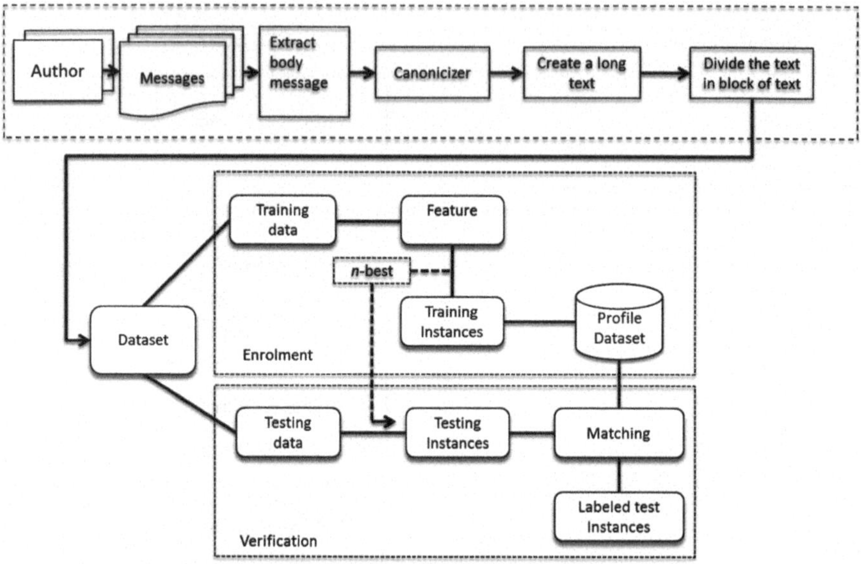

Fig. 8.2 Overview of the proposed authorship verification methodology

5 Feature Space

In this section, we present the features used in our framework and discuss our feature selection approach.

5.1 Stylometric Features

A specific user can be identified by his relatively consistent writing styles. According to Iqbal et al., the writing style of a user "contains characteristics of words usage, words sequence, composition and layouts, common spelling and grammatical mistakes, vocabulary richness, hyphenation and punctuation" [11]. In this study, we divide the features into three subsets including lexical, syntactic, and application-specific features. The list of all the features used in our work is shown in Table 8.1. A brief description of each feature subset is given below.

Lexical features indicate the preference of a user for certain group of words or symbols [29]. Lexical features can be extracted by dividing the text into tokens, where a token can be a word or a character. Word-level features may include the average sentence length in terms of words, the frequency of short and long words, the average word length, and the most frequently used words per author [22, 30–33]. These include also the vocabulary richness by quantifying the number of *hapax legomenon* and *hapax dis legomenon*, which refer to a word occurring only once or twice in a text, respectively [34–36]. Relevant character-level features include the

Table 8.1 List of common stylometric features

Features	Characteristics
1. *Lexical (character)*	
*F*1	Number of characters (C)
*F*2	Number of lower character/C
*F*3	Number of upper characters/C
*F*4	Number of white-space characters/C
*F*5	Total number of vowels (V)/C
*F*6. . .*F*10	Vowels (a, e, i, o, u)/V
*F*11. . .*F*36	Alphabets (A–Z)/C
*F*37	Number of special characters (S)/C
*F*38. . .*F*50	Special characters (e.g., "@," "#," "$," "%," "(',')," "{','}," etc.)/S
*F*51. . .*F*67	Character 5- and 6-grams (rU (b) and dU (b)) with two different values for the frequency f (i.e., $f = 1$ and $f = 2$) and for the mode of calculation of the n-grams (i.e., $m = 0$ and $m = 1$) and the discretized ru
*F*67. . .*F*192	Text-based icon (eight groups)
*F*193. . .*F*272	Unicode – Emoticons (code ranges from 1F600 to 1F64F)
*F*273...*F*528	Unicode – Miscellaneous symbols (code ranges from 2600 to 26FF)
Lexical (word)	
*F*529	Total number of words (N)
*F*530. . .*F*539	Average sentence length in terms of words/N
*F*540	Words longer than six characters/N
*F*541	Total number of short words (one to three characters)/N
*F*542	Average word length
*F*543	Average syllable per word
*F*544	Ratio of characters in words to N
*F*545. . .*F*550	Replaced words/N
*F*551. . .*F*600	The 50 most frequent words per author
*F*601. . .*F*650	The 50 most frequent 2-gram words per author
*F*651. . . *F*700	The 50 most frequent 3-gram words per author
*F*701	Hapax legomena
*F*702	Hapax dis legomena
*F*703	Vocabulary richness (total different words/N)
Syntactic	
*F*704	Total number of punctuation (P)
*F*705. . .*F*712	Single quotes, commas, periods, colons, semicolons, question marks, exclamation marks divided by P
*F*713. . .*F*824	Unicode – general punctuation (code ranges from 2000 to 206F)
*F*825. . .*F*829	Total number of conjunction, interrogative, preposition, interjection, and pronouns each one divide by N
*F*830. . .*F*1065	Ratio of functional word divided by the respective total word group

(continued)

Table 8.1 (continued)

Features	Characteristics
Application-specific	
*F*1066	Total number of sentences
*F*1067	Total number of paragraphs
*F*1068…*F*1070	Average number of characters, words, and sentences, in a block of text
*F*1071	Average number of sentences beginning with upper case
*F*1072	Average number of sentences beginning with lower case

frequency of characters comprehending upper case, lower case, vowels, white spaces, alphabets (A–Z), digits, special characters, and the writer's mood expressed in the form of icons and symbols [36–37].

Another lexical features that has proven to be efficient in capturing writing style is the measure of *n*-grams [28, 30, 38–41]. *N*-gram is a token formed by a contiguous sequence of characters or words. It is tolerant to typos including grammatical errors and misuse of punctuations. Some works showed good results by creating a vector with the most frequent words *2*-grams and *3*-grams [26]. A list of stop words was used to exclude very frequent words that convey no meaning, and only the content-specific features were retained to calculate word *n*-grams [42]. At the character level, good results are obtained when using *5*-grams and *6*-grams [43].

Syntactic features are context independent and capture author's style across different subjects. Syntactic features can be categorized in terms of punctuation and part of speech (POS) [44–46]. Punctuation is an important rule to define boundaries and identify meaning by splitting a paragraph into sentences and each sentence into tokens [12]. Punctuation includes single quotes, commas, periods, colons, semicolons, question marks, exclamation marks, and uncommon marks based on the unicode format (e.g., †, ⁝, ... ∴). The POS tagging consists of categorizing a word according to its function in the context and can be classified as verbs, nouns, pronouns, adjectives, adverbs, prepositions, conjunctions, and interjections [17, 20, 21, 47]. The weakness of this type of features is that POS is language-dependent since it relies on a language parser and also could produce some noise due to the unavoidable errors made by the parser [30].

Application-specific features capture the overall characteristics of the organization and format of a text [23, 35, 36, 47, 48]. While application-specific features can be categorized at the message level or paragraph level or according to the technical structure of the document [48], we extracted only features related to the paragraph structure because our focus is on short messages (such as e-mails and Twitter posts). Paragraph-level features include the number of sentences per block of text; the average number of characters, words, and sentences in a block of text; and the average number of sentences beginning with upper and lower case.

N-gram model is noise tolerant and effective to typos including grammatical errors and misuse of punctuations. Since online documents (e.g., e-mails, tweets) are unstructured documents, then *n*-gram can capture important features. While the approach used so far in the literature for *n*-gram modeling has consisted of computing *n*-gram frequency in a given sample document, some innovative approaches focus on analyzing *n*-grams and their relationship with the training dataset [49].

5.2 Feature Selection

Over a thousand stylistic features have already been identified and used in the literature along with a wide variety of analysis methods. However, there is no agreement among researchers on which features yield the best results. As a matter of fact, analyzing a large number of features does not necessarily provide the best results since some features provide very little or no predictive information.

Being able to keep only the most discriminating features individually per user allows reducing the size of the data by removing irrelevant attributes and improves the processing time for training and classification. This can be achieved by applying feature selection measures, which allow finding a minimum set of features that represent the original distribution obtained using all the features.

Although feature selection by an expert is a common practice, it is complex and sometime inefficient because it is relatively easy to select irrelevant attributes while omitting important attributes. Other feature selection methods include exhaustive search and probabilistic approach. Exhaustive search is a brute-force feature selection method that could evaluate all possible feature combinations, but it is time-consuming and impractical. On the other hand, the probabilistic approach is an alternative for speeding up the processing time and selecting optimal subset of features.

An effective feature selection approach is shown in Fig. 8.3 [27]. It is built on our previous works by identifying and keeping only the most discriminating features and

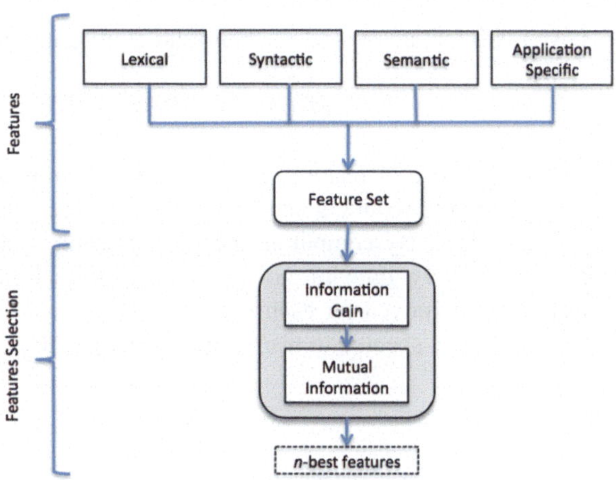

Fig. 8.3 Feature selection approach

by identifying new sets of relevant features. We derive, from the raw stylometric data, numeric feature vectors that represent term frequencies of each of the selected features. All frequencies are normalized between 0 and 1, and each user has a specific feature set that best represents his writing style.

An ideal feature is expected to have high correlation with a class and low correlation with any other features. Based on this concept, we measure the correlation between a feature and a class by computing the information gain (IG) and the correlation between a pair of features by computing the mutual information (MI).

Let $X = [x_1,x_2,...,x_n]$ denote an n-dimensional feature vector that describes our feature space. Let $S = \{X_1,X_2,...,X_m\}$ denote the set of training samples for a given user. Each training sample corresponds to a vector of feature values $X_j = [x_{ij}]_{1 \leq i \leq n}$, where x_{ij} is the value of feature x_i for sample X_j.

The information entropy of feature x_i denoted $H(x_i)$ is defined as:

$$H(xi) = -\sum_{j=1}^{m} p(xij)\log2p(xij) \tag{8.1}$$

where $p(x_{ij})$ denotes the probability mass function of x_{ij}.

Given a variable y, with samples $(y_1,...,y_M)$, the conditional entropy $H(x_i, y)$ of x_i given y is defined as:

$$H(xi|y) = -\sum_{j=1}^{m}\sum_{k=1}^{M} p(xij, yk), \log2p(xij|yk) \tag{8.2}$$

where $p(x_{ij}, y_k)$ denotes the joint probability mass function of x_{ij} and y_k.

Suppose that the dataset is composed of two classes (positive and negative). The IG for a feature x_i with respect to a class is computed as:

$$IG(\text{Class}, xi) = H(\text{Class}) - H(\text{Class}|xi) \tag{8.3}$$

Given two features x_i and x_k, their mutual information (MI) is calculated as:

$$MI(xi, xk) = H(xi) - H(xi|xk) \tag{8.4}$$

For the purpose of feature selection, it is recommended to retain only features with non-zero information gain and remove a feature when the mutual information is higher than 95%. By computing the IG for features and MI for pairs of features, features with very little or no predictive information and high correlation are identified and removed for each user. At the end, each user ends up with a subset of features that is specific to his/her individual profile.

6 Experiments

In this section, we describe in depth the experiments conducted to evaluate our approach.

6.1 Datasets

In this subsection, we describe three different datasets that were used in our experimental evaluations, namely, an e-mail corpus, a micro messages dataset based on Twitter feeds, and a forgery dataset.

1. *E-mail dataset:* The Enron corpus[1] is a large set of e-mail messages from Enron's employees. Enron was an energy company (located in Houston, Texas) that was bankrupt in 2001 due to white-collar fraud. The company e-mail database was made public by the Federal Energy Regulatory Commission during the fraud investigation. The raw version of the database contains 619,446 messages belonging to 158 users. However, Klimt and Yang cleaned the corpus by removing some folders that appeared not to be related directly to the users [50]. As a result, a cleaned version of the e-mail dataset contains more than 200,000 messages belonging to 150 users with an average of 757 messages per user. The e-mails are plaintexts and cover various topics ranging from business communications to technical reports and personal chats.
2. *Micro messages dataset:* Twitter is a microblogging service that allows authors to post messages called "tweets." Each tweet is limited to 140 characters and sometimes expresses opinions about different topics. Tweets have also other particularities such as the following:
    - The use of emoticons to express sentiments and the use of URL shorteners to refer to some external sources
    - The use of a tag "RT" in front of a tweet to indicate that the user is repeating or reposting the same tweet
    - The use of a hashtag "#" to mark and organize tweets according to topics or categories
    - The use of "@ < user>" to link a tweet to a twitter profile whose user name is "user"

    In addition, registered users can read and post tweets, reply to a tweet, send private messages, and re-tweet a message, while unregistered users can only read them. Also, a registered user can follow or can be followed by other users.

[1]Available at http://www.cs.cmu.edu/~enron/

The dataset[2] used in this study contains 100 English users and on average 3194 twitter messages with 301,100 characters per author [49]. All tweets in the dataset were posted before October 22, 2013 (inclusive).

3. *Impostor dataset:* In order to assess the robustness of an approach against forgery attempts, a novel forgery dataset was developed as part of this research. Some volunteers were invited to generate forgeries against sample tweets selected randomly from the above Twitter datasets.

Participants in the experiment consisted of ten volunteers – with seven males and three females – with ages varying from 23 to 50 years, with different background.

Tweet samples were randomly selected from ten authors, considered as legal users from the Twitter dataset. Impostor samples were collected through a simple form consisting of two sections. In the first section, tweets from a specific legal user were made available. This allows simulating a scenario where an adversary has access to writing samples. The second section involved two fields, one for participants to enter their name and the other for them to write three or four tweets trying to reproduce the writing style of the legal user. A "submit" button was used to send the sample to the database when completed as shown in Fig. 8.4. The form was sent by e-mail and made available online through a customized web page. The survey was implemented using the Google Forms platform. The only restriction was a minimum size of 350 characters per sample spread over 3–4 tweets.

During the data collection, each volunteer received a new form with different legal user information, once per every workday. All volunteers were instructed to provide one sample per day. The data was collected over a period of 30 days. The experimenters had no control over the way volunteers wrote their tweets. Collected data consisted of an average of 4253 characters per volunteer spread over 10 attacks.

6.2 Data Preprocessing

The data was preprocessed in order to normalize e-mail and tweet particularities [51, 52]. In order to obtain the same structural data and improve classification accuracy, several preprocessing steps on the data were performed.

In the Enron corpus, we used only the body of the messages from the e-mails found in the folders "sent" and "sent items" for each user. All duplicate e-mails were removed. Similarly, we removed all e-mails that contain tables with numbers when the average number of digits per total number of characters was higher than 25%. Also, we removed reply texts when present and replaced e-mail and web addresses by meta tags "e-mail" and "http," respectively.

In the Twitter and forgery corpuses, we removed all re-tweet (RT) posts and all duplicated tweets. Hashtag symbols such as "#word" and the following word were

[2]Available at http://www.uvic.ca/engineering/ece/isot/datasets/

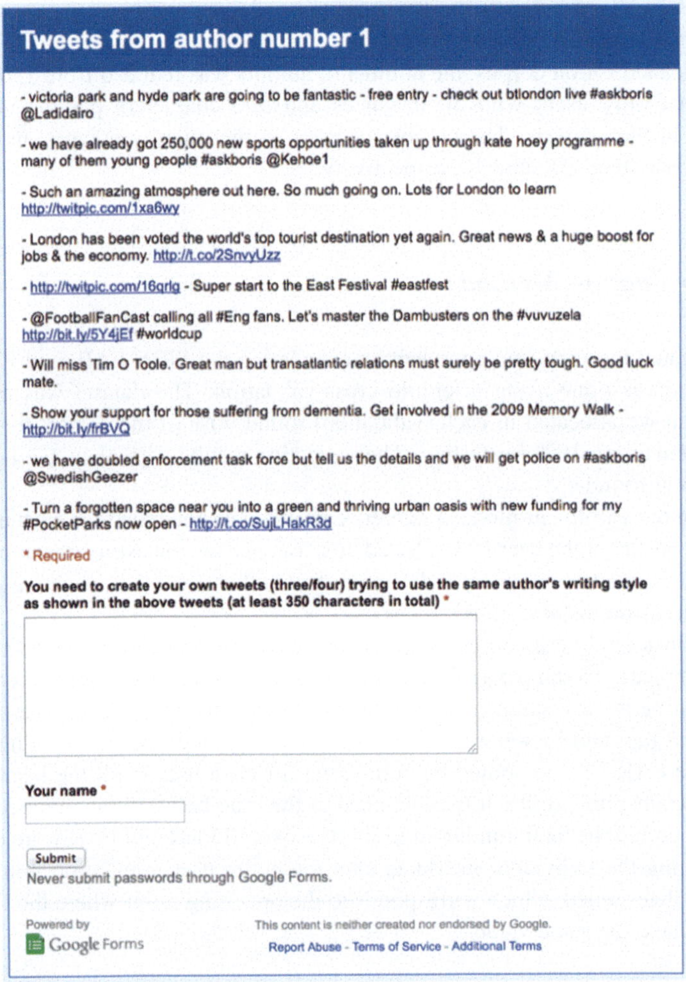

Tweets from author number 1

- victoria park and hyde park are going to be fantastic - free entry - check out btlondon live #askboris @Ladidairo

- we have already got 250,000 new sports opportunities taken up through kate hoey programme - many of them young people #askboris @Kehoe1

- Such an amazing atmosphere out here. So much going on. Lots for London to learn http://twitpic.com/1xa6wy

- London has been voted the world's top tourist destination yet again. Great news & a huge boost for jobs & the economy. http://t.co/2SnvyUzz

- http://twitpic.com/16qrlg - Super start to the East Festival #eastfest

- @FootballFanCast calling all #Eng fans. Let's master the Dambusters on the #vuvuzela http://bit.ly/5Y4jEf #worldcup

- Will miss Tim O Toole. Great man but transatlantic relations must surely be pretty tough. Good luck mate.

- Show your support for those suffering from dementia. Get involved in the 2009 Memory Walk - http://bit.ly/frBVQ

- we have doubled enforcement task force but tell us the details and we will get police on #askboris @SwedishGeezer

- Turn a forgotten space near you into a green and thriving urban oasis with new funding for my #PocketParks now open - http://t.co/SujLHakR3d

* Required

You need to create your own tweets (three/four) trying to use the same author's writing style as shown in the above tweets (at least 350 characters in total) *

Your name *

Submit

Never submit passwords through Google Forms.

Powered by Google Forms

This content is neither created nor endorsed by Google.

Report Abuse - Terms of Service - Additional Terms

Fig. 8.4 Screenshot of a form generated by a user

replaced by a meta tag "#hash"; @<user> reference was replaced by meta tag "@cite"; web addresses were replaced by meta tag "http." We also removed all messages that contain one or more of the following unicode blocks: Arabic, Cyrillic, Devanagari, Hangul syllables, Bengali, Hebrew, Malayalam, Greek, Hiragana, Cherokee, and CJK Unified Ideographs.

In both datasets, we replaced currency by a meta tag "$XX," percentage by a meta tag "XX%," date by a meta tag "date," hours by a meta tag "time," numbers by a meta tag "numb," and information between tags ("<information>") by a meta tag "TAG." Finally, the document was normalized to printable ASCII, all characters were converted to lower case, and the white space was normalized. In order to simulate CA, all messages per author were grouped, creating a long text or stream of characters that was divided into blocks.

In a cleaned Enron corpus, the number of authors was reduced from 150 to 76 to ensure that only users with 50 instances and 500 characters per instance were involved in the analysis. The number of users in the micro messages and forgery corpuses remained 100 and 10, respectively.

6.3 Evaluation Method

We evaluated the proposed approach by calculating the FRR, FAR, and EER. The evaluation was done using a tenfold cross-validation. The dataset was randomly sorted, and we allocated in each (validation) round 90% of the dataset for training and the remaining 10% for testing. The validation results were then averaged over the different rounds.

During the enrolment mode, a reference profile was generated for each user. The reference profile of the user *U* was based on a training set consisting of samples from the user (i.e., positive samples) and samples from other users (i.e., negative samples) considered as impostors.

The considered verification mode was a one-to-one matching process that consists of comparing a sample against the enrolled user profile. The FRR was computed by comparing the test samples of each user *U* against his own profile. The FRR was obtained as the ratio between the number of false rejections and the total number of trials. The FAR was computed by comparing for each user *U* all the negative test samples against his profile. It was obtained as the ratio between the number of false acceptances and the total number of trials. The overall FRR and FAR were obtained by averaging the individual measures over the entire user population. Finally, the EER was determined, which corresponds to the operating point where the FRR and the FAR have the same value.

6.4 Evaluation Results

In this section, we described the considered classification models, namely, the shallow and deep learners.

Shallow Classifiers

It has been shown that shallow classification architectures can be effective in solving many stylometric analysis problems [30, 53]. A shallow architecture refers to a classifier with only one or two layers responsible for classifying the features into a problem-specific class.

We studied three different classifiers: logistic regression (LR), SVM, and a hybrid classifier that combines logistic regression and SVM (SVM-LR).

We analyzed our feature space using the aforementioned shallow classifiers. This shows increased effectiveness of our approach compared to the existing approaches published in the literature when applied for authorship verification based on short texts. Furthermore, we investigated shorter messages, which are required for CA systems to operate with reduced window size for reauthentication.

We started our experiments involving shallow classifiers using the Enron and Twitter datasets and used different configurations for block size and number of blocks per user. Tables 8.3 and 8.4 show the performance results obtained for the Enron and Twitter datasets, respectively. Table 8.2 shows the improvement in accuracy for SVM-LR and LR over the SVM baseline classifier based on the results from Tables 8.3 and 8.4. SVM-LR and LR achieve on average 9.89% and 18.62% improvement in accuracy (i.e., EER) over SVM, respectively.

Deep Learning Classifier

We have used a deep learning classifier and assessed the robustness of our proposed approach against forgery attempts. We have investigated the use of deep models for authorship verification. More specifically, we have studied deep belief network (DBN). DBN is a type of deep neural network composed of multiple layers of restricted Boltzmann machines (RBMs) with a softmax layer added to the top for recognition tasks. Compared to the existing literature, it can be claimed that the use

Table 8.2 Accuracy improvement for SVM-LR and LR over the SVM baseline classifier

Dataset	Block size	Blocks per user	Improvement (%)	
			SVM-LR	LR
Enron	50	500	17.18	23.82
Twitter	140	100	8.68	18.90
		200	9.37	17.41
	280	50	3.47	12.51
		100	10.76	20.44

Table 8.3 Authorship verification using the Enron dataset. Authorship verification with 76 authors, block size of 500 characters, and 50 blocks per author. Feature selection was performed using the information gain and mutual information approaches

	SVM (%)	SVM-LR (%)	LR (%)
EER	12.05	9.98	9.18

Table 8.4 Authorship verification using the Twitter dataset. EER for SVM, SVM-LR, and LR using Twitter dataset involving 100 authors and varying the size of the block and the number of blocks per author. Feature selection was performed by information gain and mutual information approaches

Block size	Blocks per user	SVM (%)	SVM-LR (%)	LR (%)
140	100	23.49	21.45	19.05
	200	20.27	18.37	16.74
280	50	18.47	17.83	16.16
	100	14.87	13.27	11.83

Table 8.5 Authorship verification using DBN classifier on the Twitter dataset. EER for the Gaussian-Bernoulli DBN classifier using the Twitter dataset involving 100 authors. In the pretraining phase, the epoch was set to 100 and the learning rate was set to 0.001. In the fine-tuning phase, the learning rate was set to 0.01

Block size	Blocks per user	EER (%)
140	100	16.73
	200	16.58
280	50	12.61
	100	10.08

Table 8.6 Authorship verification by using running DBN on the forgery dataset involving ten forgery attempts against ten author profiles

Block size	Blocks per user	EER (%)
140	100	12.30
280	100	5.48

of a machine-learning method based on deep structure, in particular DBN, can help enhancing the accuracy of the authorship verification using stylometry. In the early stages of our research, we have investigated the standard DBN, which has binary neurons only. Our first approach was to normalize each input variable to binary values and run the DBN classifier. However, the results obtained did not improve those obtained from our previous work using a shallow structure. In order to strengthen the accuracy, we replaced the first Bernoulli-Bernoulli RBM layer by a Gaussian-Bernoulli RBM layer, which uses Gaussian units in the visible layer to model real-valued data. Table 8.5 shows the performance obtained by running the DBN classifier on the Twitter dataset. By running the same classifier on the Enron dataset, we obtained an EER of 8.21% with a block size of 500 characters and 50 blocks or instances per user. In both cases, DBN achieves significant improvement in accuracy over SVM-LR and LR. Although the results for DBN are very promising, there still a need to improve them in order to be comparable with other biometric systems currently used for CA (e.g., keystroke and mouse dynamics). An option could be to increase the size of the block of characters, but this goes against the need for "short authentication delay" in CA. Our future work is to investigate this challenge.

Ability to Withstand Forgery

Stylometric analysis can be the target of forgery attacks. An adversary having access to writing samples of a user may be able to effectively reproduce many of the existing stylometric features. In order to assess the ability to withstand forgery, we run the DBN classifier on the forgery dataset. Table 8.6 shows the obtained EER performance for two different block sizes, 280 and 140 characters, which are 5.48% and 12.30%, respectively. These performance results are quite encouraging. However, it is important to highlight the fact that our forgery study involved only ten attack instances on ten different user profiles. More data should be collected and analyzed to confirm these results. This will be part of our future work.

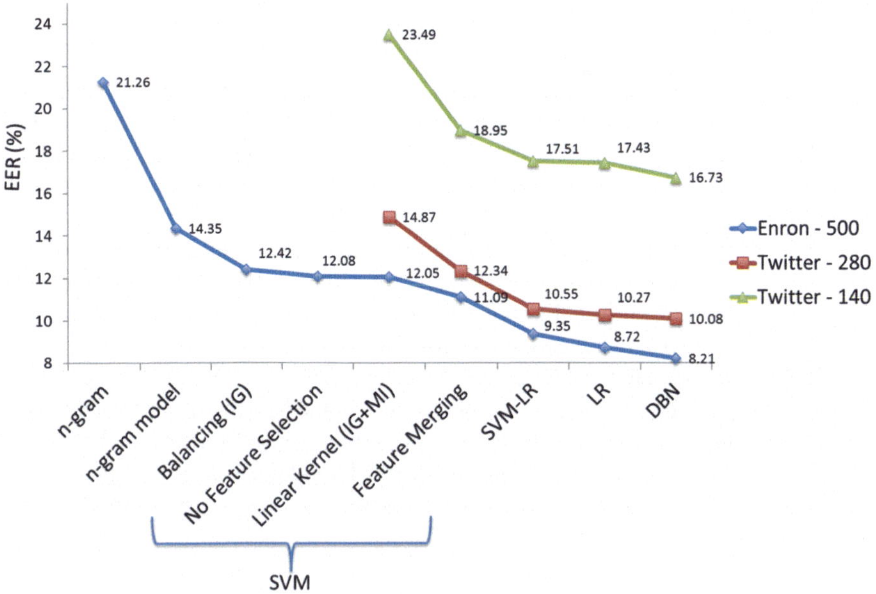

Fig. 8.5 Summary of the experiment results obtained with the Enron and Twitter datasets using shallow and deep learning classifiers. The lines represent the datasets

Using the aforementioned shallow and deep classifiers, we investigated different feature models. Figure 8.5 summarizes the performance results obtained using the Enron and Twitter datasets. The x-axis shows the different feature models and classification techniques experimented with, the y-axis shows the obtained EER, and the lines represent the datasets used.

The first two experiments were performed on the Enron dataset, and they used only the n-gram features. Our proposed n-gram model yielded an EER of 14.35%, while the baseline n-gram model yielded a EER of 21.26%. The next experiments were performed using SVM as a classifier. When the feature selection was omitted, we obtained an EER of 12.08%. Next, we extended the (information gain) feature selection technique by adding the mutual information selection approach and setting the information gain to be greater than 0. This yielded an EER of 12.05%, which indicates that our feature selection technique has a negligible impact on the accuracy of the classifier. Similar results in the literature have indicated that SVM did not benefited from the feature selection. However, the feature selection is still beneficial in terms of reduction in processing time due to the reduction in the number of features.

We investigated the impact of different SVM kernels on the accuracy. The outcome of such study revealed that SVM with linear kernel achieves better EER performances than polynomial or Gaussian. These results show that SVM-LR and LR perform much better than SVM, while DBN outperforms all the aforementioned classifiers.

7 Conclusion

Stylometry has been widely used for authorship verification and characterization, but only a few works have targeted authorship verification. Our work is distinguishable from the previous ones in this area by focusing on the challenges involved in stylometric authorship verification in the context of continuous or repeated user authentication.

We have investigated different combination of features as well as the shallow and deep learning techniques. Our experimental evaluation involving two public datasets and a forgery dataset collected in our lab yielded promising results toward the key challenges faced when using stylometry for CA.

Although the obtained results are encouraging, more work are to be done in the future to improve the accuracy of our proposed scheme by decreasing the EER and by investigating shorter authentication delays (e.g., 50 characters and below). Furthermore, there is a need to confirm the results obtained in our forgery study and further investigate the resilience of our approach to forgery by expanding the datasets used.

References

1. A.A.E. Ahmed, I. Traore, Dynamic sample size detection in continuous authentication using sequential sampling, in *Proceedings of the 27th Annual Computer Security Applications Conference*, New York, NY, USA, (2011)
2. M.E. Crosby, C.S. Ikehara, *Continuous identity authentication using multimodal physiological sensors* 5404 (2004), pp. 393–400
3. I. Traore, A. A. E. Ahmed (eds.), *Continuous authentication using biometrics: continuous authentication using biometrics: data, models, and metrics* (IGI Global, 2012), p. 385
4. R.S. Gaines, W. Lisowski, S.J. Press, N. Shapiro, Authentication by keystroke timing: some preliminary results. No. RAND-R-2526-NSF. Rand Corp Santa Monica CA, (1980)
5. A.A.E. Ahmed, I. Traore, A new biometric technology based on mouse dynamics. IEEE Trans. Depend. Sec. Comput. **4**, 165–179 (2007)
6. M. Koppel J. Schler, Authorship verification as a one-class classification problem, in *Proceedings of the 21st International Conference on Machine Learning*, Banff, (2004)
7. L. Ballard, F. Monrose, D. Lopresti, Biometric authentication revisited: understanding the impact of wolves in sheepś clothing, in *Proceedings of the 15th Conference on USENIX Security Symposium - Volume 15*, Berkeley, (2006)
8. T.C. Mendenhall, The characteristic curves of composition. Science **ns-9**, 237–246 (1887)
9. F. Mosteller, D.L. Wallace. Inference in an authorship problem: A comparative study of discrimination methods applied to the authorship of the disputed federalist papers. Journal of the American Statistical Association **58**(302), 275–309 (1963)
10. A. Abbasi, H. Chen, Applying authorship analysis to extremist-group web forum messages. IEEE Intell. Syst. **20**(5), 67–75 (2005)
11. F. Iqbal, R. Hadjidj, B.C.M. Fung, M. Debbabi, A novel approach of mining write-prints for authorship attribution in e-mail forensics. Digit. Investig. **5**, S42–S51 (2008)
12. F. Iqbal, L.A. Khan, B.C.M. Fung, M. Debbabi, E-mail authorship verification for forensic investigation, in *Proceedings of the 2010 ACM Symposium on Applied Computing*, New York, NY, USA, (2010)

13. F. Iqbal, H. Binsalleeh, B.C.M. Fung, M. Debbabi, A unified data mining solution for authorship analysis in anonymous textual communications. Inf. Sci. **231**, 98–112 (2013)
14. C.E. Chaski, Who's at the keyboard: authorship attribution in digital evidence investigations. Int. J. Digit. Evid. **4**, 1–13 (2005)
15. J. Burrows, Delta: a measure of stylistic difference and a guide to likely authorship. Lit. Ling. Comput. **17**, 267–287 (2002)
16. E. Backer, P. van Kranenburg, On musical stylometry pattern recognition approach. Pattern Recogn. Lett. **26**, 299–309 (2005)
17. Y. Zhao, J. Zobel, Searching with style: authorship attribution in classic literature, in *Proceedings of the thirtieth Australasian conference on Computer science - Volume 62*, Darlinghurst, (2007)
18. R.A.G.K. Sarawgi, Y. Choi, Gender attribution: tracing stylometric evidence beyond topic and genre, in *Proceedings of the 15th Conference on Computational Natural Language Learning*, Stroudsburg, (2011)
19. N. Cheng, X. Chen, R. Chandramouli, K.P. Subbalakshmi, Gender identification from e-mails, in *Computational Intelligence and Data Mining,* 2009. *CIDM '09. IEEE Symposium on*, (2009)
20. N. Cheng, R. Chandramouli, K.P. Subbalakshmi, Author gender identification from text. Digit. Investig. **8**, 78–88 (2011)
21. P. Juola, R.H. Baayen, A controlled-corpus experiment in authorship identification by cross-entropy. Lit. Ling. Comput. **20**, 59–67 (2005)
22. O. Canales, V. Monaco, T. Murphy, E. Zych, J. Stewart, C.T.A. Castro, O. Sotoye, L. Torres, G. Truley, A stylometry system for authenticating students taking online tests, (2011)
23. X. Chen, P. Hao, R. Chandramouli, K.P. Subbalakshmi, Authorship similarity detection from email messages, in *Proceedings of the 7th international conference on Machine learning and data mining in pattern recognition*, Berlin, (2011)
24. P. Juola, E. Stamatatos, Overview of the author identification task at PAN 2013, in *Conference and Labs of the Evaluation Forum - CLEF 2013*, Valencia - Spain, (2013)
25. S. Seidman, Authorship verification using the impostors method, in *Conference and Labs of the Evaluation Forum - CLEF* 2013, Valencia - Spain, (2013)
26. M. L. Brocardo, I. Traore, S. Saad and I. Woungang, Authorship verification for short messages using stylometry, in *Proceedings of the International Conference on Computer, Information and Telecommunication Systems (CITS)*, 2013
27. M.L. Brocardo, I. Traore, I.Woungang, M.S. Obaidat, Authorship verification using deep belief network systems. Int. J. Commun. Syst. e3259–n/a 2017
28. B. Kjell, W.A. Woods, O. Frieder, Discrimination of authorship using visualization. Inf. Process. Manag. **30**, 141–150 (1994)
29. S.M. Alzahrani, N. Salim, A. Abraham, Understanding plagiarism linguistic patterns, textual features, and detection methods, *Systems, Man, and Cybernetics, Part C: Applications and Reviews,* IEEE Transactions on, **42**, 133–149, (2012)
30. E. Stamatatos, A survey of modern authorship attribution methods. J. Am. Soc. Inf. Sci. Technol. **60**(3), 538–556 (2009)
31. J.F. Burrows, Word patterns and story shapes: the statistical analysis of narrative style. Lit. Ling. Comput. **2**, 61–70 (1987)
32. D.I. Holmes, The evolution of stylometry in humanities scholarship. Lit. Ling. Comput **13**, 111–117 (1998)
33. N. Homem, J.P. Carvalho, Authorship identification and author fuzzy Fingerprints, in *Fuzzy Information Processing Society (NAFIPS),* 2011 Annual Meeting of the North American, (2011)
34. F.J. Tweedie, R.H. Baayen, How variable may a constant be? Measures of lexical richness in perspective. Comput. Hum. **32**, 323–352 (1998)
35. O.D. Vel, A. Anderson, M. Corney, G. Mohay, Mining e-mail content for author identification forensics. Sigmod Record **30**, 55–64 (2001)

36. R. Zheng, J. Li, H. Chen, Z. Huang, A framework for authorship identification of online messages: writing-style features and classification techniques. J. Am. Soc. Inf. Sci. Technol. **57**(3), 378–393 (February 2006)
37. A. Orebaugh, J. Allnutt, Classification of instant messaging communications for forensics analysis. Int. J. Forensic Comput. Sci. **1**, 22–28 (2009)
38. M. Koppel, J. Schler, Authorship verification as a one-class classification problem, in *Proceedings of the twenty-first international conference on Machine learning*, (2004)
39. F. Peng, D. Schuurmans, S. Wang, V. Keselj, Language independent authorship attribution using character level language models, in *Proceedings of the 10th Conference on European. Chapter of the Association for Computational Linguistics - Volume 1*, Stroudsburg, (2003)
40. P. Juola, Authorship attribution for electronic documents, in *Advances in digital forensics II*, vol. 222, (Springer, New York, 2006), pp. 119–130
41. H.V. Halteren, Author verification by linguistic profiling: an exploration of the parameter space. ACM Trans. Speech Lang. Process **4**(1), 1–17 (February 2007)
42. M.R. Brennan, R. Greenstadt, Practical attacks against authorship recognition techniques, in *IAAI*, (2009)
43. M.L. Brocardo, I. Traore, I. Woungang, Toward a framework for continuous authentication using stylometry, in *The 28th IEEE International Conference on Advanced Information Networking and Applications (AINA-2014)*, Victoria, (2014)
44. H. Baayen, H. van Halteren, F. Tweedie, Outside the cave of shadows: using syntactic annotation to enhance authorship attribution. Lit. Ling. Comput. **11**, 121–132 (1996)
45. M. Koppel, J. Schler, Exploiting Stylistic Idiosyncrasies for Authorship Attribution, in *IJCAI'03 Workshop on Computational Approaches to Style Analysis and Synthesis*, Acapulco, (2003)
46. S. Argamon, S. Marin, S.S. Stein, Style mining of electronic messages for multiple authorship discrimination: first results, in *Proceedings of the ninth ACM SIGKDD international conference on Knowledge discovery and data mining*, New York, NY, USA, (2003)
47. R. Hadjidj, M. Debbabi, H. Lounis, F. Iqbal, A. Szporer, D. Benredjem, Towards an integrated e-mail forensic analysis framework. Digit. Investig. **5**, 124–137 (2009)
48. A. Abbasi, H. Chen, Writeprints: a stylometric approach to identity-level identification and similarity detection in cyberspace. ACM Trans. Inf. Syst. **26**(2), 1–29 (April 2008)
49. M.L. Brocardo, I. Traore, Continuous authentication using micro-,essages, in *Twelfth Annual International Conference on Privacy, Security and Trust (PST 2014)*, Toronto, 2014
50. B. Klimt, Y. Yang, The enron corpus: a new dataset for email classification research, in *Machine learning: ECML 2004*, (Springer, 2004), pp. 217–226
51. W.-W. Deng, H. Peng, Research on a Naive Bayesian Based short message filtering system, in *Machine Learning and Cybernetics, 2006 International Conference on*, (2006)
52. J. Cai, Y. Tang, R. Hu, Spam filter for short messages using winnow, in *Advanced Language Processing and Web Information Technology,* 2008. ALPIT'08. International Conference on, (2008)
53. M. Koppel, J. Schler, S. Argamon, Authorship attribution in the wild. Lang. Resour. Eval. **45** (1), 83–94 (March 2010)

Chapter 9
Facets and Promises of Gait Biometric Recognition

James Eric Mason, Issa Traore, and Isaac Woungang

The emerging field of behavior biometrics has prompted a re-examination of many previously overlooked human characteristics. One such characteristic that has traditionally undergone analysis in the medical realm is the gait biometric. Gait biometrics refer to the unique aspects of human locomotion that can be captured and used for recognition purposes. These biometrics offer a number of potential advantages over other traditional biometrics in their abilities to be detected at a distance and with little-to-no obtrusion to the subject of the analysis. The gait biometric also offers another potential advantage over many traditional biometrics because it is inherently difficult to spoof the complicated set of actions that compose the human gait. This chapter discusses the various approaches that have been used to perform recognition via the gait biometric and examines the performance and implications that might be expected when applying the gait biometric to a real-world scenario.

1 Approaches to Gait Biometrics

The human gait is composed of a number of repetitive characteristics coming together to form a cycle. Research by Derawi et al. [14] has suggested there are 24 distinct components that, when put together, can uniquely identify an individual's gait. This, however, presents a challenge to researchers because no single device is able to capture the entirety of motions and attributes that form the human gait. In

---

J. E. Mason (✉) · I. Traore
Department of Electrical and Computer Engineering, University of Victoria, Victoria, BC, Canada
e-mail: jemason@uvic.ca; itraore@ece.uvic.ca

I. Woungang
Department of Computer Science, Ryerson University, Toronto, ON, Canada
e-mail: iwoungan@scs.ryerson.ca

M. S. Obaidat et al. (eds.), *Biometric-Based Physical and Cybersecurity Systems*,
https://doi.org/10.1007/978-3-319-98734-7_9

research, the components examined and data extracted are largely restricted by the instrumentation used for measurement. Therefore the approaches used to accomplish gait biometric recognition relate directly to the instrumentation needed to extract gait data and fall into three categories: the machine vision approach, the wearable sensor approach, and the floor sensor approach. These three approaches together with the research that has been done in each respective area are examined in the following subsections.

1.1 The Machine Vision Approach

The machine vision approach to gait biometrics refers to the various techniques used to capture the visually observable aspects of gait. This approach has generated substantial interest from the intelligence and national security communities as it, along with facial recognition, is well suited for providing a degree of at-a-distance recognition beyond what can be accomplished with other biometric techniques. Consequently the machine vision approach to gait biometrics is the most frequently encountered gait biometric research topic, further benefiting from the availability of large public datasets such as the NIST gait database [41].

There are two primary research techniques that have been used to perform recognition via machine vision: model-free and model-based. The model-free technique, commonly called the silhouette-based technique, involves deriving a human silhouette by separating out a moving person from a static background in each frame of a video recording (Fig. 9.1). Using this silhouette-based technique, recognition can be accomplished by analyzing how the shape of the human silhouette changes over time in the sequence [62]. The silhouette-based approach exposes structural and transitional aspects of the gait such as limb length and the path taken by the feet during gait cycles. However, it presents challenges in that it may be deceived by clothing or carried objects; moreover, it may have trouble accounting for factors such as the distance or moving direction of the subject. Numerous studies have examined the potential for performing gait recognition or identification using the silhouette-based technique, and various approaches have been used to perform this

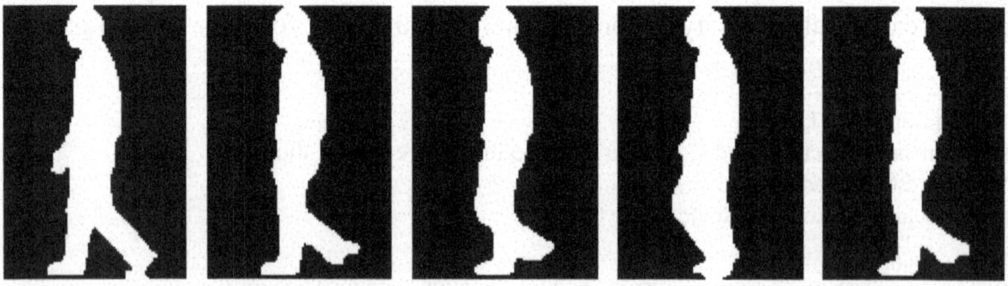

Fig. 9.1 Gait silhouette sequence. This figure presents an example of a gait silhouette sequence taken from the publically available OU-ISIR gait samples [27]

analysis including principal component analysis (PCA) for feature extraction [32], linear time normalization (LTN) to normalize gait cycles [8], and variations of the support vector machine (SVM) to perform subject recognition [34].

The other, less commonly used, technique to perform machine vision-based gait recognition is model-based [63]. This technique involves fitting a mathematical model to human movement using the features extracted from gait sequences. The model is typically composed of two distinct sub-models: one representing structural aspects of the gait subject such as the length between joints or shape of the torso and the other represents the motion kinematics or dynamics across these different parts of the body during a gait cycle (Fig. 9.2). Developing such models put constraints on each subject's range of walking motions and can be scaled to account for gait observed at varying distances or adjusted to account for alternate directions of motion. The model-based approach to gait biometric recognition also offers advantages in that it is considered to be more reliable than the silhouette approach for cases where a subject's gait may be occluded by clothing or other objects [11, 10]. In spite of these advantages, the model-based approach can be challenging to implement as it often requires a significant amount of computational power [63] to perform and is still subject to many of the physical or behavioral changes that would also affect the silhouette-based approach, such as carrying load, aging, or injury. In the research literature, several techniques have been suggested to generate gait models including the use of combined-primitive shapes to represent body structural [63] and elliptic Fourier descriptors [6] to model motion, while classification has typically been accomplished using a k-nearest neighbors (KNN) classifier [5].

One final machine vision-related approach that clearly does not fit into either the silhouette- or model-based categories of recognition involves gait recognition based on sound generated during the walking motion [26]. This audio-based approach offers many of the advantages of camera-based gait capture, as it can be

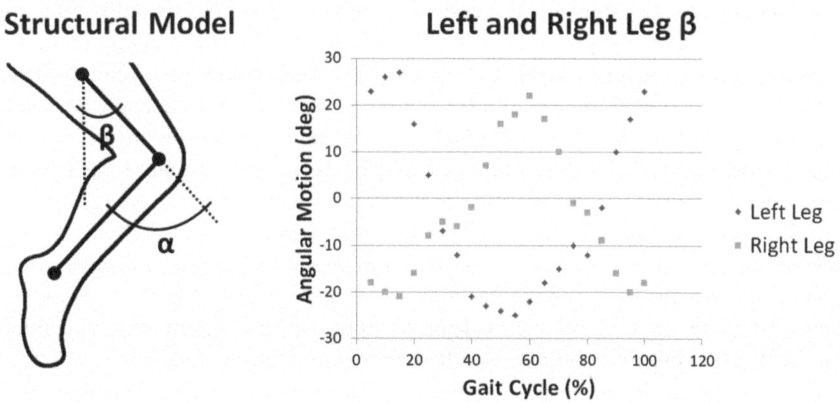

Fig. 9.2 Gait modeling. The diagram above demonstrates a simple example of the model-based gait biometric. In this case the structural model establishes two angles of rotation α and β. The dynamic model can then be established by fitting a mathematical formula to the angular motion with respect to the gait cycle (shown on the right side of the above diagram)

accomplished unobtrusively, at a distance, but has distinctly different disadvantages as it cannot be accomplished at too great a distance or in an environment with too much audible noise. This field of study has received little attention to date; nevertheless, promising results in [26] suggest it could be an exciting new realm for future gait recognition research.

1.2 The Wearable Sensor Approach

The wearable sensor approach to gait biometric recognition refers to a category of techniques that involve using sensors attached to the human body to perform recognition. These techniques derive from earlier research where gait-aware wearable sensors were primarily deployed in medical studies focusing on their usefulness for detecting pathological conditions [2]. Research into biometric applications using wearable sensors has been challenging due to the lack of publically available datasets [15] and significant implementation disparities across studies, a result of the wide variety of available sensors. In spite of the challenges, the wearable sensor approach offers distinct advantages over other gait recognition approaches, in particular the ability to perform continuous authentication, which would not always be possible with sensors fixed to a physical location.

Over the years a number of techniques have been proposed to accomplish gait recognition using wearable sensors. In one of the earliest studies, recognition was accomplished using an accelerometer sensor attached to a subject's lower leg, and data was uploaded to a computer after recording a set of gait cycles [17]. Another study used a shoe with an embedded pressure sensor, tilt angle sensor, gyroscope, bend sensor, and accelerometer [25]; in this study data was transmitted from the shoe to a nearby computer in real time. In yet another study, gait cycles were captured using a Nintendo Wii controller, with the goal of eventually obtaining the same data via mobile phone [16]. These early studies showed that gait biometric recognition could be performed with a relatively high degree of accuracy using wearable sensors, yet did not offer a full solution for more practical applications. In fact, one major weakness that had plagued the wearable sensor approach was the potential inconvenience or discomfort that may be caused by attaching sensors to the human body; another was the inability to transmit data at a distance. More recently, with the boom in smartphones and various other wearable devices, research has focused on unobtrusive gait analysis methods including shoe-based sensors [2], phone-based sensors [54], and increasingly smart watches [29] (Fig. 9.3). Moreover, many of these recent studies have made an effort to relay gait information via phone, potentially allowing for continuous authentication over large distances.

A variety of methods have been studied with respect to the implementation of biometric systems for wearable sensor-based gait analysis. These include PCA for feature extraction [25, 54] and SVM [54], KNN [14], neural network [24], and hidden Markov model (HMM) [40] techniques for classification, among others. Using these techniques, the wearable sensor approach generally produced better

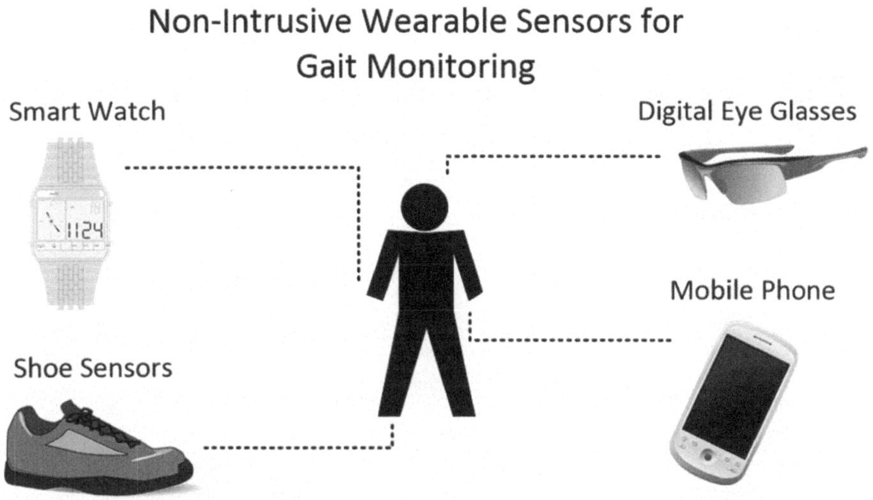

Fig. 9.3 Nonintrusive wearable sensors for gait monitoring. This figure demonstrates some of the nonintrusive devices that can be used to monitor gait via embedded wearable sensors

recognition results than the machine vision approach, albeit under conditions that may have been more favorable than those present within the larger machine vision datasets.

1.3 The Floor Sensor Approach

The floor sensor approach to gait biometrics refers to a gait analysis method by which people are recognized using signals they generate as they walk across sensor-monitored flooring. The floor sensor and wearable sensor approaches to gait biometric analysis share similar origins in that both arose from the study of biomechanical processes; much of the existing research around floor sensor technology for gait analysis focuses on its use for improving performance in athletics and discovering the effects of pathological conditions such as diabetes [52]. It was not until the late 1990s that researchers began examining this gait analysis method in the context of biometrics [1]. There are two primary types of data that can be captured using this approach: two-dimensional binary/gradient footstep frames, which represent the spatial position/force distribution of a footstep at recorded instants (Fig. 9.4), and single-dimensional force distribution plots, which represent the force exerted by the foot over a discrete time series (Fig. 9.6). When force gradient frames are captured, as was done in [59], it is also possible to acquire both frames and force distribution plots using the same capture technique.

Techniques used to capture gait via floor sensors have varied due to a wide array of available technologies as well as the fact that until recently there were no large

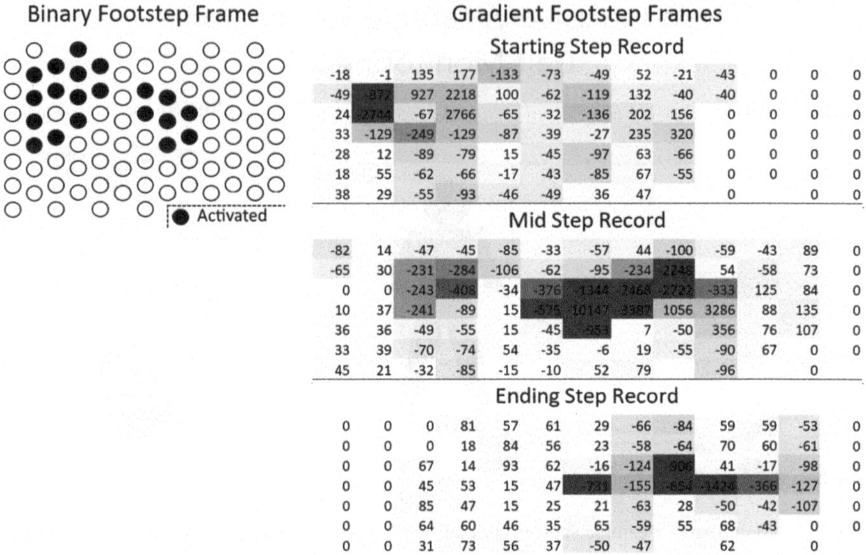

Fig. 9.4 Footstep frames. This figure demonstrates two types of data that can be acquired using a frame-based floor sensor capture method. The binary footstep frame shows it in its simplest form with sensors being activated above some given stepping force threshold, while the gradient frame demonstrates more detailed information about the spatial force distribution, in this case showing different parts of a data sample provided in [4]

publicly available datasets. One of the earliest floor sensor biometric studies used floor tiles supported in each corner by load cells to detect force signals as subjects walked over the flooring [1]. In another study, force signals were acquired using a sensor material called ElectroMechanical Film (EMFi) [56], while yet another approach was also taken in [35], this time with subjects walking over a Kistler force plate to generate step signals. An alternative capture technique implemented in [28, 55] used floor-mounted switch sensors to capture a binary frame representation of footsteps over time. To capture more feature-rich footstep gradient frames, additional approaches have been implemented including the use of floor-embedded piezoelectric sensors in [59]. The work in [59] stands out among other floor sensor-based gait recognition research in it appears to be the first to release its dataset to the public with full footstep frame sequences [4].

One of the most useful gait biometric characteristics that can be acquired with floor sensor approaches and a select group of shoe-based wearable sensor approaches is the ground reaction force (GRF) (Fig. 9.5). The GRF is a measure of the force exerted upward from the ground opposite to the foot through the course of a footstep. The GRF is composed of three perpendicular components: a vertical component representing the vertical acceleration of the body, an anterior-posterior component representing the horizontal friction between the feet and the ground, and a medial-lateral component representing friction forces perpendicular to the direction of motion. Distinctive gait behaviors can appear in these components, but it also

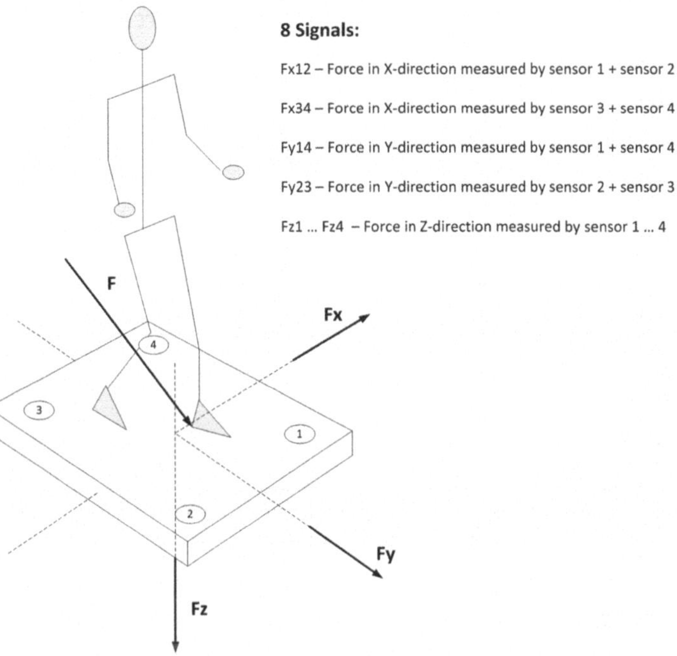

Fig. 9.5 Floor sensor signal capture. This figure, modeling the Kistler force plate [30], demonstrates how eight unique force signals can be generated using floor-mounted sensors. In this case the force "F" represents the stepping force vector, while the forces Fz (vertical), Fy (anterior-posterior), and Fx (medial-lateral) represent the three ground reaction force components that can be derived using combinations of the eight generated signals

reflects physical qualities like height and weight. Many gait recognition studies have preferred to use aggregated force of these three components; however, several have studied the three GRF components and discovered value in the individual component analysis [9, 35, 38].

Analysis of the floor sensor-based approach to gait biometrics has been accomplished using a number of different techniques. Feature extraction techniques used have included heuristic analysis (also referred to as the geometric approach) [38, 43, 49, 50, 56], PCA in both time [1, 37, 49, 50] and frequency spectra [9, 56], and a supervised feature ranking approach using fuzzy membership with the wavelet packet decomposition (WPD) of a footstep [38]. Other preprocessing techniques have included the application of normalization or scaling to footsteps, with the LTN and localized least-squares regression with dynamic time warping (LLSRDTW) obtaining slightly better recognition results in [35]. Classification techniques have also varied, with recognition or identification performed using SVM [38, 49, 50, 56], KNN [9, 37, 38, 43, 49, 56], multilayer perceptron (MLP) neural networks [56], linear discriminant analysis (LDA) [38], HMM [1], and least-squares probabilistic classifier (LSPC) [20, 35], among others.

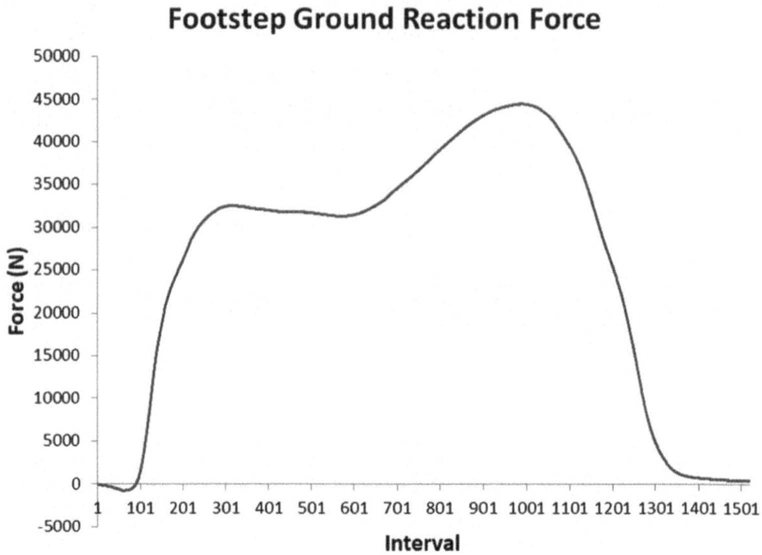

Fig. 9.6 Footstep ground reaction force. This figure demonstrates the aggregate GRF force signature generated by a footstep in [4]. The curve follows a common pattern with a local maximum occurring during the heal plant of a footstep and a second occurring as the foot pushes off the floor

1.4 Combined Approaches

A popular technique to improve biometric recognition performance involves fusing different biometric approaches together to create a more complete profile of a subject. The gait biometric, which is typically implemented in an unobtrusive manner, limits the set of additional biometrics that can be integrated without introducing obtrusiveness [7]. Moreover, as a relatively weak biometric with error rates often exceeding 1%, fusion of the gait biometric with stronger biometric technologies may actually lead to reduced performance over singular performance of the stronger biometrics [13]. Nevertheless, several studies have shown there are ways in which gait biometric approaches can be fused to improve recognition performance. In [9], a system was developed that fused data from a machine vision approach with data collected via a floor-plate sensor approach to recognize an individual walking through a monitored room and the analyzed system robustness metric increased by a factor of 5. In [59], a slightly different method was used to fuse features captured via video and those captured by floor sensors, this time achieving a 42.7% increase in performance as a result of the fusion.

In addition to fusing of gait biometric approaches, researchers have also found ways to combine gait with non-gait biometrics. One of the most convenient approaches involves the fusion of the gait and facial recognition [22], which often requires no additional sensors than would otherwise be needed to capture the gait video sequences. Other combinations of gait biometrics with other biometrics have also favored on the machine vision approach and include the fusion of gait and ear biometric, which may be better for low-quality samples where the face can be distorted [42]; a fusion of gait and speech, which in one study was proposed for human emotion recognition [31]; and the fusion of gait and body geometry [47].

2 Gait Biometric Datasets

The creation of gait biometric datasets involves a level of subject cooperation which tends to limit the number of uniquely identifiable subjects that can be enrolled; sample collection may take a considerable amount of time and require a large collection platform. Consequently, there are few publicly available gait biometric datasets (Table 9.1), and those that are available tend to favor the more easily captured machine vision data. One of the earliest publicly available datasets was the NIST-USF dataset [45] (2002), which introduced the *HumanId* challenge problem as a means to measure the results of machine vision-based gait recognition under increasingly challenging circumstances including changes to footwear and walking surface; this study achieved a maximum database enrolment of 122 subjects. A follow-up study provided a slightly larger dataset, the CASIA dataset, with 153 [12] (2006). The CASIA research group produced four different datasets with variations including records of subjects walking at different camera angles, with differing clothing and carrying condition, collection in the infrared spectrum, and with varying walking speeds. A later machine vision-based study with the goal of reducing the effects of clothing occlusion produced the TUM-IITKGP dataset [21] (2011) with several new variations of occlusion including subjects with their hands in their pockets and frames that included multiple moving subjects, to name a few. Yet, perhaps the most exciting work was done by the OU-ISIR group [27] (2012), who produced by far the largest publicly available gait biometric dataset with image

Table 9.1 Gait Biometric Datasets. This table provides a cross comparison of some of the largest publicly available datasets. Several of these research groups provide multiple datasets so the size metric in this table is simply a reflection of their largest stand-alone datasets

Research group	Name	Variant	Size
Phillips et al. [45]	NIST-USF	MV	122 subjects
Iwama et al. [27]	OU-ISIR	MV	4016 subjects
Tan et al. [12]	CASIA	MV	153 subjects
Fernández et al. [33]	AVA	MV	20 subjects
Hofmann et a. [21]	TUM-IITKGP	MV	35 subjects
Ngo et al. [39]	OU-ISIR (inertial sensor)	WS	744 subjects
Zhang et al. [64]	ZJU-GaitAcc	WS	153 subjects
Vera-Rodriguez et al. [58, 59]	ATVS	FS	127 subjects
Zheng et al. [65]	CASIA (footprint)	FS	88 subjects

The datasets here are organized as variants reflecting the collection approach: *MV* machine vision, *WS* wearable sensors, *FS* floor sensors recordings of 4012 at the time of writing. This project produced multiple smaller datasets including sets where subjects varied in clothing and walking speed.

In more recent years, a small number of research groups have started to release wearable and floor sensor-based datasets. The first large publicly available wearable sensor database was produced by the OU-ISIR group [39] (2014); in this case data was collected using an accelerometer, gyroscope, and a smartphone device placed around each subject's waist with samples collected for 744 unique subjects. Around the same time, another group produced the ZJU-GaitAcc database with 153 unique subjects whose movements were measured with accelerometers placed on the right wrist, left upper arm, right side of the pelvis, left thigh, and right ankle [64] (2015). Datasets based on the floor sensor approach have been made available by two different research groups. The CASIA group captured cumulative foot pressure images for 88 subjects [65] (2011), while a more extensive dataset was produced by the ATVS group, which included sequential footprint force measurements representing the instantaneous measurements at various points in a footstep for a total of 127 subjects [59] (2013).

3 Applications of Gait Biometrics

The implementation of gait biometrics beyond research has only recently become technically feasible, and consequently there have been few applications in industry to date. Moreover factors including privacy concerns and variability in gait as the result of aging and even emotional condition [53] present challenges for the application of gait biometrics in any real-world setting. Nevertheless, the gait biometric has some obvious advantages over other biometrics in that it can be collected unobtrusively, remotely, and potentially in a way that is less vulnerable to spoofing. This biometric can be used for security and forensic applications [36] but also has envisioned medical applications, such as gait rehabilitation for patients suffering from paralysis or sport injury, and consumer applications, such as gait-aware home automation systems [50]. With respect to security, access control and continuous monitoring are two active areas of interest.

Gait biometric recognition systems, like other biometric recognition systems, are configured according to modes of operation. The choice of operating mode, either authentication or identification, relates directly to the intended application of the biometric. Furthermore, if the system is configured in the authentication mode, it will fall under one of the three subcategories: static authentication, reauthentication, or continuous authentication. In the authentication mode, a subject will present the system with an identity and have it verified against a provided biometric sample. In identification mode only a sample will be provided, and the system will attempt to compare this against biometric profiles. With respect to gait biometric recognition, some capture approaches are more suitable than others to particular operating modes. A wearable sensor gait recognition approach could be a good choice for both reauthentication and continuous authentication as it would provide an easy means to tie an identity to data samples. And, while machine vision-based gait recognition may be suitable for continuous authentication, it may not be adequate for access control to a restricted area. In this case, static authentication would be required, and this form of authentication could better be satisfied using the floor sensor gait recognition approach to detect footsteps at an entrance. Likewise, the wearable sensor approach might make little sense in a system configured for identification mode as the sensor itself could convey the supposed identity of the subject; in this case the subject is less likely to be cooperative, and a machine vision or floor sensor approach may be best.

A number of real-world applications for gait biometrics have been envisioned (Fig. 9.7). In secure facilities or heavily monitored environments like airports, the gait biometric could be used for early identification and categorization of non-employees or travelers (for instance, using databases such as the no-fly list). It could also be used in combination with an additional factor; one example might involve combining it with a PIN to reinforce ATM security. Other gait biometric applications might include the use of GRF-based gait recognition for access control to and within buildings and homes. In this case sensors could be integrated in rugs or areas at the entrances and other locations; then based on the identity of the subject, access to specific rooms could be granted or denied. This mechanism could also be used to control access to specific devices or features in a home based on the identity of an individual. Alternatively, sensors could be embedded within shoes or clothing and report to a secured mobile device. In this case the mobile device might lock or send out an alert when it is found to be moving without the presence of its owner. Another use for such a device might be to transmit information to a receiver in a secure room allowing high-security facilities to continuously monitor access.

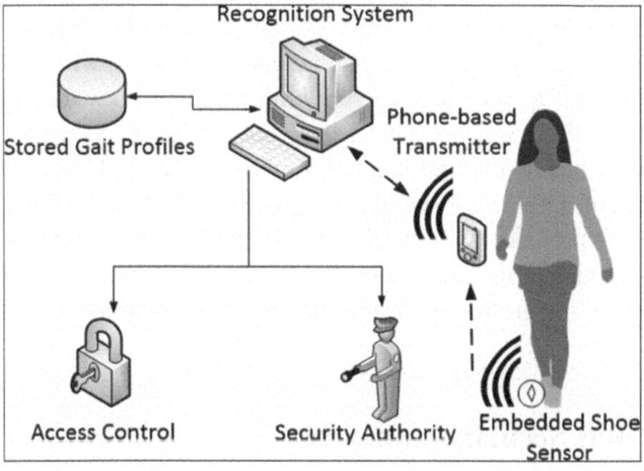

Fig. 9.7 Gait recognition application. This figure demonstrates how gait recognition might be used in a real-world setting. In this example the subject has an embedded shoe sensor which transmits gait information to a phone. It is then securely transmitted to a recognition system. The recognition system compares the provided gait profile against known matches and informs either the access control system or a security authority as to whether the subject should be authorized for access to particular location

In spite of there being many potential applications, the gait biometric has received only limited exposure in the commercial setting. In 2002, the Atlanta Business Chronicle [3] examined several potential commercial security applications for gait recognition using video and radar. Since then a number of advances have been made in the field, and a more recent publication in 2012 produced a case study [44] demonstrating a cost-time evaluation of video-based gait algorithms using Amazon Web Services for cloud computations, putting some financial figures behind the potential commercial applications of gait biometrics. The first research initiative to attract considerable commercial interest in gait-based recognition was part of a program sponsored by DARPA that established the *HumanID* Gait Challenge Problem [45]. This program set out to define a baseline by which future machine vision-based gait recognition systems could be tested. The challenge was composed of a baseline algorithm to provide a performance benchmark, a large dataset, and a set of 12 testable experiments with increasing degrees of difficulty, with special care taken to ensure the experiment reflected conditions that could be encountered in a real-world setting.

Much of the early attention surrounding commercial applications for the gait biometric focused on the machine vision approach, and substantial research has been dedicated to this area; however, the first actual commercial applications for the gait biometric have come as wearable and floor sensor approaches. These sensor-based industries have benefited from lower barriers to commercial entry including the fact that such devices are already widely being incorporated into daily living and captured data is more likely to be considered acceptable when compared with a video feed. As of the time of writing, several organizations have entered the commercial space for sensor-based gait biometrics, and their products appear to be gaining some traction. These organizations include Plantiga [46], which is developing a shoe-based gait tracking system; UnifyID [57], which is incorporating gait into a multifactor sensor-based system for personal device security; Boeing and HRL Laboratories [23], which are developing a system that continuously monitors the gait of users in contact with a specialized phone; and Zikto [66], which is developing a fitness band with built-in gait biometric tracking, to name a few. Lesser attention has been dedicated to floor sensor approaches, but ViTRAK Systems Inc. [61] has shown promise by incorporating the gait biometric as an application of its Stepscan floor sensor system. Looking forward, there is reason to believe these sensor-based gait recognition approaches will continue to outpace the machine vision approach commercially because of the rapid adoption of capable sensors into daily living and the possibilities for integration with other forms of biometrics.

4 Privacy and Security Concerns

The gait biometric shares a number of security and privacy concerns common to all biometrics as well as a few more specific to the gait itself. With respect to security, one of the greatest challenges posed specifically by gait is the fact that it has yet to achieve the reliability that would be required of a stand-alone biometric. Moreover, while the gait biometric has been considered inherently difficult to spoof, it has been shown that when an attacker knows they have a similar gait signature to an enrolled subject, spoofing becomes substantially easier [18]. Furthermore, the gait biometric like any other biometric is only as strong as the biometric system used to manage it. For instance, if an attacker were to compromise a database storing gait data, he or she could view, create, delete, or modify records and completely bypass the system. Because biometrics like gait tend to be relatively static over their intended period of use, the loss of recorded biometric data would represent a permanent compromise, unlike the loss of a password, which could simply be reset. In the same vein, the attacker could compromise a link in the system, which could potentially be more difficult to protect against with a biometric that is recorded at a distance like the gait. In this case the attacker may implement a replay attack by recording the gait signature of the subject and resending it to the system at a later point. To address such concerns, various technologies could be implemented including anti-tampering alarms, maintaining only nonreversible signatures rather than storing raw data, and encryption within the database as well as between data links. Additionally, reliability concerns might be alleviated by using the gait only as a secondary recognition method as part of a multimodal system [19].

With respect to privacy, the gait biometric carries several concerns. Biometrics in general must remain relatively static over their intended period of use to be effective, and the gait biometric is no exception, so any leak of biometric information could compromise a person's identity. The gait biometric is particularly vulnerable in this regard due to the unobtrusive nature in which it can be captured, which could allow an observer to generate biometric profiles without a person's consent or even awareness. Moreover, once a gait biometric profile has been established for an individual, that individual could be tracked using an extensive network of video cameras or floor sensors, violating the individual's right to anonymity. However, while this application of gait biometrics has attracted a great amount of attention from privacy advocates, it is still considered technically infeasible in the near future [7], and a greater potential privacy concern comes not as the result of using the gait for recognition purposes but rather than using it as a surveillance technique for tracking abnormal behaviors. For instance, if configured for surveillance, a gait monitoring system might be designed to identify persons concealing a carrying load, yet such a system could be prone to high false match rates, bringing undeserved scrutiny upon individuals suffering from injury, fatigue, or psychological conditions. Alternatively, a gait biometric analysis system could be given the dual-purpose of identifying medical conditions such as Parkinson's disease [51], violating an individual's medical privacy rights. A strong public response should be expected in the event that a gait biometric track was ever modified to take on any of the aforementioned uses and care should be taken to allay public privacy fears before deploying such a system in a real-world setting.

5 Evaluation of a Gait Biometric System

The use of the gait biometric for recognition or identification purposes requires the development of a biometric system (Fig. 9.8). This system establishes the processes that need to be performed to transform raw gait data into more distinguishable templates and later in using these templates for the generation of identity models. These two system components are often referred to as preprocessors and classifiers, respectively. The nature of a gait biometric system's preprocessors and classifier depends on the aspect of gait being analyzed. In [35], extensive research was done regarding finding suitable biometric system configurations for the footstep gait biometric. These approaches demonstrated promising results; however, the limited size and variation in the dataset (ten subjects wearing similar footwear) most likely produced better results than could be expected in a more realistic scenario. A more realistic example of the biometric recognition results that can be achieved is possible by taking the same experimental configurations and evaluating them over a far larger dataset such as the most recently released ATVS dataset [59, 60]. The ATVS dataset contains 127 unique subjects with a wide variety of footwear and carrying weights (purses, backpacks, etc.), which would be expected in a more realistic setting.

The ATVS dataset samples are composed of a left and right footstep captured as part of a single stride. Each step was recorded over a grid of 88 piezoelectric sensors, as was demonstrated in Fig. 9.4. The data was collected at a sampling frequency of 1.6 kHz with 2200 records collected per sample starting at the point the stride began. For our purposes we consider only the right-side footstep, to correspond the single-step analysis performed in [35]. In [35] the first biometric system step undertaken was the preprocessing known as sample extraction; in that case the force plate sensor data contained significant noise, and the start/end of a sample was not clearly defined, so extraction was largely based on the degree and duration of the vertical GRF force. A piezoelectric sensor-based acquisition method like that used to generate the ATVS is, under normal conditions, less susceptible to noise as it requires direct contact to produce a non-zero output. Consequently, the process of

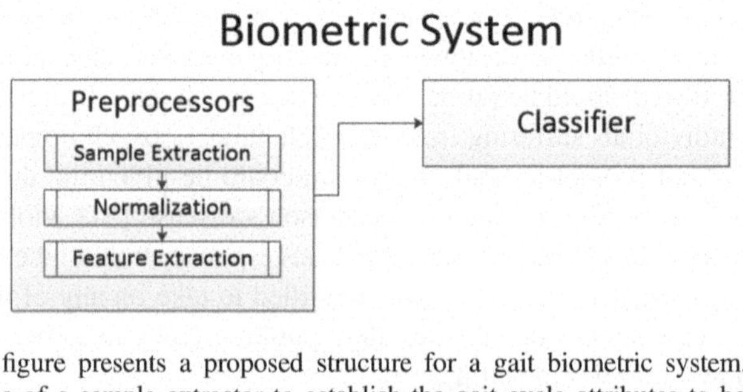

Fig. 9.8 This figure presents a proposed structure for a gait biometric system. The first step involves the use of a sample extractor to establish the gait cycle attributes to be analyzed. The second (optional) step involves normalizing the sample data, followed by a feature extraction step that reduces the sample dimensionality to a size that is easier to classify. The final step involves the use of a classifier to perform sample recognition

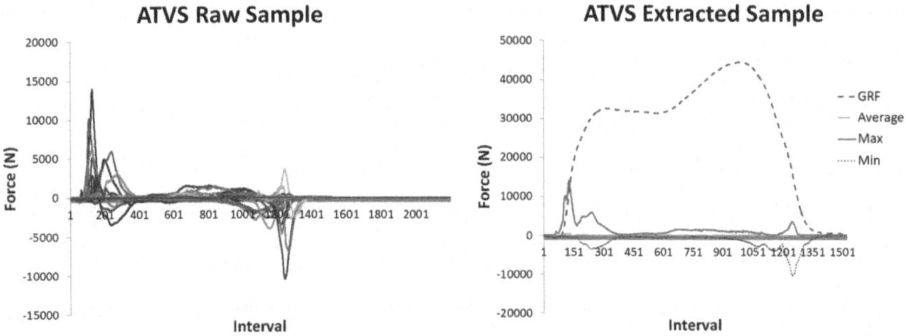

Fig. 9.9 In this figure we compare the raw data available in the ATVS dataset (left) with some of the unique temporal aspects that can be extracted from it (right). In this case the GRF, average signal value, minimum contour, and minimum contour have been extracted

sample acquisition for the ATVS is simpler; with the start of the sample being the first point, any sensor records a non-zero value and the end being the last point at which any sensor recorded a non-zero value. This provides a complete bounded footstep sample, yet the group of sensors triggered change with footsteps landing on different parts of the grid and individual sensor signals may not be particularly useful. Instead, as demonstrated in [59], several additional sample extraction techniques can be introduced to gain more useful sample information. Rather than looking at individual sensors, the ATVS researchers aggregated the data to produce the time series GRF, sensor average, minimum contour, maximum contour, and a time-invariant spatial representation (Fig. 9.9).

Taking the demonstrated ATVS time series data, further sample extraction techniques from [35] can be applied. Gait can vary in speed, and, as a result, footstep samples become misaligned from one another, presenting challenges to further preprocessing. In [35], all samples were resampled to a single length while retaining information about the total cumulative force applied in the sample's generation. Using the aforementioned approach, we are able to derive samples suitable for the further preprocessing discussed in [35]. Further preprocessing falls into two categories: normalization and feature extraction. The normalization step comes first and involves removing potentially poorly distinguishing variance between samples such as stepping speed variance or carrying condition. Feature extraction or dimensionality reduction involves reducing the effects of sample noise by reducing the sample down to only a small subset of its best discriminating attributes. To evaluate these biometric system configurations, we have taken some of the best performing configurations together with the experimental code library from [35] and applied them to the larger ATVS dataset. The results achieved by running these preprocessors with several different classifiers on the ATVS validation dataset are demonstrated in Table 9.2.

The results in Table 9.2 were acquired using the preprocessors and classifiers optimized for the [35] dataset; however, a few of the original configurations could not be translated directly between datasets. In [35], classifiers were trained with only five samples per subject, while in this evaluation of the ATVS dataset, classifiers

Table 9.2 This table demonstrates the ERR (%) results acquired by evaluating some of the best biometric system configurations from [35] over the ATVS validation dataset. To avoid bias the ATVS training dataset was used for classifier and preprocessor model generation. Note that the spectral feature extraction technique was found to produce poor results and was left out of the classifier and normalization evaluation

Evaluation of biometric system configurations				
Feature extractor	Normalizer	SVM ERR	MLP ERR	KNN ERR
Holistic		17.60	23.93	18.58
Spectral		31.87	–	–
Wavelet		18.85	23.59	20.66
Holistic	L-inf	20.95	27.93	21.36
Holistic	LTN	16.97	21.09	17.43
Holistic	LLSR	18.39	22.63	19.95
Holistic	LLSRDTW	16.97	22.06	17.99
Wavelet	L-inf	17.98	23.56	21.83
Wavelet	LTN	15.98	20.28	18.49
Wavelet	LLSR	18.64	24.40	20.68
Wavelet	LLSRDTW	17.05	20.63	18.58

were trained with 40 samples. This proved problematic for the k-nearest neighbor (KNN) classifier, where the composition of the *k*-nearest samples reflects the number of training samples. To better reflect the original KNN configuration, the value for K was chosen in proportion to the number of trained samples so the [35] best value for K of 4 became 32 in the ATVS dataset evaluation. The multilayer perceptron (MLP) classifier proved more challenging, with the large number of training samples leading to a high degree of overfitting and poor modeling when using the [35] configuration. Several changes were made to achieve results that better reflected the MLP's classification performance. Rather than using parameterized back-propagation in training, the MLP was found to perform much better when the non-parameterized resilient backpropagation [48] (Rprop) was used. To further protect against overfitting, the classifier was only trained to the point of a 1.8% training data error rate. Moreover, the [35] evaluation used a one-to-many classification strategy, but this strategy scaled poorly with more subjects and the additional imposter samples of the ATVS validation set. Instead the ATVS MLP results were acquired using a many-to-many classification strategy, whereby a different classifier model was generated for each subject; this provided far better performance.

Looking at the results in Table 9.2, it can clearly be seen that performance is worse than that found in [35]. This is expected given the larger ATVS dataset contained imposter samples and more variable sample acquisition conditions. Of interest are the areas where findings matched and diverged from the findings in [35]. With respect to feature extraction, the spectral feature extraction performance was similarly poor in comparison with the wavelet and holistic feature extractors. In contrast to the original findings, the wavelet feature extractor did not demonstrate a clear improvement in performance over the holistic feature extractor. In terms of classification, the SVM classifier clearly outperformed the KNN and MLP classifiers, again contrasting the findings in [35] where the MLP classifier was found to be

Table 9.3 This table demonstrates the ERR (%) results acquired by evaluating non-normalized, LTN, and LLSDTW-normalized ATVS training/validation data using an SVM classifier. Each biometric system configuration was evaluated using the holistic feature extraction method with a feature set containing the 120 best principal components

Holistic: 120 principal component evaluation		
Non-normalized ERR	LTN ERR	LLSRDTW ERR
15.48	15.51	14.93

the strongest of the three. Perhaps the most interesting results came from the application of normalization, where the two best stepping speed-based normalizers, linear time normalization (LTN) and localized least-squares regression with dynamic time warping (LLSRDTW), clearly outperformed non-normalized solutions. This correlates with the findings in [35], providing further evidence that a relationship useful for recognition exists between stepping speed and the shape of the GRF. However, because the best normalization conditions were subject to different optimization processes, a possible explanation for this could be another biasing factor, such as the optimal number of principal components (PCs) selected. Moreover, the results were far worse than the 14.1% achieved for the single footstep in [59], where the best 120 PCs were suggested for feature extraction and an optimized SVM was used. To analyze the time normalizer performance in the absence of potential feature extractor bias, the ATVS validation data was run for a second time using the holistic feature extractor, selecting the 120 best PCs per sample. The results shown in Table 9.3 demonstrated a considerable improvement in overall performance when using a greater number of PCs, albeit slightly worse than [59], perhaps due to differences in classifier optimization or the fact we only analyzed the right footstep. In this case LTN showed no performance improvement, but the use of LLSRDTW led to a 3.6% improvement in recognition results.

In [59] it was found that, in addition to temporal footstep signals such as the GRF, there is also value in the spatial distribution of forces through the course of a footstep and that fusing spatial features with temporal features can produce stronger recognition results. It was further shown in [59] that combining two sequential footsteps to produce a stride can produce features that substantially improve performance. In the findings of this section, we have shown it may be further improved with the application of a time-based normalizer such as the LLSRDTW. In the broader study of gait biometrics, these findings may also prove useful as other aspects of gait are also cyclic and subject to the speed at which the gait cycle was performed.

6 Conclusion

This chapter presented the challenges and benefits in using gait for biometric recognition. It was shown that gait biometrics broadly falls into three categories reflecting the approach used to capture the gait and that combining these approaches with each other or with additional biometrics such as facial recognition may improve recognition results. Additionally, this capture provided an overview of existing publicly available gait biometric datasets, demonstrating that data is now publicly available for each of the three gait recognition approaches. Furthermore, a recent assessment of current applications for gait biometrics was presented along with a discussion of potential privacy and security concerns that would accompany any real-world applications. Finally, a brief analysis was done on a floor sensor-based gait biometric, demonstrating results that may be expected for such an approach and providing further evidence to the proposition that a relationship useful to recognition exists between the speed of the gait and the shape of its temporal representation.

References

1. M.D. Addlesee, A. Jones, F. Livesey, F. Samaria, The ORL active floor [sensor system]. IEEE Pers. Comm. **4**(5), 35–41 (1997)
2. S.J.M. Bamberg, A.Y. Benbasat, D.M. Scarborough, D.E. Krebs, J.A. Paradiso, Gait analysis using a shoe integrated wireless sensor system. IEEE Trans. Inf. Technol. Biomed. **12**(4), 413–423 (2008)
3. T. Barry, Atlanta Business Chronicle. [Online]. (2002, April). http://www.bizjournals.com/atlanta/stories/2002/04/22/focus4.html?s=print
4. (2017, July) Biometric Recognition Group – ATVS. [Online]. http://atvs.ii.uam.es/sfootbd.html
5. I. Bouchrika, M.S. Nixon, Exploratory factor analysis of gait recognition. In *8th IEEE International Conference on Automatic Face and Gesture Recognition*, Amsterdam, France, 2008, pp. 1–6.
6. I. Bouchrika, M.S. Nixon, Model-based feature extraction for gait analysis and recognition. In *MIRAGE'07 Proceedings of the 3rd international conference on Computer vision/computer graphics collaboration techniques*, Rocquencourt, 2007, pp. 150–160.
7. N.V. Boulgouris, D. Hatzinakos, K.N. Plataniotis, Gait recognition: a challenging signal processing technology for biometric identification. IEEE Signal Process. Mag. **22**(6), 78–90 (2005)
8. N.V. Boulgouris, K.N. Plataniotis, H. Dimitrios, Gait recognition using linear time normalization. Pattern Recogn. **39**(5), 969–979 (2006)
9. P.C. Cattin, Biometric authentication system using human gait. Swiss Federal Institute of Technology, Zurich, PhD Thesis (2002)
10. D. Cunado, M.S. Nixon, J.N. Carter, Automatic extraction and description of human gait models for recognition purposes. Comput. Vis. Image Underst. **90**(1), 1–41 (April 2003)
11. D. Cunado, M.S. Nixon, J.N. Carter, Using gait as a biometric, via phase-weighted magnitude spectra. In *AVBPA '97 Proceedings of the First International Conference on Audio - and Video-Based Biometric Person Authentication*, London, 1997, pp. 95–102
12. D. Tan, K. Huang, S. Yu, T. Tan, Efficient night gait recognition based on template matching. In *The 18th International Conference on Pattern Recognition*, Hong Kong, 2006
13. J. Daugman, Biometric decision landscapes. University of Cambridge, Cambridge, Technical Report 1476-2986, 2000
14. M.O. Derawi, P. Bours, H. Kjetil, Improved cycle detection for accelerometer based gait authentication. In *Sixth International Conference on Intelligent Information Hiding and Multimedia Signal Processing*, Darmstadt, 2010, pp. 312–317
15. M.O. Derawi, D. Gafurov, P. Bours, Towards continuous authentication based on gait using wearable motion recording sensors, in *Continuous Authentication Using Biometrics: Data, Models, and Metrics*, ed. by I. Traoré, A.A.E. Ahmed (IGI Global, 2012, ch. 8), Hershey, PA, USA, pp. 170–192

16. M.N. Fitzgerald, Human Identification via Gait Recognition Using Accelerometer Gyro Force. [Online]. (2009). http://www.cs.yale.edu/homes/mfn3/public/mfn_gait_id.pdf
17. D. Gafurov, K. Helkala, S. Torkjel, Biometric gait authentication using accelerometer sensor. J. Comput. **1**(7), 51–58 (2006)
18. D. Gafurov, E. Snekkenes, Gait Recognition Using Wearable Motion Recording Sensors. EURASIP J. Adv. Signal Process. **2009**(1), 1–16 (2009)
19. G. Goudelis, A. Tefas, I. Pitas, Intelligent multimedia analysis for emerging biometrics, in *Intelligent Multimedia Analysis for Security Applications*, ed. by H. T. Sencar et al. (Springer, Berlin/Heidelberg, 2010, ch. 5), pp. 97–125
20. H. Hachiya, M. Sugiyama, U. Naonori, Importance-weighted least squares probablistic classifier for covariate shift adaption with application to human activity recognition. Neurocomputing **80**, 93–101 (2012)
21. M. Hofmann, S. Sural, G. Rigoll, Gait recognition in the presence of occlusion: a new dataset and baseline algorithms. In *19th International Conferences on Computer Graphics, Visualization and Computer Vision*, Plzen, 2011
22. E. Hossain, Investigating adaptive multi-modal approaches for person identity verification based on face and gait fusion. University of Canberra, PhD Dissertation, 2014
23. (2015, September) HRL Labratories, LLC. [Online]. https://www.hrl.com/news/2015/0930/
24. B. Huang, M Chen, P. Huang, Y. Xu, Gait modeling for human identification. In *IEEE International Conference on Robotics and Automation*, Roma, 2007, pp. 4833–4838
25. B. Huang, M. Chen, W. Ye, Y. Xu, Intelligent shoes for human identification. In *IEEE International Conference on Robotics and Biomimetrics*, Kunming, 2006, pp. 601–606
26. A. Itai, H. Yasukawa, Person identification using footstep based on wavelets. In *International Symposium on Intelligent Signal Processing and Communication Systems*, Totoori, 2006, pp. 383–386
27. H. Iwama, M. Okumura, Y. Makihara, Y. Yagi, The OU-ISIR gait database comprising the large population dataset and performance evaluation of gait recognition. IEEE Trans. Inf. Forensics Secur. **7**(5), 1511–1521 (2012)
28. Y. Jaeseok, User identification using gait patterns on UbiFloorII. Sensors **11**, 2611–2639 (2011)
29. A.H. Johnston, G.M. Weiss, Smartwatch-based biometric gait recognition. In *IEEE 7th International Conference on Biometrics Theory, Applications and Systems*, Arlington, 2015
30. (2017, July) Kistler force plate formulae. [Online]. http://isbweb.org/software/movanal/vaughan/kistler.pdf
31. A. Lim, H.G. Okuno, Using speech data to recognize emotion in human gait. IEEE/RSJ HBU Workshop **7559**, 52–64 (2012)
32. Z. Liu, S. Sarkar, Improved gait recognition by gait dynamics normalization. IEEE Trans. Pattern Anal. Mach. Intell. **28**(6), 863–876 (2006)
33. D. López-Fernández, F.J. Madrid-Cuevas, Á. Carmona-Poyato, J.M. Marín-Jiménez, R. Muñoz-Salinas, The AVA multi-view dataset for gait recognition, in *Activity Monitoring by Multiple Distributed Sensing*, ed. by P.L. Mazzeo, P. Spagnolo, T.B. Moeslund (Springer International Publishing, Stockohmn, 2014, ch. 3), pp. 26–39
34. J. Lu, E. Zhang, Gait recognition for human identification based on ICA and fuzzy SVM through multiple views fusion. Pattern Recogn. Lett. **28**(16), 2401–2411 (2007)
35. J.E. Mason, I. Traore, I. Woungang, *Machine Learning Techniques for Gait Biometric Recognition*, Springer International Publishing Switzerland (Springer, 2016)
36. M. Mihălcică, Gait analysis for criminal identification based on motion capture. In *The 4th International Conference "Computational Mechanics and Virtual Engineering"*, Brasov, 2011, pp. 19–20
37. A. Mostayed, S. Kim, M.M.G. Mazumder, S.J. Park, Foot step based person identification using histogram similarity and wavelet decomposition. In *International Conference on Information Security and Assurance*, Busan, 2008, pp. 307–311

38. S.P. Moustakidis, J.B. Theocharis, G. Giakas, Subject recognition based on ground reaction force measurements of gait signals. IEEE Trans. Syst. Man Cybern. B Cybern. **38**(6), 1476–1485 (2008)
39. T.T. Ngo, Y. Makihara, H. Nagahara, Y. Mukaigawa, Y. Yagi, The largest inertial sensor-based gait database and performance evaluation of gait-based personal authentication. Pattern Recogn. **47**(1), 222–231 (2014)
40. C. Nickel, C. Busch, S. Rangarajan, Using hidden Markov models for accelerometer-based biometric gait recognition. In *IEEE 7th International Colloquium on Signal Processing and its Applications*, Penang, 2011
41. M.S. Nixon, J.N. Carter, Automatic recognition by gait. Proc. IEEE **94**(11), 2013–2024 (2006)
42. M.S. Nixon, B. Imed, B. Arbab-Zavar, J.N. Carter, On use of biometrics in forensics: gait and ear. In *European Signal Processing Conference*, Aalborg, 2010, pp. 1655–1659
43. R.J. Orr, G.D. Abowd, The smart floor: a mechanism for natural user identification and tracking. In *CHI '00 Conference on Human Factors in Computer Systems*, The Hague, 2000, pp. 275–276
44. R. Panchumarthy, R. Subramanian, S. Sudeep, Biometric evaluation in the cloud: a case study with human ID gait challenge. In *IEEE/ACM 5th International Conference on Utility and Cloud Computing*, Chicago, 2012, pp. 219–222
45. J.P. Phillips, S. Sarkar, I. Robledo, P. Grother, K. Bowyer, Baseline results for the challenge problem of human ID using gait analysis. In *Fifth IEEE International Conference on Automatic Face and Gesture Recognition*, Washington, DC, 2002, pp. 137–142
46. Plantiga. [Online]. http://www.plantiga.com
47. J. Putz-Leszczynska, M. Granacki, Gait biometrics with Microsoft Kinect sensor. In *2014 International Carnahan Conference on Security Technology (ICCST)*, Rome, 2014, pp. 1–5
48. M. Riedmiller, H. Braun, Rprop – a fast adaptive learning algorithm. In *International Symposium on Computer and Information Science VII*, Antalya, 1992
49. R.V. Rodríguez, N.W.D. Evans, R.P. Lewis, B. Fauve, J.S.D. Mason, An experimental study on the feasibility of footsteps as a biometric. In *15th European Signal Processing Conference (EUSIPCO 2007)*, Poznan, 2007, pp. 748–752
50. R.V. Rodríguez, J.S.D. Mason, N.W.D. Evans, Footstep Recognition for a Smart Home Environment. Int. J. Smart Home **2**(2), 95–110 (2008)
51. A. Salarian et al., Gait Assessment in Parkinson's Disease: Toward an. IEEE Trans. Biomed. Eng. **51**(8), 1434–1443 (2004)
52. E.S. Sazonov, T. Bumpus, S. Zeigler, S. Marocco, Classification of plantar pressure and heel acceleration patterns using neural networks," in *IEEE International Joint Conference on Neural Networks (Vol. 5)*, Montreal, 2005, pp. 3007–3010
53. L. Sloman, M. Berridge, S. Homatidis, T. Duck, Gait patterns of depressed patients and normal subjects. Am. J. Psychiatry **139**(1), 94–97 (1982)
54. S. Spranger, D. Zazula, Gait identification using cumulants of accelerometer data. In *2nd WSEAS International Conference on Sensors and Signals and Visualization, Imaging and Simulation and Material Science*, Stevens Point, 2009, pp. 94–99
55. J. Suutala, K. Fujinami, J. Röning, Gaussian process person identifier based on simple floor sensors. In *Smart Sensing and Context Third European Conference, EuroSSC*, Zurich, 2008, pp. 58–68
56. J. Suutala, J. Röning, Methods for person identification on pressure-sensitive floor: Experiments with multiple classifiers and reject option. Inform. Fusion J. (Special Issue on Applications of Ensemble Methods) **9**(1), 21–40 (2008)
57. UnifyID. [Online]. https://unify.id
58. R. Vera-Rodriguez, J. Fierrez, J.S.D. Mason, A novel approach of gait recognition through fusion with footstep information. In *2013 International Conference on Biometrics (ICB)*, Madrid, 2013

59. R. Vera-Rodriguez, J.S.D. Mason, J. Fierrez, Comparative Analysis and Fusion of Spatiotemporal Information for Footstep Recognition. IEEE Trans. Pattern Anal. Mach. Intell. **35**(4), 823–834 (2013)
60. R. Vera-Rodriguez, J.S.D. Mason, J. Fierrez, J. Ortega-Garcia, Analysis of spatial domain information for footstep recognition. IET Comput. Vis. **5**(6), 380–388 (2011)
61. ViTRAK Systems Inc. [Online]. http://stepscan.com
62. W. Liang, T. Tan, H. Ning, W. Hu, Silhouette analysis-based gait recognition for human identification. IEEE Trans. Pattern Anal. Mach. Intell. **25**(12), 1505–1518 (2003)
63. C.-Y. Yam, M.S. Nixon, Model-based gait recognition, in *Encyclopedia of Biometrics*, ed. by S.Z. Li, A. Jain (Springer US, 2009), pp. 633–639
64. Y. Zhang et al., Accelerometer-based gait recognition by sparse representation of signature points with clusters. IEEE Trans. Cybern. **45**(9), 1864–1875 (2015)
65. S. Zheng, K. Huang, T. Tan, Evaluation framework on translation-invariant representation for cumulative foot pressure image. In *18th IEEE International Conference on Image Processing*, Brussels, 2011
66. Zikto. [Online]. http://zikto.com/1/w/

Chapter 10
Online Signature-Based Biometric Recognition

Sudeep Tanwar, Mohammad S. Obaidat, Sudhanshu Tyagi, and Neeraj Kumar

1 Introduction and Background

The signature is a unique and lifelong identification parameter to verify the identity of a person. Traditional banking sectors have used a grunter, person who has an account in their branch, while opening the new account by an unknown customer. A similar policy was also used, while sanctioning the loan to a person. In order to focus repayment of the loan amount, few banking systems used more than one grunter [3]. Signatures of the grunter were used as an official record to approve the opening of new banking account or sanction a loan. Banking sectors are also using the signatures very promptly for clearing the paper-based checks. Manual- or computer-based signature matching mechanisms have been used by the banks. In a manual system, the authorized person cross-examines the signatures of the account holder from database, while in the computer-based system, authentic software tools

S. Tanwar (✉)
Department of Computer Science and Engineering, Institute of Technology, Nirma University, Ahmedabad, Gujarat, India

M. S. Obaidat (✉)
ECE Department, Nazarbayev University, Astana, Kazakhstan

King Abdullah II School of Information Technology (KASIT), University of Jordan, Amman, Jordan

University of Science and Technology Beijing (USTB), Beijing, China

Fordham University, New York City, NY, USA

S. Tyagi
Department of Electronics & Communication Engineering, Thapar Institute of Engineering and Technology Deemed to be University, Patiala, Punjab, India

N. Kumar
Department of Computer Science and Engineering, Thapar Institute of Engineering and Technology Deemed to be University, Patiala, Punjab, India

M. S. Obaidat et al. (eds.), *Biometric-Based Physical and Cybersecurity Systems*,
https://doi.org/10.1007/978-3-319-98734-7_10

are used to compare the two signatures. In both the cases, checks will only be cleared after a match is found with the signature of the account holder. Such processes have been used to identify fraud checks, blocking the payments. Till now we are discussing the issues on handwriting-based signatures.

Sometimes a situation may arise by which non-repetitive nature of variation of the signatures will be observed. A couple of factors under this type of situation are an aging effect, illness, non-appropriate geographic location, and emotional state of the person, which create the problem. All these coupled together cause large intrapersonal variation. Hence, a robust system is required, which should consider the abovementioned factors and must be able to detect different types of falsifications. The system should neither be too sensitive nor too crude. It should have an acceptable trade-off between a low false acceptance rate (FAR) and a low false rejection rate (FRR). These two components are used as the quality performance measures. FRR is defined as the ratio of the number of rejected authentic test signatures to the total number of submitted authentic test signatures. On the other hand, FAR is defined as the ratio of the number of falsifications accepted to the total number of falsifications submitted. A change in decision threshold to decrease the FRR will increase FAR and vice versa.

The system which is used to verify the signatures shown in Fig. 10.1 has mainly two mechanisms: one is feature selection and another is decision-making techniques. Depending on the complete signature or partial signature, features are divided into two categories: global or local. Global features have included the testing of complete signature like what is the speed of signing, what is the size of the outer box in which signature has signed, and how much time is consumed in the signing. Global features have fixed number of measurements on a complete signature that supports signature matching on initial phase while performing the comparison. On the other side, local features are based on the individual part of the element of signature, for example, when moving from one letter to another, what is the trajectory of the signature, what is the effective change in the curvature between two successive letters, and the effective distance between two letters. Ultimately local feature works on the different vectors of dissimilar sizes. This feature performed the second level of verification; if any doubt has been identified in the global feature, then in-depth investigation can be performed by local feature. The decision-making method cares for both features:

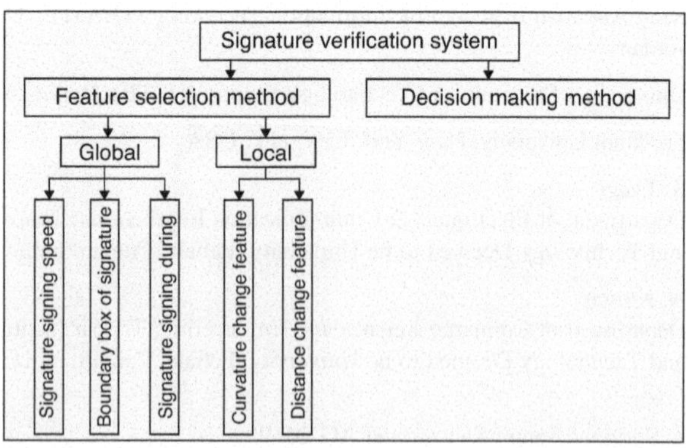

Fig. 10.1 Generalized signature verification system

global and local simultaneously. There is a chance where signatures of the same person may have dissimilar signing intervals. Possible factors are different signing speed, different locations, and the different material used for the signature, among others.

1.1 Types of Signature Verification Systems

Signature recognition and verification system (SRVS) is a growing field as it is very easy and helpful to identify a person. The signature owner can be easily identified by SRVS. Usually, signature verification systems are designed to work for both static and dynamic features. In general, there are two types of signature verification systems: online signature verification system and offline signature verification system. The next subsections will cover the two systems in detail.

1.1.1 Online Signature Verification System

Dynamic properties like the spatial and temporal characteristics of a signature can be easily obtained using online signature verification system. In addition to the dynamic properties, the following features can also be extracted: static characteristics, the speed of writing a signature, pressure points during the signature, and rushing. This information has been captured during the acquisition process. This method provides high accuracy due to an availability of dynamic characteristics, but at the same time, there is a cost increment due to complex hardware. Electronic gadgets that have the properties of pressure sensing are required to scan a signature. In online signature verification process, the user has to use the stylus (writing utensil) for signing. Its major concern is that it must be suitable for use on personal computer or handheld devices. Expected key features of the system include high accuracy, low computation, and moderate complexity; also it requires significant enrollments. The training set of an authentic signature is used to develop a model for validating the test signatures. Selection of an appropriate model is also a challenging task in designing an online signature verification system. Hidden Markov model (HMM) is one of the popular models, which is used in online signature verification system. For modeling purpose signature is formed as an array of vectors that covers every point of signature's trajectory. Hence, the selection of a perfect set of vectors in HMM can design an efficient online signature verification system. In signature verification systems, the dynamic time warping (DTW) or a distance between the model and the signature at hand is to be measured. Another interesting concept is found in [4]; the authors have used a frequency domain system for online signature verification. They have considered a fixed-length vector of 1024 complex numbers. These vectors have encoded (x, y) coordinates of every point on the signature trajectory. Then they have applied FFT with 15 Fourier descriptors, in order to select the highest amplitude feature. Fourier transform was used to replace the complicated preprocessing, normalization, and matching stages. Two experiments were performed to investigate the system.

Experiment 1: Data set (small) used in this experiment was 8 authentic signatures of a single user and 152 falsifications provided by 19 false users. Authors in [4] have achieved 2.5% error rate.
Experiment 2: In this experiment, the system was tested on a large database; 1500 signatures have been gathered from 94 users. Verification system was achieved with 10% error rate.

1.1.2 Offline Signature Verification System

This category is also known as static signature verification system [14]. In this category, simple paper- and pen-based signature is drawn and electronically captured by scanner or digital camera. Signature is stored to the biometric database under safe custody and can be used to identify the signature by biometric technology-based recognition system. The captured image indicates the unique handwriting composition of an individual. Hence, the objective of the system is to identify the forgeries, if there is any. These forgeries are related to intra- and interpersonal changeability. The simplest form of the system is to ignore the interpersonal changeability and spot as original, but, on the other hand, the system should be able to detect intrapersonal changeability and spot as forgeries. Different application areas of this technology are to validate the financial-related documents like checks, debit cards, credit cards, contracts, historical documents, etc. Several features of offline signature verification are given in [13, 15], where approximately 37 features have covered; some of them are like edge and curve. Signatures are very popular socially accepted identification technique and are being used in banking sectors, debit card and credit card transactions, smart parking system, and smart transportation system, among others. This method is very effective, especially in situations where huge database is maintained and authenticity is required. In such situations, biometric recognition-based authenticity can treat the intelligence investigations and is legally valid too. Such type of indexing is effective and can be used in the searching of the data [16]. But at the same time, offline signature verification system is very difficult to design because dynamic characteristics of the signature are not available during the verification process. Even though digital technology is growing very rapidly, a lot of departments, organizations, and banking sectors are still using paper-based working culture. The validity of appropriate person is essential in these situations and can only be verified by the offline signature verification system. Another difficulty with the offline signature recognition system is information unavailability related to the path or trajectory of the pen [17]. Further, training data are very limited as less number of signatures per subject is available. Shape matching [18] is the technique of matching, which determines the key points to avoid the difficulty with the detection and parameterization of curves. Offline signature recognition has always been the problem of shape matching. Lastly, the meaning of signature recognition technique is to recognize the novelist of a given sample, whereas the signature verification technique is to either confirm or reject the sample. Hence, in order to achieve the accurate result from offline verification system, specific design with careful handling is required due to the availability of limited features. The beauty of this system is that individual is not required personally during the verification process. In general, the key components of signature verification systems are data acquisition, preprocessing, feature extraction, comparison process, performance evaluation, and decision-making.

1.1.2.1 Features Used in Offline Verification System

The offline verification system is concerned of a signature, which has been signed by a pen. The system as shown in Fig. 10.2 contains three stages. The first stage is known as a preprocessing stage that includes some filters and preamplifiers to improve and enhance the operating quality of signature's image. The aim of this stage is to select the best quality of an image of the signature and forward to the next stage. The intermediate stage is feature extraction stage whose principle concept is based on the art of the signature of an image. Feature extraction stage consists of three different features: global feature, texture feature, and grid information feature [5]. These three features need to be calculated for the image of every signature and applied to the neural network (NN) stage, which is supposed to be the last stage. The

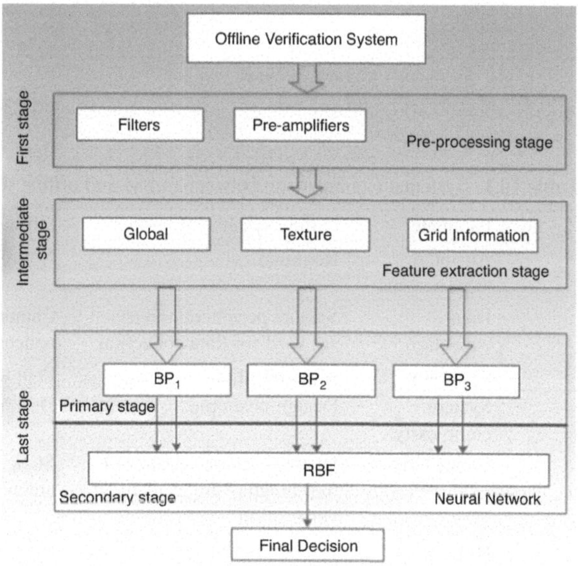

Fig. 10.2 Three-stage offline signature verification system

neural network has two-stage classifiers: primary stage that has three back propagation (BP) NN layers and a secondary stage that has two radial basis function (RBF) NN layers. Each BP paradigm has two outcomes that are applied to the second-stage classifier as an input. RBF is responsible for collecting the result from BPs and taking a final decision [6].

1.2 Comparison Between Offline and Online Signature Verification System

A systematic comparison between two signature verification systems is given in Table 10.1. Generally, for offline signature verification system, a figure of user's signature [5] is required without the knowledge of any attributes. On the other side, for online signature verification system, an array of (x, y) coordinates of user's signature is required.

The additional requirement in this system is some attributes like time, pressure, etc. Another difference between the two systems is how to get the data. In the online SRVS, data is obtained using special peripheral devices, whereas scanner or cameras are used to capture the SRVS images of the signature for offline verification systems [7]. Therefore, one can analyze that online signature verification system is better and more accurate as compared to the offline system [8]. Complex devices for computing, with high sensitivity and complicated authentication password requirement, forced to develop some simplified terminology on these devices [9, 10].

Table 10.1 Systematic comparison between online and offline signature verification system

Sr. no.	Parameter	Online system	Offline system
1	Attribute requirements	Required	Not required
2	Data collection	Special peripheral devices are used for data collection	Camera and scanners are used for data collection
3	Accuracy	More accurate	Poor as compared to the online system
4	System complexity	Design is simple	Design is very much complicated
5	Characteristics	Dynamic characteristics of the signature are available for verification	Static characteristics of the signature are available for verification
6	Human presence	Human presence is required	Human presence is not required, as verification is performed on the captured image of the signature

2 Basics of Biometric Recognition in the Context of Security

Biometrics is the automated identification of persons based on the respective behavioral and natural personality. Biometrics is a powerful tool that established an assurance that a person is dealing with a group of persons who may be already known or not known. The probability of similar fingerprints of two persons is very low. Figure 10.3 Illustrates the fundamental operations of a matching process. Biometric systems are gradually used to identify the persons and have control access to the fixed physical spaces, information, and services and to the security of international borders. The biometric technology was aimed to plummeting fraud, enhancing the safety of persons, and very importantly protecting the national security [11]. However, there are several challenges too about the effective surveillance mechanisms of biometric systems. In addition to the above, there are issues that need to be addressed: how to use, how to manage, what are the social implications, how to maintain the privacy, and how to form the legal policies are some of the quarries that need to be answered before implementation of such systems. The basic operation of a biometric system is shown in Fig. 10.3. Biometric system first captures and stores the reference samples and maintains their security too. Now, verification process involves the capturing of new biometric samples and their comparison with stored samples for the matching. The first segment of Fig. 10.3 is "capture," where electronic sensor or scanner collects biometric information of

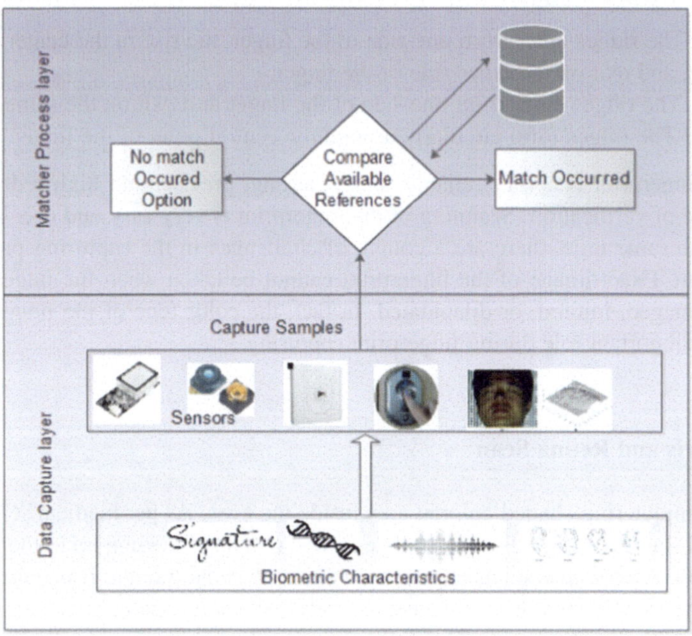

Fig. 10.3 Operation in general biometric system

individuals that need to be recognized. The second segment is "reference database," where previously enrolled information has been stored. The third segment is "matcher." This unit is very important as final identification decision can only be taken by this segment. It compares the current data with previously stored data to finalize the decision. Now the last segment of the block diagram is "action." Here, system recognition decision is revealed, and finally, actions are made based on that decision.

2.1 *Modalities*

Biometric technology provides higher amount of security and expediency than preexisting technologies of personal recognition. Nowadays, in most of applications, biometric technologies have been used, attracting customers. At the same time, public is worried about the security of biometric systems. In the next subsection, different components of biometrics have been highlighted [12].

2.1.1 Fingerprints

The fingerprint is a blueprint of ridges and valleys on the surface of fingertips of an individual. These blueprints of ridges are of three types, i.e. arch, loop, and cycle.

- Arch: The ridges enter from one side of the finger, the rise in the center forming an arc, and exit on the other side of the finger.
- Loop: The ridges enter from one side of the finger and exit on the same side.
- Cycle: The ridges form circularly around the central point of the finger.

The fingerprint is a very reliable, stable, strong, precise, and highly distinctive technique of verification. Scanning of the fingerprint is very easy and user-friendly. But at the same time, there are a couple of challenges in the capturing process of fingerprint. Exact image of the fingerprint cannot be taken when the fingertips are dirty, damaged, injured, or dilapidated. In fact, the color tone of the fingertip also plays an important role during fingerprint capturing.

2.1.2 Iris and Retina Scan

Iris is a unique ring-shaped colored area inside the eyes. As per medical science, its texture is stabilized up to the age of 2 years and remains the same up to the death of the person or until an eye has experienced some physical trauma. It is unbelievable but always true that two iris structures are different; even identical twins never have two identical iris structures. The following are the advantages of the iris-based verification system:

- Absolutely nonintrusive data collection.
- Data capturing can be done in presence of contact lenses too.
- Photo camera can be used to capture the information.
- High speed of recognition and accuracy.
- Fake irises can be easily detected.

2.1.3 Facial Recognition

Face recognition is a nonintrusive identification method, and facial images are very popular biometric characteristic to identify a person. Key approaches to face recognition are either the coordinates or shape and size of some of the facial attributes, like eyes, eyebrows, nose, lips, and chin. A facial recognition system compares the current facial image with the stored images and identifies the face if it is available in the database. Depending upon the type of accuracy, either single or series of images can be taken with a high-resolution camera. The facial image from a digital camera or from a video source is matched with the existing database.

2.1.4 Auditory Verification

The voice is also a unique feature of every individual. However, some of the persons are very expert to imitate the voice of other persons; due to this auditory verification system is less popular. Pitch of voice, time, magnitude of voice, and intensity of voice are the parameters to be used to identify the person from a data set. The acoustic pattern reflects the behavioral patterns of the individuals where a vocal signal of the person is compared with previously stored vocal signals. In some of the advanced systems, lip movement is also captured with a video camera while the person is speaking. This type of systems is a costly affair but more accurate compared to the traditional auditory verification system [1].

2.2 Comparison of Several Biometric Technologies

Several biometric components have covered in an earlier section. Table 10.2 summarizes the systematic comparison of different biometric techniques. Depending upon the application, accuracy, level of security, and requirement, either one or more components can be used for investigation purpose [1].

Table 10.2 Comparison of various biometric technologies

Biometric technology	Distinctive	Performance	Collectability	Performance	Acceptability	Uniqueness
Fingerprint recognition	High	High	Medium	High	Medium	High
Face recognition	Low	Medium	High	Low	High	Low
Signature	Medium	Medium	High	Medium	Medium	Medium
Iris and retina	High	High	Medium	High	Low	High
Audio verification	Low	Low	Medium	Low	High	Low

3 Survey of Online: Signature-Based Authentication Methods

In the revolutionary arena of information technology, the biometric technique is used to measure and analyze different characteristics of a human body for security [13]. There are a number of ways by which persons can recognize each other by different characteristics for ages. Biometric technology-based automated signature verification system is an emerging research field with several applications. The next section includes a detailed discussion on dynamic signature verification system.

3.1 Dynamic Signature Recognition

In this category, signature has been written directly to the electronic scanner, handheld device, tablet, or laptop in real time [19], which means digitations are not required here. Dynamic recognition technique is also known as online recognition technique. A transducer is required in the device to capture the signature in its true form. The beauty of this technique is involvement of time information like writing speed, pressure created by pen, and movement of pen on the device [16]. Some of the popular marketing sectors [20], where online recognition techniques have been used, are explained below:

3.1.1 Onward and Upward

According to Wayne Chodosh, Director, secure signature systems, behavioral biometrics are very helpful for the sector to move forward. Behavioral theory as per Wayne Chodosh is "Behavior is something that cannot be carried forward from one person to the other, in another way behavior can't be replicated by anybody even in presence of its demonstration." "In order to maintain the green environment paperless transactions are highly demanding in banking, insurance sectors and other applications too," highlighted by Jörg-M Lenz, manager, marketing and public relations, SOFTPRO [20]. In applications where signatures are required for confirming the intention of an individual, the dynamic signature recognition system is better for paperless demanding. In general, it is not possible to replace all paper-related documentation directly to the biometric system. However, it is a step-by-step process to replace the habit of using paper with electronic means. "In recent years, number of software vendors in the biometric sector have come up, they are increasing the number of contract announcements, publishing online case studies and in press releases – some of them are having a large number of users," says Lenz. The outcome in the nutshell is "be ready for the growth as the market is available for you."

3.1.2 Slow and Steady

Even though the technology is very popular, some challenges are still there. Signatures are playing a major role in the banking sector, and acceptance speed is very slow because several leaders are still not satisfied with the performance of software and hardware. Even they are having some legal issues too in the utilization of this technology. The financial crisis is also one of the major issues for the slow acceptance of online signature reorganization system. However, some of the developing countries like India are using this application. Aadhaar is an identity-based authentic card provided by the government of India to their residents for several applications like an identity card and a unique ID to avail the subsidy and other government policies.

3.1.3 SWOT Analysis

Dynamic signature technology helps an organization by securing electronic documents with minimum paper utilization and related cost reduction. Doing this speed of workflow increases and consequently achieves high degree automation. User acceptance can be increased by knowing the value of authenticity of their signature. The authenticated signature can be e-mailed to the remote location and works as remote electronic verified signatures. On the other side, several weaknesses are also there like the confusion in the difference between digital signatures and electronic signatures. Most of the time, users are confused and could not decide which technology to choose like biometric- or PKI-based electronic signatures. The simplest answer to this situation is just check the level of security because few persons want to achieve the security without thinking of investment.

There are several opportunities in both directions, horizontal and vertical markets. Signature capturing and verification process are handled by the banking sector, usually known as a vertical segment of the market. On the other hand, applications on the horizontal scale are adopted by several business owners and government institutions. Opportunities are good enough, but at the same time, organizations involved in this market and planning to work for a long time are worried about the threats. Several companies are capturing a signature image and storing the same to some server. The problem will be very severe, if the security is on the stake because stored image will not allow additional verification, if its authenticity is suspected.

3.2 *Function-Based Method*

Function-based methods can be classified into two categories, local and regional methods. In local methods, matching of time functions of different signatures is performed based on elastic distance measures such as dynamic time warping. In regional methods, the time functions are changed to a system of vectors describing regional properties. In most of these cases, the HMMs modeled stroke-based sequences. In the next section, we describe the application of HMMs to time sequences directly based on the dynamic functions.

3.3 HMM-Based Signature Verification

For a long time, handwritten signature has been accepted for recognition in the financial transaction and contract document, as well as being used to authenticate the materials and documents. When we use computer to collect signature by a digitizing device (electronic pen), we obtain information about the shape of the signature, and we also obtain dynamic information of the signature [21, 22]. This dynamic information generates "online" signature. This concept shows a string of sample points shipping information during the process of signing up. In other words, each dynamic information is a function according to time of signing process. Thus, the signing process generates a set of the data function over time. The online signature helps facilitate for the authentic signature because the dynamic information is more difficult to forge than the image of the signature. Hence, if anyone wants to forge signatures, he needs more efforts. However, this problem is still a challenging issue in biometrics because of the large intra-class variation and, when considering forgeries, small interclass variation. Process of online signature-based data classification is shown in Fig. 10.4.

There are many different approaches in data classification of signature. The current methods can be divided into two classes:

1. Feature-based approaches: In this approach, a signature is represented by a vector consisting of a set of global features, which are extracted from the trajectory of the signature.
2. Function-based approaches: In this approach, a signature is represented by the functions of the time, including the local properties of the signature (e.g., position trajectories (x, y), velocities, accelerations, pressures, and more).

The fusion of local and global signature verification by online mode is also possible. Global information is gathered with feature-oriented representation and documented by using Parzen Windows Classifiers (PWCs). Local information is collected on real-time basis considering different dynamic properties and documented by using hidden Markov models (HMMs) [38]. Several experiments were performed considering 16,500 different signatures from 330 users. Verification performance on random and skilled falsification was given for specific user and global decision thresholds. For small training set and user-independent thresholds, performance of global expert was found better in comparison to local analytics. A couple of theory-based approaches for combining classifiers with biometric verification are described in [39, 40]. A systematic review of recent works is given in [41, 42]. The weighted mean may also be an interesting way to combine similarity

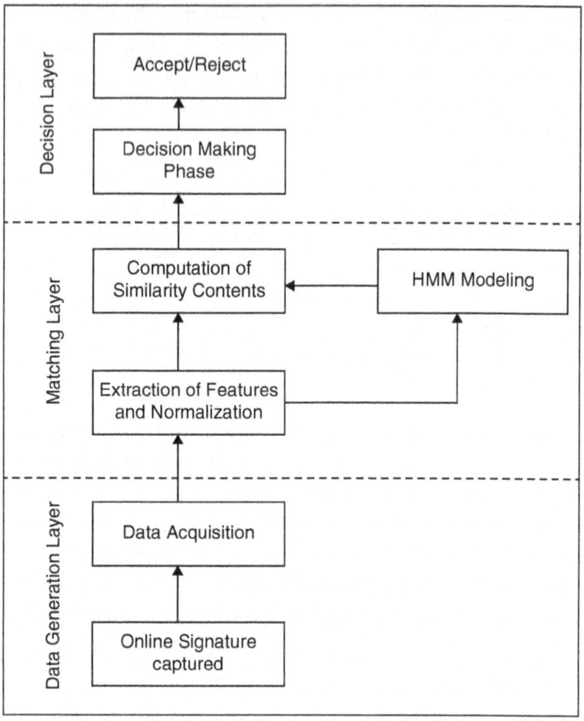

Fig. 10.4 Process of online signature-based data classification

scores given by several experts. There are several ways of doing fusion like min, max, mean, median, etc. Authors in [40] worked on fusion strategies, and they have compared max and sum functions. Similarity scores of experts of global and local environments have been collected and formulated to zero mean and unit standard deviation before fusion. This practice eases the working of the system as the model is formulated using Gaussian distribution.

3.4 Classification of Cancelable Biometrics

Function with some parameter is used to generate protected biometric templates. These parameters can be used as the key to the function [23]. The basic concept of cancelable biometric template-based on non-invertible transformations is shown in Fig. 10.5.

3.4.1 Non-invertible Geometric Transforms

The oldest method to generate cancelable biometric templates was based on the non-invertible geometric transformations. Here, the original biometric template was

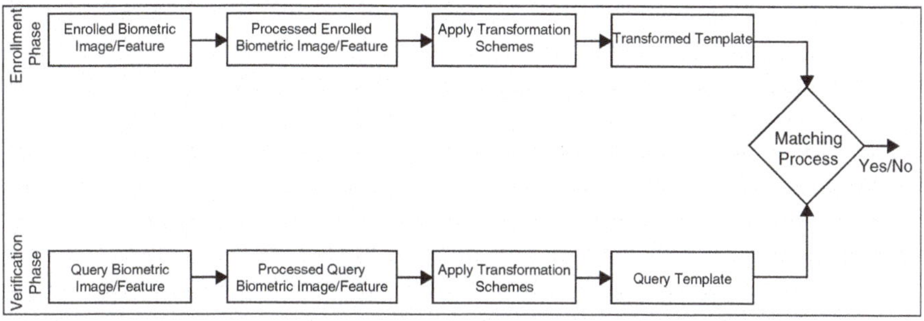

Fig. 10.5 Basic concept of cancelable biometric

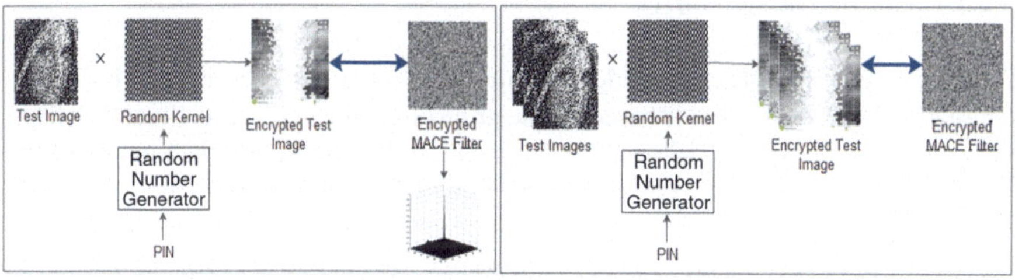

Fig. 10.6 Cancelable biometric correlation filter-based approaches: (**a**) enrollment phase for encrypted filters and (**b**) authentication phase using encrypted MACE filters

morphed by using either signal domain or feature domain transformations [24–26]. The Cartesian transformation, the polar transformation, and the functional transformation techniques were proposed for fingerprint biometric in [25, 26]. Registration is an essential phase of this method.

3.4.2 Cancelable Biometric Filters

A random convolution method for generating cancelable biometric templates was proposed in [27]. In this method biometric scheme, templates were encrypted using random user-specific convolution kernels. The training images are convolved with a random convolution kernel through random number generation. PIN is used as seed to generate the random convolution kernel. The encrypted test images are then used to generate a minimum average correlation energy (MACE) biometric filter. The output of this encrypted filter is stored and used for authentication. Figure 10.6a shows the enrollment phase for encrypted filter.

In the authentication phase, the user authentication is through the PIN and the encrypted filter, which are used to generate the convolution kernel in the enrollment phase. Now the random convolution kernel is convolved with user presenting the test face images. Then these convolved test images are cross-correlated with the encrypted MACE filter, and the resulting correlation outputs are used to authenticate the user. Figure 10.6b shows authentication phase using encrypted MACE filters. The resulting correlation output does not change while convolving the training images with any random convolution kernel prior to building the MACE filters used for biometric recognition [27]. In this way, the recognition accuracy is maintained throughout the process. Furthermore, by changing the convolutional kernel, different cancelable biometric templates can be generated from the same. There are more correlation-based cancelable biometric methods [28, 29] that exist in the literature, which include correlation-invariant random filtering (CIRF) and have almost the same accuracy as the conventional fingerprint verification based on the chip matching algorithm.

3.4.3 Knowledge Signatures

Authors in [30] used the knowledge signature for voice-based cancelable biometric templates. The key concept is based on the group signature, where any member from the group signs messages on behalf of the group. It does not reveal the identity of the member. Here, voiceprint considered as knowledge of the user and the user's voiceprint is further transmitted to a signature of knowledge template. Real signatures can only be generated by factorizing a large integer and the original feature. In this way, an individual's privacy can be protected. For more details on knowledge signatures and their uses in generating cancelable biometric templates for voiceprints, refer to [30, 31].

3.4.4 Salting Methods

The easiest way of generating cancelable biometric templates is by simply mixing in a totally artificial pattern called salting method. The mixing patterns can be one of the following: random noise, random pattern, or synthetic pattern. The authors in [32] proposed two salting methods for iris recognition, namely, GRAY-SALT and BIN-SALT. These methods used random patterns for feature gain and iris code, respectively. In these methods, it is difficult to decide how much comparative strength of the noise patterns need to be added. With the addition of strong patterns, it will further reduce the discriminative capacity of the original iris patterns and finally lead to lower recognition results. With the addition of weaker patterns, it can be easy to extract valuable information of the iris pattern from the transformed pattern.

3.4.5 Hybrid Methods

There are also several biometric template protection approaches that exist in the literature [33, 34], which were used both in cryptosystems and cancelable biometrics. The authors of [33] proposed such methods for face biometrics. In this

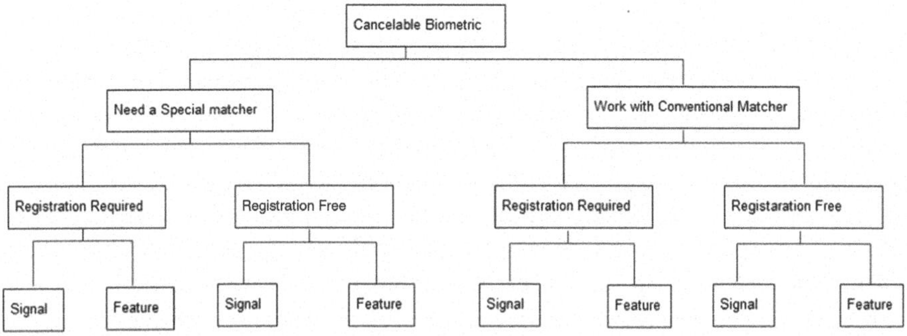

Fig. 10.7 Categorization of protection schemes of cancelable biometric

approach, bio-tokens were used, which are the revocable identity tokens produced by applying a revocable transform to biometric data, such that identity matching is done in the encoded/revocable form. This approach combines the concepts of data transformation, strong learning measures, and biometric data encryption. The scheme splits the data into two parts: the fractional part, which is used for local distance computation, and the integer part which is encrypted.

3.4.6 Summary of Cancelable Biometric Template Protection Schemes

The cancelable biometric template protection schemes, as shown in Fig. 10.7, can be divided into two main categories: one needs a special matcher and one works with conventional matcher. These schemes can be further subdivided into two categories, such as methods that do not require registration and methods that require registration of biometric samples. These methods can be further subdivided into two types of schemes: schemes where the original biometric samples are used (denoted as a signal) and schemes where the features extracted from the biometric signals (denoted as a feature) are used.

3.5 Biometric Template Protection

The number of biometric technologies has already been increased and successfully working to extract biological or non-biological features like face, iris, fingerprints, palm print, hand geometry, and signature. Fingerprints are the most interesting due to the easiness of their capturing and cost-effectiveness of respective scanner or sensor. A biometric authentication scheme, using template protection, is irreversible. One of the most critical situations of identity robbery of any person may occur if a biometric template within the database of the device of someone is negotiated. The biometric systems that rely on using single biometric are known as unimodal biometric structures [1, 35]. Alphonse Bertillon, the leader of the crook identification division of the police department in Paris, has developed an idea to pick the criminals by measuring body parts like palms, toes, and hands. After some time this idea was modified and used largely for the uniqueness of human fingerprints. The card-based documents have been prepared to store the fingerprints of the criminals. These were circulated by the police department and stored in the database, which decides the criminal's identity. After, the biometrics came into the picture and are used for law enforcement and legal function identification of criminals to provide safety for persons in sensitive jobs. Civilian and personal zone-based applications were planned that used biometrics to form the popularity in private sector [36].

Person authentication used to rely on password or tokens, like smart cards. These strategies were very problematic in the situations when passwords can be forgotten or hacked and the tokens may be misplaced or stolen. Biometrics means, however, provide a helpful way of authentication as they cannot be lost or forgotten. With the potential growth to biometric system trends in real time, biometrics is also used for authentication in programs like a laptop and smartphone log-in or accessing a building, lift, automobile, and office, among others. Currently, governmental, forensic, safety, and industrial sectors are also showing their interest to use biometric-based recognition systems. In India, for example, the government is imposing a device to capture several biometric traits from its residence, and greater than 1 billion people have been issued a complete unique identification number (UIN), also known as Aadhaar Card [36].

The biometric authentication techniques have huge potentials in creating secure systems [37] as a biometric data set of users are unique based on their identities. In general, a biometric authentication scheme has two phases. The first phase is enrollment phase, in which a user has to register his/her biometric information to a trustworthy server. A biometric template is formed for the person and is stored on some central secure server or cloud, and a smart device is provided to that person too. Now, in the second phase known as authentication phase, the same person would provide another biometric sample; the comparison has to be done with the preexisting template available in the server, cloud, or smart device. Finally, the same person is authenticated with the matching of the fresh sample to the template, and a ranking will be given by matching function.

The biometric template of the person as discussed above contains crucial information for successful authentication. Once the template of the person has been exposed, it may be captured by an attacker, misusing the personal information of that person. Therefore, it is most important to stop attackers from capturing the biometric templates of the persons. This is the major challenge to the biometric service providers to build a server or cloud that can never be compromised, and in critical situations, if compromised, then it must ensure the security of biometric templates.

In [37], Original template X is used to compute a sketch P, and to obtain another sample Y of the same biometrics, P and Y are used to recover X. Then author used similarity measures to check X and Y similarity. If both X and Y are similar, then sketch P is secure. In an authentication scheme as shown in Fig. 10.8, when the

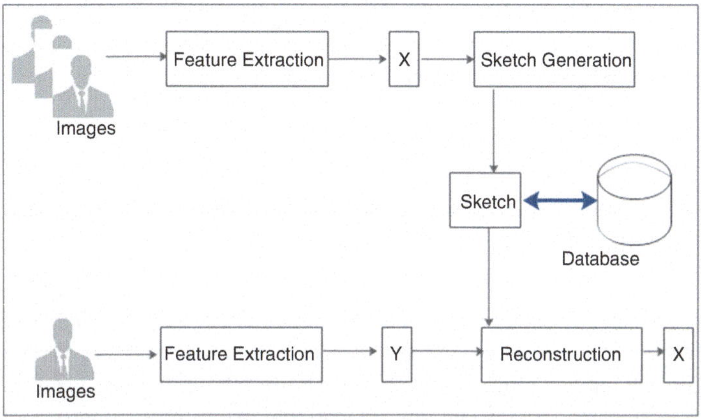

Fig. 10.8 Feature extraction, sketch generation, and reconstruction of template

original biometric template of an individual X is reconstructed from P and Y, a cryptographic key K from X can be tracked, assuming that K is uniformly distributed. For example, one can treat the key K as the "password" of that individual. If the minimum entropy of real X is bonded to the lower level, then entropy loss provides lower bound on the length of K, which can be easily tracked from a biometric data set. Hence, lower entropy loss provides longer key extraction, and it is more difficult for an attacker to copy the personal information of that individual by guessing the key K.

3.6 BioHashing for Human Authentication

The evaluation process of the biometric system is required in order to make sure that it is accurate in terms of low false rejection of legal users or low false acceptance rate of illegal users. When a legal user is denied access by biometric systems, it creates a huge impact on the usability of the system [43]. Due to this reason, the public acceptance rate of the biometric system will decrease. One of the possible solutions is to use multimodal biometrics that can reduce the probability of denial of access without compromising the false acceptance rate. Authors of [44] have proposed the solution for high false rejection rate. They proposed two-factor authentication-based model, which is based on iterated inner products between the tokenized pseudorandom number and fingerprint features of the user. Authors in [44] named BioHash code to the set of user-specific compact codes. Direct mixing of pseudorandom number with biometric data is possible and very easy and can incorporate physical tokens. The major problem with this technique [43–48] is performance degradation when a fraud "B" steals the Hash key "A" and tries to validate as "A." In such a situation, BioHashing based performance is lower as compared to the data obtained using biometric scheme alone [49]. This is an emerging area of research as no appropriate solution is given till now. BioHashing produces a vector set of biometric features set, which corresponds to the "Hash key."

4 Case Studies on Signature-Based Biometric Recognition and Challenges

Since the last few decades, signature-based biometric technologies have gained popularity due to its reliability and secure access control. There are many media to large-scale applications that exist in the literature, where they are used in applications such as airports, railway stations, educational institutions, country borders, etc. [50]. However, these applications are facing a lot of challenges in terms of security to satisfy the end users. A couple of case studies have been carried out and discussed in details, which highlight the challenges of using signature-based biometric recognition systems and their possible solutions.

4.1 Case Study on Use of Online Signature-Based Biometrics in E-Commerce

E-commerce or electronic commerce deals with financial and commercial activities over the Internet. Nowadays, E-commerce has got its place in common user purchasing, which generates an enormous amount of data due to online purchasing by online money transfer to the company account. By using any E-commerce sites, one can easily do financial transactions from home for the purchase of items. Some attractive characteristics of using it are the reduction of cost, availability, and better quality of goods. Using E-commerce system of purchasing, merchants have the option to eliminate the cost to maintain a showroom, which results in cheaper products along with reasonable online delivery costs. Since, many financial transactions that take place in online purchasing, which creates an urgent need for authentication mechanism, one of the authentication mechanism is online signature-based biometrics that are primarily used for verification of a genuine customer [1–2].

4.1.1 Objectives

The following are the objectives of this case study:

- Get a glimpse of the importance of authentication in E-Commerce.
- Understand what online signature-based biometrics is and how it works.

- Understand how online signature-based biometrics is used or can be used in E-commerce systems.
- Compare various types of authentication methods briefly.
- Understand the threats to online signature-based biometrics and examples of different types of forgeries.

4.1.2 Background

E-commerce has to deal with many insecurities regarding the theft of personal data, loss of money, and hacking of online accounts, among others. To protect customers, the various authentication techniques are used by merchants like authenticating customer's identity by using e-mail IDs, passwords, security questions, OTP, etc. But these techniques are still not adequate to handle security aspect carefully. For example, it has been observed that the passwords used by the majority of the E-commerce users are based on their names, birth dates, and cell phone numbers, and also same passwords are used to handle different accounts for the sake of simplicity and ease of remembering the passwords. This shifts the transactions done by the users in the high-risk zone because these protected passwords are easily broken by a person close to the owner of the accounts. Hence, the number of online transactions made by the users is in high-risk zone. To overcome this risk, various authentication methods based on biometrics were suggested by authors, which were proved to be more secure. They further reduced the risk of hacking accounts and stealth of valuable information of users.

One of such methods is signature-based biometrics, which uses individual's signature for authentication or access to the account. Some biometric methods, such as fingerprints, iris, and facial pattern, are unique and remain constant throughout the entire life of a person. If these are stolen or leaked, then there could be a serious threat to the identity of the concerned person. But, if online signature-based biometric is stolen or leaked, the original can be modified by the concerned person anytime, which further reduced the person identity threat.

4.1.3 Authentication System for E-Commerce

In authentication systems for E-commerce, first, the digital signature of a person is recorded. Digital signatures are recorded using specifically designed digital pad and a pen. Apps are also designed for recording of signature, in which the user uses a stylus or finger for signing. This recorded signature is converted into a digital image. The enrolled signature image is linked to the user account and is stored in the database of the online company. When the user provides a signature for authenticating an access, features of the same are compared with the features of the enrolled signature. This comparison verified the identity of the person. Moreover, this method used the concept of tokens called keys. The party, which sends the request, has a

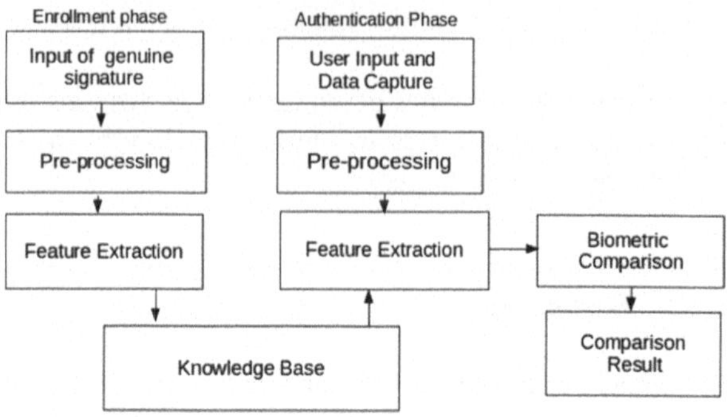

Fig. 10.9 Working of signature-based authentication

private key (for encryption), and the opposite party has the public key (for decryption). Both keys are related mathematically but in such a manner that it will not be easy to compute the relationship between them. Complete working of signature-based authentication mechanism is shown in Fig. 10.9. After obtaining the signature during the enrollment phase, special features are extracted using various feature extraction techniques or accurate mathematical algorithms. Extracted features are stored in the knowledge base, which can be further used during the time of authentication.

There are tools, like BIO SIG-ID and app BioTect-ID [51], which provide tough biometric authentication techniques that require no hardware, PIN texts, or security questions which can be easily cracked by forgers or hackers. Unlike physical biometrics, which pose a liability risk since they can never be replaced, signature-based biometric is a nonintrusive biometric that can be replaced anytime. The users are supposed to just need a password or signature to log in. It can be mainly used during the login page of online payment portals or in any service, which can authenticate and verify a digital signature based on parameters like direction, speed, length, angle, height, etc. Moreover, it even does not require any special types of equipment for the authentication process.

4.1.4 Flow of Authentication Method

Signature-based biometric method is used by merchants in various steps of transactions like logging into the account, confirmation of orders for purchase, payment of the goods, and also in confirmation of online delivery of purchased goods. At each step, the user is asked to sign digitally and the digital image is sent to the company. After proper verification of the identity, the user is given access to the next step. By this way, optimum security can be maintained, and the need of remembering and typing passwords can be reduced. Pseudo-code for online signature-based biometric authentication is given below. Figure 10.10 shows the basic block diagram of online signature-based biometric system.

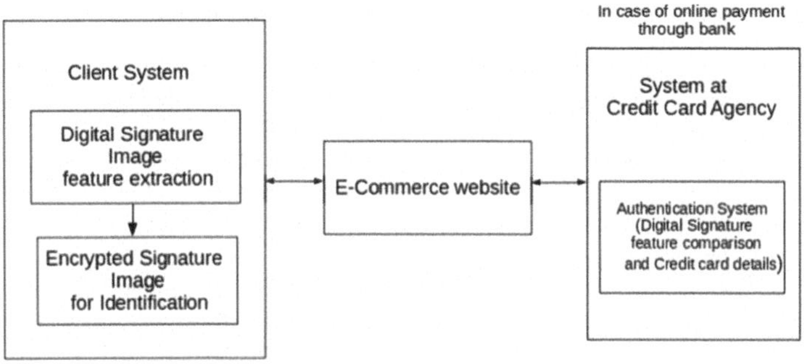

Fig. 10.10 Block diagram of flow process of authentication method

Pseudo-code for Online Signature-Based Biometric Authentication

```
Input: Online Signature-based Biometric Authentication
Output: Successful authentication
1. Go to login page of the E-Commerce website
2. Enter the signature on the digital pad with the help of
   electronic pen.
3. The processed image of signature will be sent to the
   website.
4. If (successful authentication)
{
Go ahead for shopping.
If (payment of goods through credit card)
{
Enter the signature again for authentication through credit
card system.
If (authentication successful)
{
Produce bill.
}
Else
Try again.
}
}
5. When delivery is done, enter the signature again to confirm
   the delivery.
```

4.1.5 Comparison of Various Authentication Methods

Authentication in E-commerce has several advantages like the level of security and ease to the customers. E-commerce-based company growth depends upon the comfort and trust of its customer. Table 10.3 shows the difference between various authentication methods.

Table 10.3 Comparison between different methods of authentication

Method	Details
Traditional methods (including the use of passwords, cell phone number, etc.)	If passwords are used, there may be a possible case of someone sneaking at the time of password entry, or it can be cracked by use of advanced computational algorithm which checks every possible character combinations
Signature-based biometrics	This method is more secure and it is also a cancelable biometrics. Moreover more than one characteristic of signature like curves, relative pressure, speed, etc. are compared, hence becoming more secure than other biometric-based authentication
Other biometrics like a fingerprint, iris, facial pattern, etc.	These are relatively secure, but there is a need of special instruments like the fingerprint scanner, iris scanner, etc., converting the recorded prints into a digital image by special software, etc. Also, there may be a case of stealth of the biometrics.

4.1.6 Threats to E-Commerce-Based Company

If the private or public key of the users is leaked, then it can cause unexpected verification. When the user gives access of private key to an authorized person, then the person can misuse it. Moreover, there may be cases of forgery of signatures too. Forgeries are the non-genuine signatures done by a person other than the real owner of the same signature. According to Coetzer et al. [52], basically there are three main types of forgery:

4.1.6.1 Random Forgery

The forgery in which the forger is unaware of signature even name of the genuine person and signs the documents with any random name or signature is termed as random forgery. This forgery detection is quite easy as the signature might be completely different from the enrolled signature. This type of forgery is generally performed by an unknown person, who may or may not be related to the owner of the signature.

4.1.6.2 Simple Forgery

When a forger knows the name of the person, but has not seen the genuine signature even once, and signs a document with a pattern of signature based on the name of a person, then the forgery is termed as simple forgery. There are slight chances of the signature pattern produced to be similar to that of genuine person's signature. Hence, the chances of successful forgery are increased.

Table 10.4 Comparisons between different types of forgery

Random forgery	Simple forgery	Skilled forgery
Forger is unaware of name or the signature of genuine person	Forger is aware of basic information of the genuine person like name, and others, but is completely unacquainted with his/her signature	Forger is well acquainted with basic information of genuine person like name and signature. Moreover, forger is even able to replicate the signature as it is
Chances of detection are quite low	There are good chances of detection of forgery as there remains a greater possibility of wrong pattern produced by forger even if it is dependent on forger's name	The detection of forged signature is very hard, and hence there remains a good chance of forgery to be successful
May not even need systems for detection for forgery	It may or may not require the need of advanced detection techniques	Most advanced methods are required for detection of such type of forgery

4.1.6.3 Skilled Forgery

In this type of forgery, the name as well as the original signature of a person is available with the forger. The forger also is able to replicate the signature as it is, termed as a skilled forgery. The chances of differentiating between a genuine signature and forged signature become very less, which increases the chances of successful forgery tremendously. Some advanced technologies exist in the literature, which not only compare signature pattern but also compare the speed, direction, relative force applied, and even angles in the signature, resulting in reduced rates of forgery up to certain extent. Table 10.4 shows the comparisons between different types of forgery.

There are two types of error in the system: False rejection error and false acceptance error [53]. The rate at which these errors occur are called false rejection rate (FRR) and false acceptance rate (FAR). If the rate of false acceptance is more than the false acceptance rate, it is more serious as it allows a person who is not legitimate as though he is. For FRR, the user can easily try again immediately [53], whereas a FAR of 1% is quite intolerable for the system to be even implemented. FAR of 0.1–0.001% are considered as normal.

4.1.7 Outcome from the Case Study

This case study presented a profound importance of online signature-based biometrics in any E-commerce-based company. With the rapid advancement of security systems, it has been shown that this method of authentication can be very accurate and reliable and also provide comfort to online purchasing customer through these E-commerce sites. Advance security system can verify and distinguish a forgery signature from a genuine more precisely and accurately.

4.2 *Case Study on Use of Online Signature-Based Biometrics in Retail Industry*

The online transactions in retail industries are increasing rapidly day-by-day due to increase in count of educated people, which creates high possibility of fraud. A statistical review has shown that in the United States [54], the retail industry suffers a huge loss every year as far as credit card identity theft is concerned, which nearly sums up to 50 billion of dollars. In these situations biometric signature verification can serve as the wall of protection for the identity theft and fraud as they are solely based on unique and peculiar characteristics of the humans, which may not be behaviorally copied by some other individual.

4.2.1 Objectives

The main objectives of this case study are as follows:

- As privacy of the user information is an important aspect in the retail industry, it is necessary that only the authorized owner has an access to it. At the same time, signature biometrics is a technology, which is coming up with improvements on the aspects of verification and identification day by day; it looks quite promising for use in the future.
- The combination of biometric signature and retail industry is something that cannot be neglected, to solve a number of problems that already exist or are expected to rise in the future.
- This study ultimately brings out the shortcomings, which need to be taken care of to help the large section of society associated with the retail industry.

4.2.2 Background

Proper verification has become necessary for the retail industries like shops, stores, banks, real estate, and finance. Credit cards have been used as one of the primary keys for online purchasing. But at the same time, existing security systems used for credit card are inadequate to handle high-level frauds [54]. There are possibilities of human errors in comparing the signature on the card with signature done by the customer. To handle this situation, biometric signature seems to be an efficient alternative because they are based on physical and behavioral traits of a person (which is unique for every person). It turns out to be the best source for recognition of a person identity. Figure 10.11a shows the division of signature biometrics in retail sector, and Fig. 10.11b shows the preferences of the customers.

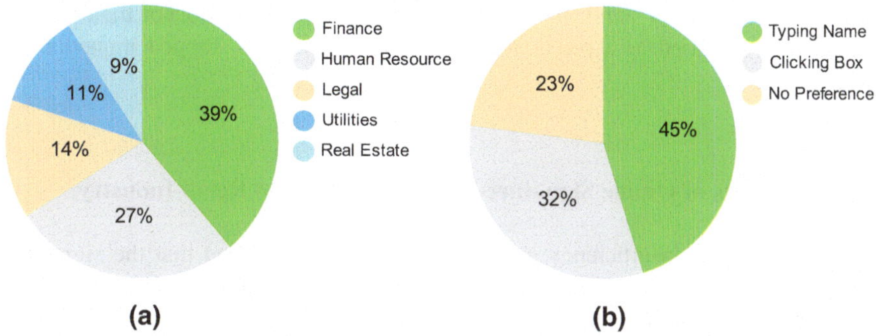

Fig. 10.11 (**a**) shows division of signature biometrics in retail and (**b**) shows the preferences of the customers

4.2.3 Cases of Identity Theft in Retail Industry

Identity theft in retail industry may occur when a thief used personal information of an individual, such as date of birth, address, and credit card number. According to some statistics of the United States, it has been noted that nearly ten million people faced the vicious web of identity theft. They lost nearly 48 billions of dollars [55]. The Federal Trade Commission (FTC), which maintains records of identity theft activity in the United States, estimated approximately ten million victims of identity theft in 2002, resulting in $48 billions of lost to businesses and financial institutions and $5 billion in out-of-pocket expenses for consumer victims [55]. Further, victims in most of the cases got to know about the fraud either by monitoring their accounts or by banks which suspected some activities to concerned account. Signature provided by the owner on the back of the card cannot solve this issue. To handle this issue, biometric signature comes into picture to prevent frauds or at least minimize the rate at which they are happening.

4.2.4 Signature Verification Process in Retail Industry

In signature verification, first the image acquisition is done. Offline signature requires paper to perform the execution, so there is a need to specify resolution, image type, and format for the scanning image. After that, it goes through preprocessing. It travels through various levels so that it can go for feature extraction later on. This is an important stage. It compares the similarity of the image with the reference that has been already taken. It increases the efficiency of the evaluation, enhances the image, and makes it noise free. These signatures are usually processed in gray. This retains the quality of signature and makes it ready for feature extraction. In this stage, decision-making is done. If the compared source and test image are similar, then signature is declared as original, and this indicates the end of matching process. On the other hand, if the source and the test image do not match, then signature is declared as a piece of forgery and the process stopped immediately [55, 56].

4.2.5 Benefits of Online Signature-Based Biometrics to Retail Industry

With the increase in efficiency and precision, it has been seen that the signature biometrics will provide a high level of security to the user and the one that manages the operations through it. It also gives the customer a new experience and can also be considered an element that increases awareness among people about the better technology. Further, promotes the very basic and the most important motto of engineering, i.e., automation, as we are all witnessing the revolution in digitization around the world, where main motto is that our reference stored data will not get lost, since in the case of manual operations, chances of data loss will be more.

4.2.6 Challenges for Retail Industries

Certain challenges need to be considered like signature variation with time (at least to some extent). There are high chances of false rejection rates. The signature biometric when compared to other methods is still in a developing stage, so improved technologies still need to come up for efficient working of retail sector. The users of retail industries may not be necessarily comfortable with the use of pad and digitizer pen for doing a signature. If the customer experience does not turn out to be satisfactory, then it can directly affect the retailers. This technology also requires high-quality hardware. All these factors need to be taken into consideration and should be worked upon so that this technology doesn't turn out to be a nightmare for those who start accepting/implementing it.

4.2.7 Outcome from the Case Study

Online signature-based biometric has started influencing the day-to-day life of people. It has its own merits and challenges during its usage in retail industry. The efficiency and accuracy of this technology still need to be made more effective so that it can be easily adopted by the larger community. Some innovation can still be thought of that can minimize the unavoidable circumstances for humans, which leads to false rejections. Though, it would not be fair to say that it does not have a promising future in the retail industry considering its advantages, advancements can make it reach new heights in this era of automation.

5 Summary and Concluding Remarks

Nowadays, reliable authentication and authorization of persons are essential tasks not only in daily activities but also in different applications like access control, financial transactions, travel and immigration, and healthcare, among others. Traditional authentication techniques, like the token-based password, are found to be less reliable because of loss and theft. These issues have been acknowledged by researchers and developers. Biometric-based techniques are considered alternative and safer methods for person authentication and identification. Biometric identifiers are unique for the individuals and found to be more reliable in comparison to token-based and knowledge-based techniques. In this chapter, we have covered the online signature-based biometric recognition system, which includes fingerprint scanning, facial geometry, footprint scanning, retina and iris patterns, DNA, and heartbeat scanning, among others. Biometrics is the science intended to perform the recognition of persons based on their biological characteristics: physiological and behavioral. Among behavior biometrics, signature verification occupies an important and a very special place in the field of biometric technology. A couple of case studies have been covered in this chapter: online signature-based biometrics in E-commerce to deal with financial and commercial activities taking place through the medium of Internet. Nowadays, E-commerce has moved to the mainstream, and an enormous amount of data and money transfer takes place due to it. Anyone can easily do his/her financial transactions or purchase required clothes or appliances while sitting at home using E-commerce systems and websites. Considering the facts discussed in the second case study that deals with the use of online signature-based biometrics in retail industry, we can conclude that by proper planning, inclusion of relevant blueprint with an appropriate, flexible, and stable biometric scheme, we can rescue numerous victims from hackers and thieves. Biometric-based schemes can serve as the wall of protection for the identity theft and fraud as they are solely based on unique and peculiar characteristics of the humans, which cannot be copied by some other individual.

References

1. M.S. Obaidat, N. Boudriga, *Security of e-Systems and Computer Networks* (Cambridge University Press, 2007)
2. M.S. Obaidat, B. Sadoun, Verification of computer users using keystroke dynamics. IEEE Trans. Syst. Man Cybernetics, Part B **27**(2), 261–269 (1997)
3. M.V. Kanawade, S.S. Katariya, Signature verification & recognition – case study. Int. J. Electron. Commun. Instrumentation Eng. Res. Dev. (IJECIERD) **3**(1), 77–86 (2013)
4. B. Zhang, M. Fu, H. Yan, Handwritten signature verification based on neural 'gas' based vector quantization. in *IEEE International Joint Conference on Neural Net-works*, 1998, pp. 1862–1864
5. Sansone, Vento, Signature verification: Increasing performance by a multi-stage system. Pattern Anal. Applic. **3**, 169–181 (2000)

6. E.J.R. Justino, F. Bortolozzi, R. Sabourin, Off-line signature verification using HMM for random, simple and skilled forgeries. in *ICDAR 2001, International Conference on Document Analysis and Recognition*, vol. 1, 2001, pp. 105–110
7. J.F. Vélez, Á. Sánchez, A.B. Moreno, Robust off-line signature verification using compression networks and positional cuttings. in *Proceedings of 2003 I.E. Workshop on Neural Networks for Signal Processing*, vol. 1, 2003, pp. 627–636
8. A. Fallah, M. Jamaati, A. Soleamani, A new online signature verification system based on combining Mellin transform, MFCC and neural network. DSP **21**(2), 404–416 (2011)
9. L. Findlater, J.O. Wobbrock, D. Wigdor, Typing on flat glass: examining ten-finger expert typing patterns on touch surfaces. in *Proceedings of the 2011 annual conference on Human factors in computing systems, CHI '11*, 2011, pp. 2453–2462
10. N. Sae-Bae et al., Biometric-rich gestures: a novel approach to authentication on multi-touch devices. in *CHI '12*, 2012, pp. 977–986
11. S. Prabhakar, S. Pankanti, A.K. Jain, Biometric recognition: Security and privacy concerns. IEEE Secur. Priv. Mag. **1**, 33–42 (2003)
12. S. Kulkarni, N. Takawale, A comparative study of biometric techniques. Int. J. Innovative Res. Comput. Commun. Eng. **4**(1), 879–883 (2016)
13. Z. Riha, V. Matyas, Biometric authentication system. FIMU, 4–44 (2000)
14. S.D. DiptiVerma, Static signature recognition system for user authentication based two level cog, Hough transform and neural network. Int. J. Eng. Sci. Emerg. Technol. **6**(3), 335–343 (2013)
15. G. Agam, S. Suresh, Warping based offline signature recognition. Inf. Forensics and Secur. **2** (3), 430–437 (2007)
16. F.Z. Marcos, signature recognition state–of–the–art. IEEE A&E Syst. Mag., 28–32 (2005)
17. J. Wen, B. Fang, and T. Jhang, Offline signature verification: a new rotation invariant approach, pp. 3583–3586
18. B. Jayasekara, A. Jayasiri, L. Udawatta, An evolving signature recognition system. in *First International Conference on Industrial and Information Systems, ICIIS 2006*, 2006, pp. 529–534
19. J. Zheng, G. Zhu, On-Lin handwriting signature recognition based on wavelet energy feature matching. in *Proceedings of the 6th World Congress on Intelligent Control and Automation*, 2006, pp. 9885–9888
20. Survey on Dynamic Signature Recognition: A sign of the times – Part 2, Biometric Technology Today, 2009, pp. 1–2
21. J. Fierrez, HMM-based on-line signature verification: Feature extraction and signature modeling. Pattern Recogn. Lett. **28**, 2325–2334 (2007)
22. T.Q. Ton, T. Nhu, HMM based online signature verification. Int. J. Adv. Res. Comput. Eng. Technol. (IJARCET) **4**(4), 1448–1451 (2015)
23. V.M. Patel, N.K. Ratha, R. Chellappa, Cancelable biometrics: A review. IEEE Signal Proc. Mag. 2015
24. N.K. Ratha, J.H. Connel, R. Bolle, Enhancing security and privacy in biometrics-based authentication systems. IBM Syst. J. **40**(3), 614–634 (2001)
25. N. Ratha, S. Chikkerur, J. Connell, R. Bolle, Generating cancelable fingerprint templates. IEEE Trans. Pattern Anal. Mach. Intell. **29**(4), 561–572 (2007)
26. R.M. Bolle, J.H. Connel, N.K. Ratha, Biometrics perils and patches. Pattern Recogn. **35**(12), 2727–2738 (2002)
27. M. Savvides, B. Kumar, P. Khosla, Cancelable biometric filters for face recognition, in *International Conference on Pattern Recognition*, vol. 3, (2004), pp. 922–925
28. K. Takahashi, S. Hirata, Cancelable biometrics with provable security and its application to fingerprint verification. IEICE Trans. Fundam. Electron. Commun. Comput. Sci. **94-A**(1), 233–244 (2011)

29. S. Hirata, K. Takahashi, Cancelable biometrics with perfect secrecy for correlation-based matching. in *Advances in Biometrics, ser. Lecture Notes in Computer Science,* ed. by M. Tistarelli, M. Nixon, vol. 5558, (Springer, Berlin/Heidelberg, 2009), pp. 868–878
30. W. Xu, Q. He, Y. Li, T. Li, Cancelable voiceprint templates based on knowledge signatures, in *International Symposium on Electronic Commerce and Security*, (2008), pp. 412–415
31. J. Camenisch, M. Stadler, Efficient group signature schemes for large groups, in *International Cryptology Conference on Advances in Cryptology*, (Springer, London, UK, 1997), pp. 410–424
32. J. Zuo, N. Ratha, J. Connell, Cancelable iris biometric. in *International Conference on Pattern Recognition*, 2008, pp. 1–4,
33. T. Boult, Robust distance measures for face-recognition supporting revocable biometric tokens. in *International Conference on Automatic Face and Gesture Recognition*, 2006, pp. 560–566
34. T. Boult, W. Scheirer, R. Woodworth, Revocable fingerprint biotokens: accuracy and security analysis. in *IEEE Conference on Computer Vision and Pattern Recognition*, 2007, pp. 1–8
35. M.M. Ashish, G.R. Sinha, Biometric template protection. J. Biostat. Biometric App. **1**(2), 202 (2016)
36. Unique Identification Authority of India. Multipurpose National Identity Card
37. Y. Sutcua, Q. Lib, M. Memon, *How to Protect Biometric Templates*, pp. 1–11
38. T. Kanade, A. Jain, N.K. Ratha, *AVBPA 2005, LNCS* (Springer, Berlin/Heidelberg, 2005), pp. 523–532
39. E.S. Bigun, J. Bigun, B. Duc, S. Fischer, Expert conciliation for multi modal person authentication systems by bayesian statistics. in Proceedings of AVBPA, Springer, LNCS-1206, 1997, pp. 291–300
40. J. Kittler, M. Hatef, R. Duin, J. Matas, On combining classifiers. IEEE Trans. Pattern Anal. Machine Intell. **20**, 226–239 (1998)
41. A.K. Jain, A. Ross, S. Prabhakar, An introduction to biometric recognition. IEEE Trans. Circuits Syst. Video Technol. **14**, 4–20 (2004)
42. R. Plamondon, G. Lorette, Automatic signature verification and writer identification – the state of the art. Pattern Recogn. **22**, 107–131 (1989)
43. A. Lumini, L. Nanni, An improved BioHashing for human authentication. Pattern Recogn. **40**, 1057–1065 (2007)
44. A.T.B. Jin, D.N.C. Ling, A. Goh, Biohashing: Two factor authentication featuring fingerprint data and tokenised random number. Pattern Recogn. **37**(11), 2245–2255 (2004)
45. A.T.B. Jin, D.N.C. Ling, Cancelable biometrics featuring with to kenised random number. Pattern Recogn. Lett. **26**(10), 1454–1460 (2005)
46. A.T.B. Jin, D.N.C. Ling, A. Goh, Personalised cryptographic key generation based on face hashing. Comput. Secur. J. **23**(7), 606–614 (2004)
47. T. Connie, A. Teoh, M. Goh, D. Ngo, Palm hashing: A novel approach for dual factor authentication. Pattern Anal. Appl. **7**(3), 255–268 (2004)
48. Y.H. Pang, A.T.B. Jin, D.N.C. Ling, Cancelable palmprint authentication system. Int. J. Signal Process. **1**(2), 98–104 (2005)
49. D. Maio, L. Nanni, MultiHashing, human authentication featuring biometrics data and to kenised random number: a case study FVC2004. Neurocomputing, 2006, to appear
50. I. Nakanishi et al., DWT Domain On-Line Signature Verification, 2011, pp. 184–196
51. https://www.biosig-id.com/industry/biosig-id-banking-or-ecommerce
52. J.A. Du Preez, B. Herbst, J. Coetzer, Offline signature verification using the Discrete Radon Transform and a Hidden Markov Model. EURASIP J. Appl. Signal Processing **4**, 559–571 (2004)
53. M.C. Fairhurst, Signature verification revised: Promoting practical exploitation of biometric technology. Electron. Commun. Eng. J. **9**(6), 273–280 (1997)
54. T.G. Zimmerman et al., Retail applications of signature verification. Biometric Report
55. J.B. Ciulla, Ethics effectiveness: The nature of good leadership, in *The Nature of Leadership*, 2nd edn., (SAGE, Los Angeles, CA, 2012), pp. 508–540
56. U. Pal et al., Signature-based biometric authentication. in *Computational Intelligence in Digital Forensics: Forensic Investigation and Applications*, 2014, pp. 285–314

Chapter 11
EEG-Based Biometrics

Florian Gondesen, Matthias Marx, and Dieter Gollmann

Acronyms

ANN	Artificial neural network
BCI	Brain-computer interface
BSS	Blind source separation
EEG	Electroencephalography
EMG	Electromyogram
EOG	Electrooculogram
ERP	Event-related potential
fMRI	Functional magnetic resonance imaging
ICA	Independent component analysis
k-NN	k-nearest neighbors
LDA	Linear discriminant analysis
MEG	Magnetoencephalography
PCA	Principal component analysis
RSVP	Rapid serial visual presentation
SSEP	Steady-state evoked potential
SSVEP	Steady-state visually evoked potential
SVM	Support vector machines
VEP	Visually evoked potential

F. Gondesen (✉)
Hamburg University of Technology, Hamburg, Germany
e-mail: gondesen@tuhh.de

M. Marx
University of Hamburg, Hamburg, Germany

D. Gollmann
Hamburg University of Technology, Hamburg, Germany

SCSE, Nanyang Technological University, Singapore, Singapore

M. S. Obaidat et al. (eds.), *Biometric-Based Physical and Cybersecurity Systems*,
https://doi.org/10.1007/978-3-319-98734-7_11

1 Introduction

Biometrics require a feature that is unique, permanent over a long period, observable, and sufficiently difficult to imitate. A source for such a feature could be the human brain. It is a very complex organ, containing approximately 86 billion neurons. Their interconnections are affected by a person's experience and determine his or her behavior. From this alone, the brain should provide a source of individuality sufficient for biometrics. However, as the brain constantly adjusts while processing experiences, permanence is expected to be limited.

Due to its complexity and the fact that it is an internal organ, capturing the brain's structure is difficult. However, there are side channels leaking information of the brain's processing. From the blood oxygen levels that can be measured by functional magnetic resonance imaging (fMRI) or near-infrared spectroscopy, spatial activity patterns can be inferred. As the neurons work electrochemically, it is possible to measure brain activity electrically or magnetically. The noninvasive methods are called electroencephalography (EEG) and magnetoencephalography (MEG), respectively. Due to the weak magnetic fields, MEG requires properly shielded rooms and uses superconductors for the sensors. This makes it very expensive compared to EEG, where electrical potentials are measured on the scalp. Recently, consumer-grade EEG headsets have appeared, making EEG the method of choice for practical biometric applications based on brain activity measurement. An EEG-based biometric system would require the subject to wear an EEG headset while executing a certain task. From the recorded EEG data, features have to be extracted that fulfill the requirements for distinctiveness and permanence under that task.

2 Fundamentals

In this section, we will briefly explain EEG. We will introduce EEG recording devices and techniques that can be used for brain-computer interfaces. Then, we will cover data collection, signal processing, and signal analysis. Finally, we will present software that lets you build and analyze experiments.

2.1 Electroencephalography

EEG is a method of recording neural activity of the brain. The first EEG was recorded by Hans Berger in 1924 [1]. When neurons are activated, synaptic currents are produced. Those currents generate an electrical field. Because of the attenuation from multiple layers of tissue and bones, a large number of neurons have to be activated synchronously so that the signal can be measured by electrodes which are placed along the scalp [2, p. 7]. Voltages are in the range of a few microvolts. The

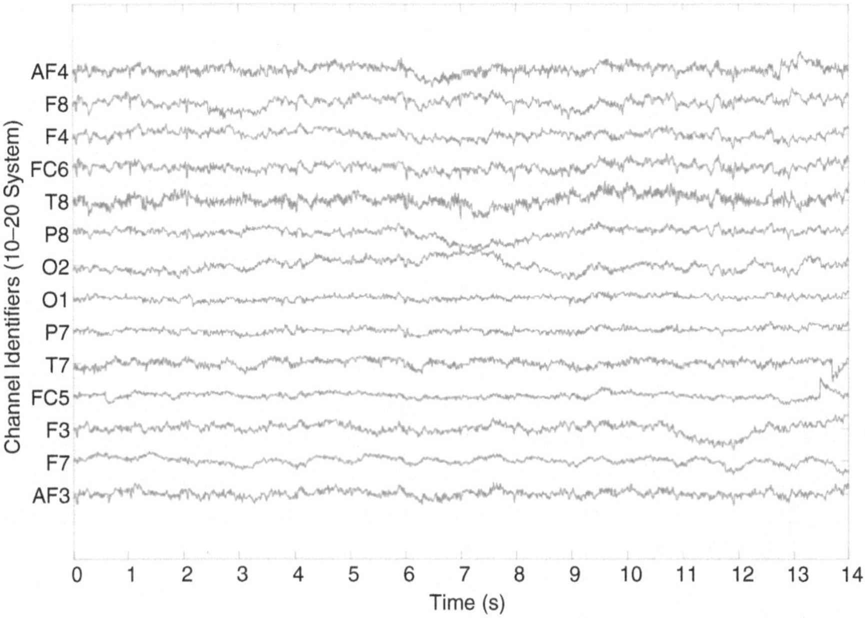

Fig. 11.1 EEG recorded with an EMOTIV Epoc EEG headset with 14 electrodes positioned in the 10–20 system

spatial resolution depends on the number of electrodes used and is in the range of a few centimeters. EEG devices typically have sampling rates between 128 Hz and 2048 Hz. Higher rates are usually not necessary, as the human EEG has a spectrum similar to pink noise. Figure 11.1 shows the EEG of a subject in resting state with eyes open. The corresponding power spectral density is depicted in Fig. 11.2. The frequency range below 21 Hz contains 90% of the total power. Figure 11.3 shows the time-frequency representation.

To describe rhythmic activity, the EEG is often divided into certain frequency bands. Activity in those bands is associated with different mental states [3, 2]. The different bands are shown in Table 11.1.

EEG is clinically used in diagnostics, for example, of epilepsy or brain death. It can also be used to identify different sleep states, as those are associated with activity in certain frequency bands. There are many approaches using EEG for biofeedback systems that aim to improve cognitive performance, like memory or attention, or to treat mental, cognitive, and behavioral disorders [4]. EEG can be used as a data acquisition method for brain-computer interfaces (BCIs), which are especially important for people suffering the locked-in syndrome.

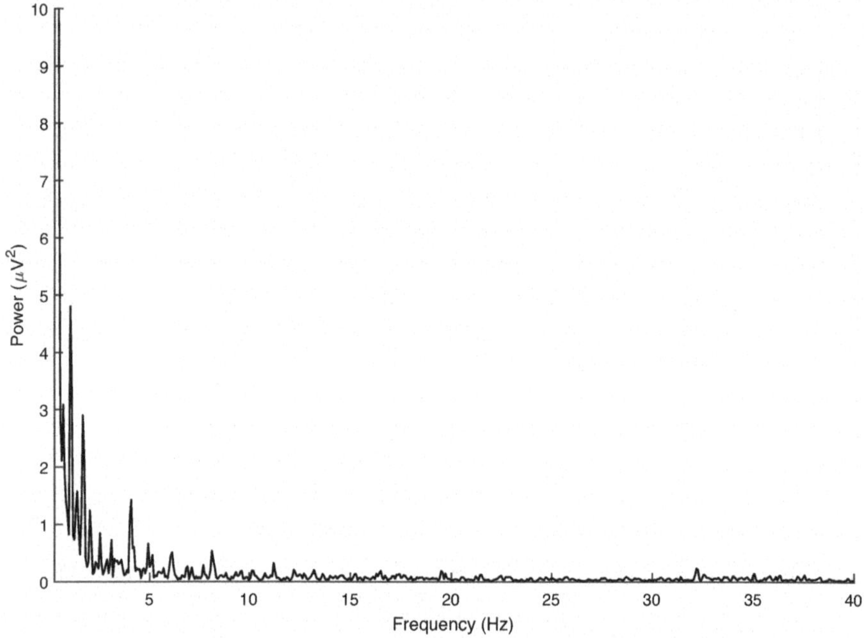

Fig. 11.2 Spectrum, averaged over all channels of the EEG shown in Fig. 11.1

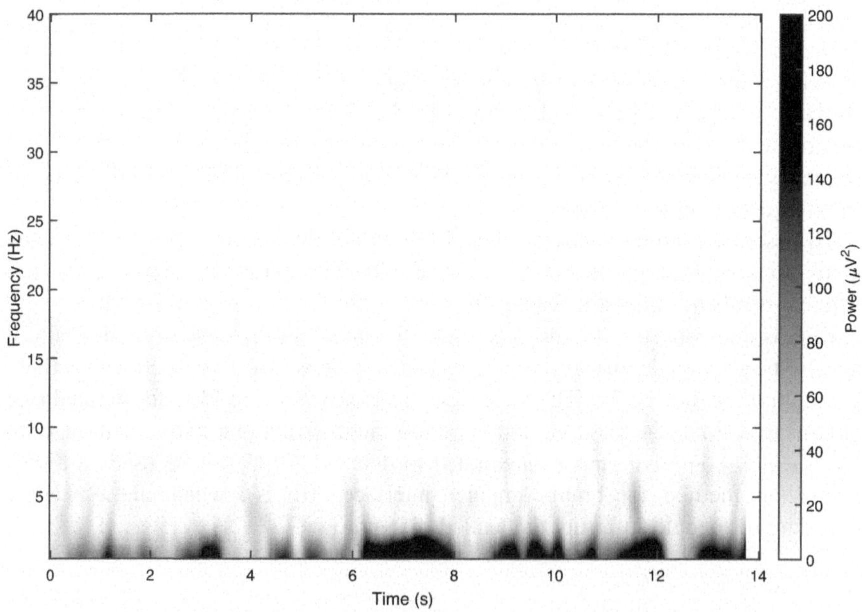

Fig. 11.3 Time-frequency representation, averaged over all channels of the EEG shown in Fig. 11.1. Most black areas reach power beyond scale

Table 11.1 EEG frequency bands

Band	Bandwidth (Hz)	Region	Associated with
Delta (δ)	0.5–4	Varies	Deep (dreamless) sleep
Theta (θ)	4–7.5	Varies	Creative inspiration, deep meditation, drowsiness
Alpha (α)	8–13	Occipital and parietal	Physical and mental relaxation
Beta (β)	14–26	Frontocentral	Active thinking, active attention
Gamma (γ)	>30	Frontocentral	Active information processing, processing of sensory stimuli

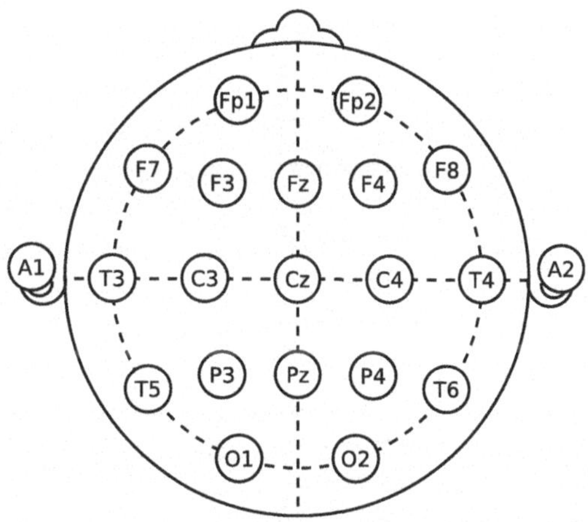

Fig. 11.4 21 electrodes of 10–20 system for EEG [5].

2.1.1 Electrodes and Electrode Placement

EEG recording systems use different types of electrodes. An electrolytic gel or saline solution can serve as conductor between electrode and skin. Dry or semidry polymer electrode pads have shorter preparation times but may provide recordings with higher noise.

The 10–20 system (see Fig. 11.4) is a method of describing the locations of the electrodes. Capital letters *F*, *P*, *O*, *T*, or *C* identify lobes of the brain (frontal, parietal, occipital, temporal, or central); see Fig. 11.5. Odd numbers refer to electrodes on the left, even numbers to electrodes on the right hemisphere. Electrodes placed on the midline are referred to by the letter *z*. Earlobe positions are identified by a capital *A*, frontal polar positions by *Fp* [2, pp. 15–17].

2.1.2 Artifacts

Artifacts are signals that arise from sources other than the brain. They can be physiological or extraphysiological in origin. Physiological artifacts arise from

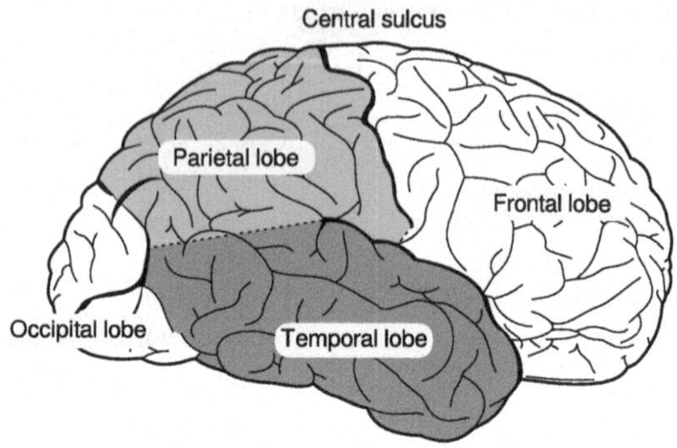

Fig. 11.5 Lobes of the brain [6, Fig. 728]

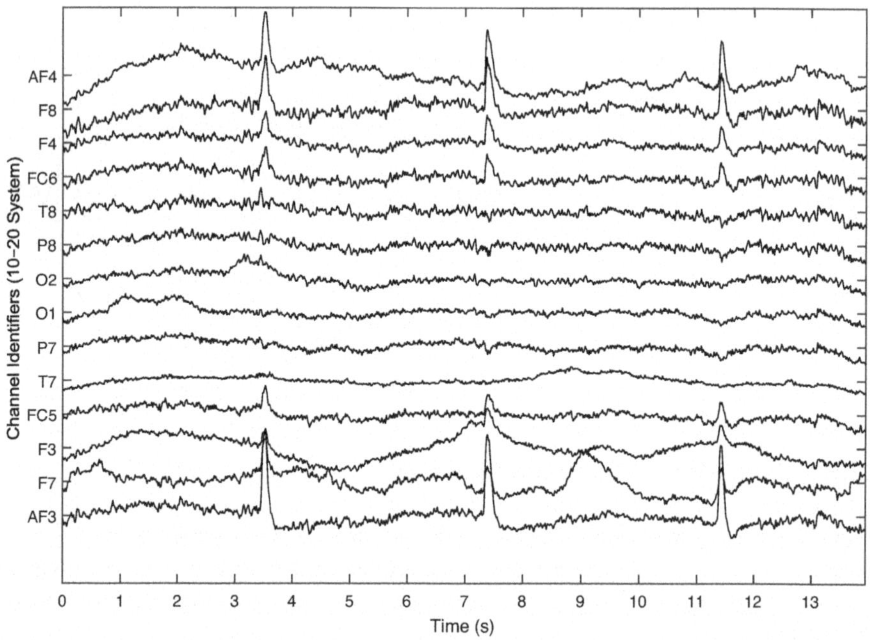

Fig. 11.6 EEG recorded with an EMOTIV Epoc EEG headset with 14 electrodes positioned in the 10–20 system. The subject was instructed to rest and blink after visual cues which appeared at seconds 3, 7, and 11

body activities. Eye blinks, eye movements, and head movements lead to strong physiological artifacts in the recorded data. Artifacts that arise from eye blinks or eye movements are referred to as ocular artifacts and are visible as spikes in frontal electrode locations. Figure 11.6 shows the EEG of a subject instructed to blink after visual cues. The EEG contains three spikes prominent in the set's most frontal electrodes *AF3* and *AF4*. They occur approximately half a second after the subject was cued to blink. The corresponding time-frequency representation is depicted in Fig. 11.7.

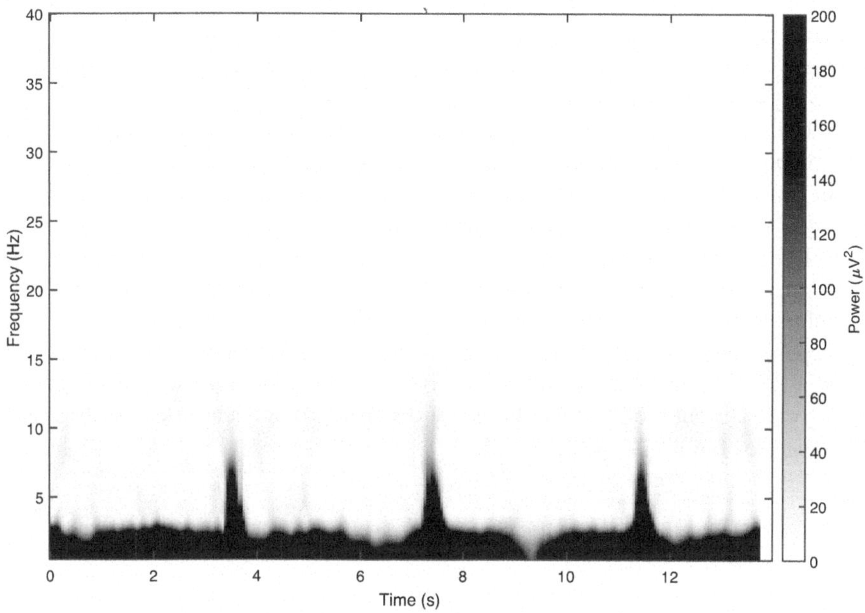

Fig. 11.7 Time-frequency representation, averaged over all channels of the EEG shown in Fig. 11.6. Most black areas reach power beyond scale

Swallowing or neck contraction can lead to smaller artifacts [7]. Artifacts originating from the pharyngeal muscles have a broad frequency distribution and affect many electrode sites. Their strength depends on the level of contraction [8]. The channel *T8* in the EEG depicted in Fig. 11.1 appears to be noisier.

The position *T8* is close to facial muscles; thus muscle tension is a possible explanation for the noise. Extraphysiological artifacts arise from electrical interference from power sources or from the EEG recording system. They are visible as spikes in several electrode locations [7].

Contamination with artifacts makes the analysis of EEG signals more difficult [9]; hence it is beneficial to eliminate sources of artifacts in advance. Here, the instructions to the subjects play an important role (see Sect. 2.4.1). Artifacts can also be filtered out or detected, to exclude the contaminated data from further analysis. We further discuss handling artifacts in Sect. 2.4.2.

2.1.3 Event-Related Potentials

Event-related potentials (ERPs) are the electrical response of the brain to sensory, affective, or cognitive events. They are generated in response to external stimuli. The amplitude of ERP waves ranges from 1 μV to 30 μV and is relatively small compared to the background EEG activity [2, p. 125].

Latencies, amplitudes, and topographies vary with stimulus or response features but are not limited to that. The ERP is also influenced by the subject's reactions or attitudes toward a stimulus as well as the design of the experiment. The subject's physiology can play a role as well.

The components of an ERP are named starting with the letter P or N depending on the component being positive (P) or negative (N). Digits typically indicate the peak latency in milliseconds after event onset. Alternatively, digits are used to enumerate the positive and negative deflections separately.

If the event is an external stimulus, the ERPs are often named by its modality. For example, a visually evoked potential (VEP) is the ERP corresponding to a visual stimulus.

Depending on the latency, components can be attributed to different areas of the brain. ERP components occurring within 100 ms after stimulus onset are influenced by physical attributes of the stimulus. Later occurring components are nonobligatory responses to stimuli [2, pp. 127–128].

2.1.3.1 P300

The P300 is a positive deflection in the ERP. Its peak latency is about 300 ms after stimulus onset. The window for the P300 ranges between 220 ms and 500 ms after stimulus onset. The amplitude of the P300 wave is about 10 μV, measured relative to a pre-stimulus baseline [10]. An ERP with a P300 component is depicted in Fig. 11.8. P300s can be elicited through a so-called oddball paradigm, where a rare and relevant target stimulus is shown in a sequence of nontarget stimuli. When stimuli are presented, the P300 is most prominent with stimuli that the subject is paying attention to. Therefore, the experimenter controls the direction of attention and requires the subject to perform a task that involves the target stimuli and not the nontarget stimuli. Generally, P300s occur only in response to task-relevant stimuli [10].

The P300 is best recorded over the midline centro-parietal regions of the brain. Multiple occurrences can be averaged, and information from multiple electrode locations may be combined. Amplitude and latency may vary, so that averaging may not deliver the desired results [10]. When the target stimulus is more improbable, the P300 wave becomes larger. The amplitude becomes smaller and the latency longer when discriminating the target stimuli from the nontarget stimuli becomes difficult. But also when the task is too easy, the amplitude may become smaller. Many other factors influence the P300 wave. Its latency is related to the age of subjects. In general, the amplitude is smaller when the subject is older. Also its scalp distribution changes significantly with age. The P300 becomes more frontal in older subjects. Extroverts have smaller P300 waves than introverts. Alcohol and other drugs increase the latency and reduce the amplitude. Dementia, schizophrenia, and autism influence amplitude or latency [10, 2].

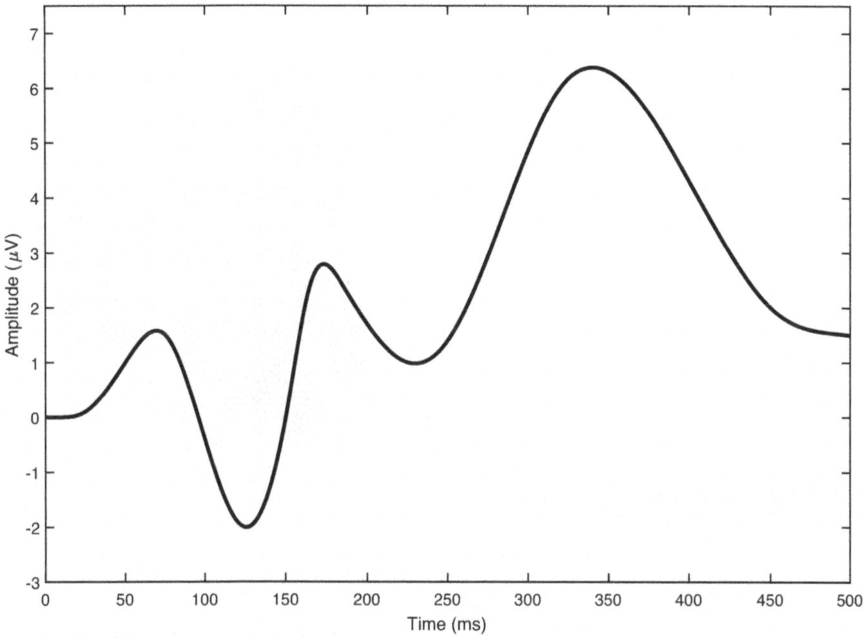

Fig. 11.8 ERP with P300 component

2.1.3.2 N400

The N400 component of the ERP is a negative deflection peaking around 400 ms after stimulus onset. The time window of negativity ranges from 200 ms to 600 ms. The voltage is not necessarily negative; it is only negative relative to the ERP of a similar stimulus not eliciting an N400. The variance of the latency is very low, though age, diseases, and language proficiency can influence the timing. The modulation of the amplitude is related to semantic processing of the stimulus. It is, for example, correlated to the cloze probability when sentences with odd endings are used as stimuli: "take coffee with cream and dog." Though the topographies vary with the stimulus type, the N400 can be recorded over centro-parietal regions.

2.1.4 Steady-State Evoked Potentials

Observing a flickering light can evoke activity at the same frequency in the visual cortex. This phenomenon is called steady-state visually evoked potential (SSVEP) or sometimes visual steady-state response. SSVEP is typically measured at a frequency range of 3.5–75 Hz [11, 12]. In different frequency ranges, state evoked potentials exist also for somatosensory and auditory stimulation. The effect of steady-state evoked potential (SSEP) at different frequencies varies among individuals.

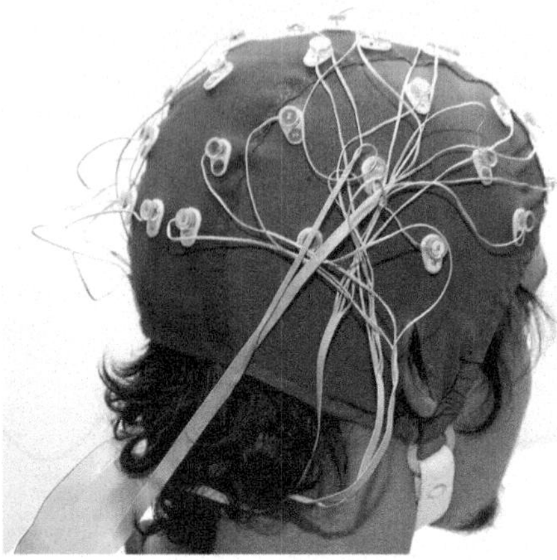

Fig. 11.9 Electrode cap with 32 electrodes

2.2 EEG Recording Devices

The following section provides a short overview of two very diverse EEG recording devices: the BioSemi ActiveTwo and the EMOTIV Epoc. They differ in price, performance, and usability. The ActiveTwo was introduced in 2002 and has been further developed since then [13]. The Epoc was released in 2009. A slightly improved version, the Epoc+, followed in 2013. Google Scholar lists several thousands of results for both devices [14, 15].

2.2.1 BioSemi ActiveTwo

The BioSemi ActiveTwo is an EEG recording device for research applications by BioSemi B.V. It costs approximately 13.500 € with 8 electrodes and approximately 75.000 € for 256 + 8 electrodes [16]. The electrodes are mounted in an electrode cap (see Fig. 11.9). Its sampling rate ranges from 2 kHz to 16 kHz. It differs from low-cost EEG devices in that it supports more electrodes and each channel comes with its own 24 bit ADC. For a 128-channel headcap, a preparation time of 30 min is realistic [17].

2.2.2 EMOTIV Epoc

The low-cost EEG headset by EMOTIV Inc. is designed for research and brain-computer interface applications. It has 14 channels and costs approximately 799$. The reusable electrodes consist of gold-plated discs and felt pads (see Fig. 11.10) that have to be soaked in a saline solution before use. Its sampling rate is 128 Hz or, for the newer Epoc+, 256 Hz. The specified bandwidth for both versions is 0.2 Hz–43 Hz. The EEG headset comes with a single 14-bit ADC [18]. Due to its headset design, it can be mounted in a few seconds. But the preparation can take several minutes as some electrodes might not have good contact until slight position adjustments or have not been sufficiently soaked. A person wearing an Epoc EEG headset is shown in Fig. 11.11.

Grummett et al. [19] evaluated the Epoc comparing it to more expensive EEG systems. While showing a higher noise floor, it was found suitable to measure the two of the selected EEG phenomena. ERPs could not be tested due to timing problems with the supplied recording software Testbench. Time-locked experiments require to store the time of the events synchronously to the EEG data. Different from research-grade EEG systems, the Epoc does not have a hardware trigger channel. The trigger signals, called markers by EMOTIV, are added in software. Testbench can receive them via a serial port. When no serial port is available, virtual serial ports or USB to serial adapters can be used. They might further influence the timing. Nonetheless, the Epoc has been used in ERP biometrics studies [20–22].

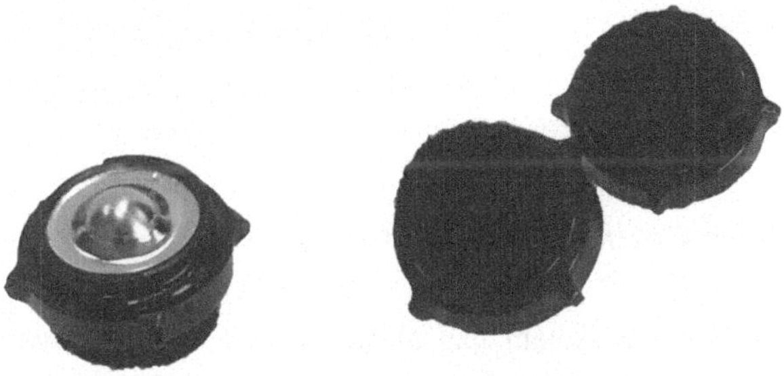

Fig. 11.10 EMOTIV Epoc's reusable electrode pads

Fig. 11.11 Person wearing an EMOTIV Epoc EEG headset

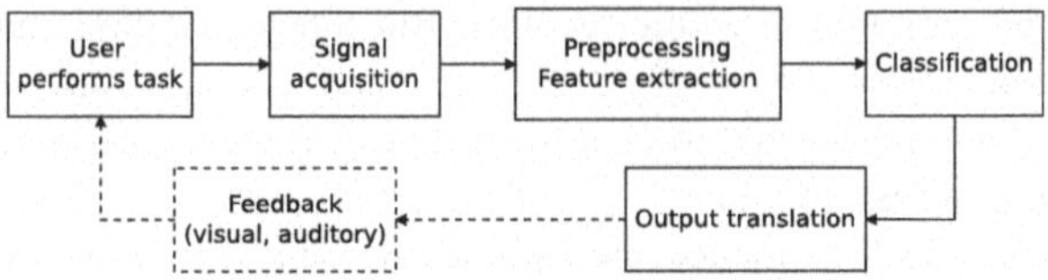

Fig. 11.12 Components of a BCI [24]. The methods for the individual components can be selected, by and large, independently of each other. Feedback is optional

Fig. 11.13 Screen of a P300 speller. The letters are lit up column- and row-wise

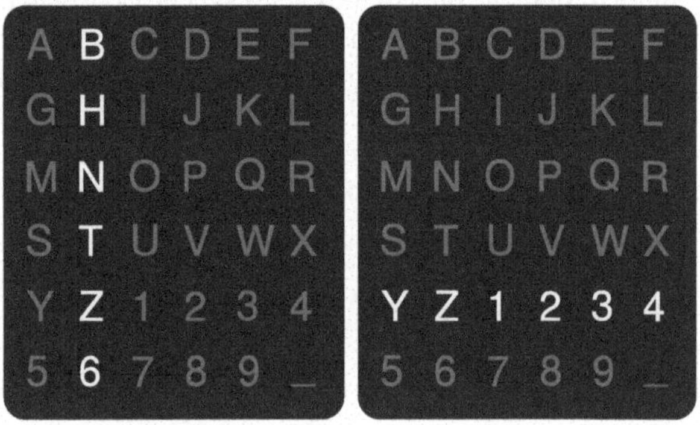

2.3 *Brain-Computer Interfaces*

A BCI is a system that allows a user to input data to a computer without the activation of efferent nerves [23]. This requires the user to do a mental task that deterministically changes features of the brain activity that can be captured by an acquisition device. This feature has to be extracted, classified, and translated to a suitable output, which can, for example, be a letter typed on a virtual keyboard. The appearance of the typed letter on a screen is a commonly used feedback to the user, but BCIs without feedback are possible. The components of a BCI are shown in Fig. 11.12. For every component there exist several different approaches or methods. Though they can be selected, by and large, independently of each other, adjacent components need compatible parameters. The feature extraction method is required to extract a feature that is actually measurable by the acquisition system and influenced by the task. Otherwise the classifier will not be able to distinguish between the different conditions of the task.

2.3.1 P300 Speller

An example of a BCI is the P300 speller. A screen (depicted in Fig. 11.13) shows a matrix of letters that are lit up column- and row-wise. The user focuses on the letter to be typed, thus making the rare event of that letter lighting up relevant, which elicits an ERP with a P300 component. The system recognizes the time of the P300 event and thus identifies the letter.

2.4 Data Collection, Signal Processing, and Analysis

Building an EEG-based biometric system is very similar to a BCI. It consists of the same components (shown in Fig. 11.12) with one major difference: instead of using a feature that allows to deduce the user's will, it has to facilitate the conclusion who the user is. In this section we will discuss how to design the components in relation to the application of biometric identification.

2.4.1 Conducting EEG Experiments

The center of an EEG experiment is the subject, who has to perform a certain task. The task has to be known prior to the experiment. The experimenter can usually explain the task, while the EEG system is being set up. It helps to demonstrate at least a part of the experiment before commencing it. When the EEG is set up, the subject can be asked to clench teeth or blink while watching the live EEG signals to explain the severity of artifacts. This helps reducing activity generating them [25]. However, urging the subject not to blink or move their eyes changes the nature of the task. The subject might pay attention to the eyes and might lose focus on the stimuli. This can, for example, significantly affect the P300 [10].

When visual stimuli are presented, a cross can serve as a fixation point to look at (see Fig. 11.14). It can help in reducing eye movements.

Many experiments require the subject to pay full attention. EEG experiments can be very lengthy and boring so that subjects might lose motivation and pay less attention. Including a score on how well the subject performs can help to compensate this. If this is not compatible with the task, it might be possible to introduce slightly deviant trials, often referred to as "target" that has to be detected. But this creates an

Fig. 11.14 Fixation cross on gray background

oddball paradigm, which might not be wanted. The target trials have to be excluded from the analysis then.

If the sequence of the experiment follows patterns that can be learned, the subject's brain might adapt to this which might influence the EEG. It is best practice to randomize the order of the stimuli. Constant timing can also create a bias.

When dealing with visual stimulation, care has to be taken because of the risk of epileptic seizures. These can even occur to subjects who have no history of epilepsy.

2.4.1.1 Rapid Serial Visual Presentation

Rapid serial visual presentation (RSVP) is a method for presenting visual stimuli. In RSVP, visual stimuli appear serially in the same location of a screen. The rate at which the stimuli appear can vary between less than 50 ms and several hundred milliseconds. RSVP has been used to address questions about reading and language processing [26]. In 1975, Potter explored that an observer can detect an expected scene even when it is presented so briefly that it would not otherwise be remembered [27]. RSVP can be applied to speed up experiments in which the oddball paradigm is used to elicit P300s.

When a subject has to identify multiple targets within a continuous stream of stimuli, the subject might miss one or more of the target stimuli. Interference is produced when the presentation of another stimuli starts before target-identification processes of previous target stimuli are complete. Further target stimuli are poorly detected when they are presented during a 270 ms interval beginning 180 ms after the first target stimulus has been shown [28]. This phenomenon is known as *attentional blink*. Another phenomenon observed in RSVP is *repetition blindness*. Subjects have been shown to have difficulty detecting repeated occurrences of target stimuli when multiple target stimuli appear within a time frame of 500 ms [29].

2.4.2 Improving Signal-to-Noise Ratio

The EEG signal components of interest are not only often contaminated with artifacts but also superimposed with uncorrelated background activity of the brain. A general approach dealing with those two kinds of noise is averaging. EEG experiments are usually structured into smaller units, referred to as trials that are repeated several times. Assuming signal components unrelated to the experiment to be zero-mean, averaging over the trials will cancel out the noise. As artifacts are often significantly stronger than EEG signals, the occurrence of a spike in one of ten trials can still be prominent in the average and can be confused with an expected spike. Thus artifacts are often removed in advance.

2.4.2.1 Handling Artifacts

Some artifacts can be easily handled by filtering. One example is the ubiquitous 50 Hz or 60 Hz signal from the power grid. EEG devices typically have built-in notch filters for these frequencies.

Artifacts originating from the eyes, the electrooculogram (EOG), show distinctive spikes in EEG channels next to the eyes. It is also possible to attach additional electrodes below the eyes to measure EOG activity. Muscle artifacts, the electromyogram (EMG), are detectable on many electrode sites and have a broad frequency range. Due to the pink noise-like property of the EEG, it can be detected at high frequencies with low EEG activity, for example, 110–140 Hz. The detection can be automated according to certain thresholds or done by visual inspection by the experimenter. The latter poses the risk of introducing a bias into the data, as trials containing artifacts are usually excluded from further analysis.

2.4.2.2 Handling Background Activity

ERPs are time-locked signals. Their signal-to-noise ratio profits greatly from time-locked averaging. Non-time-locked components are canceled out. Some ERP components, like the N400, are very weak. To detect them it is necessary or at least beneficial to compare the average of trials containing the components to the averaged trials not containing it. Comparing two different conditions is not limited to the analysis of ERP, and it can be done in time or frequency domain. One of the conditions is often referred to as baseline. The comparison can be absolute, by subtraction, or relative by division. Figure 11.15 shows a baseline-corrected version of the time-frequency representation shown in Fig. 11.3. The power of each frequency was divided by the corresponding average power of a 5-second timespan where the subject was resting and not cued to blink.

2.4.3 Selecting and Extracting Features

An EEG-based biometric system has to discriminate EEG data of different subjects. The data is typically organized in trials or averages of multiple trials that have a length of few seconds. Depending on the sampling rate, the data consist of hundreds to thousands of samples per channel. Directly using these data as a feature vector for any machine learning approach will probably have poor results, as training sets are typically small. Recording more trials consumes time and may be exhausting for the subject. Thus, it is reasonable to apply a feature extraction method prior to classification that minimizes the size of the feature vector while minimizing loss of discriminant information. Where to find discriminant information in the EEG may largely depend on the subject's task. If, for example, the experiment aims to elicit a known ERP component like the P300 or N400, it might be sufficient to only regard the EEG data of the corresponding timespan. As these ERP components are rather

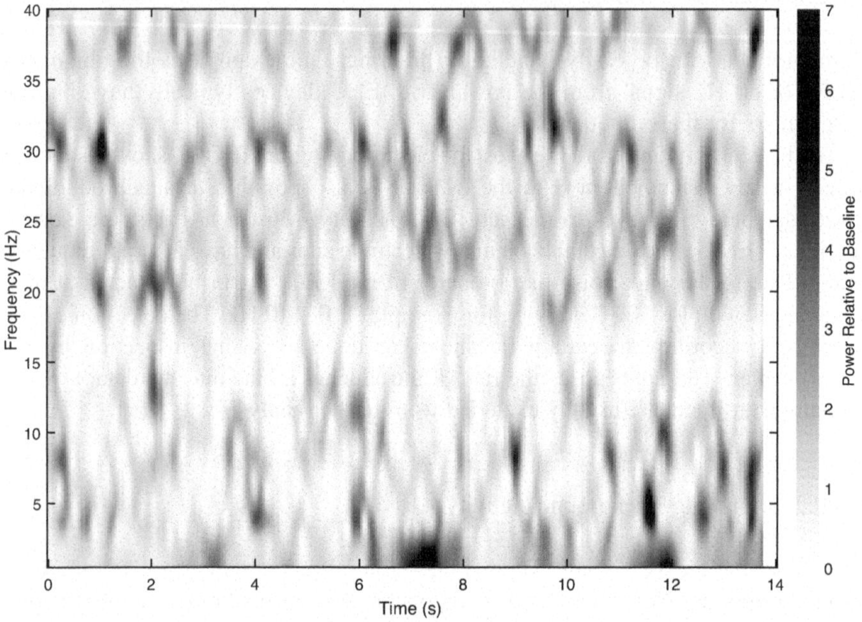

Fig. 11.15 Baseline-corrected version of the time-frequency representation (shown in Fig. 11.3) of the EEG shown in Fig. 11.1. The timespan of 2.5 seconds before the plotted EEG was used as relative baseline

slow, downsampling might be applicable too. If the topography of the component is known, channels can be averaged or excluded. The resulting feature vector would cover the characteristics of the ERP component, but other discriminant information would be reduced.

The effects of different tasks on the EEG, especially with regard to their discriminant information is not completely understood and thus subject to biometrics research. To search for suitable features in the EEG, it helps to transform the data. As tasks can influence the oscillatory patterns, it is helpful to transform the time signal to the frequency domain [30]. The spectrum of a 14-second resting state EEG is shown in Fig. 11.2. Using the spectrum of a whole trial compresses the signal into a single time bin, thus reducing the size of the feature vector. But it lacks information about time-dependent variations. To keep temporal information, several spectra of different timespans of the trial can be computed by the short-time Fourier transform (STFT) or similar techniques. A time-frequency representation for the 14-second resting state EEG is shown in Fig. 11.3. This helps in identifying changes in EEG due to tasks but also adds another dimension to the data. A popular method to structure multivariate data is to apply principal component analysis (PCA). PCA transforms the data to an orthogonal basis, whose vectors are sorted by the largest variance. It is possible to reduce the data, by dropping the vectors of least variance.

Under the assumption that the EEG signal is a mix of statistically independent signals of different origin, independent component analysis (ICA) can be used for blind source separation (BSS). This decomposes the EEG into signals of different origin. Hence the signal correlated to the task and different noise signals are separated.

2.4.4 Classification

In biometric identification, data have to be classified corresponding to the users. Thus multiclass classifiers are required. Generally, multiple binary classifiers can be coupled to solve multiclass problems [31]. This is often done with support vector machines (SVM) that are popular in EEG-based biometrics research. Other popular classifiers are different forms of artificial neural network (ANN), linear discriminant analysis (LDA), and k-nearest neighbors (k-NN) [32]. Some classifiers require a Gaussian distribution. This cannot generally be assumed for any feature vector derived from EEG data.

2.5 *Software*

Manufacturers of research-grade EEG systems offer software to visualize the live signal and record it. Interfaces to access the live data are typically provided, too. Consumer-grade devices typically have interfaces that allow access to predefined features of the EEG for using it in end-user application, while access to raw EEG data is restricted. Additional to data acquisition, EEG experiments require different steps of data analysis and often the presentation of precisely timed stimuli. We briefly describe a few software packages that facilitate these tasks.

2.5.1 FieldTrip

FieldTrip is an open-source MATLAB toolbox developed by the Donders Institute for Brain, Cognition and Behavior, in Nijmegen, the Netherlands. Among others, it offers functions for EEG analysis, covering preprocessing, time-locked analysis, time-frequency analysis, and plotting. It is compatible with different EEG data formats and offers real-time acquisition methods [33].

2.5.2 EEGLAB/BCILAB

EEGLAB is a MATLAB toolbox for processing electrophysiological data licensed under the GPL. It features an interactive graphical user interface, functions for artifact removal, ICA and time-frequency analysis, and plotting [34]. BCILAB is a toolbox based on EEGLAB, adding BCI-specific features [35].

2.5.3 Psychtoolbox

Psychophysics Toolbox is a free MATLAB toolbox allowing to deliver precisely timed visual and auditory stimuli. It also contains functions to facilitate processing of inputs [36].

2.5.4 BCI2000

BCI2000 is a software system for BCI research. Its C++ sources are free for nonprofit research and education. It features integration of MATLAB scripts and a network interface for live interactions with arbitrary software [37].

3 Methods for EEG-Based Biometric Systems

An EEG-based biometric system requires the subject to perform a task while his or her EEG is recorded. The signal then has to be preprocessed, and features have to be selected that allow a classifier to distinguish if the recorded data matches to a stored feature set. For all stages shown in Fig. 11.12, different methods can be used. The selection of methods and their fine tuning largely depends on the methods used in the other stages. A set of methods can be referred to as paradigm.

We distinguish between *event-related paradigms* where the task is connected to a certain stimulus that creates an ERP and *non-event-related paradigms* where the aim of the tasks is not to elicit an ERP. Some paradigms combine biometrics (*what you are*) with some sort of shared secret (*what you know*). Furthermore, BCIs can be used to enter passwords, which is generally not a biometric approach. In contrast to traditionally typed passwords, BCI password spellers resist shoulder surfing attacks.

3.1 Event-Related Paradigms

Due to the time-locked nature of ERPs, it is important to know the exact moment of the event. Thus examining the ERP of an internal event, for example, a certain thought, is difficult. Typical approaches of ERP biometrics therefore use external stimuli. They can in principle be applied to any of the five classical senses, but creating gustatory, olfactory, or somatic stimuli is difficult in practice. Audition and vision are technically easier to use; stimuli can easily be delivered by a PC or smartphone. To date, most publications on ERP-based biometrics use VEPs. Visual stimulation allows a high variety of different stimuli that can be delivered quickly via RSVP (see Sect. 2.4.1.1), which helps to create a sufficient number of trials without consuming too much time. To create a biometric system based on the ERP paradigm, components of the ERP that are highly depending on personal traits should be chosen. The task, the type of stimuli, as well as the selection of the actual stimuli can have a big impact on certain ERP components. For ERP analysis, trials of the same condition are typically averaged. This reduces components related to background activity and noise (see Sect. 2.4.2). Although ERPs are typically analyzed in the time domain, analysis in the frequency domain is also possible. The features to extract from this signal are the positions and amplitudes of the wave components, sometimes referred to as landmarks. Which landmarks are suitable depends largely on the complexity of the stimulus and of the task. For earlier ERP components that are associated with sensory processing, simpler stimuli are sufficient. Later ERP components are associated with higher levels of processing (see Sect. 2.1.3) and thus require a more complex experimental protocol. The experimental protocol also influences the frequency bands (See Table 11.1). While even simple ERP experiments that require the subject to pay attention to a stimulus should affect the beta band, only protocols that involve active information processing should affect the gamma band activity.

3.1.1 Visually Evoked Potentials

Visual stimulation offers a broad range of stimulus complexity. The components of ERP directly depend on the type of stimulus. A picture probably entails more cognitive processing than a simple flash. In this section we try to structure the VEP paradigms by the presumed effect of the stimuli selected.

3.1.1.1 Early Visually Evoked Potential Components

In 2006, Power et al. [38] studied the VEPs produced by inverting a black and white checkerboard every second on 13 subjects. Besides different sets of electrodes in the occipital region, a variety of features was evaluated:

- The amplitudes of the early ERP components P100, N75, and N135 (latencies were disregarded because they are assumed not to vary much between individuals).
- Spectral power in the alpha and beta bands, because VEP stimulation was found to increase activity in these bands.
- Autoregressive models of order 3, 4, 5, and 6.

Autoregressive models reached classification accuracies between 25% and 63.5%. The spectral power in the beta band yielded accuracies around 50% to 60%, superior to the alpha band with accuracies below 20%. The amplitudes of the P100, N75, and N135 yielded accuracies between 36.5% and 67.3%. The highest accuracy of 86.5% was reached by combining P100, N75, and the beta band on the channel set *O1*, *Oz*, and *O2*.

Singhal et al. also used a checkerboard as stimulus for their study with 10 subjects in 2007 [39]. Arguing that an onset/offset VEP is more variable than a VEP elicited by pattern reversal, the checkerboard was shown for a flash of 100 ms at a rate of one per second. Only 300 ms of post-stimulus data were analyzed with a mother wavelet due to similarity to the VEP waveform. Out of 343 wavelet coefficients, the 35 highest energy coefficients were used as feature vector. A k-NN classifier yielded a peak accuracy of 42%. An alternative method using phases, latencies, and amplitudes of the onset/offset VEP components P75, N125, and P150 yielded 78% accuracy when using a two-dimensional Gaussian kernel.

In 2015, Barbosa et al. [40] conducted a study with 16 infant subjects using looming stimuli. A VEP is expected approximately 800 ms before a seemingly approaching object hits. A detector for VEPs was used to segment the data into VEP candidates of 200 ms. The classification with an ANN had an accuracy of 62.5%. This was argued to be comparable to other research, as more subjects were tested, but fewer trials recorded.

3.1.1.2 Recognition and Memory

Palaniappan et al. [41] started to use VEP for biometrics in 2002 using pictures of the Snodgrass and Vanderwart set that consists of 260 standardized line drawings of single easily recognizable objects [42]. The stimuli were presented for 300 ms, followed by a 5.1 s interstimulus interval. The feature extracted from data segments of 1 s per trial was the total power in the gamma band. This band was argued to be "related to higher brain functions like perception and memory" and "the level of perception and memory access between individuals are generally different." Half of the 40 artifact-free trials per each of the 10 subjects were used to train an ANN. With the other half used as test set, depending on the parameters, the average classification rate was 90.95%, peaking at 95%. In subsequent research, the experimental protocol was kept, but the analysis parameters were changed.

A more complex task was used by Das et al. in 2009 [43] in a study with 20 subjects. Pictures of cars and faces with added white noise were shown for 40 ms, preceded by a fixation screen and followed by a blank screen each for a random period of 0.5–1.5 s. The subjects had to decide after each presentation to which class the picture belonged. 200 ms pre-stimulus and 500 ms post-stimulus were analyzed. SVM and LDA classifiers were applied. By applying a method they called "Fisherbrains," the most discriminative spatial and temporal features were identified. As expected, electrodes over the visual cortex were optimal. The period from 120 ms to 200 ms contained most discriminant information. Classification on pre-stimulus data was only slightly above chance level. With the post-stimulus data, SVM performance peaked at 94%.

In 2015, Armstrong et al. [44] used acronyms as stimuli on 45 subjects. The idea behind this is that people know different sets of acronyms and their processing in semantic networks results in an N400 (see Sect. 2.1.3). Four different classifiers were tested, with accuracies between 82% and 92%. A combination to a meta-classifier reached 97%. To assess the permanence, the experiment was repeated with a smaller number of subjects after about 2 weeks and about half a year, showing similar performance.

Based on this, Ruiz-Blondet [45] evaluated polarizing foods and celebrities and compared it to other visual stimuli as well as different tasks: oddball, pass-thought, and resting state. As it was the best performing classifier in the previous study, only cross correlation was used. When using just the optimal sensor, color food achieved an accuracy of 86% on 50 subjects. Voting classification on three sensors reached 100%. It was found that the topography of discriminant information depends on the type of stimulus. While it is strongest over the occipital cortex for foods, it is more central for celebrities.

3.1.1.3 P300

Das et al. [46, 47] used an oddball paradigm (see Sect. 2.1.3.1) in two slightly different protocols, one using geometric shapes and one using letters and digits. The parameters of the data analysis were tuned to find optimal classification results. Inspecting the frequency band from 0.5 Hz to 8 Hz of 600 ms from stimulus onset in 50 trials, EERs around 14% could be achieved. Nontarget stimuli outperformed the target stimuli. The 50 subjects were tested up to three times over a couple of weeks to verify the permanence of the features.

3.1.2 Secret Components

ERP components can significantly differ depending on the meaning a stimulus has to the subject. This can be, for example, related to familiarity or relevance to the task. Picking a set of stimuli with a high variety of meaning to the population helps to create an efficient paradigm. If stimuli are selected that are known to have a certain meaning for a subject and uses the effect on the ERP as a feature, a shared secret is introduced. If this secret is, for example, the familiarity of faces, an imposter able to obtain the information, could familiarize with the pictures of faces and increase his chance of a match.

Touyama and Hirose conducted a VEP study with five subjects in 2008 [48]. Nine different images were presented, of which the subject could select an arbitrary number as targets in order to elicit a P300 component in the ERP. PCA was used for dimension reduction of the temporal data. By using target and nontarget images, an LDA classifier had a peak rate of 97.6%.

In 2016 Koike et al. used a similar protocol with one out of five Zener cards [20]. PCA was combined with several machine learning algorithms, achieving a peak accuracy of 96.7% on the 25 subjects. Data acquisition was done with the EMOTIV Epoc EEG headset (see Sect. 2.2.2).

Also in 2016, Harshit et al. [22] conducted an ERP study with the EMOTIV Epoc on five subjects. The protocol consisted of stimuli in different modalities: familiar voice, unfamiliar voice, self-voice, and self-face. They showed that the response to a familiar voice has a much higher amplitude. The mean classification accuracy was 79.73%.

3.2 Non-Event-Related Paradigms

Non-event-related paradigms take the brain's response to certain longer-lasting tasks into account. In contrast to event-related paradigms, there is no short-duration event, which can be used for time locking. To our knowledge, this paradigm was presented first by Poulos et al. in 1998 [49]. The authors demonstrated that person identification based on EEG-extracted features is possible.

In general, the procedure for identification or authentication is always the same. Subjects perform one or multiple tasks that are described below while their EEG is being recorded. After preprocessing and feature extraction, classification follows. A classifier has to be trained beforehand using training data.

3.2.1 Tasks

A wide variety of tasks have been used in non-event-related paradigms. We describe them briefly in the following. Two selected tasks will be described in more detail. It is not always clear why authors chose certain tasks, even though the task may influence many things, for instance, the regions of the brain that are being activated or the EEG activity in certain frequency bands. It is also possible that certain tasks have better repeatability [50].

Poulos et al.'s subjects were instructed to *rest with closed eyes* [49, 51]. The majority of papers on non-event-related paradigms considers this task [52–61].

Keirn and Aunon let subjects perform various mental tasks (*resting with eyes open, resting with eyes closed, math, object rotating, letter composing and visual counting*) while their EEG was recorded. The authors demonstrated that it is possible to distinguish these tasks in the recorded data [62]. They suggested that people with locked-in syndrome could use their thoughts to communicate with their surroundings. These tasks were taken up by Palaniappan [53, 55] to build a biometric identifier. Paranjape et al. also instructed their subjects to *rest with closed eyes* and *with open eyes* [52].

A new task has been discussed by Thorpe et al. They introduced *pass-thoughts*. If different pass-thoughts by the same subject and if same *pass-thoughts* by different subjects could be distinguished, *pass-thoughts* may be utilizable for an authentication scheme.

Motor imagery tasks have been used by Marcel and Millán [63] and Ashby et al. [56]. Marcel and Millán also asked their subjects to *generate words* beginning with the same random letter.

Brigham and Kumar presented an identification scheme which is based on *imagined speech* [64].

Chuang et al. let their subjects perform few tasks that have been used before (*resting with eyes closed, pass-thoughts, motor imagery*) and introduced new tasks (*object counting, recitation, eye and audio tone*) [65].

Eye and Audio Tone
Subjects close their eyes and listen for an audio tone. Upon hearing the tone, the subjects open their eyes and stare at a dot.

Generate Words
Subjects generate words beginning with the same random letter.

Imagined Speech
Subjects imagine speaking without performing any overt actions.

Letter Composing
Subjects compose a letter without vocalizing. If the task is repeated several times the subjects are told to try to pick up where they left off in the previous task.

Math
Subjects solve nontrivial and nonrepeating multiplication problems and are instructed not to vocalize or make overt movements.

Motor Imagery
Subjects imagine that they are moving a finger, limbs, or a specific motion from sport. These simulated movements activate similar regions of the brain which are involved in motor preparation [66].

Object Counting
Subjects count objects of a previously chosen color.

Object Rotating
Subjects imagine the rotation of an object that was shown before.

Pass-Thought
Subjects focus on a previously chosen thought.

Recitation
Subjects imagine that they are singing a song or reciting a passage.

Resting
Subjects close their eyes, focus on their breathing and remain still. A variant is resting with open eyes.

Visual Counting
Subjects close their eyes and imagine numbers being written on a blackboard sequentially.

The acceptance of several mental tasks has been surveyed by Chuang et al. Their subjects ($n = 15$) found some tasks difficult (*motor imagery, pass-thoughts*). All tasks were found boring or would not be repeated by parts of the surveyed group [65].

3.2.1.1 Resting with Eyes Closed

The *resting with eyes closed* task is often used in non-event-related paradigms. In the resting state, the alpha activity is dominant (see Table 11.1). The maximum voltage of the alpha wave can be measured over the occipital region, but it also appears in parietal and posterior temporal regions of the brain (see Fig. 11.5). The amplitudes vary from subject to subject and in a given subject also from moment to moment [67]. The influx of light, for instance, due to eye opening, temporarily blocks posterior alpha waves and was already mentioned by Berger in 1929 [1]. Mental tasks or other stimuli do also temporarily block alpha waves, but their influence is not as high as the influence of light [67].

3.2.1.2 Pass-Thoughts

The interesting approach of *pass-thoughts* was introduced by Thorpe et al. They described a system where a user's thought (the pass-thought) acts as a password. That thought has to be recorded with enough accuracy to distinguish between different thoughts of the same user and between same thoughts of different users. The physiological uniqueness of a user's brain would act as biometric component [68].

If it were possible to extract single thoughts from the brain, one could not only build a shoulder surfing resistant authentication system. It would also revolutionize all areas of human-machine interaction. However, as of today, there is no BCI available that is capable of extracting thoughts. That is why Thorpe et al. presented a feasible alternative: a password speller that utilizes event-related potentials.

Chuang et al. use pass-thoughts as one task among many [65]. However, they do not investigate whether different thoughts of the same user can be distinguished. In a follow-up paper, Johnson et al. investigate the extent to which both factors, the secret pass-thought and the biometric component, are necessary for successful authentication. Their result is that Chuang's system is more dependent on biometric components than on the chosen secrets [69].

3.3 Steady-State Evoked Potentials

As the SSEP responses of different frequencies are individual, they could be used as biometric feature if permanence criterion is fulfilled. In 2015 Phothisonothai [70] used SSVEP as biometric feature. The true acceptance rate varied between 60% and 100% on a population of five.

4 Discussion

Various aspects need to be taken into account when designing EGG-based biometric systems. In this section, we will first discuss performance of EEG devices and their usability in general, followed by an overview on attacks and countermeasures. After that, we will address ethics, acceptance, and privacy. Finally, we will discuss future perspectives.

4.1 Performance

At a first glance, EEG-based biometrics is a promising approach. There are several papers claiming classification accuracies close to 100%. But many studies have been conducted on very small populations between 4 and 20 subjects or used a publicly available dataset that is believed to be biased [40]. Depending on the application, it is desirable to be able to sufficiently discriminate much larger populations. Without doubt, the brain itself has got enough uniqueness. But how much of it can be obtained by EEG readings?

Ruiz et al. achieved an accuracy of 100% on a higher population of 50 subjects by combining different types of stimuli in an ERP protocol. But this comes at the cost of longer recording sessions. Per subject, EEG data of 35 minutes were recorded, in contrast to about 5 minutes in other studies. Thus it can be reasoned that by combining multiple features and increasing the recording time, the decrease of the accuracy in higher populations can be compensated. This would, of course, increase the conflict with collectability. Even with an EEG headset with dry electrodes, setting it up takes longer than the total process of conventional biometric systems like fingerprint or iris scans. Using simpler EEG devices with shorter setup times decreases signal quality, thus decreasing accuracy. A small number of electrodes can also decrease accuracy, as the spatial information is reduced.

Many studies rely on a single EEG recording session that is then separated into trials for training and test sets. This poses the risk that the result is not reproducible because it relies on the discriminant information of a transient feature. Such a transient feature could depend on the affective state of the subject. Permanence in EEG-based biometrics has only been assessed in few works [44, 47]. The features used here were sufficiently stable for at least half a year.

EEG-based biometric authentication systems can be deployed under laboratory conditions. There is no evidence that event-related paradigms outperform non-event-related paradigms, or vice versa. In both paradigms, there are tasks that are practically applicable, e.g., *recognition of pictures* (event-related) or *resting with eyes closed* (non-event-related). However, due to the long setup phase and long recording times, EEG-based biometric authentication systems may only be attractive in applications where users already wear EEG devices for a longer period of time. In that regard, tasks that do not rely on a user's attention are preferable and would allow continuous authentication.

4.2 Attacks

Attacks on EEG-based biometric systems can be grouped in different families: *direct attacks* which target the sensor and *indirect attacks* that focus on other parts of the biometric system. In addition, biometric systems can be used as a *side channel*.

Johnson et al. describe a direct attack on Chuang et al.'s [65] non-event-related authentication system. In that system, users can choose their own secret in some of the tasks, for example, which song to recite. The authors address the question of whether an attacker gains advantage from information about a given user's secrets [69].

Maiorana et al. discuss an indirect attack against biometric systems. They assume that an attacker can access the scores produced by a classifier. The attacker uses them to generate EEG templates by means of a hill-climbing procedure until achieving successful authentication [71].

Martinovic et al. showed that it is possible to detect which of the presented stimuli in image-based experiments are related to the user's secret information, including information related to the user's credit cards or the persons known to the user [72]. The problem with their attack is that the stimuli were noticeable by the users. Frank et al. were able to improve this attack by decreasing the stimuli's display time below 13.3 ms. This short period of time is usually not sufficient for conscious perception [73].

4.3 Ethics, Acceptance, and Privacy

EEG data allow to draw conclusions about a user's illnesses, addictions, or moods. For that reason, we will discuss ethics and user acceptance. In addition, storage and evaluation of personal health data are subject to regulations of data protection and security laws. This will be covered at the end of this subsection.

Farah presents a broad overview of practical and philosophical issues in the application of neurotechnology and reviews key examples for each type of issue [74]. Kubler et al. and Haselager et al. discuss ethical aspects of BCI as a medical tool (e.g., with respect to locked-in patients) [75, 76]. Vlek et al. go a step further and imagine a scenario in which a BCI system reveals to an employer the fact that the BCI user might have been drinking the night before [77]. In general, BCIs can be used to obtain personal information. They may be able to reveal psychological states and traits, sexual attraction, or personality characteristics like extraversion, neuroticism, risk aversion, pessimism, persistence, and empathy [74]. That is why research proposals should be reviewed by an ethics committee. As a consequence, the protection of this personal information raises research questions, such as, how to control which features are extracted from the EEG. These and other questions are discussed by Bonaci et al. [78] and Takabi [79].

Acceptance of BCIs has been studied with respect to users with severe motor disabilities [80]. Guger et al. investigated how many users would be able to operate an EEG-based BCI [81]. We can imagine that users might be rather skeptical because a device is reading their minds and, as mentioned above, illnesses, addictions, or moods may be revealed. On the other, the technical innovation could make users curious. Shoulder-surfing resistance or continuous authentication may be features that could outweigh the aforementioned concerns.

EEG data contain information about the user's health and the collection of medical data raises special legal concerns. A comprehensive overview of legal implications of brain imaging is provided by the Committee on Science and Law of The Association of the Bar of the City of New York [82]. Bonaci et al. [78] and Takabi [79] make suggestions for privacy-friendly EEG-based biometrics. However, in a large number of papers, the legal implications of brain imaging with regard to EEG-based biometrics receive too little attention. It is important to ensure, if EEG-based biometrics are used widely, that these do not disclose more information than other biometrics or a traditional password.

4.4 Perspectives

EEG-based biometrics are still far from being significant for real-world applications. Though recording devices and data analysis methods keep improving, it is unclear if it will ever become as common as fingerprint based biometrics. If it becomes common to wear EEG headsets, for example, to use it as a game controller, the game provider might apply biometric methods to identify and track the users. This could be interesting for marketing purposes.

Other than fingerprints, we are not leaving our *brainprints* anywhere behind; thus it is unlikely that EEG-based biometric identification will become important in crime investigations. Even if a *brainprint* was available, recording an artifact-free sample from a reluctant subject might be difficult and it might be even possible to practice countermeasures. Countermeasures have been examined in regard to a P300 based lie detector [83]. On the other hand, not leaving *brainprints* behind and being able prevent sampling are welcome features for authentication.

The extensive time used for setting up the recording device and collecting the data is a main obstacle for a wide deployment of EEG-based authentication. Only in high-security environments such delays can be tolerated. In high-security environments, continuous authentication is a welcome feature. To avoid distraction from the actual task, the authentication should run in the background. This would require authentication schemes that are either task independent or directly tied to the real work of the person to authenticate. The feasibility of distraction-free continuous EEG-based authentication is an open research field.

5 Conclusion

The availability of consumer-grade EEG devices has increased the interest in EEG-based biometrics. As the brain itself holds very much discriminant information, high accuracies are expected. The EEG is a behavioral characteristic that cannot be captured remotely, preventing it from being copied easily. Taking EEG samples is a main disadvantage as it consumes at least minutes compared to conventional biometrics requiring a few seconds or less. Speeding that process up without losing too much accuracy is a main challenge in research. It is not clear how much can be gained by better algorithms or improved EEG devices. EEG might generally be the limiting factor and might have to be replaced by other, possibly invasive, measuring techniques. The long setup and measuring times can be disregarded in applications, where an EEG is worn anyway. This would allow continuous authentication schemes.

As the EEG is not fully understood, it is possible that future research enables extraction of information from recorded EEG samples that should be kept private. The resulting ethical and legal questions have hardly been addressed by biometrics research yet. This risk, accompanied by the usability aspects, is an obstacle for mass deployment for biometrics based on the EEG.

Glossary

Affective state psycho-physiological construct combining subjective evaluation, arousal, and motivation. 23.

Brain-computer interface input device usable without activation of efferent nerves. 2, 33.

Efferent nerves nerves carrying information from the brain to peripheral effector organs. 11.

Electroencephalography measurement of the brain's electrical activity on the scalp. 1, 33.

Event-related potential EEG response to a certain event. 7, 33.

Steady-state evoked potential EEG response to a periodical stimulus. 9, 33.

Steady-state visually evoked potential EEG response to a visual flicker stim-ulus. 9, 33.

Visually evoked potential EEG response to a visual stimulus. 7, 33.

Zener card deck of five cards with simple symbols, used in extrasensory perception experiments. 19.

References

1. H. Berger, Über das Elektrenkephalogramm des Menschen. Eur. Arch. Psychiatry Clin. Neurosci. **87**(1), 527–570 (1929)
2. S. Sanei and J. A. Chambers, *EEG Signal Processing* (Wiley, 2013)
3. X. Jia, A. Kohn, Gamma Rhythms in the Brain. PLoS Biol. **9**(4), e1001045 (2011)
4. J. Rogala, K. Jurewicz, K. Paluch, E. Kublik, R. Cetnarski, A. Wróbel, The do's and don'ts of neurofeedback training: a review of the controlled studies using healthy adults. Front. Hum. Neurosci. **10**, 301 (2016). https://www.frontiersin.org/articles/10.3389/fnhum.2016.00301/full
5. Asanagi, "21 electrodes of International 10-20 system for EEG." https://commons.wikimedia.org/wiki/File:21_electrodes_of_International_10-20_system_for_EEG.svg, 2010. [Online; accessed16-April-2017]
6. H. Gray, *Anatomy of the Human Body* (Lea & Febiger, Philadelphia (US), 1918)
7. B.J. Fisch, R. Spehlmann, *Fisch and Spehlmann's EEG Primer: Basic Principles of Digital and Analog EEG* (Elsevier Health Sciences, Amsterdam (Netherlands), 1999)
8. I.I. Goncharova, D.J. McFarland, T.M. Vaughan, J.R. Wolpaw, EMG contamination of EEG: spectral and topographical characteristics. Clin. Neurophysiol. **114**(9), 1580–1593 (2003)
9. S.S. Spencer, P.D. Williamson, S.L. Bridgers, R.H. Mattson, D.V. Cicchetti, D.D. Spencer, Reliability and accuracy of localization by scalp ictal EEG. Neurology **35**(11), 1567–1567 (1985)
10. T.W. Picton, The P300 wave of the human event-related potential. J. Clin. Neurophysiol. **9**(4), 456–479 (1992)
11. C.S. Herrmann, Human EEG responses to 1–100 Hz flicker: resonance phenomena in visual cortex and their potential correlation to cognitive phenomena. Exp. Brain Res. **137**(3–4), 346–353 (2001)
12. S. Amiri, R. Fazel-Rezai, V. Asadpour, A review of hybrid brain-computer interface systems. Adv. Hum. Comput. Interact. **2013**, 8 (2013)
13. BioSemi B.V., BioSemi news. https://www.biosemi.com/new.htm, 2016. [Online; accessed 16-April-2017]
14. Google Inc., Google Scholar Search Results for Biosemi ActiveTwo. https://scholar.google.de/scholar?q=BioSemi+ActiveTwo, 2017. [Online; accessed 16-April-2017]
15. Google Inc., "Google Scholar Search Results for Emotiv Epoc." https://scholar.google.de/scholar?q=Emotiv+Epoc, 2017. [Online; accessed 16-April-2017]
16. BioSemi B.V., BioSemi products prices. https://www.biosemi.com/faq/prices.htm, 2016. [Online; accessed 16-April-2017]
17. BioSemi B.V., BioSemi pin active electrode. https://www.biosemi.com/pin_electrode.htm, 2016. [Online; accessed 16-April-2017]
18. EMOTIV Inc., EMOTIV EPOC+ 14 Channel Mobile EEG. https://www.emotiv.com/product/emotiv-epoc-14-channel-mobile-eeg/, 2016. [Online; accessed 16-April-2017]
19. T. Grummett, R. Leibbrandt, T. Lewis, D. DeLosAngeles, D. Powers, J. Willoughby, K. Pope, S. Fitzgibbon, Measurement of neural signals from inexpensive, wireless and dry EEG systems. Physiol. Meas. **36**(7), 1469 (2015)
20. T. Koike-Akino, R. Mahajan, T.K. Marks, Y. Wang, S. Watanabe, O. Tuzel, P. Orlik, High-accuracy user identification using EEG bio-metrics. In *International Conference of the IEEE Engineering in Medicine and Biology Society*, 2016, pp. 854–858
21. R.J. Rodriguez, Electroencephalogram (EEG) based authentication leveraging Visual Evoked Potentials (VEP) resulting from exposure to emotionally significant images. In *IEEE Symposium on Technologies for Homeland Security*, 2016, pp. 1–6
22. R.S. Harshit, K.P. Thomas, K. Smitha, A. Vinod, Online Electroencephalogram (EEG) based biometric authentication using visual and audio stimuli. In *IEEE EMBS Conference on Biomedical Engineering and Sciences*, 2016, pp. 454–459
23. J.R. Wolpaw, N. Birbaumer, D.J. McFarland, G. Pfurtscheller, T.M. Vaughan, Brain-computer interfaces for communication and control. Clin. Neurophysiol. **113**(6), 767–791 (2002)

24. M. Bensch, Examination and comparison of methods to increase communication speed of paralysed patients by brain-computer interfaces. PhD thesis, University of Tübingen (2010)
25. S.J. Luck, *An Introduction to the Event-Related Potential Technique* (MIT Press, Massachusetts (US), 2014)
26. M.C. Potter, Rapid serial visual presentation (RSVP): a method for studying language processing. New Meth. Reading Comprehens. Res. **118**, 91–118 (1984)
27. M.C. Potter, Meaning in visual search. Science **187**, 965–966 (1975)
28. J.E. Raymond, K.L. Shapiro, K.M. Arnell, Temporary suppression of visual processing in an RSVP task: an attentional blink? J. Exp. Psychol. Hum. Percept. Perform. **18**(3), 849 (1992)
29. N.G. Kanwisher, Repetition blindness: type recognition without token individuation. Cognition **27**(2), 117–143 (1987)
30. S. Sanei, *Adaptive Processing of Brain Signals* (Wiley, Chichester, West Sussex, 2013)
31. T. Hastie, R. Tibshirani, Classification by pairwise coupling. Proc. 1997 Conf. Adv. Neural Inform. Proces. Syst. **26**(2), 451–471 (1998)
32. M.D. Pozo-Banos, J.B. Alonso, J.R. Ticay-Rivas, C.M. Travieso, Electroencephalogram subject identification: A review. Expert Syst. Appl. **41**(15), 6537–6554 (2014)
33. R. Oostenveld, P. Fries, E. Maris, J.-M. Schoffelen, FieldTrip: open source software for advanced analysis of MEG, EEG, and invasive electrophysiological data. Comput. Intell. Neurosci. **2011**, 1 (2011)
34. A. Delorme, S. Makeig, EEGLAB: an open source toolbox for analysis of single-trial EEG dynamics including independent component analysis. J. Neurosci. Methods **134**(1), 9–21 (2004)
35. C.A. Kothe, S. Makeig, BCILAB: a platform for brain-computer interface development. J. Neural Eng. **10**(5), 056014 (2013)
36. M. Kleiner, D. Brainard, D. Pelli, A. Ingling, R. Murray, C. Broussard, et al., What's new in Psychtoolbox-3. Perception **36**(14), 1 (2007)
37. G. Schalk, D.J. McFarland, T. Hinterberger, N. Birbaumer, J.R. Wolpaw, BCI2000: a general-purpose brain-computer interface (BCI) system. IEEE Trans. Biomed. Eng. **51**(6), 1034–1043 (2004)
38. A.J. Power, E.C. Lalor, R.B. Reilly, Can visual evoked potentials be used in biometric identification?. In *International Conference of the IEEE Engineering in Medicine and Biology Society*, 2006, pp. 5575–5578
39. G.K. Singhal, P. RamKumar, Person identification using evoked potentials and peak matching. In *Biometrics Symposium*, IEEE, 2007, pp. 1–6
40. I.B. Barbosa, K. Vilhelmsen, A. van der Meer, R. van der Weel, T. Theoharis, EEG biometrics: on the use of occipital cortex based features from visual evoked potentials. *Norsk Informatikkonferanse* (2015)
41. R. Palaniappan, P. Raveendran, Individual identification technique using visual evoked potential signals. Electron. Lett. **38**(25), 1634–1635 (2002)
42. J.G. Snodgrass, M. Vanderwart, A standardized set of 260 pictures: norms for name agreement, image agreement, familiarity, and visual complexity. J. Exp. Psychol. Hum. Learn. Mem. **6**(2), 174 (1980)
43. K. Das, S. Zhang, B. Giesbrecht, M.P. Eckstein, Using rapid visually evoked EEG activity for person identification. In *Engineering in Medicine and Biology Society, 2009. EMBC 2009. Annual International Conference of the IEEE*, 2009, pp. 2490–2493
44. B.C. Armstrong, M.V. Ruiz-Blondet, N. Khalifian, K.J. Kurtz, Z. Jin, S. Laszlo, Brainprint: assessing the uniqueness, collectability, and permanence of a novel method for ERP biometrics. Neurocomputing **166**, 59–67 (2015)
45. M.V. Ruiz-Blondet, Z. Jin, S. Laszlo, CEREBRE: a novel method for very high accuracy event-related potential biometric identification. IEEE Trans. Inf. Forensics Secur. **11**, 1618–1629 (2016)

46. R. Das, E. Maiorana, D. L. Rocca, P. Campisi, EEG biometrics for user recognition using visually evoked potentials. In *International Conference of the Biometrics Special Interest Group*, Sept 2015, pp. 1–8
47. R. Das, E. Maiorana, P. Campisi, EEG biometrics using visual stimuli: a longitudinal study. IEEE Signal Process. Lett. **23**(3), 341–345 (2016)
48. H. Touyama, M. Hirose, Non-target photo images in oddball paradigm improve EEG-based personal identification rates. In *International Conference of the IEEE Engineering in Medicine and Biology Society*, Aug 2008, pp. 4118–4121
49. M. Poulos, M. Rangoussi, E. Kafetzopoulos, Person identification via the EEG using computational geometry algorithms. In *European Signal Processing Conference*, IEEE, 1998, pp. 1–4
50. T. Harmony, T. Fernández, J. Silva, J. Bernal, L. Díaz-Comas, A. Reyes, E. Marosi, M. Rodríguez, M. Rodríguez, EEG delta activity: an indicator of attention to internal processing during performance of mental tasks. Int. J. Psychophysiol. **24**(1), 161–171 (1996)
51. M. Poulos, M. Rangoussi, V. Chrissikopoulos, A. Evangelou, Person identification based on parametric processing of the EEG. IEEE Electron. Circuits Syst. **1**, 283–286 (1999)
52. R.B. Paranjape, J. Mahovsky, L. Benedicenti, Z. Koles, The electroencephalogram as a biometric. Canadian Conf. Electr. Comput. Eng. **2**, 1363–1366 (2001)
53. R. Palaniappan, Electroencephalogram signals from imagined activities: a novel biometric identifier for a small population. In *Intelligent Data Engineering and Automated Learning*, Springer, 2006, pp. 604–611
54. A. Riera, A. Soria-Frisch, M. Caparrini, C. Grau, G. Ruffini, Unobtrusive biometric system based on electroencephalogram analysis. EURASIP J. Adv. Signal Process. 18 (2008)
55. R. Palaniappan, Two-stage biometric authentication method using thought activity brain waves. Int. J. Neural Syst. **18**(01), 59–66 (2008)
56. C. Ashby, A. Bhatia, F. Tenore, J. Vogelstein, Low-cost electroencephalogram (EEG) based authentication. In *IEEE EMBS Conference on Neural Engineering*, 2011, pp. 442–445
57. D.L. Rocca, P. Campisi, G. Scarano, EEG biometrics for individual recognition in resting state with closed eyes. In *International Conference of Biometrics Special Interest Group*, 2012, pp. 1–12
58. E. Maiorana, D.L. Rocca, P. Campisi, EEG-based biometric recognition using EigenBrains. In *IEEE International Conference on Multimedia Expo Workshops*, 2015, pp. 1–6
59. C. Mao, B. Hu, M. Wang, P. Moore, EEG-based biometric identification using local probability centers. In *International Joint Conference on Neural Networks*, 2015, pp. 1–8
60. Y. Wang, L. Najafizadeh, On the invariance of EEG-based signatures of individuality with application in biometric identification. In *International Conference of the IEEE Engineering in Medicine and Biology Society*, 2016, pp. 4559–4562
61. K.P. Thomas, A. Vinod, EEG-based biometric authentication using gamma band power during rest state. Circuits Syst. Signal Proces. **37**(1), 1–13 (2017)
62. Z.A. Keirn, J.I. Aunon, A new mode of communication between man and his surroundings. IEEE Trans. Biomed. Eng. **37**(12), 1209–1214 (1990)
63. S. Marcel, J. del R. Millán, Person authentication using brainwaves (EEG) and maximum a posteriori model adaptation. IEEE Trans. Pattern Anal. Mach. Intell. **29**(4), 743–752 (2007)
64. K. Brigham, B.V. Kumar, Subject identification from electroencephalogram (EEG) signals during imagined speech. In *IEEE BTAS*, 2010, pp. 1–8
65. J. Chuang, H. Nguyen, C. Wang, B. Johnson, *I Think, Therefore I Am: Usability and Security of Authentication Using Brainwaves*, Springer, 2013, pp. 1–16.
66. M. Jeannerod, Mental imagery in the motor context. Neuropsychologia **33**(11), 1419–1432 (1995)
67. E. Niedermeyer, F.L. da Silva, *Electroencephalography: Basic Principles, Clinical Applications, and Related Fields* (Lippincott Williams & Wilkins, Philadelphia (US), 2005)
68. J. Thorpe, P.C. van Oorschot, A. Somayaji, Pass-thoughts: authenticating with our minds. In *Proceedings of the 2005 Workshop on New Security Paradigms*, NSPW '05, (New York, NY, USA, 2005), pp. 45–56, ACM

69. B. Johnson, T. Maillart, J. Chuang, My thoughts are not your thoughts. In *ACM UbiComp*, 2014, pp. 1329–1338
70. M. Phothisonothai, An investigation of using SSVEP for EEG-based user authentication system. In *IEEE APSIPA*, 2015, pp. 923–926
71. E. Maiorana, G.E. Hine, D. La Rocca, P. Campisi, On the vulnerability of an EEG-based biometric system to hill-climbing attacks algorithms' comparison and possible countermeasures. In *IEEE BTAS*, 2013, pp. 1–6
72. I. Martinovic, D. Davies, M. Frank, D. Perito, T. Ros, D. Song, On the feasibility of side-channel attacks with brain-computer interfaces. In *USENIX Security*, 2012, pp. 143–158
73. M. Frank, T. Hwu, S. Jain, R. Knight, I. Martinovic, P. Mittal, D. Perito, D. Song, Subliminal probing for private information via EEG-based BCI devices. *arXiv preprint arXiv:1312.6052 [cs.CR]*, 2013
74. M.J. Farah, Neuroethics: the practical and the philosophical. Trends Cogn. Sci. **9**(1), 34–40 (2005)
75. A. Kubler, V. Mushahwar, L.R. Hochberg, J.P. Donoghue, BCI meeting 2005–workshop on clinical issues and applications. IEEE Trans. Neural Syst. Rehabil. Eng. **14**(2), 131–134 (2006)
76. P. Haselager, R. Vlek, J. Hill, F. Nijboer, A note on ethical aspects of BCI. Neural Netw. **22**(9), 1352–1357 (2009)
77. R.J. Vlek, D. Steines, D. Szibbo, A. Kübler, M.-J. Schneider, P. Haselager, F. Nijboer, Ethical issues in brain-computer interface research, development, and dissemination. J. Neurol. Phys. Ther. **36**(2), 94–99 (2012)
78. T. Bonaci, J. Herron, C. Matlack, H.J. Chizeck, Securing the exocortex: a twenty-first century cybernetics challenge. In *IEEE Conference on Norbert Wiener in the 21st century*, 2014, pp. 1–8
79. H. Takabi, Firewall for brain: towards a privacy preserving ecosystem for BCI applications. In *IEEE CNS*, 2016, pp. 370–371
80. S. Blain-Moraes, R. Schaff, K.L. Gruis, J.E. Huggins, P.A. Wren, Barriers to and mediators of brain-computer interface user acceptance: focus group findings. Ergonomics **55**(5), 516–525 (2012)
81. C. Guger, G. Edlinger, W. Harkam, I. Niedermayer, G. Pfurtscheller, How many people are able to operate an EEG-based brain-computer interface (BCI)? IEEE Trans. Neural Syst. Rehabil. Eng. **11**(2), 145–147 (2003)
82. New York City Bar Association et al., Are your thoughts your own, in *Neuroprivacy and the Legal Implications of Brain Imaging*, (The Committee on Science and Law, New York, 2005)
83. J.P. Rosenfeld, M. Soskins, G. Bosh, A. Ryan, Simple, effective countermeasures to P300-based tests of detection of concealed information. Psychophysiology **41**(2), 205–219 (2004)

Part III
Hybrid Technologies

Chapter 12
Multimodal Biometric Invariant Fusion Techniques

P. Viswanatham, P. Venkata Krishna, V. Saritha, and Mohammad S. Obaidat

1 Introduction

Recent advancements in technology have given scope for more threats to personal data and national security due to large amount of stored data. The information transmitted through online can be easily hacked and override the authorized user by the hackers. There are many traditional methods such as password-, watermarking-, and cryptography-based systems to protect the data from the hackers. But these methods are not sufficient to handle new generation applications [1–4].

The biometric based authentication was introduced to avoid the brute force attack. Here, the authentication process is performed by the unique physical features of humans like fingerprint [5], iris, retina, hand geometry, etc. They provide high-secured systems than the traditional methods. Initially, the mono-biometric [6] authentication systems were used to authenticate users and secure systems. Fingerprint verification system is the one of the biometric authentication systems that is highly reliable and is being extensively used by forensic experts. Fingerprint applications include entrance control, door-lock applications, fingerprint identification

P. Viswanatham
School of Information Technology and Engineering, VIT University, Vellore, India

P. Venkata Krishna (✉)
Department of Computer Science, Sri Padmavati Mahila Visvavidyalayam, Tirupati, India
e-mail: dr.krishna@ieee.org

V. Saritha
Department of Computer Science and Engineering, Sri Padmavati Mahila Visvavidyalayam, Tirupati, India

M. S. Obaidat
Department of Computer and Information Science, Fordham University, New York, NY, USA
e-mail: m.s.obaidat@ieee.org

M. S. Obaidat et al. (eds.), *Biometric-Based Physical and Cybersecurity Systems*,
https://doi.org/10.1007/978-3-319-98734-7_12

mouse, and fingerprint mobile phones, among others. The biometric fingerprint means allow authorized users access to multiple clinical, financial, and other systems. It also avoids forgery of certificates, conveying of false information, threats, and crimes.

There are three stages in the fingerprint verification system. These are the enhancement, feature extraction, and comparison. Image enhancement is the preprocessing stage where the quality of the edges is improved and contrast level is increased. The poor-quality images will have low-contrast edges and also the boundaries are not well defined which reduce the ratio of FAR and FRR to about 10% [7].

In case of biometrics, the huge number of images needs to be maintained irrespective of the features, as the population is typically high. Hence, an effective compression technique is required in order to utilize the storage space efficiently. But the disadvantage of using the compression technique is loss of data, which leads to inaccurate matching. In this chapter, the Morlet wavelet algorithm is discussed for fingerprint enhancement and compression during the preprocessing stage of fingerprint verification system [8].

Minutiae-based methods [9, 10] and image-based methods [11–13] are the two variations in fingerprint verification systems. Minutiae is defined as the points of interest in the fingerprint. Minutiae are used as features in minutiae-based methods, and the position of the minutiae, their orientation, and type are stored as sets. The disadvantage is that they may not utilize rich discriminatory information and may have high computation complexity, whereas the image-based methods utilize ridge pattern as feature. Tico et al. [14] proposed transform-based method using digital wavelet transform (DWT) features, while Amornraksa et al. [15] proposed using digital cosine transform (DCT) features. These transform methods show a high matching accuracy for inputs which are identical to the one in its own database. However, these methods have not considered the invariance to an affine transform to deal with different input conditions.

To satisfy the variability condition, integrated wavelet and Fourier-Mellin transform (WFMT) [16] using multiple WFMT features is used. However, this scheme is not suitable for all types of fingerprint images since it chooses core point as a reference point.

To overcome these methods, the simple binaries method is introduced to extract the core reference point. The Zernike and invariant moments are calculated from the reference point invariant to translate, rotate, and scale. The feature is evaluated by the range of correlation between the moments, which reduces the number of features required for comparison during authentication. In this, the authentication is performed by single biometric system [11], which results in high error rates when many similar features exist in the database.

In order to overcome the high error rates, the multimodal biometric system has been developed. It means that more than one biometric [17] is used simultaneously in order to authenticate and validate the user as well as to maintain more information for security purpose. The multimodal biometric system leads to having more information for authentication so it takes more time for authentication and consumes more storage. It results in high complexity, storage, and execution time. The new fused biometric systems have been introduced to solve the above constraints where the features of the multiple biometrics are combined into a single feature and the authentication is performed using predefined threshold value.

The multimodal biometric fusion system leads in an increase in the error rate for authentication due to the more similar features. There are many fusion methods based on decision, score, and feature level that are used in biometric authentication system. These techniques differ upon what biometric information is going to be fused and how the fusing is done. In decision-level fusion techniques [18], the biometric image was divided into equal small squares from which the local binary patterns are fused to single global features pattern. The performance of these techniques leads to 95% of accuracy. The score level fusion technique [19] is fusing the PCA analysis of the face and fingerprint into single identification system, and in this case the error rate reaches more than 11%. The feature level fusion techniques [20] fuse the feature points of the fingerprint and the face and provide 97% efficiency, but none of the previous fusion techniques provide zero error rates.

In this chapter, a new simple and robust fusion technique called the multimodal biometric invariant moment fusion authentication system has been introduced, and it provides better adaptation of genuine and imposter among various test data sets. The fused algorithm gives a single identification decision (data sets) using coefficients which solve the problem of timely constraints and storage space [21]. This approach provides better results than score, feature, and decision-level fusion technique.

2 Multimodal Biometric Invariant Moment Fusion Authentication System

In multimodal biometric system, more than single biometric is used for authentication purpose. Usually, both mono- and multimodal systems perform the two major operations, namely, enrolment and authentication. During enrolment, the distinct information of the biometric is stored in the database which is used for verification. After enrolment, the authentication is performed by comparing the information with the stored information. Depending upon the ratio of similar or non-similar data, the genuine or imposter must be identified.

2.1 Invariant Moment Fusion System

The binaries method extracts the core reference point in which the Zernike and invariant moments are calculated. Translation, rotation, and scaling are performed on invariants. The final features for authentication are evaluated by the range of correlation between the moments to reduce the amount of storage.

2.2 *Fingerprint*

2.2.1 Morlet Enhancement and Compression

The Morlet fingerprint image enhancement and compression [8] consists of two-stages in processing. They are wavelet analysis and smoothening. In wavelet analysis, the Fourier transforms are applied on the 2D Morlet wavelet and the original image separately. The transformed images are then obtained from these transformed functions. The corrected two-dimensional continuous wavelet transform (2D CWT) is obtained by applying the inverse Fourier transform in the transformed image. During the smoothing process, the orientation and the frequency image [22] of the 2D CWT image are estimated and applied in the Gabor filter in order to remove noise.

The steps involved in the algorithm are as follows:

1. The image is decomposed using Morlet wavelet.
2. Ridge segmentation is done to identify the broken ridges.
3. The ridge orientation is estimated.
4. The frequency is estimated using orientation image.
5. The final image is reconstructed based on adjoining chosen filtered blocks.

2.2.2 Morlet Wavelet

2.2.2.1 2D Continuous Wavelet Transforms

2D CWT is performed by convolving a wavelet function and image. For $f(x, y) \in L_2R$, 2D CWT in time domain is given as:

$$cwt(s, a, b) = \frac{1}{\sqrt{s}} \int\int f(x, y)\psi\left(\frac{x-a}{s}, \frac{y-b}{s}\right)dxdy \tag{12.1}$$

where s is the "dilation" parameter used to change the scale and a, b are the translation parameters used to slide in time. The factor of $s^{1/2}$ is a normalization factor to keep the total energy of the scaled wavelet constant.

The 2D CWT in frequency domain is given as:

$$cwt(s, w_1, w_2) = \sqrt{s}F(w_1, w_2)\Phi(sw_1, sw_2) \tag{12.2}$$

where w_1 and w_2 refer to the frequency of the image, $F(w_1, w_2)$ is the low-frequency spectrum, and $\phi(w_1, w_2)$ is the phase modulation, which defines the spectrum of deformed image. The Fourier transform in Morlet wavelet is applied to the image, which calculates the discrete points depending on the scale and displays the real part of the inverse Fourier transform.

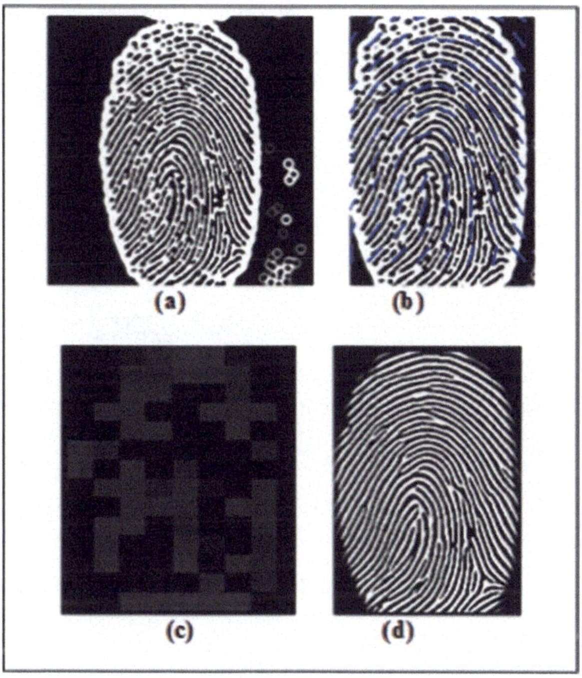

Fig. 12.1 The resultant phases of the enhancement: (**a**) Morlet image, (**b**) orientation image, (**c**) frequency image, and (**d**) enhanced image

$$\psi(kx, ky) = \sqrt{2\pi}\left(e^{-\frac{1}{2}(2\pi kx - k) + (2\pi ky)^2} - e^{-\frac{1}{2}k^2\psi}\right)e^{-\frac{1}{2}\left(2\pi kx^2 + 2\pi ky^2\right)} \tag{12.3}$$

The decomposition of the fingerprint image by 2D Morlet wavelet is shown in Fig. 12.1a. The resultant transformed image has good contrast and enhanced ridges with compression.

2.2.3 Ridge Orientation

The orientation image represents an intrinsic property of the fingerprint image and defines invariant coordinates for ridges and furrows in a local neighborhood as shown in Fig. 12.1b. A ridge center maps itself as a peak in the projection. The projection waveform facilitates the detection of ridge pixels. The ridges in the fingerprint image are identified with the help of eight different masks. The ridges are separated from the fingerprint image by the following equations:

$$I(x, y) = I(x, y) - \text{mean} \tag{12.4}$$

$$S(x, y) = I(x, y)/\sigma \tag{12.5}$$

where σ is the standard deviation and $I(x, y)$ is an integrated image.

By viewing ridges as an oriented texture, a number of methods have been proposed to estimate the orientation field of fingerprint images [22]. Given a transformed image, N, the main steps for calculating dominant directions are as follows:

1. Divide N into blocks of size w × w.
2. Compute the gradients and apply Gaussian filter G_{xy}. The gradient operators are simple Sobel operators and Gaussian filter is applied as follows:

$$G_{xy} = \frac{1}{2\pi\sigma^2} e^{-\frac{x^2+y^2}{2\sigma^2}} \tag{12.6}$$

3. Estimate the local orientation of each block centered at pixel *(i, j)*

$$O(x, y) = \frac{\pi}{2} \times \tan\left(\left(\frac{G_{xy} - G_{yy}}{G_{xy}}\right)/2\right) \tag{12.7}$$

where the degree of smoothening is governed by the variance σ^2.

2.2.4 Frequency Image

The frequency of the fingerprint image is estimated using the orientation image $O(x, y)$ by Eq. 12.7, and it is shown in Fig. 12.1c. The block is rotated and cropped based on the orientation. The median filtering is then applied for smoothening.

$$F(x, y) = \frac{F(u, v)W(u, v)I(u, v)}{W(u, v)I(u, v)} \tag{12.8}$$

where $W(u, v) = \frac{u}{\sqrt{2}}, \frac{v - \frac{u}{\sqrt{2}}}{2}$

$F(u, v)$ is the wavelet transformed image and $I(u, v)$ ensures that the valid ridge frequency is non-zero. The ridge of 3–25 pixels is the valid range.

2.2.5 Enhanced Image

The Gabor filter optimally captures both local orientation and frequency information to smoothen the fingerprint image. By tuning a Gabor filter to specific frequency and direction, the local frequency and orientation information can be obtained, which will be used for extracting texture information from images, which gives smoothing as a part of enhancement by removing the noise shown in Fig. 12.1d.

$$E(x,y) = \frac{1}{2\pi\sigma_x\sigma_y} e^{\left[-\frac{1}{2}\left(\left(\frac{x^2+y^2}{\sigma_{x,y}^2}\right)/2\right)\cos 2\pi fx\right]} \quad (12.9)$$

where σ_x and σ_y determine the shape of the filter envelop and f represents the frequency of the image.

2.2.6 Determination of Reference Point and Regions of Interest (ROI)

The reference point is determined in order to evaluate the ROI of the fingerprint image, which are used to extract the Zϕ moments. This process simplifies the process of the extraction by reducing its complexity.

The Otsu method is used to define the threshold to the binaries of the image. Intra-class variance is defined as a weighted sum of variances of the two classes:

$$\sigma_\omega^2(t) = \omega_1(t)\sigma_1^2(t) + \omega_2(t)\sigma_2^2(t) \quad (12.10)$$

Weights ω_1 and ω_2 are the probabilities of the two classes separated by the threshold of variance and $\sigma_1{}^2$*and* $\sigma_2{}^2$ variances of these classes, respectively. Minimizing the intra-class variance is the same as maximizing interclass variance:

$$\sigma_b^2(t) = \sigma^2 - \sigma_\omega^2(t) = \omega_1(t)\omega_2(t)[\mu_1(t) - \mu_2(t)]^2 \quad (12.11)$$

which is expressed in terms of class probabilities ω_I and class means μ_i. The class probability ω_i(t) is computed from the histogram t:

$$\omega_i(t) = \sum_{i=0}^{t} p(i) \quad (12.12)$$

while the class mean $\mu_i(t)$ is:

$$\mu_i(t) = \left[\sum_{i=0}^{t} p(i)x(i)\right]/\omega_i \quad (12.13)$$

where $x(i)$ is the value at the center of the ith histogram. Similarly, we can compute $\omega_2(t)$ and $\mu_2(t)$ on the right-hand side of the histogram.

The algorithm to binaries detects and crops the ROI of fingerprint:

1. Compute histogram(t) and probabilities of each intensity level
2. Set up initial$\omega_i(0)$ and $\mu_i(0)$.
3. For $t = 1$ to maximum intensity, do:

3.1 Update ω_i and μ_i
3.2 Compute $\sigma_b{}^2(t)$ using Eq. 12.11
3.3 Thresholds $\sigma_{b1}{}^2(t)$ greater and $\sigma_{b2}{}^2(t)$ equal or minimum is defined

$$\text{Threshold } T = \frac{\sigma_{b1}^2(t) + \sigma_{b2}^2(t)}{2} \tag{12.14}$$

$$\text{Binaries image } E_b(x, y) =^{1 \text{ if } E(x,y)>T}_{0 \text{ if } E(x,y)>T} \tag{12.15}$$

4. The region labeled with four connected components is chosen which determines the high curvature region used to determine ROI.
5. The median of the region is taken as reference point, and image is cropped into size of 120 × 120. It is shown in Fig. 12.2.

2.2.7 Invariant and Zernike Moment Analysis

The algebraic invariants and Zernike moment are calculated from the reference point of the fingerprint and are invariant to scale, position, and rotation. Algebraic invariants are applied to the moment generating function under a rotation transformation. Nonlinear centralized moment and absolute orthogonal moment invariants are calculated with reference. Fingerprint ZΦ invariants [18] are shown in Table 12.1.

Fig. 12.2 The resultant phases of the fingerprint enhancement with singular point detection

Table 12.1 Fingerprint Zϕ invariants

Data sets	Train image database						
Fing1	6.6739	24.1707	30.6781	30.3368	66.175	42.5687	60.8585
Fing2	6.6439	21.8419	26.9747	30.2023	60.5443	41.5152	58.8209
Fing3	6.6444	14.9212	28.2185	28.0322	57.1951	35.5803	58.8439
Fing4	6.5548	14.7008	29.5278	28.9722	59.2708	37.7285	58.6214
Fing5	6.6496	23.3503	30.7699	31.9627	64.0907	48.2492	63.8234
Fing6	6.6524	23.6642	30.9556	30.0366	62.4862	43.1547	60.855

Invariant Moments

Central moments of order 3 or less are for translational invariance. For a 2D continuous function $f(x, y)$, the moment of order $(p + q)$ is defined as:

$$m_{0,0} = \sum_{i,j=1}^{n} f \quad \bar{x} = \frac{m_{1,0}}{m_{0,0}} \bar{y} = \frac{m_{0,1}}{m_{0,0}} m_{1,0} = \sum_{i=1}^{n} x \cdot f$$

$$m_{0,1} = \sum_{j=1}^{n} y \cdot f m_{1,1} = \sum_{i,j=1}^{n} x \cdot y \cdot f$$

$$m_{2,0} = \sum_{i=1}^{n} x^2 \cdot f m_{0,2} = \sum_{j=1}^{n} y^2 \cdot f m_{1,2} = \sum_{i,j=1}^{n} x \cdot y^2 \cdot f m_{3,0} = \sum_{i=1}^{n} x^3 \cdot f m_{0,3}$$
$$= \sum_{j=1}^{n} y^3 \cdot f m_{2,1} = \sum_{i,j=1}^{n} x^2 \cdot y \cdot f$$

Second-order central moment for image orientation for scaling invariant:

$$\xi_{1,1} = \frac{\left(m_{1,1} - \bar{y} \cdot m_{1,0}\right)}{m_{0,0}^2} \xi_{2,0} = \frac{\left(m_{2,0} - \bar{x} \cdot m_{1,0}\right)}{m_{0,0}^2} \quad \zeta_{0,2} = \frac{\left(m_{0,2} - \bar{y} \cdot m_{0,1}\right)}{m_{0,0}^2}$$

$$\xi_{3,0} = \frac{\left(m_{3,0} - 3\bar{x} \cdot m_{2,0} + 2 \cdot \bar{x}^2 \cdot m_{1,0}\right)}{m_{0,0}^{2.5}} \quad \zeta_{0,3} = \frac{\left(m_{3,0} - 3\bar{y} \cdot m_{0,2} + 2 \cdot \bar{y}^2 \cdot m_{0,1}\right)}{m_{0,0}^{2.5}}$$

$$\xi_{2,1} = \frac{\left(m_{2,1} - 2\bar{x} \cdot m_{1,1} + \bar{y} \cdot m_{2,0} + 2\bar{x}^2 \cdot m_{0,1}\right)}{m_{0,0}^{2.5}}$$

$$\xi_{2,1} = \frac{\left(m_{1,2} - 2\bar{y} \cdot m_{1,1} - \bar{x} \cdot m_{0,2} + 2\bar{y}^2 \cdot m_{1,0}\right)}{m_{0,0}^{2.5}}$$

A set of seven invariant moments derived from the second and third moments is a set of absolute orthogonal moment invariants proposed by Hu [23].

Rotational invariant moments: $\varphi(1) = \xi_{2,0} + \zeta_{0,2}$.

Moment of inertia (pixel intensity to physical density for rotation invariant).

$$\varphi(2) = \left(\xi_{2,0} + \zeta_{0,2}\right)^2 + \left(4\xi_{1,1}^2\right)\varphi(3) = \left(\xi_{3,0} - 3\xi_{1,2}\right)^2 + \left(3\xi_{2,1} - \zeta_{0,3}\right)^2 \varphi(4)$$
$$= \left(\xi_{3,0} - \xi_{1,2}\right)^2 + \left(\xi_{2,1} + \zeta_{0,3}\right)^2$$

$$\varphi(5) = (\xi_{3,0} - 3\xi_{1,2})(\xi_{3,0} + \xi_{1,2})\left((\xi_{3,0} + \xi_{1,2})^2 - 3(\xi_{2,1} + \xi_{0,3})^2 + (3\xi_{2,1} - \xi_{0,3})\right.$$
$$\left.(\xi_{2,1} + \xi_{0,3})\left(3(\xi_{3,0} + \xi_{1,2})^2 - (\xi_{2,1} + \xi_{0,3})^2\right)\right)$$

$$\varphi(6) = (\xi_{2,0} - \xi_{0,2})\left((\xi_{3,0} + \xi_{1,2})^2 - (\xi_{2,1} + \xi_{0,3})^2 \right.$$
$$\left. +4\xi_{1,1}(\xi_{3,0} + \xi_{1,2})(\xi_{2,1} + \xi_{0,3})\right)$$

$$\varphi(7) = (3\xi_{2,1} - 3\xi_{0,3})(\xi_{3,0} + \xi_{1,2}) \times \left((\xi_{3,0} + \xi_{1,2})^2 - 3(\xi_{2,1} + \xi_{0,3})^2\right)$$
$$+(3\xi_{1,2} - \xi_{3,0})(\xi_{2,1} + \xi_{0,3})\left(3(\xi_{3,0} + \xi_{1,2})^2 - (\xi_{2,1} + \xi_{0,3})^2\right))$$

Skew invariants distinguish between mirror and identical images.

Zernike Moments

The Zernike moment is a set of complex polynomials $\{V_{nm}(x,y)\}$, which form a complete orthogonal set over the unit disk of $x^2 + y^2 \leq 1$ from the polynomial in polar coordinates, where n is the +ve integer or 0, n-$|m|$ is even, $|m| \leq n$ and $\theta = \tan(y/x)$.

The radial polynomial:

$$R_{nm}(r) = \sum_{s=0}^{(n-|m|)/2} \frac{(-1)^s (n-s)!}{s!\left[\frac{n+|m|}{2} - s\right]!\left[\frac{n-|m|}{2} - s\right]!} r^{n-2s} \tag{12.16}$$

The Zernike moment is:

$$Z_{nm}(x, y) = \frac{n+1}{\pi} \sum_{x=0}^{N} \sum_{y=0}^{M} f(x, y) V_{n,-m}(x, y) \tag{12.17}$$

2.3 *Face Fusion System*

The architecture of the face fusion system is shown in Fig. 12.3. The eigen faces are extracted from the face and used for authentication [17]. Initially, the mean and difference of each image in the training set is computed by using Eqs. 12.18 and 12.19. Then the entire centralized image T is merged using mean to obtain the result A. The merged value is used for computing the surrogate covariance matrix L using

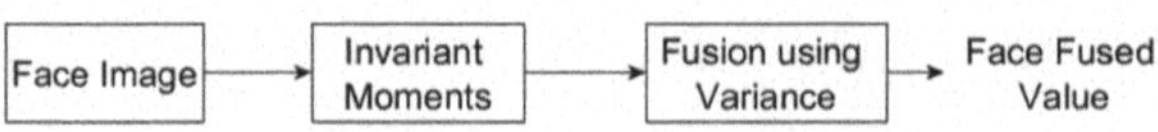

Fig. 12.3 Block diagram of face fusion

Table 12.2 Face Zϕ invariants

Data sets	Train image database					
Face 1	−0.0861	0.0292	0.2199	0.0595	−0.1391	−0.0263
Face 2	0.1025	−0.0871	0.0046	0.0363	−0.1580	0.0161
Face 3	−0.0021	−0.2707	0.0512	−0.0392	0.0847	0.2199
Face 4	0.3195	−0.0552	0.1880	−0.3034	−0.1184	−0.1025
Face 5	−0.3618	−0.0130	0.3020	−0.2350	0.4339	−0.2700
Face 6	−0.4902	0.8825	−0.2266	−1.0756	−0.1895	1.0297

Eq. 12.20. The diagonal elements of covariance matrix are taken as eigen faces using Eq. 12.21. Eigen elements are sorted and are eliminated if their values are greater than 1. Finally, the six invariant features are extracted from the faces using Eq. 12.22.

The high dimensionality makes a good face recognition algorithm. The sample tested face features fusion is shown in Table 12.2.

$$\text{mean} = \frac{1}{n}\sum_{i=1}^{n} X_i \tag{12.18}$$

$$A_i = T_i - \text{mean} \tag{12.19}$$

$$L = A' \times A(X_i - \text{mean}) \tag{12.20}$$

$$[V \times D] = \text{Eig}(L) \tag{12.21}$$

$$\text{Variant} = L \times A \tag{12.22}$$

2.3.1 Fusion

The data sets are independently computed by the described variants of face and fingerprint [18]. The variation distance of the moments is calculated using Eqs. 12.23 and 12.24. It is used for enrolment and comparison during authentication.

$$d_1 = \mu(\varphi_i), \mu(\sigma(\varphi_i)), \mu\left(\sigma^2(\varphi_i)\right), \frac{\mu(\varphi_i)}{\mu(\sigma(\varphi_i))} \tag{12.23}$$

$$d_2 = \mu(Z_i\varphi_i), \mu(\sigma(Z_i\varphi_i)), \mu\left(\sigma^2(Z_i\varphi_i)\right), \frac{\mu(Z_i\varphi_i)}{\mu(\sigma(Z_i\varphi_i))} \tag{12.24}$$

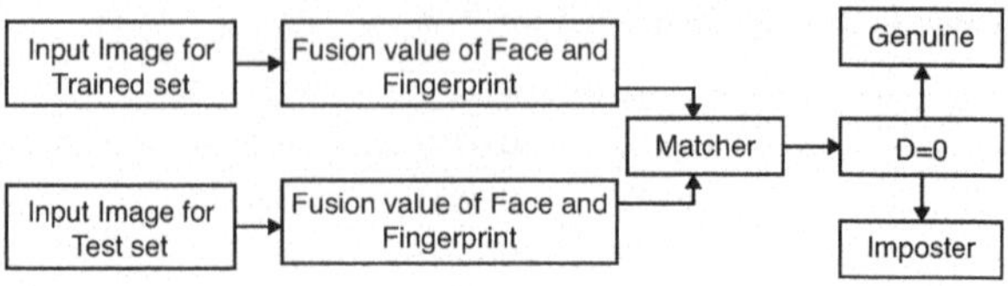

Fig. 12.4 Block diagram of the face and fingerprint fusion authentication

2.3.2 Authentication

The multimodal biometric authentication is one of the new breeds of authentication system performed by means of more than one biometric in order to validate/authenticate the user. The overall architecture of our authentication system is shown in Fig. 12.4. The trained set of inputs in which invariant moment is extracted is fused and enrolled in the database. Now during authentication, the test data input image of fingerprint and face scanned by the user is fused and compared with the fused value in the database. Then matching is performed by calculating the correlation r between the distance d_i of enrolled moments α and verification moments β by Eq. 12.25. The correlation between the fused values computed using Eq. 12.25 and variation using Eq. 12.26 determine whether the user is legitimate or not.

The resultant difference value is compared with the threshold value to validate the user using Eq. 12.22. The threshold value is based upon the sensitivity of the system. If the difference is low, then the similarity will be higher and it crosses the threshold limit to authenticate the user. Otherwise, the user is not authenticated. This multimodal biometric authentication system performed well and provides more than 99% accuracy.

$$r = \frac{2C_{rf}}{C_r + C_f} \text{ where } C_r = \sum_{i=0}^{N} \alpha(i)^2$$

$$C_f = \sum_{i=0}^{N} \beta(i)^2 \text{ and } C_{rf} = \sum_{i=0}^{N} \alpha(i)^2 \beta(i)^2 \tag{12.25}$$

$$D = Fused_{scanned} - Fused_{Enrolled} \tag{12.26}$$

$$A = \begin{cases} \frac{100 - D}{100} \times 100 < \text{Th} = \text{Notauthehticated} \\ \frac{100 - D}{100} \times 100 > \text{Th} = \text{Authenticated} \end{cases} \tag{12.27}$$

3 Experimental Results

The fingerprint image database used in this experiment is the FVC2002 database, which contains four distinct data set DB1, DB2, DB3, and DB4.

The performance is evaluated in terms of false acceptance rate (FAR) and false reject rate (FRR).

$$FAR = \frac{\text{Number of accepted imposter}}{\text{Total number of imposter}} \times 100 \tag{12.28}$$

$$FRR = \frac{\text{Number of rejected genuine}}{\text{Total number of genuine}} \times 100 \tag{12.29}$$

The FAR means imposter accepted as a genuine user, and FRR means the genuine user is rejected as imposter. They are calculated using the Eqs. 12.28 and 12.29, respectively.

The equal error rate (EER) is used as a performance indicator, which indicates the point where FRR and FAR are equal and for evaluating the performance in terms of recognition rate.

The receiver operating characteristic is used as another performance indicator (ROC). It plots the genuine acceptance rate (GAR = 1-FRR) against FAR. The missing probability and alarm probability are evaluated.

Finally, EER is evaluated and results are shown in Figs. 12.5, 12.6, and 12.7, where it is shown that the performance of the proposed system works well in comparison with other image-based approaches.

The DCT coefficient used by Amorniska in [15] and Jimin [16] used WFMT features; Sha [13] with Gabor filter and Ju [24] with invariants using BPNN are compared, and results shown in Table 12.3 with the proposed method provided more accuracy.

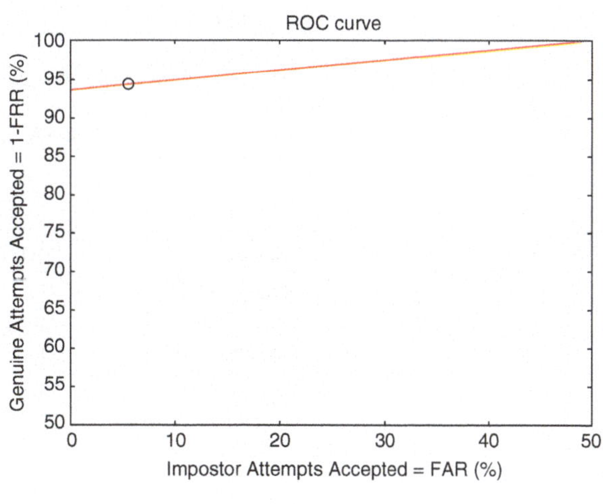

Fig. 12.5 The performance evaluation of the proposed method shows the ROC determines the GAR against FAR

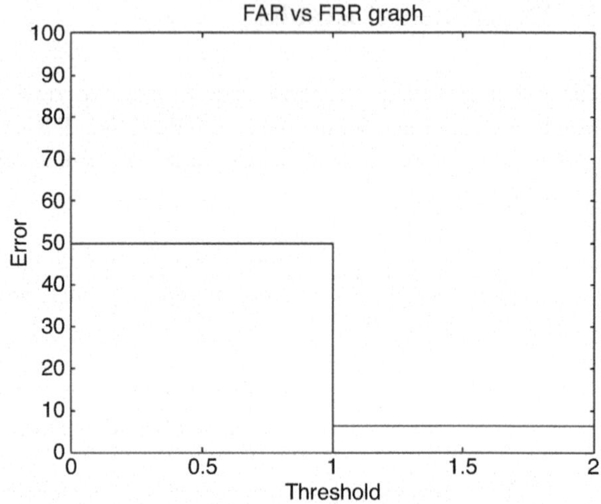

Fig. 12.6 Probability of missing and alarm

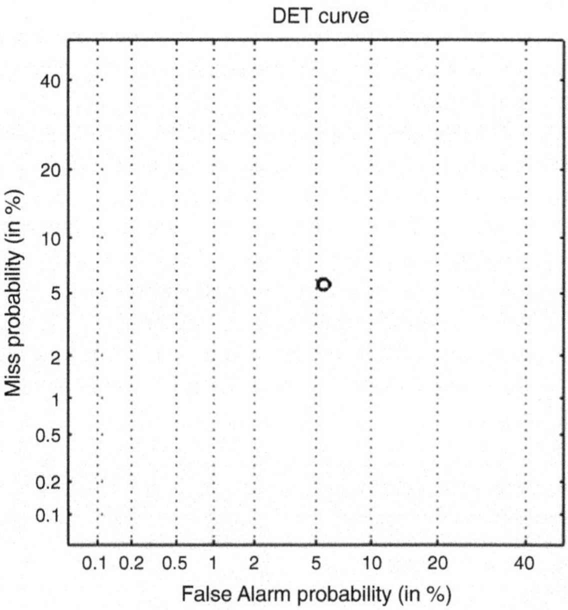

Fig. 12.7 Threshold and the equal error rate (ERR) between FAR vs FRR

Table 12.3 The GAR% against FAR% of the proposed method compared with other methods

Methods	DB1	DB2	DB3	DB4
Amorniska [13]	91.4	85.7	82.1	92.6
Sha [15]	92.7	88.9	85.3	93.2
Jin [14]	94.3	92.4	90.6	94.9
Ju [25]	96.4	95.8	94.2	97.3
Multimodal fusion	98.7	97.2	95	98.2

4 Conclusion

The combined Morlet enhancement with fusion of Zernike and invariant moment features of fingerprint and face is fused by evaluating the distance, mean, and correlation. The combined Morlet enhancement with fusion of Zernike and invariant moment features reduces the storage of features and error rate. The binaries approach using high curvature region accurately determines the reference point used to extract the moments. It is invariant to affine transformations on various input condition. The combined feature maintained for authentication into single identification data reduces the amount of biometric features. The analysis on multimodal biometric using $Z\varphi$ moment's invariant improves the verification accuracy up to 97% as compared to other approaches. The maximum FAR and FRR were maintained at less than 1%. This system demonstrates high reliability, robustness, and good performance in personnel authentication systems.

References

1. M.S. Obaidat, N. Boudriga, *Security of e-Systems and Computer Networks* (Cambridge University Press, Cambridge, UK, 2007)
2. M.S. Obaidat, B. Sadoun, Verification of computer users using keystroke dynamics. IEEE Trans. Syst. Man Cybern. B **27**(2), 261–269 (1997)
3. M.S. Obaidat, B. Sadoun, Keystroke dynamics based identification, in *Biometrics: Personal Identification in Networked Society*, ed. by A. Jain et al. (Springer, Kluwer, 1999), pp. 213–229
4. W. Stallings, *Cryptography and Network Security- Principles and Practices* (Prentice-Hall, Upper Saddle River, 2003)
5. T. Jea, V. Govindaraju, A minutia-based partial fingerprint recognition system. Pattern Recogn. **38**(10), 1672–1684 (2005)
6. T. Jea, V.K. Chavan, V. Govindaraju, J.K. Schneider, Security and matching of partial fingerprint recognition systems. Proc. SPIE **5404**, 39–50 (2004)
7. D. Maio, D. Maltoni, A.K. Jain, S. Prabhakar, *Handbook of Fingerprint Recognition* (Springer, Berlin, 2003)
8. P. Viswanathan, P. Venkata Krishna, Fingerprint enhancement and compression method using Morletwavelet. Int. J. Signal Imaging Syst. Eng. **3**(4), 261–268 (2010)
9. S. Prabhakar, J. Wang, A. K. Jain, S. Pankanti, R. Bolle. Minutiae verification and classification for fingerprint matching. In *Proc. 15th International Conference Pattern Recognition*, Vol. 1, Barcelona, September 3–8, 2000, pp. 25–29
10. J. Liu, Z. Huang, K. Chan, Direct minutiae extraction from gray-level fingerprint image by relationship examination. Proc. Int. Conf. Image Process. **2**, 427–430 (2000)
11. P. Viswanathan, P. Venkata Krishna, Morlet Wavelet fingerprint invariant automated authentication system. Int. J. Recent Trends Eng. **4**(1), 1–5 (2010)
12. C. Chen, Decision level fusion of hybrid local features for face recognition. In Neural networks and signal Processing, 2008 International Conference on (pp. 199–204). IEEE (2008).
13. L.F. Sha, F. Zhao, X.O. Tang, Improved finger code for filter bank-based fingerprint matching. Proc. Int. Conf. Image Process. **2**, 895–898 (2003)
14. M. Tico, E. Immonen, P. Ramo, P. Kuosmanen, J. Saarinen, Fingerprint recognition using wavelet features. Proc. IEEE Int. Symp. Circuits Syst. **2**, 21–24 (2001)

15. T. Amornraksa, S. Achaphetpiboon, Fingerprint recognition using DCT features. Electron. Lett. **42**(9), 522–523 (2006)
16. A.T.B. Jin, D.N.C. Ling, O.T. Song, An efficient fingerprint verification system using integrated wavelet and Fourier-Mellin invariant transform. Image Vis. Comput. **22**(6), 503–513 (2004)
17. D. Maio, D. Maltoni, Direct gray scale minutia detection in fingerprints. Trans. PAMI **19**(1), 27–40 (1997)
18. P. Viswanathan, P. VenkataKrishna, Multimodal biometric invariant moment fusion authentication system. Information Management Processing, BAIP 2010, Springer CCIS, vol 70, 2010, pp. 136–144
19. G.L. Marcialis, F. Roli, Score-level fusion of fingerprint and face matchers for personal verification under "stress" conditions. In *14th International Conference on Image Analysis and Processing (ICIAP 2007)* 0-7695-2877-5/07 $25.00 © 2007 IEEE
20. A. Rattani, D.R. Kisku, M. Bicego, M. Tistarelli, Feature level fusion of face and fingerprint biometrics 978-1-4244-1597-7/07/$25.00 ©2007 IEEE
21. T.-Y. Jea, V. Govindaraju, A minutia-based partial fingerprint recognition system. Pattern Recogn. **38**(10), 1672–1684 (2005)
22. C.I. Watson, G.T. Candela, P.J. Grother, Comparison of FFT fingerprint filtering methods for neural network classification. NISTIR **5493** (1994) Available: https://ws680.nist.gov/publication/get_pdf.cfm?pub_id=900727
23. M.K. Hu, Visual pattern recognition by moment invariants. IRE Trans. Info. Theory **IT-8**, 179–187 (1962)
24. J.C. Yang, D.S. Park, Fingerprint verification based on invariant moment features and nonlinear BPNN. Int. J. Control. Autom. Syst. **6**(6), 800–808 (2008)
25. L. O'Gormann, J.V. Nickerson, An approach to fingerprint filter design. Pattern Recogn. **22**(1), 29–38 (1989)

Chapter 13
Biometrics Based on Healthcare Sensors

Mohammad S. Obaidat, Tanmoy Maitra, and Debasis Giri

1 Introduction

Miniaturization of wireless sensor network into body-centric network makes huge advantages in the healthcare monitoring system. Some heterogeneous psychological healthcare sensors are deployed into a wearable device or planted to the different segments of the human body so that they can collect data in terms of present health condition data such as blood pressure, heartbeat rate, glucose level in blood, and others. This is in addition to many related psychological data from the human body [1, 2]. After collecting the data, the healthcare sensors transmit the data to a controller, which may be a gateway node or mobile device. By applying the application through mobile devices, the human can get medical facilities from the remote places through the Internet. Figure 13.1 depicts network architecture for an e-medical system using healthcare system, where a patient can get different medical facilities by the help of communications among sensors, controller, and different service provider servers. As communications are done through insecure channel like Internet, data security along with proper authentication is highly required in such

M. S. Obaidat
ECE Department, Nazarbayev University, Astana, Kazakhstan

King Abdullah II School of Information Technology (KASIT), University of Jordan, Amman, Jordan

University of Science and Technology Beijing (USTB), Beijing, China

Fordham University, New York City, NY, USA

T. Maitra
School of Computer Engineering, KIIT University, Bhubaneswar 751024, Odisha, India

D. Giri (✉)
Department of Information Technology, Maulana Abul Kalam Azad University of Technology, Nadia 741249, West Bengal, India

M. S. Obaidat et al. (eds.), *Biometric-Based Physical and Cybersecurity Systems*,
https://doi.org/10.1007/978-3-319-98734-7_13

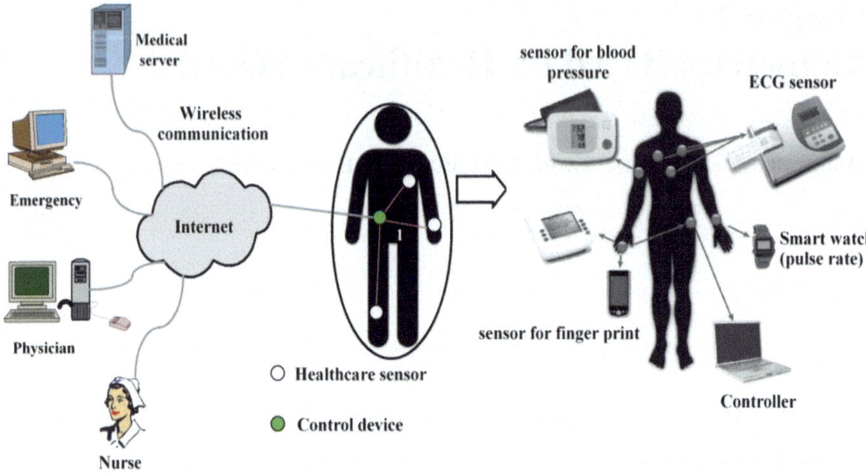

Fig. 13.1 A schematic view of network architecture using healthcare sensors in medical systems

system. Biometric-based authentication is highly demanding in order to increase the security level into the system [3]. Figure 13.1 also shows the different body sensors placed in different segments of the body in order to capture the psychological parameters as biometric features of the patient.

For the biometric-based solution, sensor is the key thing by which a user can interact. In biometric ecosystems, different devices like high-definition cameras to catch face biometrics, ultrasound devices to imprint multiple layers of a fingerprint, infrared cameras to recognize irises biometrics, and subdermal imaging devices to depict palm and finger veins are required. For this purpose, specialized hardware is needed for the sensors so that they can capture the information of distinctive biological features, which is used to identify and verify in biometric mechanism. Recently the consumers became familiar with biometric security measures. For an example, Apple attached a fingerprint sensor to its iPhone. Nowadays, in many industrial applications like e-banking and healthcare, biometric technologies are being accounted rapidly to fulfill the twofold demands, i.e., data security and physical security. For healthcare systems, biometric-based security can go a long way in improving operations and trust in clinics, physician offices, and pharmacies.

The use of biometrics in remote care applications provides both demands, freeing up care staff time tables as well as bed space normally taken up by chronic patients while still offering critical around-the-clock observation. In the future, the healthcare and health insurance industries may depend on biometrics for patient recognition purpose. Fingerprint scans have been used by the blood banks so that the donors feel easy to obey the Federal regulations, which are required for supportive identification of every donation. This mechanism reduces the overhead to store likelihood of duplicate donor records. Other advantages of recognizing patients using biometrics are as follows:

- *Enhancing the reaction capacity to medical emergencies*: Unconscious patients or the patients who are unable to talk can still be quickly identified along with relevant medical history like drug allergies and recent treatments by scanning fingerprint of the patients.

- *Avoiding errors in medication*: Whenever nurses govern medication at the bedside, biometrics can prospectively substitute patient's wristbands and barcodes as identifiers.
- *Decreasing billing fraud*: Biometric identification can be replaced with paper insurance cards. By this identification, a common type of imposture can be protected, which may happen when a patient poses as another to get benefits in medical insurance. It also prevents dishonest providers from applying phantom assert. This is because a biometric scan supplies the proof of a patient's physical existence in the healthcare system.
- *Beneficial on health information exchange*: In healthcare, multiple databases may exist for the same patient. In this case, there is a strong likelihood that one or all will be insufficiently putting the patient at risk for drug interactions or improper services. Patient identification using biometrics can overcome the need for a patient to fill multiple forms at time of each provider's visit, and it also ensures that all information is stored into the same patient's database. However, when digitized patients' information is shared with the providers, perfect patient identification is also required.

2 Real-World Applications: From Marketing Viewpoint

Biometric-based secure systems in healthcare facilities have rapidly grown up worldwide in the past several years because authentication using biometric template of a human can achieve less error in full identification of a patient as well as can provide a secure transmission of sensitive medical data. In today's struggling economy, countries all over the world are showing their great interest in biometric-based security in order to keep privacy of patients' data. Therefore lots of money is invested to build such biometric-based secure system [4]. North America dominates the global healthcare biometrics market by building up a healthcare fraud controlling system to legitimize the patients using their biometric features [4].

According to Technavio's market research analysis [5], the market of global biometric sensors is projected to increase at a CAGR of around 10% between 2016 and 2020. Figure 13.2 shows the usage of biometrics in medical systems from past several years, which is reported by 6Wresearch in 2015 [6]. They also reported that in biometrics market, maximum market revenues have been accrued by fingerprint biometrics technology. Low price of fingerprint-based biometric devices and ease of usage make it most popular in the family of biometrics.

Olea Sensor Networks [7], a company for Internet of Things and wireless sensor networks, is one of the companies bringing advanced vital biometrics to the world of remote healthcare. This company designed a healthcare sensor named as OS-3008 sensor for health monitoring. The healthcare sensor can measure cardiac and respiratory metrics from the patients. For this purpose, this sensor can be placed in the front pocket and there is no need to have skin-to-sensor interaction directly.

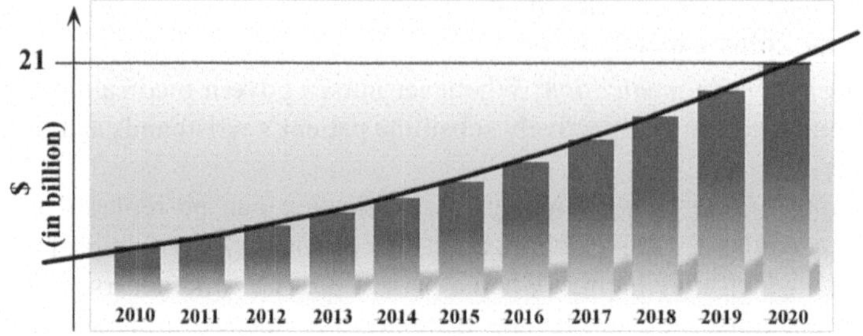

Fig. 13.2 Marketing growth of biometric-based healthcare sensor in medical systems [6]

Phelps Memorial Hospital Center, New York, USA, has developed biometric telehealth system in partnership with Visiting Nurse Association of Hudson Valley [8]. The system can measure biometrics such as patient's weight, pulse, and blood pressure readings in his/her home and can report to the hospital securely so that medical facility can access them from remote places.

3 Constraints of Healthcare Sensor

For the health monitoring applications, different healthcare sensors are implanted into patient's body or positioned on body or embedded into the wearable devices so that they can capture the physiological data like EEG, blood pressure, glucose level in blood and pulse rate of humans. Different biosensors like PillCam Capsule Endoscopy, blood-glucose biosensor, potentiometric biosensor, and thermometric biosensor are implanted into the body. An analytical biosensor device contains the immobilized biological components (i.e., nucleic acid, enzyme, antibody, hormone, and organelle), which interact with analyte and make physical or electrical signals that can be captured and visualized. The propagation loss of signal for biosensors is quite high when they communicate with any on-body device like mobile or laptop. This is because the signal has to propagate into the different layers of fascia, and thus, signal loses its strength. Biosensors can be digested as well as they are easily diluted in the blood. Furthermore, the biosensor provides the services for the little time span as it has short lifetime. Figure 13.3 shows different biosensors used in healthcare system.[1]

On the other hand, different physiological sensors are deployed on patient's body, which sense the physical condition and transmit their data to a mobile device or any gateway nodes. Body-centric communications are taken place to propagate the transmitted signal. The physiological sensors have limited battery power, low computational power, limited memory, and limited range of communication,

[1]http://science.dodlive.mil/2014/04/18/the-science-of-sweat-skin-biosensors/

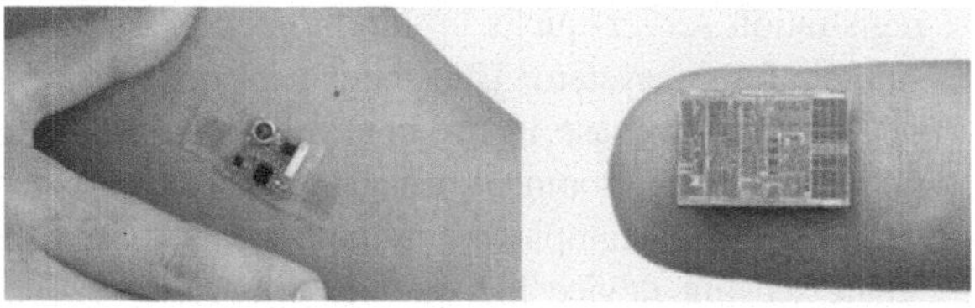

Fig. 13.3 Different types of biosensors. (See footnote 1)

Table 13.1 Purpose to use of different types of healthcare sensors

Sensor type	Usage
Thermostat, bimetallic thermostat, thermistor, resistive temperature detectors (RTD), thermocouple	To capture any physical changes of body temperature that are produced in the form of analog or digital output
Continuous glucose monitoring (CGM)	To timely (day and night) measure the sugar in blood for better diabetes management
Pulse oximetry	To detect oxygen saturation (SO_2) of human to measure percentage of hemoglobin and protein in blood that carries oxygen
Electrocardiogram sensor (ECG or EKG)	To provide report on the electrical activity of heart over a periodic time span using electrodes attached to the skin
Electroencephalography sensor (EEG)	Sleep disorders, coma, encephalopathies, brain death, tumors, stroke, and other focal brain disorders can be diagnosed. For this purpose, voltage fluctuations produced from ionic current within the neurons of the brain is measured.

which is upper bounded by human height. Table 13.1 provides the purpose to use of biosensors as well as on body sensors in healthcare system.[2]

4 Biometric-Based System: An Overview

A biometric system [9] is a pattern recognition system. It produces particular biological and/or behavioral features, known as trait possessed by a human during the fill up of a form to get service. Then it matches the features with the available template of the same trait in the biometric record to identify the person during the service providing time. There are mainly two stages in biometric-based system: (a) registration phase and (b) authentication phase.

1. *Registration phase*: In this stage, each person (patient) has to register his/her biometric characteristics either in controlled environment or in uncontrolled environment so that the biometric characteristics can be stored into his/her corresponding database. In a controlled scenario, the registration party (i.e.,

[2]https://en.wikipedia.org/wiki/List_of_sensors

registration server) picks up the biometrics features of a patient, which will be suitable for the system. Thus the patient needs to be more cooperative at the time of registration phase in the controlled environment. On the other hand, in an uncontrolled environment, a patient provides his/her biometrics by his/her own choice. For an example, when the patients submit their biometrics through their smart sensing device like mobile phone, there does not exist any controlling party. However, for both scenarios, the physical presence of the patients/person in front of the biometric sensor is a must. The biometric sensor collects a sample of biometric trait of the patients. Then the collected biometric sample is transformed into a reference format known as template. After that the processed template is compared with the existing biometric sample templates, which are previously stored into the database of other patients. If it does not find any match with existing templates in the database of other patients, then only the sample will be stored into the database corresponding to that patient.

2. *Authentication phase*: After performing registration phase, the patient again submits his/her biometrics into the sensor, which is same as authentication process. However, the patient will be authenticated to get authorized service if the captured template matches with the stored biometric samples in the database. Otherwise, the patient will be refused to get medical service.

4.1 Biometric Feature Extraction Using Fuzzy Extractor

Generally, to extract a unique feature of a human from biometric information, different fuzzy extractor functions are used. A schematic representation is given in Fig. 13.4 to demonstrate the basic mechanism of feature extraction and authentication. From Fig. 13.4, a generic view of fuzzy extractor [10] is described. Fuzzy extractors [10] map biometric data into random strings, which make it easy to build cryptographic techniques in biometric-based security system. As a biometric key, the

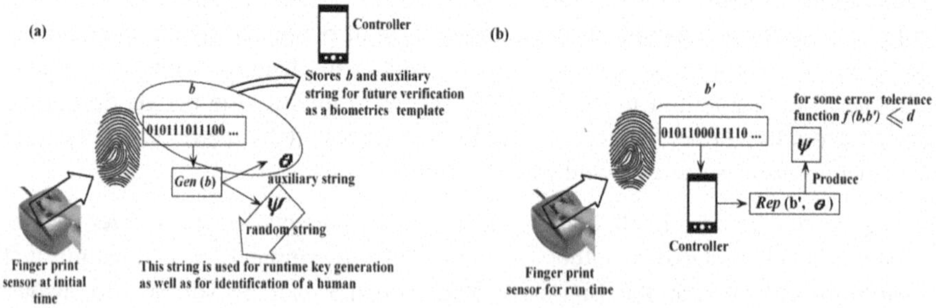

Fig. 13.4 Basic mechanism to features' extraction (**a**) and authentication (**b**) a schematic representation

random strings are utilized to encrypt/decrypt information as well as to verify users' records.

A fuzzy extractor can be explained as a black box system, which has two procedures: (1) The generating function *GEN*(.) takes a binary string b (which is basically the features of biometrics) of metric space M as an input, where $M \in \{0,1\}^n$, for some n bits and produces a random string, say $\psi \in_R \{0,1\}^{a_1}$, for some a_1 bits and an auxiliary string, say $\theta \in \{0,1\}^{a_2}$, for some a_2 bits, where $a_2 = a_1$ or n bits. Moreover it can be written as $GEN : M \rightarrow \psi \times \theta$, where *GEN*(.) is the mapping function. (2) Another procedure known as reproduction function *REP*(.) which takes a binary string, say $b^{'}$ of the metric space $M \in \{0,1\}^n$, where $b \neq b^{'}$ and a uniform distributed binary string, say $\theta \in \{0,1\}^{a_2}$, produces the random string $\psi \in_R \{0,1\}^{a_1}$. It can be represented as $REP : M \times \theta^{'} \rightarrow \psi$, where *REP*(.) is the mapping function.

Before going to use healthcare sensors in medical system, in the registration procedure, the healthcare sensor follows the below procedure, which may be done offline by using any dedicated secure channel.

In Fig. 13.4a, sensor produces the biometric information or psychological information b after scanning the patient or human who needs medical facility [10]. Then the sensor extracts the features (θ, ψ) from the biometric information b using GEN function and transmits the biometric information b and the auxiliary string θ to the controller using secure channel. The controller stores (b, θ) as a biometric template for the user so that the controller can recognize the legitimacy of the human in the future. The sensor uses the random string ψ to make a runtime secret key during the communication in the future. The above phase is addressed as registration phase of the medical system.

According to Fig. 13.4b, during the data communication online, sensor produces the biometric information $b^{'}$ and generates $(\theta^{'}, \psi^{'}) \leftarrow GEN(b^{'})$. Using some cryptographic technique, the sensor sends $(b^{'}, \psi^{'})$ to the controller. After obtaining $(b^{'}, \psi^{'})$, the controller fetches the biometric template (b, θ) of the user and compares $des(b^{'}, b) \leq \partial$, where *des*(.) is an error tolerable distance function and ∂ is the threshold value of the tolerable error of the system. If the said condition holds, the controller reproduces $\psi^{''} \leftarrow REP(b, \theta)$ and checks whether computed $\psi^{''}$ and obtained $\psi^{'}$ are equal or not. If they are equal, the controller assumes that the patient or human is registered and valid; otherwise, discard the human. After that the controller and the healthcare sensor make a common secret key using the random string ψ for secure data communication. The said procedure is known as authentication procedure in medical system.

4.2 Difficulty to Design Biometric-Based Identification

To design a biometric-based system, the following measurements should be checked as the performance metrics [3], [11, 12]:

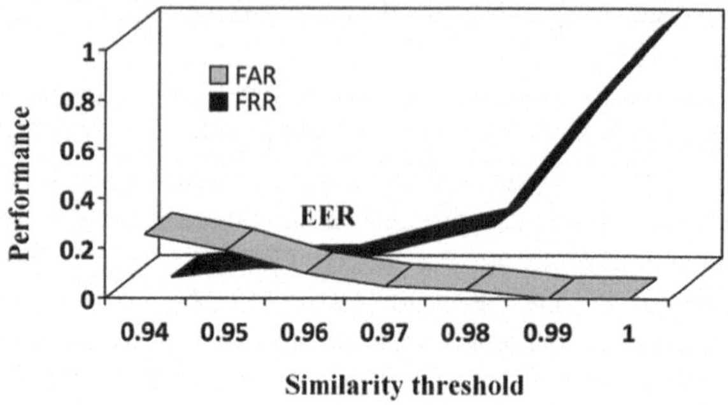

Fig. 13.5 FAR and FRR versus the identification threshold [13]

- *False acceptance rate (FAR)*: If the produced biometric information is faulty and it is incorrectly matched with matching stored template during authentication procedure with some probability, which is more than the threshold value, then an impersonator can be treated as valid.
- *Faulty enrollment*: During the registration procedure, a healthcare sensor may incorrectly produce faulty biometric information due to the low-quality inputs. As a result, during authentication process, a valid patient can be rejected due to the wrong stored biometric template.
- *False rejection rate (FRR)*: If the produced biometric information is accurate and it is incorrectly rejected due to nonmatching with stored template during authentication procedure with some probability which is more than the threshold value, then a valid patient may be rejected due to fault in the system.
- *Equal error rate (EER)*: It is the rate at which both acceptance and rejection errors are equal. Generally, a system having lowest equal error rate is considered as efficient and tolerable system.

An identification threshold (0–1) creates the boundary between acceptance and rejection in a matching mechanism. Figure 13.5 depicts a graphical representation of FAR and FRR as a function of several thresholds values. The converging point between FAR and FRR curves is known as EER [12]. For the lower values of threshold, if FAR increases, then it can be said that the biometric system is more tolerable with respect to input variations and noise.

5 Necessity of Biometric-Based Security Systems

Personal medical data has to be preserved against errors and misconduct in healthcare where sensors are responsible to transmit and receive data from each other through wireless insecure channel [2]. The lightweight security mechanism has to be stated so that it can meet the boundary of limited resources like bounded energy and low computational facility of sensors. After investigation for the suitable solution, it has been endorsed that the intrinsic capability of a human body to exchange information is a unique and also resource-saving solution, which may provide security in wireless communications. For example, physiological attributes of a human can be captured to generate unique identifiers for verifying sensors during the data transmission time. Moreover, to protect against various types of threat, we should address the following matters [14]:

Circumvention: If an adversary may get access to the system safeguarded by biometrics, then he/she can read the sensitive data such as medical records of a patient. Besides this, by breaking the privacy of the registered patients, the adversary can also change the sensitive data.

Repudiation: A valid patient may get access on the facilities provided by a service provider and then he/she can claim that an adversary had circumvented the system. For an example, a hospital clerk alters the medical records of a patient and then he/she can deny the responsibility by saying that an attacker has altered the records.

Covert acquisition: An adversary may sneakily get the raw biometric data of a patient to get access on the system. For example, the latent fingerprints of a patient can be raised from an object by an adversary. In the future, the adversary may use the latent fingerprints to build a digital or physical artifact of that patient's finger.

Collusion: An employee of the system (like an administrator) may intentionally alter the system parameters so that an adversary can enter to the system.

Coercion: A masquerader may create pressure on a legitimate user by some enforcement to grant him/her access to the system.

Denial of service (DoS): An adversary may flood the system resources to the point where the access of legal patients will be denied. For instance, a server which provides access requests can be flooded with a large number of fake requests; therefore computational resources are overloaded, and processing of valid requests will be refused.

A biometric-based system needs to be efficient in terms of computation when it is concerned with medical system. This is because, in medical system, real-time data with low delay is needed in order to monitor the patient. Different cost of biometrics and their security levels are presented in Table 13.2.

6 Different Attacks on Biometric-Based Security System

There are various security threats like chosen plaintext attack, chosen ciphertext attack, impersonation attack, man-in-the-middle attack, software module attack, and insider attack on the template database which are discussed here.

- *Chosen ciphertext attack* (CCA): An adversary adaptively chooses the ciphertext C_i and sees the corresponding plaintext P_i by decrypting under an unknown key for some unknown number of tries i. If the number of tries is limited, then this kind of attack is called *lunchtime attacks* or *CCA1*; otherwise, it is called as

Table 13.2 Purpose to use of different types of healthcare sensors: an overview [9], [14–16]

Biometrics	Characteristic		Vulnerability	Solution
Fingerprint recognition	High	Distinctiveness, performance, permanence	Synthetic or disjoint fingers	For synthetic fingers, thermal scanners are used to observe the temperature. Moisture pattern detection over the fingertip skin can detect the vitality of a fingerprint
	Medium	Collectability, acceptability,		
	Low	Circumvention		
Facial recognition	High	Collectability, acceptability, circumvention	Another person can be characterized trying to imitate	Combined with other biometric technology (multimodality). The use of facial termographies
	Medium	Permanence		
	Low	Distinctiveness performance		
Iris recognition	High	Distinctiveness, permanence performance	Prosthetic eye	Infrared system captures the flow of warm blood in veins of iris
	Medium	Collectability		
	Low	Acceptability, circumvention		
Hand geometry	High	Collectability	Synthetic or dismembered hand	Blood vessel of veins carries blood to the heart. Blood flow through veins for each person has unique behavioral and physical traits
	Medium	Distinctiveness, permanence, acceptability, circumvention, performance		
Voice recognition	High	Acceptability, circumvention	A tape recording of an authorized user	To use a text-dependent system (different for each trial)
	Medium	Collectability		
	Low	Distinctiveness, permanence, performance		

adaptive chosen ciphertext attack or *CCA2*. In such an attack, an attacker has a probability to submit one or more known ciphertexts (captured communication messages) into a system. The system executes an algorithm *ALGO* on the basis of submitted ciphertexts as inputs and produces the corresponding plaintexts just like a black box mechanism. From the produced pieces of information (plaintexts), the attacker can try to extract or recover the hidden secret key required for decryption of the submitted ciphertexts.

- *Chosen plaintext attack* (CPA): An attacker may have *P* number of unrestricted plaintexts, which is controlled by him/her. The attacker then supplies *P* number of plaintexts to the encryption oracle (a black box system). The encryption oracle provides *P* number of ciphertexts to the adversary after encrypting the plaintexts. After receiving *P* number of ciphertexts in such a way that the adversary knows each plaintext-ciphertext combination, i.e., he/she knows each cipher corresponding to plaintext, the attacker can try to extract the secret key used by encryption oracle to encrypt the plaintexts. The computational complexity may be minimized, since the adversary has no restriction to choose the plaintexts to match his/her needs.

- *Impersonation attack*: If an adversary gets secret information like identity of patient, secret key from the captured transmission messages, then he/she may successfully enter into the system as the legitimate party. The aim of a strong entity authentication protocol is to construct such a way so that any party B, distinct from A, cannot play the role of$A$.
- *Man-in-the-middle attack*: In this attack, an adversary stealthily relays and modifies the communication messages between two communicators who think that direct communication is going on. Moreover, in active eavesdropping, the adversary builds self-governing connections with the victims and relays messages between the victims to ensure them that they are communicating directly to each other through a private connection. But, the entire communications are run by the adversary. The adversary intercepts all the relevant messages passing between the two victims.
- *Software module attack*: The executable program at a module can be altered in a way so that it always produces the results desired by the attacker. Trojan-horse attack is the common attack in this scenario. To overcome this attack, specialized hardware or secure code execution practices can be used to impose secure execution of the software. Algorithmic integrity is another solution, which also imposes the component of software integrity. Algorithmic integrity can be defined as follows: software should manage any provided input in a desirable way. For an example of algorithmic drawback, assume a matching module in which a particular input B_0 is not managed properly, and when B_0 is supplied to the matcher, it always produces the acceptance (yes) decision. However, this loophole may not affect the biometric-based functioning system. This is because the probability of generating such B_0 from a real biometric data can be negligible. But, an attacker may utilize this drawback to easily break the security without being detected.
- *Attacks on the template database*: An adversary from inside the system may try to alter or can share the biometrics template with the other untrusted parties. Therefore, this attack produces very much damage for the biometric-based security system. Attacks on the template can lead to the following three security threats: (a) a template can be substituted by a masquerader's template to achieve unauthorized access; (b) after stealing the template of a valid person, physical presence of the valid person can be mounted from the template as a replica to get success on unauthorized entry to the system as well as the other systems, which follow the same biometric template; and (c) the stolen template may be replayed to the matcher to obtain uncertified entrance. For an example, a biometric template stolen from a healthcare's database can be used to search criminals' biometric records or cross-link to person's banking records.

7 A Survey of Biometric-Based Communication Protocols

Different biometric-based secure communication protocols are reviewed in this section, where different psychological data like fingerprint, heartbeat rate, EEG, or ECG of human sensed by healthcare sensors are used as biometric features. By taking these features, secure key establishment between healthcare sensors and controller devices as well as verification of legitimacy of the patients can be achieved.

Poo et al. [17] explored the use of security approaches in body-centric network. In this study [17], a biometric approach that uses the inherent features of the human body as identity for the authentication purpose or the means of protecting the distribution of a cipher key, which ensures the security in communication, Bao et al. [18] analyzed the performance of a biometric like the inter-pulse intervals (IPIs) of heartbeats that have been computed from electrocardiogram and photoplethysmogram of 99 subjects.

By getting the benefit from wireless technology, microelectronics and material science have employed to the evolution of sensors that can be utilized for monitoring of unreachable environments. Health monitoring and telemedicine are the common applications where sensors are used. Health monitoring involves collection of data about crucial health information from different parts of the body and making decisions based on them. This information is of personal nature and is needed to be secured because insecurity may also lead to serious consequences. Due to the extreme constraints of energy, memory, and computation, securing the communication among the sensors is not a trivial problem. However, key distribution is central to any security mechanism. Biometrics extracted from the body has been used in a study by Cherukuri et al. [19].

Miao et al. [20] presented a key distribution system by which two sensors placed on a human body can agree on a variable cryptographic key. The authors of the study [20] argued that simulations based on ECG data from MIT PhysioBank database are giving a minimum half total error rate (HTER) of 0.65%. Next, we will present briefly a study that describes some biometric-based systems using healthcare sensors in more details.

7.1 *Biometric-Based Authentication Using Body Sensors*

Wang et al. [21] devised a biometric-based secure authentication mechanism for wireless body area networks. In the mechanism [21], biometric information has been distributed by healthcare sensors, which are placed at different segments of a person's body. In this study, the security has been preserved in the data communications among these sensors by using authentication and selective encryption technique. The authors showed that the proposed scheme occupied low computational power in the environment of limited resources like bandwidth and battery.

Wavelet-domain hidden Markov model (HMM) classification method [22] has been used for correct authentication. The biometric feature used in this work is the ECG. The basic feature of this is given below.

Wavelet-domain hidden Markov model (HMM): Two sensors are placed on the patient's body, which collect critical ECG data for a certain time duration. After collecting different ECG signals as a training dataset, Wavelet-domain hidden Markov model [22] is applied for each class C_i to derive parameter θ_m, which is basically a statistical characteristic of ECG signal. This statistical characteristic parameter is required to determine the most likelihood observation of ECG signal during authentication phase. However, if a new ECG observation w comes, then using θ_m for each class C_i, maximum likelihood for w is calculated as [22]:

$$\begin{aligned} f_W(w) &= \sum_{m=1}^{M} f(w|\theta_m) \\ &= \sum_{m=1}^{M} \Pr_s(m) f_{W|s}(w|S=m) \end{aligned} \tag{13.1}$$

where, for a random variable W, a M-state mixture model consists of (a) a discrete random state variable S and takes the values $s \in [1, M]$ with pmf $\Pr_S(s)$ and (b) the conditional pdfs $f_{X\,|\,s}(x|\,S=m)$, $s \in [1, M]$.

Authentication: Figure 13.6 shows the authentication procedure. A healthcare sensor S_1 (sender) collects ECG data *MSG*. Then S_1 performs hashing operation on *MSG* as *h*(*MSG*). For each wavelet coefficients w_i, it determines the suitable parameter θ that best characterizes the ECG. Then S_1 performs key hashing using the key θ and produces *MAC*. Next, it sends $\langle MSG, MAC\rangle$ to the other sensor S_2 for authentication.

After getting $\langle MSG, MAC\rangle$, it performs hash operation on *MSG* to get *h*(*MSG*). Then S_2 finds the maximum likelihood estimation on both *MAC* and *h*(*MSG*). The

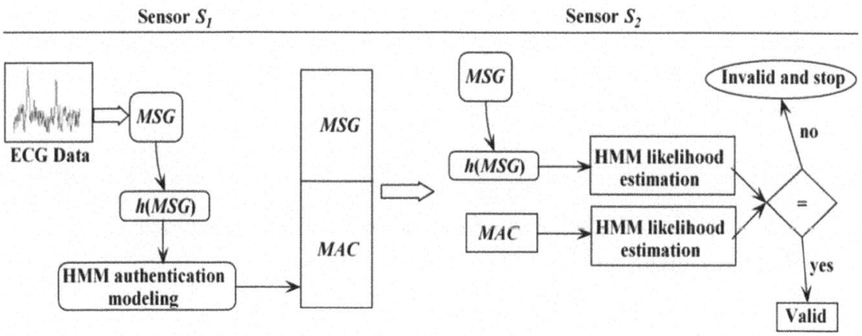

Fig. 13.6 A biometric-based security mechanism using HMM [21]

estimation will be closed enough for the same value of θ. If it is close to each other, S_2 accepts S_1 as authentic.

7.2 *Beat-to-Beat Heartbeat Interval: A Biometric Approach for Authentication*

As discussed earlier, in the m-Health applications, to monitor and provide treatment of diseases, several heterogeneous biomedical sensors are implanted (or placed) in (or on) an individual's body. The sensors form a body-centric network through wireless connectivity, where how to maintain security on private data transmission is a major concern. A study by Bao et al. [23] proposed a solution to handle the problem of authentication of sensors in body area sensor network for m-Health. For this purpose, this study [23] took physiological signals detected by biomedical sensors due to its dual functionalities: (a) it is suitable for specific medical application like healthcare, and (b) sensors in the same body can identify each other by biometrics of the patient. The basic procures in [23] are briefly reviewed next.

Registration: A common secret key s is embedded into the memory of all healthcare sensors before going to implant into or place on the body of a patient.

Authentication: Whenever a sensor wants to transmit its data via multi-hop fashion, the receiver verifies the sender each and every time through the biometrics of the patient. The sender generates a random number r and extracts the biometric feature I, i.e., heartbeat from the patient. Then the sender encrypts the biometric feature I using the secret key s as $Enc[I]_s$ and sends $\langle r, Enc[I]_s \rangle$ to the receiver. For this purpose, symmetric key cryptosystem can be used here.

After getting $\langle r, Enc[I]_s \rangle$ from the sender, the receiver decrypts $Enc[I]_s$ using the same key s to get biometric feature I. Receiver also extracts the biometric feature I' from the patient's body at runtime and then compares $des(I, I') \leq \theta$, where $des(\cdot)$ is a function to measure hamming distance between two features. If the condition holds, the receiver replies $\langle r, Enc[I']_s \rangle$ to the sender; otherwise, it rejects the sender.

After getting $\langle r, Enc[I']_s \rangle$ from the receiver, the sender decrypts $Enc[I']_s$ using the same key s to get biometric feature I'. Then it compares the hamming distance between I and I'. If the distance is within the threshold value θ, then it accepts the receiver as authentic; otherwise, it rejects the message.

7.3 *Fingerprint-Based Authentication and Secure Data Transmission*

Nowadays, body-centric wireless sensor network technologies are growing rapidly in healthcare application due to its easy access and self-configuration [2]. Providing data integrity and protecting secret information of the messages are the most serious issues when sensors transmit their data in wireless sensor networks [1]. The authors in [24] proposed a biometric-based security mechanism for healthcare applications using body sensors. The proposed mechanism [24] has three objectives. First, to elevate the level of security, a key has been computed from the receiver's fingerprint biometrics, and using the computed key, the actual data is encrypted. Second, to minimize the transmission-based attack like replay attack and chosen cipher/plain-text attack, a genetic operator has been used to randomize the fingerprint-based cryptographic key. Third, only authenticated receiver can decrypt the message as well as the proposed mechanism ensures the data integrity as biometric cryptographic key has been used. Different phases of fingerprint-based authentication and secure data transmission [24] are as briefly reviewed below.

Registration: In [24], patients maintain a biometric template of doctors into their smart devices, and similarly, doctors maintain a biometric template of patients into their smart devices in the registration procedure. For this purpose, only fingerprint is used as the biometric. However, templates are maintained to produce biometric key during the secure data transmission between doctor and the patients.

The patents use the biometric template (stored in smart devices of patient during registration phase) of doctor to produce runtime key for encryption data whenever the patients want to get e-health service from the doctor. At the receiver end, the doctor can get the patients' data after submitting his/her biometrics to his/her smart device by producing the same biometric key. Furthermore, when doctors want patients' data, they use patients' biometric template to produce runtime key for encryption. In the following, key generation mechanism proposed in [24] is briefly reviewed where a patient sends his/her health-related data to a doctor.

Key generation: During data transmission from a patient to doctor, the smart device of the patient produces a runtime key to encrypt the data by following procedure shown below:

(a) A patient retrieves the biometric features of the doctor from his stored biometric template and applies a genetic two-point crossover operation with the help of his own identity. Figure 13.7 shows the basic operation.
(b) After crossover operations, the smart device produces a 64 bit key in order to encrypt the data.

Fig. 13.7 Key generation procedure

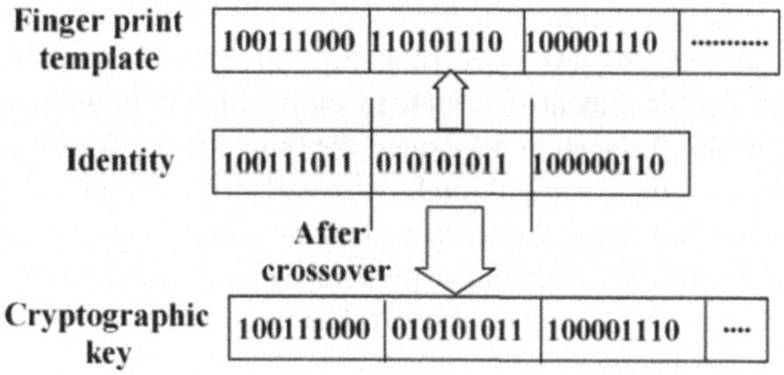

However, at the receiver end, i.e., doctor inputs their biometric into their smart device and then the smart device produces the same key if the input biometric information is correct by applying the genetic two-point crossover operation.

Encryption/Decryption: In the approach reported in [24], both receiver and sender produce a common secret key for the correct biometric features of the receiver. However, after generating key, smart device encrypts the data sensed by different healthcare sensors using the key. For this purpose, symmetric key cryptography (block cipher) like AES or DES is used to encrypt the data. In addition, decryption procedure is reverse of the encryption procedure as private key cryptography is used.

8 Open Questions: Future Direction

Though current studies try to solve the issues in biometric-based security systems, there are some open questions that need addressing by researchers including the ones given below:

Prone to fraud the system: Is it easy to fool the system by unlawful mechanisms? Is it feasible to steal a biometric feature or template of other persons?

Uniqueness and distinctiveness of the biometric feature: What is the probability to get different persons with same biometric features? Can biometric template be accurately distinguishable?

Changing of the biometric features: How fast do the biometric features of a person change over time? What will happen if a user loses the used biometric features, in case of losing eyesight or losing a finger? Can a face be identified after 20 years?

Accuracy of the system: How accurate is the system while taking the biometric sample? Is the authentication system producing comfort to the user?

Cooperation of the user: How much cooperation is required from the user while collecting the biometric sample? Are the method used to collect biometrics accepted by local culture?

On the other hand, if biometric template database is stolen or compromised from smart devices like a mobile phone or a healthcare sensor (i.e., sink), then whole security system may be hampered. However to overcome this issue, template-free biometric system is highly required. In addition, many people hang back to submit either raw or processed biometric information to centralized system. Furthermore, the people may also hesitate to supply his/her biometrics to an untrustworthy system because the system may share the biometric data with other service providers. Thus, decentralized recognition facility will be more reasonable for the current applications while delivering the services. In such scenario, a biometric system should reserve the biometric information of patient not in a centralized manner, but in decentralized manner along with encrypted databases in a way so that the individual and distinct servers can have full control.

9 Conclusion

Patient safety continues to be one of healthcare's most pressing challenges. There are many perspectives from which patient safety can be addressed. The elimination of duplicate medical information and the elimination of medical identity theft stand out as two of the main issues in the healthcare industry. Costing the industry millions of dollars per year in administrative costs, legal expenses and liabilities, in addition to placing patient safety at risk, the core cause of these problems is generally inaccurate patient identification; a problem that can be rectified by the adoption of biometric technology. Biometrics also eliminates the challenges for recognizing patients. For an example, for each time, when a patient visits a healthcare organization, he/she has to submit his/her first name and surname multiple times. Furthermore, we may have some patients with same first name and surname. Therefore, there is a chance to build multiple medical records against the different or same patient. Basically, biometrics uses physiological features of the human body for recognizing the patient to eliminate the requirement of other information of the patient like insurance card, social security number, or date of birth during registration. A biometric template can be directly linked to an electronic medical database for perfect credentialing on subsequent visits. Thus biometrics ensure that medical records of each patient are perfectly registered no matter what variation of their names is submitted, and no duplicate medical database can be generated.

This study focused on biometrics-based on healthcare sensors. We also explained the trends and challenges regarding security and privacy in healthcare system using medical sensors. It has been exhibited that a well-planned security technique must be designed for the prosperous deployment of such applications. The chapter further discussed the different difficulties that are faced in order to design a biometric-based security system using healthcare sensors. Different possible security attacks in biometric-based systems were also reviewed. Furthermore, some solutions to design biometric feature-based key distribution as well secure data transmission are also addressed and reviewed.

References

1. M.S. Obaidat, S. Misra, *Principles of Wireless Sensor Networks* (Cambridge University Press, Cambridge/New York, 2014)
2. S. Misra, V. Tiwari, M.S. Obaidat, LACAS: learning automata-based congestion avoidance scheme for healthcare wireless sensor networks. IEEE J. Sel. Area Commun. **27**(4), 466–479 (2009)
3. M.S. Obaidat, B. Sadoun, Verification of computer users using keystroke dynamics. IEEE Trans. Syst. Man Cybernet. Part B **27**(2), 261–269 (1997)
4. http://www.idg.com.au/mediareleases/27490/healthcare-cyber-security-market-to-benefit-from/
5. Technavio: Global Biometric Sensors Market 2016–2020. https://www.technavio.com/report/global-sensors-global-biometric-sensors-market-2016-2020

6. 6Wresearch partnering growth. http://www.6wresearch.com/press-releases/global-biometric-bio-metric-market.html
7. Olea Sensor Networks. http://www.oleasys.com/
8. Health in the home. http://findbiometrics.com/biometric-healthcare-in-the-home-302190/
9. S. Prabhakar, S. Pankanti, A.K. Jain, Biometric recognition: security and privacy concerns. IEEE Secur. Priv. **1**(2), 33–42 (2003)
10. D. Giri, R.S. Sherratt, T. Maitra, A novel and efficient session spanning biometric and password based three-factor authentication protocol for consumer usb mass storage devices. IEEE Trans. Consum. Electron. **62**(3), 283–291 (2016)
11. M.S. Obaidat, D.T. Macchairllo, An on-line neural network system for computer access security. IEEE Trans. Ind. Electron. **40**(2), 235–242 (1993)
12. I. Traore, I. Woungang, B. Khalilian, M.S. Obaidat, A. Ahmed, Dynamic sample size detection in learning command line sequence for continuous authentication. IEEE Trans. Syst. Man Cybern. B **42**(5), 1343–1356 (2012)
13. M.A. Zahhad, S.M. Ahmed, S.N. Abbas, Biometric authentication based on PCG and ECG signals: present status and future directions. SIViP **8**(4), 739–751 (2014)
14. M.S. Obaidat, N. Boudriga, *Security of e-Systems and Computer Networks* (Cambridge University Press, 2007)
15. M.S. Obaidat, B. Sadoun, in *Biometrics: Personal Identification in Networked Society*. Keystroke dynamics based identification (Kluwer, Boston, 1999), pp. 213–229
16. M.F. Zanuy, On the vulnerability of biometric security systems. IEEE Aerosp. Electron. Syst. Mag. **19**(6), 3–8 (2004)
17. C.C.Y. Poon, Y.-T. Zhang, S.-D. Bao, A novel biometrics method to secure wireless body area sensor networks for telemedicine and m-health. IEEE Commun. Mag. **44**(4), 73–81 (2006)
18. S.D. Bao, C.C.Y. Poon, Y.T. Zhang, L.F. Shen, Using the timing information of heartbeats as an entity identifier to secure body sensor network. IEEE Trans. Inf. Technol. Biomed. **12**(6), 772–779 (2008)
19. S. Cherukuri, K.K. Venkatasubramanian, S.K.S. Gupta, Biosec: a biometric based approach for securing communication in wireless networks of biosensors implanted in the human body, in *Proc. of International Conference on Parallel Processing Workshops*, Taiwan, pp. 432–439, 2003
20. F. Miao, L. Jiang, Y. Li, Y.T. Zhang, Biometrics based novel key distribution solution for body sensor networks, in *Proc. of Annual International Conference of the IEEE Engineering in Medicine and Biology Society*, Minneapolis, pp. 2458–2461, 2009
21. H. Wang, H. Fang, L. Xing, M. Chen, An integrated biometric-based security framework using wavelet-domain HMM in wireless body area networks (WBAN), in *Proc. of IEEE International Conference on Communications (ICC)*, Kyoto, pp. 1–5, 2011
22. M.S. Crouse, R.G. Baraniuk, R.D. Nowak, Hidden Markov modelsfor wavelet-based signal processing, in *Signals, Systems and Computers, the Thirtieth Asilomar Conference*, pp. 1029–1035, vol. 2, 1996
23. S.-D. Bao, Y.-T. Zhang, L.-F. Shen, Physiological signal based entity authentication for body area sensor networks and mobile healthcare systems, in *Proc. of IEEE Engineering in Medicine and Biology 27th Annual Conference*, Shanghai, pp. 2455–2458, 2005
24. B. Shanthini, S. Swamynathan, Genetic-based biometric security system for wireless sensor-based health care systems, in *Proc. of International Conference on Recent Advances in Computing and Software Systems*, Chennai, pp. 180–184, 2012

Chapter 14
Biometric Authentication for Wearables

Harin Sellahewa, Nasiru Ibrahim, and Sherali Zeadally

1 Introduction

The first wearable computer could be ascribed to the invention of Edward O. Thorp in 1961. Edward and Claude Shannon built a small computer that computes and predicts where a roulette wheel ball would land [1]. The device had 12 transistors and microswitches which easily fit in a shoe. Inputs were registered in the shoe, and an ear attachable speaker was used as the output. The system was tested in Shannon's basement before going to the casinos in Las Vegas. One person (Shannon) needed to wear the shoe with the small computer in and the other person (the bettor – Edward) to wear the speaker. When the ball is rolled, Shannon would press one of the switches in the shoe, and a sound would be transmitted to the speaker indicating where the ball could land, and consequently the bet would be placed there. The system would only work when bets are placed after the ball has been set in motion. The device prediction caused a gain of 44% which was in line with their basement test outcome. However, the system was short-lived because of hardware problems, but others went on to perfect the device and built similar wearables throughout the 1970s, all for the purpose of cheating the roulettes [2].

Further, there were several other inventions over the course of time leading up to the millennium. One of the famous ones is the creation of the consensual first wearable computer, a backpack-mounted head camera by Steve Mann in the 1980s as he described in [2]. His device, an early Google Glass-like style, had the capability

H. Sellahewa (✉) · N. Ibrahim
Department of Applied Computing, University of Buckingham, Buckingham, UK
e-mail: harin.sellahewa@buckingham.ac.uk

S. Zeadally
College of Communication and Information, University of Kentucky, Lexington, KY, USA

M. S. Obaidat et al. (eds.), *Biometric-Based Physical and Cybersecurity Systems*,
https://doi.org/10.1007/978-3-319-98734-7_14

of taking photographs unobtrusively. He followed through with the development of a webcam that wirelessly transmits images from the backpack-mounted device to the web in 1994 [3]. Another wearable computer invented in the 1990s is the Forget-Me-Not device which was developed to help with everyday memory problems such as finding a lost document, remembering somebody's name, send reminders, etc. [4]. The device recorded interactions with people and devices and stored them in a database for subsequent retrieval.

This section presents a background to wearable technologies, including their developments over the past decade, the current landscape, and trends. A brief introduction to biometric authentication is also given. The rest of the chapter is organized as follows: Sect. 2 looks at a wide variety of wearable technologies and their capabilities in detail; a review of literature on the use of wearable technologies within "traditional" biometric modalities is presented in Sect. 3; the use of wearable technologies to enable new types of biometrics and their application are explored in Sect. 4. Challenges in biometric authentication for wearables are discussed in Sect. 5. This chapter concludes with final remarks in Sect. 6.

1.1 Wearable Technologies

The term wearables or wearable technology has become popular recently because of recent advances in mobile technologies and the realization of their potential benefits. Wearable technology could be referred to as devices or electronics that are equipped with microchips and other sensory features to perform one or many tasks and that are comfortably worn on the body as clothing or accessories [5]. Wearable devices are often integrated with or designed to replace existing accessories such as a watch. These devices have more mobility and flexibility and could in some functions perform better than the traditional handheld and mobile devices such as tablets and laptops due to their sole function specification. One of the most important features of this technology is the communication capability it has with the wearer as well as other devices. It can communicate and give feedback in real time.

Wearable technology includes devices such as watches, glasses, bands, fabrics, jewelry, and so on. Wearable technology has had significant impact in the fields of health, fitness, and sports as well as education, entertainment, military, music, and even law enforcement. However, a field that wearable technology has not actually been exploited but would be practical is the field of biometrics. Wearable technology tracks or monitors the physiological, behavioral, and biomedical functioning of the human body, and that information could be used to identify and verify identities.

1.1.1 Developments of the Past Decade in Wearable Technologies

Microsoft was one of the early companies to join the smartwatch race with the introduction of the Microsoft SPOT Watch in 2003 [6]. It was thought of as revolutionary, smart, and sexy at least at that time by early adopters. It had features of alerting Windows Messengers, news headlines, stock updates, and weather forecast. However, the device was short-lived as the product development ceased 2 years after, and it was officially discontinued in 2008. SPOT Watch was a major commercial failure, and that could be attributed to a few factors. First is incorrect timing as it clashed with the booming of smartphones. It was not just the fast adoption of smartphones but also the competitive disadvantage it had as the smartphone evolved and offered more features. Second is the incompatibility with other devices around, and that was due to the network technology used. Microsoft decided to use the FM signal which meant limited coverage and compatibility. Another factor was the battery power which was very short and limited to 3 days (depending on usage). Microsoft SPOT died quietly, but its legacy and revolutionary ideas did not as more companies undertook developments to overcome the limitations and shortcomings of Microsoft SPOT.

Garmin launched their first in a series of the Garmin Forerunner GPS sports watches in 2003 [7]. These are watches designed to tell the time, but they also tracked information such as steps, speed, distance travelled, altitude, and pace, most innovatively mark movement paths, and navigate through coordinates. Garmin has continued to innovate and produce GPS watches, smart bands, and smartwatches.

In 2006, Nike and Apple collaborated to introduce the Nike+ iPod Sports Kit which consists of a small sensor that fits into a Nike shoe and a receiver plugged into an iPod device that receives the data/communication. The device tracks activities and workouts by measuring distance, the pace of run or walks, and calories burnt. This kit has opened the floodgate for major companies to have their own devices and innovations as in the following year, Fitbit debuted their first personal fitness recording device, Fitbit Zip. It is a small device that could be clipped at various places on clothing such as belt, pocket, bra, and others. It has the features of tracking activities such as workouts, running, calories burnt, and log food and most importantly has a memory that could track up to 7 days of detailed motion data and daily totals of 23 days. The battery lasts between 4 and 6 months and has a 3-axis accelerometer [8]. Since then, Fitbit has expanded, and it is now a major player in the smart wearable technology, especially for smart bands and watches.

It can be seen that early smartwatches and smart bands (also known as activity trackers) had restricted functionality and were limited to performing electronic pedometers and few other measures [7, 8]. However, after 2010, devices not only became advanced in their processing powers but also contain enhanced biometric sensors capable of monitoring and storing heart rate, galvanic skin response, skin temperature, heat flux, sleep patterns, and movement by employing high-precision 3-axis accelerometers, gyroscopes, and other sensors. These devices are also becoming more advanced in that they could potentially do or have the same functions (i.e., a smartwatch having activity tracking, health monitoring, navigation, biometric sensing, and many other features).

1.1.2 Future Trends in Wearable Technology

The future of wearables could be looked at from two perspectives: first, the features and the capabilities of the devices and, second, the functions or the applications or use resulting from the features and capabilities of the devices.

The increasing number of manufacturers and technological companies developing new wearable devices would suggest that wearable technology would become mainstream and be integrated into our everyday life just like the smartphone.

The wearable industry has seen a steady growth in terms of adoption and the number of wearable devices sold. In the first quarter of 2016, the total shipment volumes reached 19.7 million units, an increase of 67.2% from the 11.8 million units shipped in the first quarter of 2015. In the third quarter of 2016, the total wearable shipments reached 23 million units with smartwatch having the most growth of 3.1% year over year according to the International Data Corporation (IDC) Worldwide Quarterly Wearable Device Tracker [9]. This growth did not slow down as evidenced in the report by IDC in 2017. In 2016, the wearable device shipment increased by 25% with 102.4 million devices shipped [10].

However, despite the increase in shipments and sales, there was a notable market share decrease in the same year particularly for Fitbit whose shares dropped from 26.8 percent to 22 percent in 2016. The struggle made Fitbit rethink its strategy and decided to reduce its global workforce by 6% (approximately 110 jobs) [11]. This could be attributed to the device's lack of personalization and the incapability of running third-party applications. It is not only a Fitbit issue but a wide wearable challenge.

Wearable devices also suffer from abandonment after a period of use due to people not finding them useful or boredom or fault or even lack of incentives and gamification. According to Gartner, smartwatch abandonment was 29% and for fitness trackers 30% in 2016 [12]. Hence, in the future, wearable devices are not only going to be personalized, but they will also monitor and understand the dynamic behavior and changing interests of the user, thereby providing adequate and customized feedback and experience.

Wearable devices that track or monitor movement and gestures do not always provide accurate readings. Due to the critical nature of the data these devices record, the industry will have to enforce a guarantee and standard on the accuracy for every measurement. There are many health and fitness applications of wearables that are based on biometric sensors, and their accuracies are paramount to their successful application.

1.1.3 Current Market and Growth Forecasts in Wearable Technologies

The worldwide wearable market continued to grow in the first quarter of 2017 according to IDC press release [13], with 24.7 million wearable shipment (up to 17.9% from 20.9 million in Q1 2016).

The wearable landscape is evolving with new companies and products emerging and companies venturing into other wearable products; hence, the market leadership might not be sustainable [13]. The forecast is bright as adoption is expected to rise (despite the brief slowdown in 2016), and new devices that are personable and smarter and offer more than just tracking information are expected in the next 5 years. The wearable market is forecasted to grow at a compound annual growth rate (CAGR) of 18.4% to reach $37.39 billion by 2022 from $11.43 billion accounted for in 2015 [14]. Further, the worldwide wearable device shipments are expected to rise from 102.4 million in 2016 to 237.5 million in 2021, a 18.3% CAGR [15].

1.2 Biometrics

In our daily interactions, we recognize people with their faces, body shapes, way of walking, voice, and even by their smell. In the field of forensics, fingerprints, footprints, and DNA are used to identify the criminals. Biometrics can be used in two ways: identification and identity verification (also referred to as verification or authentication) [16]. In identification, a person is being recognized among many people (one-to-many matching), while in verification, the task is to verify the claimed identity of a person (one-to-one matching). The system tries to verify if the identity is correct or incorrect by comparing the claimed identity with some previously stored genuine record.

1.2.1 Operation Modes

Before any identification or verification could take place, the system has to be trained to recognize the genuine person resulting in two modes of operation, namely, the training mode and the identification/verification mode.

Training mode The mode where the system is trained with the genuine user's data by providing as many inputs as needed to generate a unique profile/template. This mode consists of four stages: data acquisition, preprocessing, feature extraction, and data storage. *Data acquisition* requires the user to present to the system the appropriate biometric characteristic, i.e., the finger, iris, face, palm, etc., and the system scans and collects all the necessary raw data. The collected raw data is then *preprocessed* (using techniques such as normalization, discretization, etc.) before the required features for identification or verification are extracted (*feature extraction*) and eventually stored in the database as profile/template (*data storage*). These steps are repeated until the appropriate number of samples is obtained.

Identification/verification mode Once the system is trained by presenting several inputs and a unique profile has been created and stored, the system is set for identification or verification tasks. In this mode, all the processes in the training mode are performed in the same manner except the *data storage*. The extracted features do not need to be stored as they are only used for identification or verification. The templates are retrieved from the storage and compared with the newly extracted features in the *feature extraction* process, for similarity measurement. A similarity score is obtained based on a matching function, and that score is compared with a predefined threshold. If the score is higher than the threshold, then the verification status will be true or positive (i.e., accepted); else the status will be false or negative (i.e., rejected). In the task of identification, the system will output the identity of the person.

1.2.2 Types of Biometrics

There are two types of biometrics: physiological and behavioral biometrics [17]. Physiological biometrics use metrics related to the physical and biological characteristics of the human body. These characteristics include face, fingerprint, iris, palm and hand, and so on. Behavioral biometrics, on the other hand, uses metrics related to the behavioral characteristics of the human. Some of the characteristics include how the user walks (gait) and talks (voice) and the user's typing rhythm (keystroke) and signature. The premise is that all of those characteristics have unique metrics that can distinguish individuals and cannot be stolen, forgotten, or lost, hence their application in identification and verification. In verification, the characteristics could be used for static and continuous authentication. In the case of static authentication, the user is authenticated only once before being granted access to a system, whereas with continuous authentication the system authenticates and reauthenticates the user after every interval of time.

1.2.3 Multimodal Biometrics

Multimodal biometrics employ the use of data or information from two or more different biometric modalities (characteristics), for instance, the use of data from face and speech for identification or verification. Most wearable devices have multi-interaction types (touch and voice command) and also track many types of biometric data. They are independently used, and they are also not necessarily used for verification or identification. Multifactor biometrics is suitable in this case because data from sensors on multiple body parts could be integrated to provide a more accurate recognition rate, since a motivation of employing multimodal biometrics is to increase the identification and recognition accuracies. Furthermore, the deployment and integration of multimodal biometrics in wearable could provide a new dimension to authenticating individuals in a seamless and unobtrusive manner.

2 Current State of Wearable Technologies

Wearables have been receiving increasing attention since the launch of Google Glass in 2012. Today, smart devices such as fitness trackers and smartwatches are leading the consumer market. The key players in the smartwatch business are Samsung, Apple, Garmin, and Sony while in the smart band race include Xiaomi, Fitbit, and Garmin.

Interests in eyewears have slowed down since the failure of Google Glass, but new vendors are reviving the market and releasing new products especially for sports [18, 19] and entertainment [20]. Smart clothing has experienced significant growth since 2015 especially in fitness, biometrics, and health monitoring. The vendor and customer bases are gradually growing with products ranging from smart sock [21], bras [22], and T-shirts [23]. These types of clothing could be utilized in tracking athletic performance or vital information and communicate with workers such as first responders heading into dangerous situations. In 2016, smart clothing had a market share of 1.2% and a shipment of 1.3 million units [15]. IDC's 5-year forecast has the wearable market growing by 18.3% CAGR and shipping by 237.5 million units. Among those units, smartwatches would be 152.0 million (64.0% market share), 57.5 million in smart bands (24.2%), and 22.3 million in smart clothing (9.4%) [15].

2.1 Types of Wearable Technologies

2.1.1 Smartwatches

Smartwatches (similar to regular watches) are devices that are worn on the wrist but are capable of performing more complex and computational processes. Early smartwatches had limited functionality and were able to carry out elementary calculations and basic data storage and retrieval. However, current smartwatches have more sophisticated features and capabilities than the regular watches which tell/

display the time and date. The devices typically possess enhanced data processing functionality similar to a smartphone or tablet and can run apps. Modern smartwatches have special features, such as biometric monitoring, GPS, and mapping capability, and can even function independent of a smartphone.

The first smartwatch that dominated the market was the Pebble smartwatch which was released in 2012. The smartwatch was operated by the Pebble OS and can connect to both iOS- and Android-operated phones in order to display email, text, and social media notifications. The smartwatch had a screen of 144 $\times$ 168 pixels, no microphone or sound, 128 KB of RAM, 130mAh battery power, four buttons, 3-axis accelerometer, magnetometer, light sensor, and Bluetooth connectivity. Later versions of the watch had more advanced features such as voice control, heart rate monitor, step counter, and calorie and sleep tracking [24]. Until the shutdown of production and the company, Pebble has sold over a million units of smartwatches. Looking forward to the advancement in smartwatches today, the Pebble would be considered as bulky and ineffective.

2.1.2 Smart Bands/Activity Trackers

Smart bands, also referred to as activity trackers, have become a legitimate means of tracking exercises or activities and daily calorie burns via various motions via sensors. These devices are also worn on the wrist just like the smartwatches and GPS watches but are tailored toward health and fitness goals and tracking. They have specific functions and usually have limited customization. Typically, they are not cumbersome and seamless in terms of capturing and distributing exercise, health, and even biometric data without being intrusive.

2.1.3 Smart Clothing and Jewelry

Smart clothing and jewelry are relatively new in the broad field of wearable technology. It is, however, attracting a lot of attention as more companies are exploring the concept of connected garments and jewelries. Smart clothing is similar to regular textiles and fabrics but connected by embedding electronics and sensors and is also able to collect and communicate various types of data and information. Much more than strapping devices on the wrist, face, feet, or ear, smart clothing can continuously track heart rate, monitor emotions, and much more. For instance, they could provide a more precise reading of the heart rate as they are closer than a device strapped to the wrist or feet. Smart clothing follows the trends of the other wearables in that they track and monitor workout or performance-enhancing exercises which will have an enormous impact on athletics, extreme sports, and military industries. Some examples of smart clothing include Owlet Smart Sock, Hexoskin Smart, and OMsignal Bra.

All of the technologies highlighted above have the ability to capture and monitor biometric data that could be used to identify and verify individuals. They could use not only traditional biometric modalities such as gait, voice, and so on but also novel biometric traits and modalities such as heart rate (electrocardiogram – ECG) and brain signals (electroencephalogram – EEG).

2.1.4 Smart Eyewear

Smart eyewear includes the different classes of head-mounted displays (HMDs) and glasses (smart). They are devices with computing capabilities and are worn over the head with a small optic display lens which could allow users to see through them (in the case of augmented reality). These devices are also capable of augmenting the physical environment or displaying projected images. Such devices can run self-contained applications (without external dependencies) and can connect to the Internet via touch buttons and voice commands. They can work as a stand-alone device or connect with some other devices around for better personalization and efficient usage and experience.

However, there are differences between the HMDs and the smart glasses in terms of usage and the projection of information. HMDs such as the augmented reality (AR) and virtual reality (VR) devices are typically used only to provide or add information into what the user perceives. For instance, in the case of a VR device, it provides an immersive experience where the user feels his/her physical presence in the virtual world because he/she can control his/her interactions by using data captured and sent to the system via motion sensors. Some VR devices include Oculus Rift [25] and HTC Vive [27]. In AR, the display only adds information to the real-world objects. For instance, an AR device could project directions and navigate the wearer on the road. Some AR devices are the Microsoft HoloLens [26], Steve Mann's EyeTap [28], Google Glass [29] (which has been discontinued), Vuzix [30], and Solos [19]. Other glasses are just for photo- and video-capturing purposes and are therefore not categorized as any of the above. An example of this device is the Spectacles by Snap [20].

Table 14.1 shows the specifications and features of some recent devices in the wearable categories discussed above.

2.2 Technical Specifications

Table 14.1 Latest smart wearable devices and their features

Name/ manufacturer	Storage	Communication	Processing	Sensors	Battery	Features
Smartwatches						
Samsung Gear S3 by Samsung Classic and Frontier [31]	768 RAM and 4GB internal memory	Bluetooth Wi-Fi NFC GPS LTE with eSIM	Tizen OS Exynos 7270 Dual 1.0GHz processor chipset	Accelerometer Gyroscope Barometer ambient light Heart rate monitor (HRM) Altimeter Speedometer	380mAH (3–4 days) Wireless charging	IP68 water and dust resistant Samsung Pay, activity and fitness tracking Support third party apps Multi-touch screen Android and iOS compatible Make/receive calls Voice command.
Apple Watch 2 by Apple, Inc. [32]	8GB with limited music and photo spaces	Bluetooth NFC Wi-Fi GPS	WatchOS 3.0 Apple S2 Dual-Core	Accelerometer Gyroscope HRM Ambient light	Built-in rechargeable lithium-ion battery (up to 18 h)	SIRI voice command IPX7 water resistant Health and fitness tracking Support third-party apps
Garmin Forerunner 935 by Garmin Inc. [33]	64 MB	Bluetooth ANT+ GPS Wi-Fi	Garmin OS	Accelerometer Gyroscope Thermometer Barometer HRM	Rechargeable lithium-ion battery	iPhone, Android, and Windows compatible Multisport and activity tracking feature – workout, golf, swimming
Smart bands						
Fitbit Charge 2 by Fitbit [34]	• 7 days (motion data) • 30 days HR and exercise tracking .	Bluetooth Wi-Fi	–	• Optical heart rate tracker. • Accelerometer. • Altimeter. • Vibration motor.	Lithium polymer (up to 5 days)	Activity and auto sleep tracking Multisport mode iOS, Android, and Windows compatible
Xiaomi Mi Band 2 By Xiaomi [35]	–	Bluetooth	–	Accelerometer HRM	70mAh Lithium-polymer (up to 20 days)	Smartphone unlocking Smartphone alerts IP67 splash and dust resistant

						iOS and Android compatible Steps and sleep tracking Idle alert
Garmin vívosmart 3 by Garmin Inc. [36]	14 days of 7 timed activities	Bluetooth ANT+	–	HRM Barometer Accelerometer Ambient light	Rechargeable lithium-ion (up to 5 days)	Activity tracking and HR features Stress monitoring
Smart clothing and jewelry						
Athos [23]	–	Bluetooth Low Energy	–	18 sEMG (10 on shirt and 8 on short) 4 HRM on shirt	10 h – full charge	Muscle activity (pectoral, biceps, triceps, lats, deltoids, inner/outer quad, hamstring, and glute) Heart rate Machine washable iPhone/iPad/iPod compatible
Owlet Smart Sock 2 [21]	–	Bluetooth 4.0 Wi-Fi	–	HRM Pulse oximeter	Rechargeable sock lasts up to 18 h	Heart rate Oxygen level iPhone5+/iOS9+ compatible
Ringly [37]	–	Infrared	–	Accelerometer Five-color LED light	Up to 24–48 h	Activity tracking (steps, distance, calories burned) Compatible with iPhone 5+ andAndroid 4.3+ Mobile alerts Water resistant Third-party app support

3 Traditional Biometrics Using Wearable Technologies

Wearable devices are becoming very popular, with smartphones being the most adopted and widely used device. Studies have shown that users are within proximity of their smartphones almost 90% of the time [38]. The smartphone's features and capabilities offer ample opportunity to monitor and track users and their activities and recently their physiological and behavioral characteristics. These smartphone features have made it possible for smartphones to be used in various application areas from health monitoring to detecting and tracking a person's activity.

Traditional biometrics aim to identify and verify the identity of individuals using their physiological and behavioral modalities. Physiological modalities relate to but not limited to fingerprint, palm veins, face recognition, deoxyribonucleic acid (DNA), palm print, hand geometry, iris, and retina. Behavioral modalities include keystroke dynamics, gait, voice, and signature.

3.1 Gait Recognition Using Wearable Technologies

Gait recognition has been an active topic of investigation using wearable sensors and mobile devices. With the widespread advances and installation of sensors in modern wearable devices such as smartwatches and smart bands, movement measurement has become much more significant. These devices (which are worn almost all of the time) are incorporated with accelerometer and gyroscope sensors that are used for data collection in gait recognition.

The use of smartphones to recognize the movement of individuals and consequently identify or verify them for access control and activity recognition has been well researched. Fernandez-Lopez et al. [39] used a smartphone (Samsung Galaxy 4) for gait recognition. They recruited 15 people aged 21–38 and asked them to walk in a hallway six times while carrying a smartphone on their hip (belt). The data collection was divided and done in 2 days with each participant contributing 12 samples. They used four classifiers for evaluation, Euclidean distance, dynamic time warping (DTW), cycle rotation metric (CRM), and Rotation Manhattan, and they obtained error equal rate (EER) of 26.67%, 28.07%, 16.38%, and 16.49%, respectively. Another study [40] used a Google G1 phone which was put on the hip as well. The methodology is similar to the previous study we have described, but in [40], 51 persons were involved and reported an error rate of 20.1% using DTW. Furthermore, Shi et al. [41] used multiple sensors to implicitly and continuously authenticate a user. They used accelerometer (for the gait), touch screen (for gestures), microphone (for voice), and location history (for environmental change). Sensors are activated and data collected depending on the individual's motion and setting (i.e., context-aware). For instance, when the user starts to walk, the accelerometer starts taking readings. The study indicated that the use of multisensor was suitable as source for implicit user identification.

At present, only a few researches use smartwatch or bands for gait recognition. One challenge in the use of smartphones for gait recognition is that the users keep their smartphones in different places (bag, pocket, and waist) and in different orientations, which ultimately affects the accuracy and the performance of the recognition. However, challenges such as location and orientation are not present in smartwatches because they are always worn on the wrist and in same orientation. Johnston and Weiss (2015) [42] carried out a study to show the feasibility of using smartwatches for gait-based biometrics by demonstrating the high levels of accuracy that can result from smartwatch-based gait recognition. The experimental procedure requires users to wear a smartphone on the wrist of the nondominant hand and then walk for about 5 min with a few turns. The data collected (by the accelerometer and gyroscope) from 57 participants are then transmitted to a smartphone which is placed in the participant's pocket and to a server. They extracted 30 features to which four classifiers were used: multilayer perceptron (MLP), random forest, rotation forest, and naive Bayes in WEKA. The highest authentication accuracies (EER) obtained with accelerometer were 1.4% (random forest) and with gyroscope 6.3% (MLP).

3.2 Handwriting/Gesture Recognition Using Wearable Technologies

Lewis et al. [43] conducted a feasibility experiment by using free-form gesture data collected from a smartwatch to verify the wearer of the device. They evaluated their system's capability in countering attacks by simulating random and mimicry attacks. Random attacks assume that the attacker has no knowledge of the legitimate gesture, while mimicry attack assumes that the attacker knows the legitimate gesture by observation. The study used Samsung Galaxy Gear smartwatch and collected 300 samples from 5 participants with each participant providing 15 random and mimicry attacks. After evaluating the two attacks, no random forgery was successful, and 45 mimicry forgeries were falsely accepted as genuine gestures. The experiment shows a promising result, but more analysis with more data is needed in order to validate its effectiveness.

3.3 Multimodal Biometrics Using Wearable Technologies

Peng et al. [44] proposed a noninvasive continuous authentication with touch and voice behavioral biometrics on a wearable glass. The wearable glass used was Google Glass that has a touch pad and a microphone/speaker for voice commands. The Google Glass device supports multimodal biometric authentication and potentially better security against traditional attacks such as smudge, observational, mimicry, and replay attacks. They used six touch types: single tap, swipe backward (forward), swipe down, and two-finger swipe forward (backward), and voice commands such as “OK, Glass! Take a picture!” The features extracted for both input types were duration, distance (*x*,*y*), speed (*x*,*y*), and pressure (mean, median, standard deviation, min, max, first, last, and max pressure portion) from both one- and two-finger swipes (each) and Mel frequency cepstral coefficients (MFCC) for voice commands. In addition, they used accelerometer, gyroscope, and magnetometer sensors from which they extracted mean, median, and average change in readings before and after a gesture. Thirty-two subjects participated, and data was collected in multiple sessions over a 2-h time frame. Seven classifiers (one-class support vector machine) were used corresponding to six touch types and one voice command. The classifiers output a binary result depending on their measurements (1, genuine, and 0, imposter) and then are sent to an aggregator called tree re-weighted (TRW) to make the final decision based on its computed likelihood and confidence. The proposal (GlassGuard) achieved 93% detection rate and 3% false alarm rate on less than 5 events for single events and 99% detection and 0.5% false alarm rate on 3.46 user events with mixed events.

Activity monitoring and recognition are other applications of wearables, which have been well studied on smartphones. Monitoring daily activities through smartphones is highly accurate and has been mostly used in primarily clinical and medical environments to ensure that patients are performing the recommended and desired actions [45–47].

An interesting commercial implementation of smart wearables and its interconnectivity with the surrounding devices is the BIOWATCH [48]. BIOWATCH is a device that could be attached to a watch, smart band, or bracelet. It scans the vein and communicates with paired devices using Bluetooth and near-field communication (NFC). As long as the device is on the user and the user is authenticated, access to any paired device (automobile, electronic (phones, laptops), building access control, online payments, e-banking, and so on) becomes seamless (i.e., no keys, passwords, and other modes of access will be required). The only drawback is that this device is external to the wearable, and we believe in due time smartwatches, bands, and bracelets would have that capability of verifying wearer’s identity and communicating to the connected and paired devices.

A wearable that provides secure authentication as long as the device is worn is the Nymi Band [49] which is shown in Fig. 14.1. The Nymi Band is a wristband that uses multimodal biometric to provide seamless authentication to applications, devices, and services. The modalities utilized are heart beat for ECG authentication

Fig. 14.1 Nymi Band [49]

and touch ID (optional). The device is also capable of recognizing gestures and tracking activities. When the user has been verified, that verification status is communicated to any Nymi-enabled device within via Bluetooth and NFC. The active authentication state would remain until the device is taken off the wrist or the wearer is beyond proximity. This device provides a good way for enterprises to have smooth access to devices, locations, and resources and could be extended to various other settings. The biometric authentication provided by the band could become mainstream in the near future of wearable devices.

Smart bands have also been studied in the context of stress detection and monitoring. Mental stress is one of the growing problems of the present society. The number of people experiencing mental stress is increasing day by day. This scenario is significant when it comes to actions or professions that are very critical such as pilots, medical doctors, and so on. Zubair et al. [50] use a smart band that continuously monitors and measures the skin conductance of the wearer using two electrodes inside of the band to measure the stress level of the wearer. The band measures the response of DC voltage request sent to the skin and passes it to a smartphone for classification and prediction. This approach ultimately increases the efficiency and effectiveness of caring and well-being and also organizations to better understand how to get more productivity from their employees, for example, military in combat scenarios.

4 Emerging Biometrics Using Wearable Technologies

4.1 Emerging Biometrics

Large companies and upcoming start-ups are realizing the potential of wearable technologies along with their interactions and experiences. Today we have wide variety of products ranging from smartwatches to bands and even clothing among numerous others. These products are equipped with sensors that have the capability of interacting and communicating with the body and the environment. So far, these devices have made the greatest impact in health, sports, and fitness areas where they have been used to measure various user conditions such as movement activities, health information, and food intake patterns, thereby making suggestions on improving activities and general well-being.

Emerging wearable technology can have an even greater impact as the data collected could be extended to various applications that require authentication or identification but with new types of biometric data. The technology collects biometric data such as heart rate, but such data is mostly utilized for health diagnosis and could easily extend to authentication. These capabilities would allow biometric authentication and identification of not only the traditional modalities (gait, touch, voice) but also new modalities that would provide high accuracy and recognition rates with distinguishing and distinctive features.

Several researches have been carried out on the feasibility of emerging biometrics with wearable technology, two of which are electrocardiogram (ECG) and electroencephalogram (EEG). ECG measures the electrical activities and rhythm of the heart [51]. These signals are unique (depending on the anatomic structure of the individual's heart and body) and detected by electrodes (sensors) attached to the skin [52]. Traditionally, to collect or detect the ECG signals, bulky external sensors have to be attached to the body (arm, leg, and chest) and the data transmitted to a central server station for processing. With ECG, the sensors are not ideally integrated into smartphones because there should be skin contact. What devices do we put on and interact with every day that would be more practical to integrate the sensors? Most probably none other than our accessories (i.e., watches, bands, bracelets, and even clothes).

Choi et al. [52] proposed a method of authentication on mobile devices using ECG and a sensor. They collected noisy ECG signals but cancelled using a band-pass filter before extracting eight features. The proposal was tested on 175 people (each collecting 1 heartbeat and 15s data). The proposed method achieved 4.61% EER on a single pulse and 1.87% of EER on 15s data, using RBF-kernel SVM. Although the device was a CardioChip, they argued that wearable mobile devices could also be used, which is a reasonable argument as the same data is collected.

Electroencephalogram (EEG) is another modality that is utilized in human identification. EEG is the use of electrodes (sensors) attached to the scalp to record signals and activities of the brain [53]. Bashar et al. [54] carried out a preliminary study to measure the feasibility of using a wearable headset for authentication. They experimented on nine EEG records (from nine people) using a band-pass filter to remove any noise. Three features were extracted, and a supervised error-correcting output code (ECOC) multiclass model classifier was applied, and they obtained a true positive rate (TPR) of 94.44%. The study showed the potential in using wearable headset for EEG human authentication. However, more features and data need to be collected and investigated to validate its effectiveness and accuracy. An attractive proposition was made by Thorpe et al. [55] who used the what they called "pass-thoughts" for authentication. In their proposed approach, a wearable headset would be worn by the user who would think of a password. That process of thinking is captured, transmitted, and analyzed as the password – "authenticating with our minds" [55]. The idea is still in theory as no experiment has been conducted to investigate its feasibility. Although, at the moment, the idea exists only as fiction, with the rapid advancements in technologies and especially wearable, it is within our technological reach.

Could human daily activity information such as walking, running, climbing, and jumping be used to authenticate implicitly? The answer is yes, as validated by the study of Zeng et al. [56]. This study demonstrated two results: first, human daily activity information could be used for authentication, and, second, sensors worn at different places can communicate and provide a higher recognition accuracy. The researchers asked 30 participants to perform 30 min of action which involves running, jumping, climbing, and running while wearing different devices with sensors on their arm, wrist, thigh, waist, and ankle. Accelerometer data was collected, and the validation achieved a high authentication accuracy of 97% (1% false positive rate) using a single wearable device and up to 99.6% with fusion of data from multiple devices. Authentication using the aforementioned approach can be used in two ways: first, to make sure that the device is being used by the legitimate user, and, second, to provide seamless authentication with other devices around by communicating the authentication status of the user, thereby leaving the devices always unlocked. In the first case, a conventional approach to static and continuous authentication can be implemented where, at first, the user uses an instantaneous modality to authenticate himself/herself and then monitors an implicit modality for continuous assurance that the user has not changed or the device has not been removed. Ojala et al. [57] described a similar process where explicit fingerprint was used as clearance to make sure the owner is the one to use the wristband. The wristband then implicitly and continually monitors various signals (heart rate, motion) and uses them to measure and ensure that the device is still worn by the legitimate user.

Other applications of biometrics on wearable devices include 3D signature verification using multisensor smart jewelry (finger ring) [58] and micropayment with mobile and wearable that utilized two-factor (password and biometrics) authentication [59].

A third trend is the maturity and richness of the technology and tools available in current wearable products. While topics such as implicit authentication and biometrics have long been studied in laboratories, a lack of commercial sensing platforms has restricted the impact of laboratory results on real products and services. With current wearables, this is no longer true as devices typically feature a range of advanced sensors and associated software support such as dedicated application programming interfaces (APIs) that provide access to the information they gather. This means that product designers and software developers, as well as researchers, can experiment and implement authentication techniques leveraging these capabilities. Based on these current trends, we anticipate that novel wearable authentication schemes will reach the market more rapidly than with previous technological platforms such as mobile devices.

4.2 Use Cases of Wearable Biometrics

Emerging wearable devices with their capabilities of tracking and measuring new biometric modalities present an innovative way of verifying and authenticating the identity of an individual. Enabling biometric verification on wearable devices further increases the usefulness of these devices from mere medical application, sports, and recreation to military, criminal justice system, finance, and even the corporate field.

Healthcare Wearables have had the greatest impact on the medical sector, and virtually all wearable devices have a sensor that tracks a medical parameter such as ECG or EEG, skin quality, and so on. Today, doctors are even capable of tracking the movement, body reactions, and patients' medical condition in real time without physical interaction via different types of wearables. The wearable could be used by the elderly to communicate with their doctors or emergency response in the case of an emergency such as a dangerous fall or irregular heart or brain activity. This leads to improved efficiency in medical diagnosis, reduced health costs, and an increase in life span and gross domestic product (GDP) [60]. All these are possible with current wearable devices that do not use biometric verification. Therefore, one question is what benefit could the addition of biometric authentication to wearable device offer? When the patient's information is communicated to the doctor, how does the doctor know that the information is from his/her patient? Hence, before communication, the device being worn could authenticate the wearer and if a genuine person then grants access to all features. Otherwise, we should block or limit access to all features. This way we can ensure that any information communicated only came from a genuine patient and the diagnosis would be accurate.

Privacy Control On our smartphones, tablets, and computers, we can set up device access control using PINs, passwords, patterns, or biometrics to ensure our data confidentiality. We need that to control who we allow access to the private, social, and confidential information that may have been stored on device. Smart wearable devices are no exception as they also store an enormous amount of different types of information such as personal, medical, and so on. If we protect our smartphones and other devices, we should also protect our wearables. For instance, Google Glass is activated by the words "OK Glass," and all features would be active [61].

Real-Time Information Access Imagine a scenario where a doctor in the operation room has all of the information he requires about the patient, the operation, research information, or even access to another expert in real time. Although there are good benefits with such a system which provides access to all kinds of resources, a related concern is the focus of the doctor: would the technology overwhelm or distract him/her from focusing on the main operation?

Another scenario is with law enforcement. It would be interesting for an officer to access some database in real time for face identification without logging into their system. This could be possible in real time, while a police officer speaks to a person with the resulting information being projected onto the officer's field of view. The smart glasses could take a picture of that person and send it to a cloud-based police service where facial recognition technology would match the picture with an entry in a police database. The police officer might not know who he was talking to, but this technology would be able to alert the officer if the person was a suspect in an ongoing case and had a criminal record. However, the use of wearable, in this case, might also present a lot of privacy concerns.

Military Wearables have for long been used in military applications to assess and measure the soldier's physical and internal condition for optimal performance and readiness for deployment. Several sensors have been created to allow soldiers and their superiors to monitor various conditions and assess injuries and their severity, for instance, tiny biosensors developed by the US military that are bandage-like and are placed on the skin to track what flows in the sweat, to monitor their health and improve their performance. The data obtained from the sensors (heart rate, respiration rate, and hydration) is used to learn how best to deploy soldiers and how to enable them to operate at peak performance [62].

There have been recorded casualties in the US Army due to cases of heat-related injuries, i.e., the environment getting too hot while soldiers engage in intensive operations. The US Army Research Institute of Environmental Medicine collaborated with a team from MIT Lincoln Laboratory and a Marine Corps expeditionary rifle squad to develop a wearable device that would detect early symptoms of heat stress. This way the soldiers and their supervisors would know when to cool down for better physical and psychological conditions which would positively affect the efficiency and thought process of the soldiers [63].

Soldiers in the field battling cannot afford an infiltration from the enemy as that would cause devastating effects or even loss of lives. This infiltration could happen when an enemy gets hold of the army's uniform and wears it to impersonate some officer. To prevent impersonation, a wearable device or, more specifically, smart clothing could be used to identify any impersonation attempt. As officers would be verified continuously, any failed verification would alert the central command office to necessary actions. The same approach could be applied to security personnel assigned to protect heads of states and very important people (VIP).

Criminal Justice System Judges may put lawbreakers under house arrest, travel restriction, or parole condition depending on the offense committed, the pending litigations, health condition, and time spent in jail to reduce overcrowding in prisons. In that case, the lawbreakers' movements need to be monitored to ensure that they adhere to their restrictions. For that reason, they are required to wear an ankle monitor or bracelet at all times which communicates mainly the wearer's location via Global Positioning System (GPS) or radio frequency to a receiver (that is monitored by the law enforcement) [64]. Any movement outside of the allowed range or tampering with the device would automatically alert the law enforcement. However, the device has been shown to have a vulnerability that allows it to be hacked without triggering an alert [65]. In addition, it could be removed and strapped on another person or left in the house, while the offender escapes unknown to the law enforcement. A wearable biometrics employed in this scenario would not only monitor the location and the movement of the device but would identify a change in the wearer of the device or even when the device has been taken off. For instance, if a smart band with biometric identification (e.g., ECG) is used, it would continuously verify the identity of the wearer while also tracking and communicating location information. Further, the wearable biometrics could also transmit health data, so any inconsistency or abnormal activity of the wearer would be detected. This provides many ways by which the offender can be tracked and makes it difficult to compromise the system.

4.3 System Architectures

The amount of data generated by individuals is growing at an exponential rate. Wearable technology is further accelerating the production of more data. Humans

Fig. 14.2 Current wearable architecture

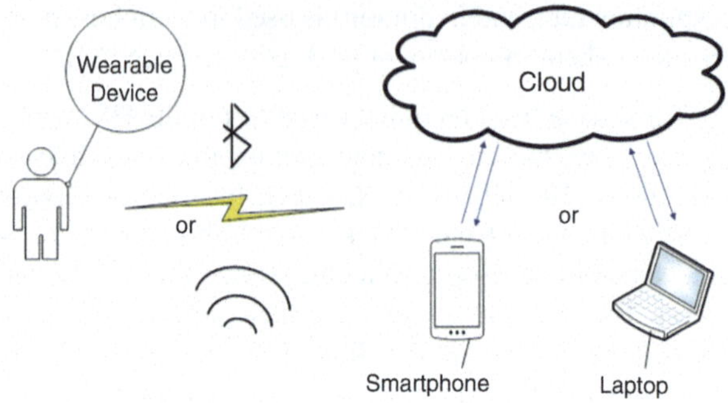

produced a total of 4.4 zettabytes in 2013 and would create 44 zettabytes by 2020 (1 zettabyte is equal to 44 trillion gigabytes). Daily, 2.5 exabytes (10^18) of data are produced which is equivalent to 530,000,000 songs, 90 years of HD videos, and 10 million Blu-ray discs which when stacked would measure the height of the Eiffel Tower fourfold [66]. This data, often referred to as big data, and a whole field deal with the challenges of analyzing, processing, and storage of this information.

As wearable technology develops further, the amount and type of data that the devices collect would expand. The devices already generate massive quantities of data every day ranging from biomedical information, exercise and workout, social interaction, geographical location, photos, and videos for and from social media platforms. At present, the challenges are not in the generation of the data but in the analysis and storage of that data. For instance, biomedical information (e.g., heart rate) is recorded over time, and it needs to be stored and analyzed to give feedback on the state of the heart. The information is voluminous, and the device has a limited storage capacity and processing power which means no complex processing or algorithm can be executed because it would use so much power that would require the charging of the device every couple of hours. How can the data be analyzed and stored for historical tracking and monitoring? Currently, one proposed architecture described in [67] works as follows: the wearable device collects the data and transmits wirelessly (usually Bluetooth and Wi-Fi) to a paired node (laptop, smartphone, or server) where the data would be processed and analyzed. After the analysis, the resulting information is either displayed on the laptop and smartphone or sent back to the wearable devices. Figure 14.2 shows the wearable architecture inspired by [67].

The current architecture only works when biometric authentication is not applied. When biometric authentication is used, a different architecture has to be implemented due to the nature of the operation of authentication. There are two modes in biometric system: training (enrolment) and verification (authentication) (as presented in Sect. 1.2). These modes have similar working stages with a few extended activities in verification mode as shown in Fig. 14.3.

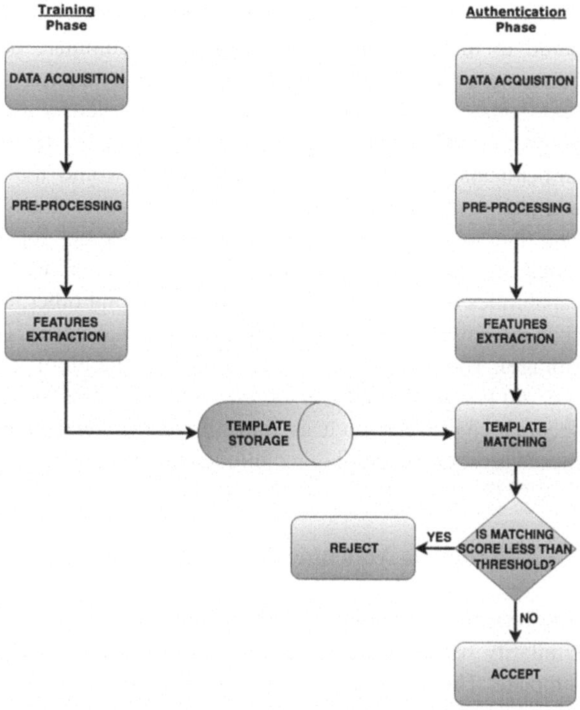

Fig. 14.3 Biometric system architecture

In any biometric authentication system, the architecture in Fig. 14.3 is applied, and in some cases, the data/template storage, template matching, and decision are done on device itself.

On device		Cloud	
Pros	Cons	Pros	Cons
Faster storage and retrieval	Data lost on device loss	Ubiquitous	High latency
Absolute data controls	Limited processing power	Scalability	High data transfer risk (man-in-the-middle attack)
		More processing power required	Reliability issue depending on service
		Strong security measures needed	Limited control over data

4.3.1 Data/Template Storage

Wearable devices are by design non-cumbersome and small which require the use of very small batteries. The size limits the battery life of the devices and consequently affects their processing power. Further, the device storage is relatively small compared to the amount of data they can accumulate over time. For instance, the Samsung Gear S3 smartwatch has the ability to run applications, track workout, play songs, and measure the distance travelled with only a limited memory of 768 MB (RAM) and 4GB (internal storage) [31]. If the smartphone has biometric authentication capability, how and where would the data template be stored in addition to the other information? In this case, there are two possible solutions. The first solution is to explicitly allocate memory for biometric templates and the remaining storage for the other features. This solution would, however, limit the capabilities of the device in terms of supporting other features and information. But such information could be moved to an external storage (i.e., the cloud or expandable storage). The second solution is to off-load the biometric data template to the cloud (i.e., no template would be stored on the device). However, this solution has some security challenges such as transmission channel security and cloud security. In these cases, cryptographic techniques might be of use.

4.3.2 Template Matching

The process of template matching involves the use of algorithms to measure the similarity between the stored biometric templates and newly presented biometric template. These algorithms are sometimes computationally complex and expensive to run. Hence, they would require high computing capabilities. However, wearable devices having tiny batteries mean they are limited in terms of the complexity of processing they could do. Besides, the device would need to be recharged frequently to compensate for the power consumption and keep the device active. In terms of usability, users would not want a device that requires charging every few hours in addition to the fact that it offers limited functionality that other devices (smartphones) could provide. Now, the question is how would the template matching work on wearable devices? The best possible and practical solution is to send the collected data to an external node (i.e., server, cloud, smartphone, tablet, or laptop). The node would then perform the template matching with the implemented algorithm, make a decision, and communicate only the outcome to the wearable device. This way, the complex processing part is done on a more powerful node, and only a small memory is needed for the results.

4.3.3 Cloud-Based Solutions

The cloud could provide the best solution for providing biometric authentication on wearable devices as all of the burdens of processing that are memory and computationally intensive are off-loaded from the device. This adds an additional phase in the architecture of the biometric system on wearable devices.

How would such a biometric system work in enrolment and authentication modes in a cloud-based solution? In the enrolment mode, the raw biometric data is captured on the wearable device and transmitted to the cloud via a secure channel. The data is then preprocessed, and vital features that are potentially distinguishable are extracted and stored in a database in the cloud as templates. The template needs to be in an encrypted form so that it cannot be viewed even if the cloud storage is compromised. Here, the only role that the device plays is the collection and transmission of the modality information.

In the authentication mode, the raw biometric data is captured and transferred to the cloud via the same channel as the previous mode. The same activities (preprocessing and feature extraction) are carried out but are not saved. The template obtained and the retrieved stored template are compared, and a similarity score is computed. A threshold (user dependent or independent) is determined by the system and compared with the similarity score. If the score is greater or equal to the threshold, the obtained template is considered to be genuine else would be considered as imposter. The decision would then be transmitted back to the device via the secure channel. In the case that the decision is positive, not only that the device full functions would be accessible but can communicate and provide seamless and unobtrusive usage of other connected devices. The device-to-device communication could be established via several communication protocols (Bluetooth, Wi-Fi, infrared, 3G, 4G). For instance, a smartwatch that utilizes biometric authentication could communicate its status to a laptop, and that would allow seamless usage for the user as he/she would not need to put in their passwords whenever they want to use the laptop. Apple Inc. has implemented a similar functionality of unlocking MacBook with Apple Watch. The automatic unlocking works when a user is wearing their Apple Watch and is near the Mac computer [68]. The Apple Watch has to be unlocked with a four-digit passcode before the auto unlock could work even when near the computer.

Nymi Inc. implements a similar architecture, but instead of cloud, they used the user's smartphone [69]. The Nymi Band captures ECG data and then passes it to a smartphone which runs Nymi Companion Application (NCA) where the data is saved. Template matching is done upon receipt of a verification template on the NCA in real time, and the status is communicated back to the smart wristband. The status is communicated to other devices and systems that have Nymi-enabled applications (NEAs), and they remain unlocked as long as the smart wristband remains on the user's wrist.

The medium of transfer and the storage of data in the cloud possess few security and privacy issues. First, how secure is the transmission channel? If the medium does not utilize strong encryption mechanism, then, an eavesdropper, after interception, would be able to decode the information with relative ease. Therefore, the transmission has to be on a secure channel from the device to the cloud and vice versa. Second, how are the data (especially sensitive) stored in the cloud? Does it utilize a strong cryptographic method? Storing the information in the cloud does not mean it is secure without a protection method. Cloud storage could be compromised, and when this happens, the attacker should not be able to make sense of the information, and that is achieved by a strong encryption method. Third, by storing the information in the cloud, the owner of the data has shifted to the cloud service providers.

However, users should know how much control they have on the data and who has access to their data and who and can modify the data. These are important issues to consider when using a cloud-based architecture for biometric systems. Section 5.4 presents a further discussion on privacy issues.

5 Challenges and Limitations

Over the years, wearable technology has evolved in its design, functions, and features and has increased its user base. Nevertheless, it is yet to become mainstream because of various challenges that still need to be addressed which include device limitations as well as technical and social challenges.

5.1 Limited Battery Life

Wearable devices are small by design because of their application and place worn on the body. The batteries mounted on wearable devices are small resulting in limited battery life. Devices such as smartwatch and smart band are equipped with features to enhance usability and user experience, but they often consume a lot of power.

Currently, wearable devices' battery power lasts only a few days, and even that depends on the intensity and duration of usage. For instance, the Samsung Gear S3 has a 380mAh battery which they claim to last 3–4 days [31], while the Apple Watch 2 battery lasts up to 18 h [32]. Many companies make theoretical claims of low energy-consuming devices that make the battery life much longer. Xiaomi Mi Band 2 claims 20 days of battery life for their smart band [35]. However, the device is equipped with one accelerometer and heart rate monitor and only tracks sleep, steps, time, alert, and heartbeat.

The battery technology advancement is lagging behind compared to other technologies such as the chip technology. This increasing gap could be attributed to cost and battery chemistry [70]. New wearable devices work around the battery limitation by adopting multi-battery usage, fast-charging technology, wireless charging, and energy-efficient communication channels (such as Bluetooth 4.0) to keep batteries alive for longer, for example, Moto 360 [71], Samsung Gear S3 [31], and Recon Jet [18]. Further, other companies adopt new processor and sensor technologies that are more powerful and less power consuming [72].

Until a major battery technology breakthrough, wearables would continue to use workarounds, while the battery longevity and power consumption would continue to be challenges that need to be addressed for widespread use.

5.2 *Limited Data/Template Storage*

The storage capacities of wearable devices vary from temporary storage of a few days to months (e.g., Fitbit Charge 2, 30 days), while others have permanent storage with capacities in gigabytes (e.g., Apple Watch 2, 8GB). Such storage capacities allow information from sensors, applications, entertainment, and even personal to be tracked and stored. Over time the collected information would exceed the capacity of the device, and some data would have to be removed or transferred to external storage.

Storage of information on the wearable device raises both security and privacy concerns. As these devices are small, they have a high chance of being misplaced or stolen. If this happens, all of the data stored (including personal information) is at risk of being misused, and this consequently would cause a breach of privacy. To secure device's data, access has to be controlled by the use of strong authentication methods.

The external storage of user's data poses more threats and concerns. The information is also within the reach of the company providing the storage services (in the case of cloud service). New concerns and issues emerge for which innovative solutions must be sought for: who owns the data? Who else has access to the data? How is the data stored? Is the data encrypted or in plaintext? Does the user have control over the information (i.e., can they edit, delete, and add)? Are there any third-party disclosures? Who is the third party? Does the user consent to disclosures? These privacy and security concerns need to be addressed in the future as they are critical to the industry and organizational growth.

5.3 *Limited Processing Power*

The difficulty of wearable devices running complex algorithms or performing complex processes is directly related to battery life because the complex algorithms require intensive processing which in turn needs high computational power. Current wearable devices track predominantly health, exercise, fitness, and location information from which they calculate metrics such as steps, distance, speed, heart rate, and so on. Depending on the wearable device, some of the metrics are measured and displayed on the device. The calculations are not necessarily compute intensive and would require low power to operate. However, the cumulative features' calculations could be compute intensive, and in that case, the metrics would be analyzed and displayed on an external device with more processing power and storage capacity such as a smartphone or a laptop.

Presently, only a few wearable devices have biometric authentication capabilities, and even for them, the authentication processes are not executed on the device. The Nymi Band captures the heartbeat rate and sends it to an application on the paired smartphone which performs all of the biometric authentication processing and makes a decision. The smartphone then sends the decision status (whether genuine or an imposter) to the band where it is stored and communicated to other compatible systems for seamless authentication. The authentication process cannot be done on the wearable device because the device needs to track other features (e.g., exercises) as well which would need a lot of processing power. Further, no one would want to use a device that would require recharging every couple of hours.

For wearable devices to have the capability of running complex and computationally intensive algorithms, first, the battery technology would require tremendous improvement. Second, the processor and communication channels need to implement low energy transmission techniques (such as ANT+ and Bluetooth Low Energy) [73].

5.4 Security and Privacy

Wearable devices are becoming increasingly popular largely due to their capabilities and features they can support such as personal health, exercise information, sleep patterns, calories burned, blood pressure, heart rate, and other biometric information. Despite all the hype about wearables, there are considerable concerns about the devices and the huge amount of personal data it generates about their users.

One of the main concerns is user privacy: the fear of covert observation, data ownership, confidentiality, and disclosure [74]. These fears are fueled by the nature of some of the devices, for instance, the concern with smart glasses that users would capture images of people covertly without their consent and knowledge [75]. This particular scenario has both an external concern and an internal concern. The external concern is the person that is being observed covertly while internal to the observer.

Steve Mann, the inventor of EyeTap Digital Eye Glass, maintained that the digital eye glass is the least privacy-invasive camera as it is placed in the same location where a natural camera with a storage space has existed (i.e., the human eye and mind [76]). He added that surveillance cameras are more privacy invasive as they are mounted on unusual vantage points and record every action of the people nearby. Some cameras are mounted at points that could see over balconies and windows (as shown in Fig. 14.4) invading the homes all in the name of security. Mann's argument for EyeTap is if higher authorities, large organizations, or entities would do the watching from above, then ordinary people should have the ability to also do the watching from below, hence the term "surveillance with sousveillance."

Another privacy concern that arises with wearable devices is the storage of data. Significant amounts of data are collected, and in most cases individuals would not necessarily want to be put in a public domain. Hence, the data should be stored in a secure place and form. With wearable devices, we have entrusted the vendors with our data, and they store the data in the cloud or on device or an external node [69]. However, telling the user's where the data is saved is only half the story. Security measures must be enforced by the company to secure the data, and they

Fig. 14.4 CCTV cameras looking right into house windows [77]

must be made known to users (e.g., some pertinent issues include is the data encrypted? Is the storage access controlled? How about periodic review and monitoring?).

Data disclosure or misuse is also a major privacy concern, especially by unauthorized third parties. This is a nefarious act that could have tremendous impact on the wearer of the device. Imagine a situation where data is sold, stolen, or leaked via a breach (unintentional or intentional) for some gains to third parties such as insurance companies, pharmaceuticals, marketing firms, and so on. The mere fact that the companies may be able to identify a wearer's data through well-known techniques such as data mining or otherwise could result in the wearer incurring higher health costs, life, or other insurance rates. The disclosure, when consented to specific parties, could be advantageous (e.g., for the elderly and their doctors). The doctors could continuously see and monitor the health of their patients remotely and make necessary medication adjustments and recommendations. Still, this disclosure will have to be explained, consented, and agreed to by the patient.

How much ownership and control does a wearable device owner have over the data generated by the device? As highlighted in the system architecture section, many wearable devices do not store data on the device but at the manufacturer's cloud storage facilities. This provides the users ubiquitous access to the data on many platforms but at the expense of giving the cloud storage provider some control over the stored data. Therefore, some relevant questions arise in this case: who owns the data generated by the wearable device? Is it the entity that owns the device and the storage platform or the entity who generates the data being stored? It is worth pointing out that owning the data does not necessarily mean having control of the data. How much control does the wearer have on the data that he/she creates? Can the wearer manipulate the data (i.e., edit, delete any data)? The answers should be explicit and transparent to users in the company's privacy policy statements. However, this issue leads to further questions: how can the user get control over their data? How can the companies be compelled to state such data management issues/ rights in the privacy statements explicitly? In the future, data protection and confidentiality regulations need to extend to the wearable domain in order to protect the privacy of the data and the wearer of the device. While regulations, standards, and enforcements should be ensured by the relevant authorities, the users should also bear some responsibility and read the policy statements carefully so that they know what they are signing up for and also be able to make well-informed decisions on the use of the wearable device.

6 Conclusion

In this chapter, we have presented an overview of biometrics for wearable technology. The types of wearable devices currently on the market keep growing and are being used in many application areas. We have highlighted some of the key vendors, products, and technologies of wearables in this work. Increasing performance and new features together with decreasing costs are key drivers that are accelerating the adoption of wearable devices by end users and professionals in the future.

We have also presented the new area of wearable biometrics for authentication purposes. We have discussed how wearable technology has enhanced the use of traditional biometric modalities and is enabling new modalities to be used for authentication. In particular, wearable technology has shown to be the most promising in providing frictionless and continuous authentication.

We have also presented a conventional wearable architecture, and we have discussed how it would evolve with the authentication factors because of some of the inherent limitations of current wearable devices. Limited storage, battery life, and processing power along with security and privacy are some of the challenges that need to be addressed in the future in order to increase the adoption rate of wearables and enable them to be deployed and used to their fullest potential.

References

1. E.O. Thorp, The invention of the first wearable computer, in *Digest of Papers. Second International Symposium on Wearable Computers (Cat. No.98EX215)*, (1998), pp. 4–8
2. S. Mann, An historical account of the "WearComp" and "WearCam" inventions developed for applications in "personal imaging", in *Digest of Papers. First International Symposium on Wearable Computers*, (1997), pp. 66–73

3. WearTech.org, WearComp.org, WearCam.org, EyeTap Digital Eye Glass, WearTech (TM) Wearable Technologies/research/consulting/public speaking/inventing/philosophy. [Online]. Available: http://wearcam.org/. Accessed 04 Apr 2017
4. M. Lamming, M. Flynn, Forget-me-not: Intimate computing in support of human memory, in *Procedings of FRIEND21, 1994 International Symposium on Next Generation Human Interface*, (1994), p. 4
5. Wearable technology and wearable devices: everything you need to know, wearable devices [Online]. Available http://www.wearabledevices.com/what-is-a-wearable-device/
6. Microsoft presents Smart Personal Objects Technology (SPOT)-based wristwatches at CES, News Center, 09-Jan-2003 [Online]. Available https://news.microsoft.com/2003/01/09/microsoft-presents-smart-personal-objects-technology-spot-based-wristwatches-at-ces/
7. Garmin and G. L. or its subsidiaries, 'Forerunner® 101', *Garmin*. [Online]. Available: https://buy.garmin.com/en-US/US/p/231. Accessed 07 Apr 2017
8. Fitbit Zip™ Wireless activity tracker. [Online]. Available: https://www.fitbit.com/uk/zip. Accessed 07 Apr 2017
9. Fitness trackers in the lead as wearables market grows 3.1% in the third quarter, according to IDC, www.idc.com. [Online]. Available: http://www.idc.com/getdoc.jsp?containerId=prUS41996116. Accessed 12 Apr 2017
10. Wearables aren't dead, they're just shifting focus as the market grows 16.9% in the Fourth Quarter, According to IDC, www.idc.com. [Online]. Available: http://www.idc.com/getdoc.jsp?containerId=prUS42342317. Accessed 12 Apr 2017
11. C. McGoogan, Fitbit cuts staff as interest in wearables stalls, The Telegraph, 30-Jan-2017 [Online]. Available http://www.telegraph.co.uk/technology/2017/01/30/fitbit-cuts-staff-interest-wearables-stalls/
12. Gartner survey shows wearable devices need to be more useful. [Online]. Available: http://www.gartner.com/newsroom/id/3537117. Accessed 13 Apr 2017
13. Xiaomi and Apple tie for the top position as the wearables market swells 17.9% during the first quarter, according to IDC, www.idc.com. [Online]. Available: http://www.idc.com/getdoc.jsp?containerId=prUS42707517. Accessed 13 Apr 2017
14. Admin, 'Wearable technology market 2017 global trend, segmentation and opportunities forecast To 2022 | AB Newswire'
15. Wristwear dominates the wearables market while clothing and earwear have market-beating growth by 2021, according to IDC', *www.idc.com*. [Online]. Available: http://www.idc.com/getdoc.jsp?containerId=prUS42371617. Accessed 15 Apr 2017
16. A.K. Jain, A. Ross, S. Prabhakar, An introduction to biometric recognition. IEEE Trans. Circuits Syst. Video Technol. **14**(1), 4–20 (Jan. 2004)
17. L. Wang, X. Geng (eds.), *Behavioral Biometrics for Human Identification: Intelligent Applications* (Medical Information Science Reference, Hershey, 2010)
18. Recon Jet. Smart eyewear for sports., Recon Instruments. [Online]. Available: https://www.reconinstruments.com/products/jet/. Accessed 02 Jun 2017
19. Head mounted display sunglasses for cycling, Solos Wearables. [Online]. Available: http://www.solos-wearables.com/. Accessed 05 May 2017
20. Spectacles by Snap Inc. (United Kingdom). [Online]. Available: https://www.spectacles.com/uk/. Accessed 12 May 2017
21. Owlet smart sock & baby care | track heart rate & oxygen levels. [Online]. Available: http://www.owletcare.com/. Accessed 13 May 2017
22. OMbra - The ultimate running bra that just happens to be smart, OMsignal. [Online]. Available: https://omsignal.com/. Accessed 13 May 2017
23. Athos. [Online]. Available: https://www.liveathos.com/. Accessed 1 Jun 2017
24. Pebble 2 + heart rate | Pebble smartwatch | Smartwatch for iPhone & Android, Pebble. [Online]. Available: https://www.pebble.com/pebble 2-smartwatch-features. Accessed 23 Apr 2017
25. Oculus Rift | Oculus. [Online]. Available: https://www.oculus.com/rift/. Accessed 23 Apr 2017

26. Microsoft, 'Microsoft HoloLens', Microsoft HoloLens. [Online]. Available: https://www.microsoft.com/en-gb/hololens. Accessed 24 Apr 2017
27. Vive | Discover virtual reality beyond imagination. [Online]. Available: https://www.vive.com/uk/. Accessed 24 Apr 2017
28. EyeTap personal imaging lab. [Online]. Available: http://www.eyetap.org/research/eyetap.html. Accessed 27 May 2017
29. Google glass, Wikipedia. 07-Jun-2017 [Online]. Available https://en.wikipedia.org/wiki/Google_Glass
30. Vuzix | view the future. [Online]. Available: https://www.vuzix.eu/. Accessed 12 Jun 2017
31. Samsung Gear S3 highlights, The official Samsung Galaxy Site. [Online]. Available: http://www.samsung.com/global/galaxy/gear-s3/. Accessed 12 May 2017
32. Apple Watch Series 2, Apple (United Kingdom). [Online]. Available: https://www.apple.com/uk/apple-watch-series-2/. Accessed 12 May 2017
33. Garmin and G. L. or its subsidiaries, 'GPS Running Watch | Forerunner 935', Garmin. [Online]. Available: https://buy.garmin.com/en-GB/GB/p/564291. Accessed 12 May 2017
34. Fitbit Charge 2™ Heart rate + fitness wristband. [Online]. Available: https://www.fitbit.com/uk/charge2. Accessed 13 May 2017
35. Mi Band 2 - Mi Global Home. [Online]. Available: http://www.mi.com/en/miband2/. Accessed 13 May 2017
36. Garmin and G. L. or its subsidiaries, 'vívosmart 3 | Garmin | Fitness Tracker', Garmin. [Online]. Available: https://buy.garmin.com/en-GB/GB/p/567813. Accessed 14 May 2017
37. RINGLY, 'Ringly Luxe - Smart Rings', RINGLY. [Online]. Available: https://ringly.com/products/smart-ring. Accessed 13 Jun 2017
38. A.K. Dey, K. Wac, D. Ferreira, K. Tassini, J.-H. Hong, J. Ramos, Getting closer: an empirical investigation of the proximity of user to their smart phones, in *Proceedings of the 13th international conference on Ubiquitous computing*, (2011), pp. 163–172
39. P. Fernandez-Lopez, J. Liu-Jimenez, C. Sanchez-Redondo, R. Sanchez-Reillo, Gait recognition using smartphone, in *2016 I.E. International Carnahan Conference on Security Technology (ICCST)*, (2016), pp. 1–7
40. R. Ferrero, F. Gandino, B. Montrucchio, M. Rebaudengo, A. Velasco, I. Benkhelifa, On gait recognition with smartphone accelerometer, in *2015 4th Mediterranean Conference on Embedded Computing (MECO)*, (2015), pp. 368–373
41. W. Shi, J. Yang, Y. Jiang, F. Yang, Y. Xiong, SenGuard: Passive user identification on smartphones using multiple sensors, in *2011 I.E. 7th International Conference on Wireless and Mobile Computing, Networking and Communications (WiMob)*, (2011), pp. 141–148
42. A.H. Johnston, G.M. Weiss, Smartwatch-based biometric gait recognition, in *2015 I.E. 7th International Conference on Biometrics Theory, Applications and Systems (BTAS)*, (2015), pp. 1–6
43. A. Lewis, Y. Li, M. Xie, Real time motion-based authentication for smartwatch, in *2016 I.E. Conference on Communications and Network Security (CNS)*, (2016), pp. 380–381
44. G. Peng, G. Zhou, D.T. Nguyen, X. Qi, Q. Yang, S. Wang, Continuous authentication with touch behavioral biometrics and voice on wearable glasses. IEEE Trans. Hum. Mach. Syst **47** (3), 404–416 (Jun. 2017)
45. Y. Chen, C. Shen, Performance analysis of smartphone-sensor behavior for human activity recognition. IEEE Access **5**, 3095–3110 (2017)
46. N. Alshurafa, J. Eastwood, S. Nyamathi, W. Xu, J.J. Liu, M. Sarrafzadeh, Battery optimization in smartphones for remote health monitoring systems to enhance user adherence, in *Proceedings of the 7th international conference on PErvasive Technologies Related to Assistive Environments*, (2014), p. 8
47. S. Majumder, T. Mondal, M.J. Deen, Wearable sensors for remote health monitoring. Sensors **17**(1), 130–2017
48. BIOWATCH | Biometric authentication [Online]. Available https://www.biowatch.ch/web/. Accessed 21 May 2017

49. Product overview, Nymi. [Online]. Available: https://nymi.com/product_overview. Accessed 25 May 2017
50. M. Zubair, C. Yoon, H. Kim, J. Kim, J. Kim, Smart wearable band for stress detection, in *2015 5th International Conference on IT Convergence and Security (ICITCS)*, (2015), pp. 1–4
51. N. H. S. Choices, Electrocardiogram (ECG) - NHS choices, 24-Nov-2016. [Online]. Available: http://www.nhs.uk/Conditions/electrocardiogram/Pages/Introduction.aspx. Accessed 27 May 2017
52. H.S. Choi, B. Lee, S. Yoon, Biometric authentication using noisy electrocardiograms acquired by mobile sensors. IEEE Access **4**, 1266–1273 (2016)
53. N. H. S. Choices, Electroencephalogram (EEG) - NHS choices, 24-Nov-2016. [Online]. Available: http://www.nhs.uk/Conditions/EEG/Pages/Introduction.aspx. Accessed 27 May 2017
54. M.K. Bashar, I. Chiaki, H. Yoshida, Human identification from brain EEG signals using advanced machine learning method EEG-based biometrics, in *2016 I.E. EMBS Conference on Biomedical Engineering and Sciences (IECBES)*, (2016), pp. 475–479
55. J. Thorpe, P. C. van Oorschot, and A. Somayaji, 'Pass-Thoughts: Authenticating with our Minds', 121, 2005
56. Y. Zeng, A. Pande, J. Zhu, P. Mohapatra, WearIA: Wearable device implicit authentication based on activity information
57. S. Ojala, J. Keinanen, J. Skytta, Wearable authentication device for transparent login in nomadic applications environment, in *2008 2nd International Conference on Signals, Circuits and Systems*, (2008), pp. 1–6
58. Multi-sensor finger ring for authentication based on 3D signatures (PDF download available)', ResearchGate. [Online]. Available: https://www.researchgate.net/publication/264083695_Multi-sensor_Finger_Ring_for_Authentication_Based_on_3D_Signatures. Accessed 29 May 2017
59. B.-R. Cha, S.-H. Lee, S.-B. Park, G.-K. Lee, Y.-K. Ji, Design of Micro-payment to strengthen security by 2 factor authentication with mobile & wearable devices. Adv. Sci. Tech **109**, 28–32 (2015)
60. '8 Mind-blowing Uses of Wearable Technology (Seriously...)'. [Online]. Available: http://www.govtech.com/fs/news/8-Mind-blowing-Uses-of-Wearable-Technology-Seriously.html
61. "ok glass," - Google glass help. [Online]. Available: https://support.google.com/glass/answer/3079305?hl=en. Accessed 08 Jun 2017
62. E. H. L. com J. 4, 2014, and 10:26 Am, 'U.S. military biosensors could reveal science of sweat'. [Online]. Available: http://www.cbsnews.com/news/u-s-militarys-skin-sensors-could-reveal-science-of-sweat/. Accessed 05 Jun 2017
63. C4ISRNET, 'Wearable sensors could prevent troops from overheating', C4ISRNET. [Online]. Available: http://www.c4isrnet.com/articles/mit-marine-team-developing-sensors-to-avoid-overheating-troops. Accessed 05 Jun 2017
64. Ankle monitor, Wikipedia. 04-Jun-2017 [Online]. Available https://en.wikipedia.org/wiki/Ankle_monitor
65. Security flaw allows criminals to hack and escape from house arrest devices, The Stack, 14-Aug-2015 [Online]. Available https://thestack.com/security/2015/08/14/security-flaw-allows-criminals-to-hack-and-escape-from-house-arrest-devices/
66. M. Khoso, How much data is produced every day?, Level Blog, 13-May-2016 [Online]. Available http://www.northeastern.edu/levelblog/2016/05/13/how-much-data-produced-every-day/
67. Working with wearables and Bluemix. [Online]. Available: https://www.ibm.com/developerworks/library/ba-bluemix-wearables/index.html. Accessed 09 Jul 2017
68. Automatically unlock your Mac with your Apple Watch, Apple Support. [Online]. Available: https://support.apple.com/en-gb/HT206995. Accessed 12 Jun 2017
69. Nymi Whitepaper 08-Nov-2015 [Online]. Available https://nymi.com/sites/default/files/Nymi%20Whitepaper.pdf. Accessed 01 Jun 2017

70. L. Goode, Don't expect battery life miracles in wearables anytime soon, The Verge, 15-Jan-2016. [Online]. Available: https://www.theverge.com/2016/1/15/10775660/fitness-tracker-smartwatch-battery-problems-apple-motorola-lumo. Accessed 30 May 2017
71. Moto 360 by Motorola - Smartwatch powered by android wear, Motorola. [Online]. Available: https://www.motorola.com/us/products/moto-360. Accessed 27 Apr 2017
72. A. R. M. Ltd, 'Markets | Wearables – ARM', ARM | The architecture for the digital world. [Online]. Available: https://www.arm.com/markets/wearables. Accessed 10 Jun 2017
73. S. Zeadally, S.U. Khan, N. Chilamkurti, Energy-efficient networking: Past, present, and future. J. Supercomput. **62**(3), 1093–1118 (Dec. 2012)
74. A. Perez, S. Zeadally, Privacy issues and solutions for consumer wearables. IT Prof **99**, 1–1 (2017)
75. A.J. Perez, S. Zeadally, S. Griffith, Bystanders' Privacy'. IT Prof **19**(3), 61–65 (2017)
76. S. Mann, 'Eye am a camera: Surveillance and Sousveillance in the Glassage', *Time* 02-Nov-2012 [Online]. Available http://techland.time.com/2012/11/02/eye-am-a-camera-surveillance-and-sousveillance-in-the-glassage/. Accessed 30 May 2017
77. Bree~commonswiki. (2006) [online]. Available https://commons.wikimedia.org/wiki/File:Bewakingsbollen_station_Aarschot.JPG. Accessed 30 May 2017

Chapter 15
Cognitive Biometrics for User Authentication

Ahmed Awad and Yudong Liu

1 Introduction

Biometric systems are used to detect human characteristics presenting them in the form of metrics that could be used in user identification. Physiological and behavioral characteristics are utilized for such purpose. Physiological biometrics describe human characteristics related to the human body such as hand geometry, fingerprint, retina, and face. Such characteristics are not expected to change over the course of the person's lifetime. Behavioral biometrics, on the other hand, focus on detecting characteristics related to patterns in human behavior. These patterns are always related to specific set of actions performed by the user.

User behavior relies on two important factors, the mechanical action itself and how the user performs it and the habit or knowledge or understanding the human builds from interacting with such environments.

Cognition is a process performed by the human brain involving knowledge acquisition, reasoning, and understanding through experience. In contrast to biometrics, such process is hard to measure and capture since there is no straightforward way to detect its characteristics.

Figure 15.1 illustrates the different steps involved in a human-computer interaction process. The human brain receives signals from the eye as a result of changes taking place on the computer screen. The brain translates these signals into a concept or set of concepts and decides about how to react to such changes. The brain, controlling the human body parts, sends signals to order the body part to react to

A. Awad (✉)
University of Washington, Bothell, WA, USA
e-mail: ahmedaa@uw.edu

Y. Liu
Western Washington University, Bellingham, WA, USA

M. S. Obaidat et al. (eds.), *Biometric-Based Physical and Cybersecurity Systems*,
https://doi.org/10.1007/978-3-319-98734-7_15

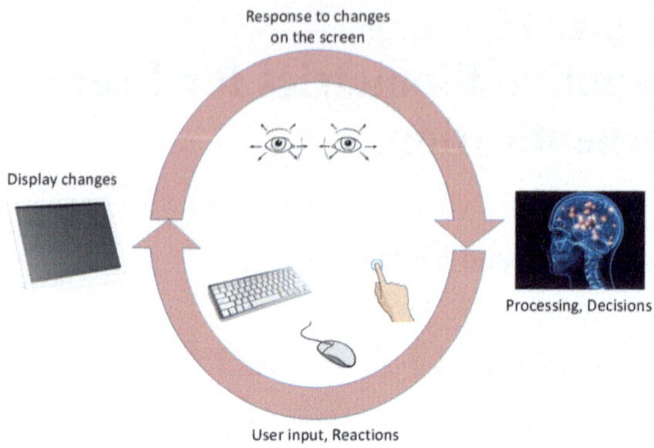

Fig. 15.1 The cycle of actions and reactions: user interaction with a computer and the different steps involved

such concepts. The reaction could be as simple as a face expression or an action to change the status of the surrounding environment.

The user will apply a mechanical action on the computer using one of the peripheral devices. The operating system and the running application will process this input and apply changes on the computer screen in a form of communication with the user. Such changes will be detected by the human eye and processed again with the brain. This sequence forms a cycle of actions and reactions between the user and the computer as shown in Fig. 15.1.

Behavioral sensing devices could be implanted in this cycle to detect the user actions, such as mouse dynamics and keystroke dynamics, or to monitor the user's eye movements. Cognitive activities performed by the brain are hard to detect. A sophisticated setup is required to detect signals generated by the heart, the brain, and the autonomic nervous system. The electrocardiogram (ECG), the electroencephalogram (EEG), and the electrodermal response (EDR) are used by [11] to build a biometric signature that can be used for authentication. Such sensors are very intrusive and not applicable for implementation in passive authentication environments.

The approach we propose in this chapter is to try to capture the cognitive factors through the behavioral biometric sensors and the eye tracking sensor. The goal is to model the cycle of actions and reactions presented above in order to capture the relations between the different biometric factors involved.

2 Eye Tracking Biometrics

Human eyes provide a rich source of information to identify who the person is. Eye tracking biometrics that analyze the complex and unique characteristics of the eye were long conventionally divided into two fields: iris biometrics and retina biometrics. The iris is a muscle within the eye, which regulates the size of the pupil and thus controls the amount of light that enters the eye. An iris recognition system automatically identifies the identity of a subject by analyzing the random pattern of the iris. It first acquires images of an iris using near-infrared light. Once the image is captured and found, the set of pixels that cover the iris on the image are then transformed into a digital form for future matching/verification. Retina recognition captures the patterns of blood vessels on the thin nerve on the back of the eyeball that processes light entering through the pupil. Since both iris and retina patterns are highly distinctive and remain relatively stable throughout one's lifetime, both are considered as highly accurate biometrics [12–16].

Compared to iris biometrics and retina biometrics, eye movement-based biometrics is a relatively new field of research. Previous research has shown a solid correlation between eye movement and underlying cognitive processes [17, 18]. Information about which objects have been visually inspected and in which order and context and the duration the eyes were staying on those objects can help researchers to infer what cognitive processes were involved to perform a task related to these objects. Some of the pioneering work on eye movement investigated their functionality in the task of reading [19, 20] and picture viewing [21, 22]. More recently, more research on human identification based on eye movement has been done due to the availability of more affordable and less obtrusive eye tracking devices and accessibility of machine learning tools [23–27]. Eye movement features, such as number of fixations, fixation duration, saccadic velocity and duration, gaze position, pupil diameters, etc., provide a novel data resource for biometric recognition.

When collecting eye movement data through recording the eye movement by an eye tracking device, some main questions that need to be considered include what type of visual stimuli to use, how the experiment is set up, what data quality to achieve, how to assess such a quality, and what preprocessing operations (such as data cleaning) need to apply on the raw data.

Different types of visual stimuli have been adopted related to the aspects of oculomotor characteristics under consideration. They are generally classified into the following categories [28]:

- "Jumping" point of light (horizontal, vertical, and random)
- Cognitive dot patterns
- Text excerpts
- Face images
- Various-content (natural) image
- Video sequences

Figure 15.2 [28] shows a general demonstration of some common types of visual stimuli that can be used in an eye movement experiment, along with examples of performed eye movement scan paths. The first one is the random "jumping" point stimulus. As observed, it induces saccades of various amplitude, which can be used for the inspection of the physiological characteristics of the oculomotor system. More details can be found in [28]. The chosen stimuli need to allow for the extraction of various eye movement features which should be used by the machine learning method for the biometric recognition task.

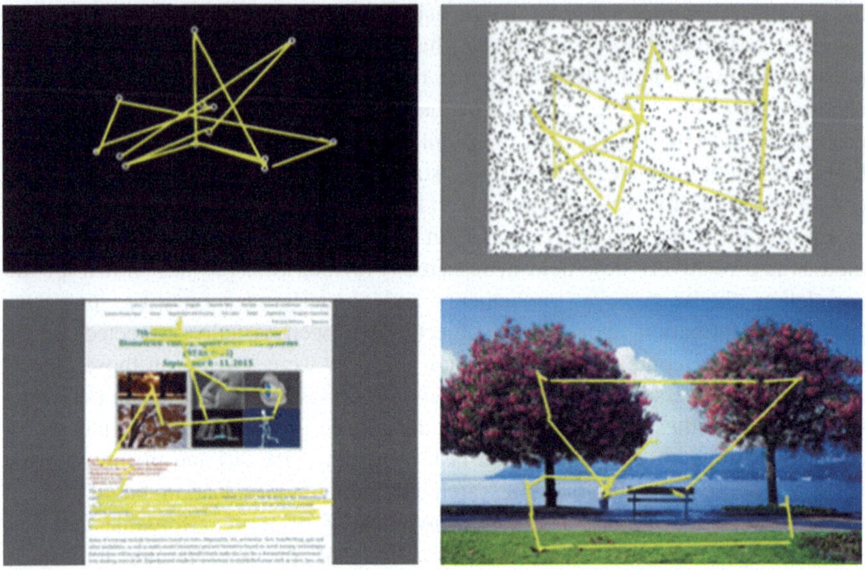

Fig. 15.2 Graphical demonstration of different types of visual stimuli and the performed eye movement scan paths (Images are taken from [28])

Eye movement recording can be conducted either in a lab-controlled environment or in an open-free environment, depending on the required data quality. Generally, lab-controlled environment is used when high-quality dataset is required. The selection of an eye tracking device is dictated by the goal of the research. The factors such as whether it is desk-mounted, remote, or head-mounted/wearable and what the temporal resolution, spatial resolution, and spatial accuracy are needed to be considered when such a device is acquired and calibrated. Another important parameter in collecting the eye movement data is the subjects from whom the data is collected, the adequate number of subjects, the balanced composition of subjects in age, race, gender, with or without corrected lens, etc. In a typical eye movement recording, the subject is positioned in front of a computer screen where the visual stimulus is presented, while the eye tracker captures the performed eye movements (see Fig. 15.3 [28]). Some important setup parameters that need to be configured for ensuring a clear and comfortable visual field during the experiments are the stimulus screen dimensions and resolution and the distance of the subject from the stimulus screen. Other experimental parameters that need to be decided are the calibration setup, the duration of the experimental trials, and the time interval between the experimental trials [28].

To extract features from the eye movement data, the common approach is to initially classify the eye movement samples into fixations and saccades, and then, a large variety of basic features can be extracted from the corresponding amplitude, velocity, and acceleration profiles. A set of more complex features such as various ratios of amplitude, velocity, and acceleration characteristics can be further

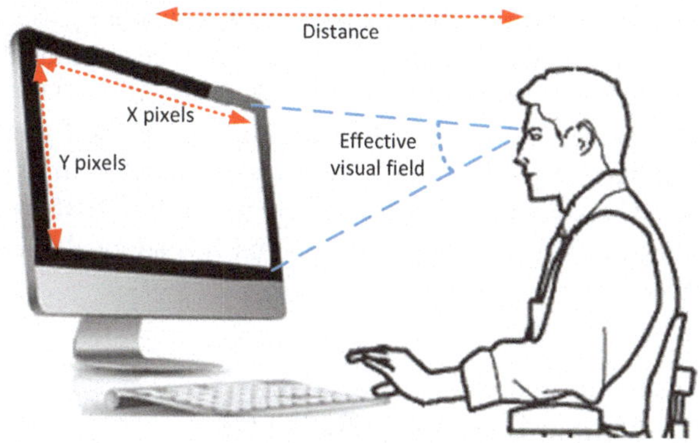

Fig. 15.3 A typical eye movement recording setup

calculated using these basic features. To summarize, the common features that are used in an eye movement biometric recognition systems generally fall into the following categories [28]:

- Amplitude (or position)
- Duration
- Velocity
- Acceleration
- Direction
- Frequency
- Scan path
- Blink rate

Different machine learning methods have been adopted where the features are used either directly or by building histograms over them. Some common classification methods include neural networks, KNNs, and SVM.

One of the pioneering works [23] that used eye movements for biometric purpose used stimuli type of "jumping" point of light and achieved a 1.36% false acceptance rate (FAR) and a 12.59% false rejection rate (FRR) for a group of nine subjects. Later works [24, 30–37] have adopted different types of visual stimuli, used a variety of feature combinations and machine learning methods, and reported performance improvements on varied number of subjects from 12 to 200. The BioEye 2015 (https://bioeye.cs.txstate.edu/) competition was organized as an attempt to advance the research field of eye movement biometrics, by giving the scientists and researchers opportunity to use a common dataset that contains eye movement recorded using high-quality standards and captured using different visual stimuli and with different time intervals. The system that ranked first [28] used a large number of statistical features extracted from position, velocity, and acceleration profiles of fixation and saccades. The redundant and correlated features extracted in this stage were removed via a feature selection process. Several algorithms were tested, and the best IR (identification rate) was achieved by a neural network-based framework. The same authors reported in [27] that equal error rate (EER) is improved up to 2.59% and rank-1 accuracy of 89.54% in RAN (random 30-min dataset of BioEye 2015 database [29] containing 153 subjects). According to the authors, Template aging effect has also been studied using data taken after an interval of 1 year. The average EER obtained is 10.96% with a rank-1 accuracy of 81.08% with 37 subjects. In their proposed method, eye movement data is classified into fixations and saccades. Features extracted from fixations and saccades are used by a Gaussian radial basis function network (GRBFN)-based method for biometric authentication. A score fusion approach is adopted to classify the data in the output layer.

3 Mouse Dynamics Biometrics

Mouse dynamics is a behavioral biometrics that could be used in different security applications. The purpose of this biometrics is to model the behavior of the user and find the distinguishing factors that establish the user's identity. Mouse dynamics biometrics was pioneered in 2003 by Ahmed and Traore of the University of Victoria. The new biometrics was introduced as a tool to identify and verify user's identity based on how he/she interacts with the computer using the mouse input device. The new biometrics is utilized in static/dynamic authentication, intrusion detection, and forensics.

Mouse dynamics can be described as the characteristics of the actions received from the mouse input device for a specific user while interacting with a specific graphical user interface. The first step in understanding the actions received from the input device is to identify the categories where those actions fall. A mouse action can be classified into one of the following categories:

- Mouse move (MM): general mouse movement
- Drag and drop (DD): mouse button down, mouse move, then mouse button up
- Point and click (PC): mouse movement followed by a click or a double click
- Silence: no movement

Data collected from the mouse device require processing and filtering before it can be analyzed to model the biometric behavior. Figure 15.4 [1] shows an example of captured raw data. Each point on the curve represents a detected mouse action. The figure shows the relation of traveled distance and the elapsed time for a set of captured actions. Such representation is not suitable for user classification and behavior comparison. It is obvious that there is a need to process the collected data more in order to model the user's behavior.

The characteristics of mouse dynamics can be described by a set of factors generated as a result of analyzing the recorded mouse actions. These factors represent the components of what the mouse dynamics signature for a specific user is and which can be used in verifying the identity of the user.

Different approaches can be used in each category to collect the factors characterizing it. Some examples of the type of factors collected from each analysis include the following:

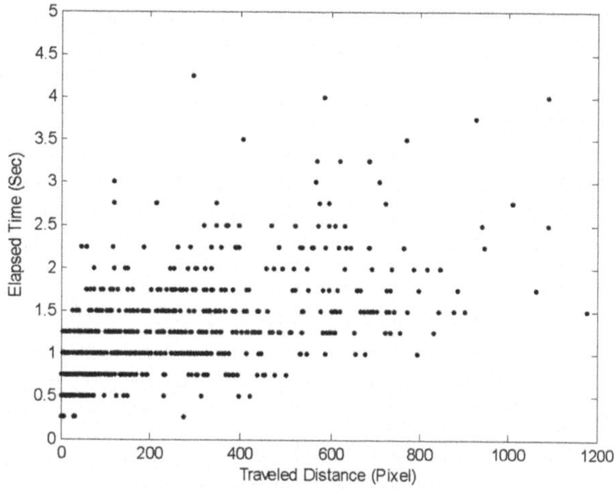

Fig. 15.4 Raw mouse dynamics data. Each point represents a mouse action described with the distance traveled and the elapsed time

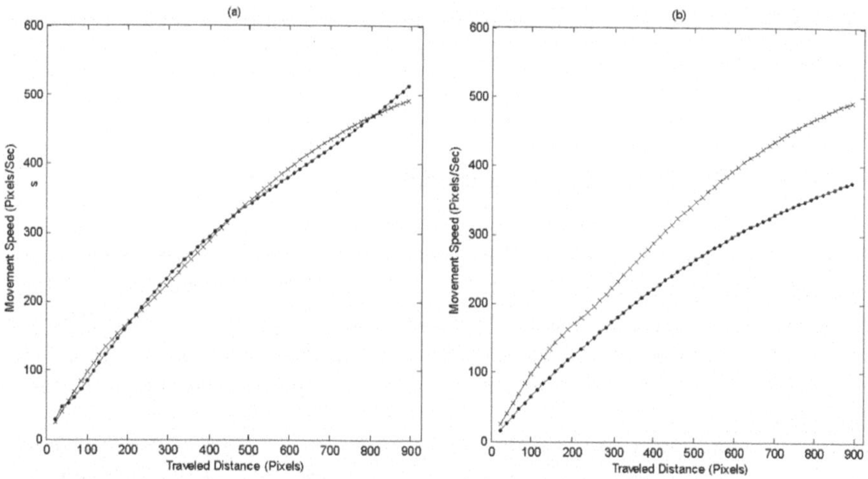

Fig. 15.5 MSD curves for two sessions (**a**) belonging to the same user, (**b**) belonging to two different users

1. Movement Speed compared to traveled Distance (MSD)
2. Average Movement speed per movement Direction (MDA)
3. Movement Direction Histogram (MDH)
4. Average movement speed per Types of Actions (ATA)
5. Action Type Histogram (ATH)
6. Traveled Distance Histogram (TDH)
7. Movement elapsed Time Histogram (MTH)
8. Silence Histogram (SH)

Each of these factors is represented with a set of numbers that are used to construct the mouse dynamics signature for a specific user. Figure 15.5 shows one of these factors, the Movement Speed compared to Traveled Distance (MSD), for two different users. Sessions belonging to the same user are similar with very little deviation, while sessions belonging to two different users deviate from each other confirming an expected difference in user behavior.

Figure 15.5 [2] shows an example of a detectable deviation in behavior between two different users. The first curve shows two sessions belonging to the same users: very little deviation is noticeable between the two curves. The second curve shows two sessions belonging to two different users where very high detectable deviation is measurable between the two different curves.

A classification system should be designed to be able to calculate the user's signature from raw training data and differentiate between different behaviors based on the calculated signatures. Ahmed and Traore used a neural network for such purpose [2]. The neural network used in the detection process is a feed-forward multilayer perceptron network consisting of three layers. The input layer consists of 39 nodes, the total number of inputs representing the factors involved in the mouse dynamics signature. The hidden and output layers consist, respectively, of 40 nodes and 1 node. Experimental evaluation produced an accuracy of a FAR of 2.4649% and a FRR of 2.4614%.

4 Keystroke Dynamics Biometrics

Keystroke dynamics is a well-established biometric. Since the early 1980s, a rich body of research has been produced in this area [4–9]. Keystroke dynamics recognition systems measure the dwell time and flight time for keyboard actions. The raw data collected for keystroke includes the time a key is depressed and the time the key is released. Figure 15.6 [3] illustrates the different types of data collected from a keyboard action.

Based on this data, the duration of keystroke (i.e., length of time a key is depressed) and the latency between consecutive keystrokes are calculated and used to construct a set of monographs, digraphs, trigraphs, or n-graphs producing a pattern identifying the user.

Figure 15.7 [4] gives an example on how a calculated signature is used to identify a user based on data provided in a given session. Figure 15.7a compares calculated

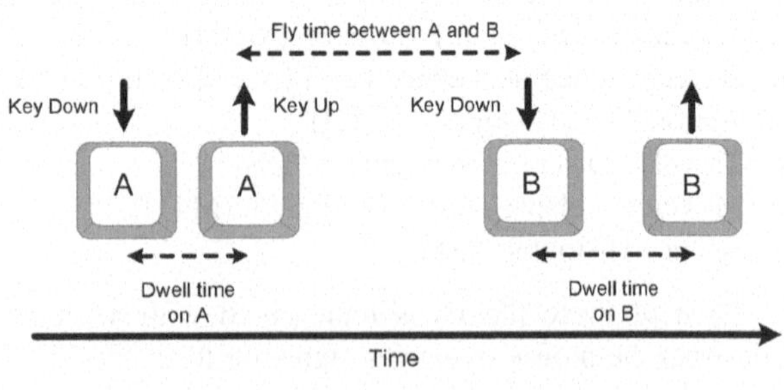

Fig. 15.6 Data collected from a keyboard action. Dwell time and fly time are used to construct monographs and digraphs

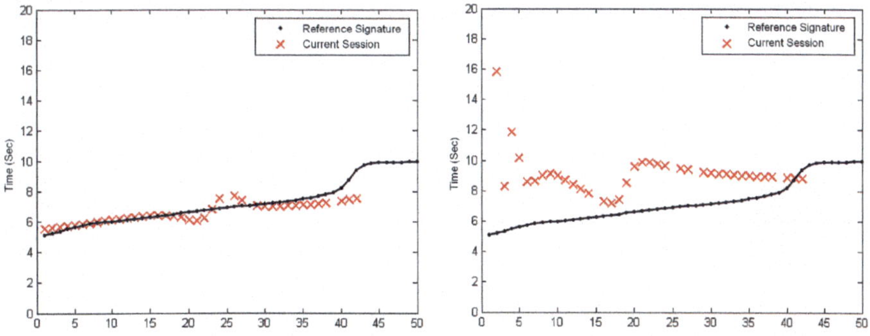

Fig. 15.7 Data collected from a keyboard action. Dwell time and fly time are used to construct monographs and digraphs. The figure shows two comparisons between the dwell times for actions, performed in (**a**) a legitimate user's session and (**b**) an attacker's session, and the reference signature of the user

monographs collected from a user's session to his/her own signature. The dwell time of the collected monographs falls on or close to the signature curve. Figure 15.7b performs the same comparison but using data collected from a different user. As expected, and since the behavior is different, session's monographs deviate from the signature curve of the legitimate user.

Most of the research work done in this area target fixed text detection and focus on using this technology for static authentication or access control [4, 5, 10]. In static authentication the user is asked to enter a specific text on the keyboard, e.g., a password, a passphrase, or a predefined word or a set of words. During the enrollment process, the user is required to enter the same fixed text several times in order to get reasonable amount of data for the identification.

For dynamic and passive monitoring, it is necessary to be able to detect the user without requiring him to enter a predefined message or text.

To study the cognitive behavior, it is expected that the user will respond differently to what is displayed on the screen. So, free text detection is essential for our purpose. However, free text detection presents huge challenges, which explain the limited number of related work available in the literature.

One of the important works in this field is [3] where a model was built to enable the creation of a biometric signature using data collected from fully passive free text sessions. The model approximates the relation between the different monographs and digraphs in order to fill in missing factors required for user enrollment. Experimental evaluation involving 53 users in a heterogeneous environment yielded a false acceptance ratio (FAR) of 0.0152% and a false rejection ratio (FRR) of 4.82%, at an equal error rate (EER) of 2.46%.

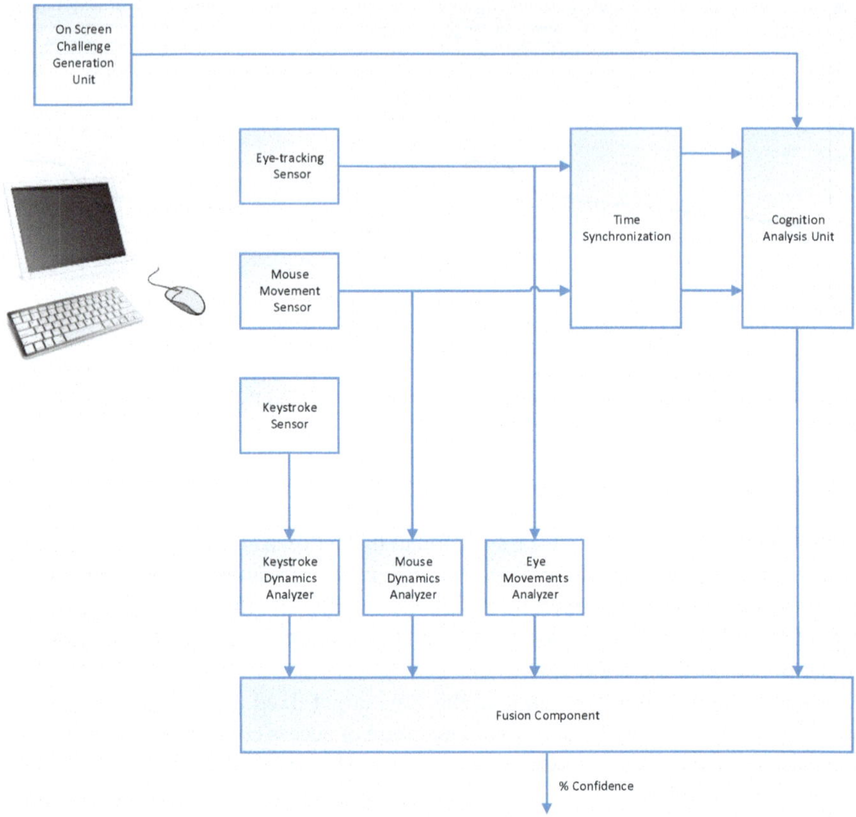

Fig. 15.8 A hybrid model correlating features collected from three biometric sensors with cognitive features representing how the user reacts to changes on the screen

5 Hybrid Model for Authentication

A typical model for user authentication relies on the processing of user's input in order to create confidence on his/her identity before granting access. Such models rely fully on data collected from biometric sensors. A decision is made based on biometric features collected from the sensors and how far they are from expected user's behavior. Such models completely ignore the cognition factor and the relation between the user's actions captured by different biometric sensors.

The hybrid model presented in Fig. 15.8 takes this relation in consideration. In order to capture the different relations, it is necessary for the system to be able to synchronize the captured activities to determine which activities are done is response to which and which are affected by which.

Three biometric sensors (Fig. 15.8) are used to capture the user behavior, mouse dynamics, keystroke dynamics, and eye tracking sensor. The sensors will provide raw data which could be analyzed independently to produce three decisions represented by % confidences which could be fused to produce the final decision. Cognitive analysis unit will also process captured raw data after synchronizing the time differences between the eye tracking and the mouse dynamics sensors.

The unit models the relation between the two dynamics by detecting behavioral pattern similarities and producing factors representing such relations, for example, the relation between the mouse movement speed and eye movement speed for cursor movement in a specific direction.

Authentication systems could include a user challenge in the form of prompting the user for input or reacting to the user's input by displaying something on the screen or changing the colors or the shape of specific areas on the screen. This is applicable to both passive and active authentication systems. In passive systems the changes on the screen are dependent on the operating system or the application in use.

The hybrid model introduces a component responsible for detecting the changes on the screen and synchronizing them with the detected user's reactions. The unit could go beyond that in case of active authentication and generate on-screen challenges to trigger user's reactions leading to the capturing of specific cognitive features.

The cognition analysis unit analyzes and correlates these features and generates a % confidence that contributes to the overall % confidence calculated using a tunable fusion technique.

6 Conclusion

With today's increasing demand on dependable and secure systems, stronger and more secure authentication mechanisms are required. Designers of such systems strive to incorporate technologies that are usable, less intrusive, and more secure. Biometric systems are good candidates for such purpose. The accuracy of biometrics relies on the maturity of the model used and how accurate it is in capturing different human characteristics.

In this chapter, we presented a model that can be used to integrate the different factors representing behavioral biometrics and the cognitive characteristics of a human. We believe that the inclusion of such analysis will add to the accuracy of the authentication system since the decision will be made based on more features that are unique for each user. The goal is to lower the false acceptance and false rejection by incorporating the cognitive dimension in the analysis. Future work will focus on modeling the cognitive behavior, building an intelligent on-screen challenge component that is capable of triggering the required biometric feedback, and enhancing the eye tracking sensor by developing new factors that better describe the behavior.

References

1. A.A.E. Ahmed, I. Traoré, Behavioral biometrics for online computer user monitoring, in *Image Pattern Recognition*, (World Scientific Publishing, Singapore, 2007), pp. 243–263
2. A.A.E. Ahmed, I. Traore, A new biometric technology based on mouse dynamics. IEEE Trans. Dependable Secure Comput. **4**(3), 165–179 (2007)
3. A.A. Ahmed, I. Traore, Biometric recognition based on free-text keystroke dynamics. IEEE Trans. Cybernet. **44**(4), 458–472 (2014)
4. A.A.E. Ahmed, Employee surveillance based on free text detection of keystroke dynamics, in *Handbook of Research on Social and Organizational Liabilities in Information Security*, (IGI Global, Hershey, Pennsylvania, 2009), pp. 47–63. Web. 22 May 2017
5. F. Bergadano, D. Gunetti, C. Picardi, User authentication through keystroke dynamics. ACM Trans. Inform. Syst. Secur. **5**(4), 367–397 (2002)
6. M. Brown, S.J. Rogers, User identification via keystroke characteristics of typed names using neural networks. Int. J. Man-Mach. Stud. **39**(6), 999–1014 (1993)
7. P. Dowland, S. Furnell, and M. Papadaki, Keystroke analysis as a method of advanced user authentication and response, in *Proceedings of the IFIP TC11 17th International Conference on Information Security: Visions and Perspectives*, 7–9 May 2002, pp. 215–226
8. D. Gunetti, C. Picardi, Keystroke analysis of free text. ACM Trans. Inform. Syst. Secur. **8**(3), 312–347 (Aug. 2005)
9. F. Monrose, A. Rubin, Authentication via keystroke dynamics, in *Proceedings of the 4th ACM Conference on Computer and Communications Security*, Apr 1997, pp. 48–56
10. M.S. Obaidat, B. Sadoun, Verification of computer users using keystroke dynamics. IEEE Trans. Syst. Man Cybernet. Part B **27**(2), 261–269 (1997)
11. K. Revett, F. Deravi, K. Sirlantzis, Biosignals for user authentication - towards cognitive biometrics?, in *2010 International Conference on Emerging Security Technologies*, Canterbury, 2010, pp. 71–76
12. R.P. Wildes, Iris recognition: An emerging biometric technology. Proc. IEEE **85**(9), 1348–1363 (1997)
13. Y. Zhu, T. Tan, Y. Wang, Biometric personal identification based on iris patterns, in *Pattern Recognition, 2000. Proceedings 15th International Conference on*, vol. 2. IEEE, 2000
14. L. Ma et al., Personal identification based on iris texture analysis. IEEE Trans. Pattern Anal. Mach. Intell. **25**(12), 1519–1533 (2003)
15. A. Cavoukian, *Privacy and Biometrics* (Information and Privacy Commissioner, Ontario, 1999)
16. A. Jain, R. Bolle, S. Pankanti, *Biometrics: Personal Identification in Networked Society*, vol 479 (Springer Science & Business Media, New York, 2006)
17. M.A. Just, P.A. Carpenter, Eye fixations and cognitive processes. Cogn. Psychol. **8**(4), 441–480 (1976)
18. K. Rayner, Eye movements in reading and information processing: 20 years of research. Psychol. Bull. **124**(3), 372 (1998)
19. G.W. McConkie et al., Eye movement control during reading: I. The location of initial eye fixations on words. Vis. Res. **28**(10), 1107–1118 (1988)
20. E.D. Reichle et al., Toward a model of eye movement control in reading. Psychol. Rev. **105**(1), 125 (1998)
21. J.M. Henderson, A. Hollingworth, Eye movements during scene viewing: An overview. Eye Guid. Read. Scene Percept. **11**, 269–293 (1998)
22. G.R. Loftus, N.H. Mackworth, Cognitive determinants of fixation location during picture viewing. J. Exp. Psychol. Hum. Percept. Perform. **4**(4), 565 (1978)
23. P. Kasprowski, J. Ober, Eye Movements in Biometrics, in *International Workshop on Biometric Authentication*, Springer Berlin Heidelberg, 2004
24. R. Bednarik, et al., Eye-movements as a biometric, in *Image Analysis* 2005, pp. 16–26
25. C. Holland, O.V. Komogortsev, Biometric identification via eye movement scanpaths in reading, in *Biometrics (IJCB), 2011 International Joint Conference on*. IEEE, 2011

26. C.D. Holland, O.V. Komogortsev, Complex eye movement pattern biometrics: analyzing fixations and saccades, in *Biometrics (ICB), 2013 International Conference on*. IEEE, 2013
27. A. George, A. Routray, A score level fusion method for eye movement biometrics. Pattern Recogn. Lett. **82**, 207–215 (2016)
28. I. Rigas, O.V. Komogortsev, Current research in eye movement biometrics: an analysis based on BioEye 2015 competition. Image Vis. Comput. **58**, 129–141 (2017)
29. O.V. Komogortsev, I. Rigas, Bioeye 2015: competition on biometrics via eye movements, in *Biometrics Theory, Applications and Systems (BTAS), 2015 I.E. 7th International Conference on*. IEEE, 2015
30. T. Kinnunen, F. Sedlak, R. Bednarik, Towards task-independent person authentication using eye movement signals, in *Proceedings of the* 2010 *Symposium on Eye-Tracking Research & Applications*. ACM, 2010
31. O.V. Komogortsev, et al., Biometric identification via an oculomotor plant mathematical model, in *Proceedings of the* 2010 *Symposium on Eye-Tracking Research & Applications*. ACM, 2010
32. O.V. Komogortsev, et al., Biometric authentication via oculomotor plant characteristics, in *Biometrics (ICB),* 2012 *5th IAPR International Conference on*. IEEE, 2012
33. Z. Liang, F. Tan, Z. Chi, Video-based biometric identification using eye tracking technique, in *Signal Processing, Communication and Computing (ICSPCC),* 2012 *IEEE International Conference on*. IEEE, 2012
34. I. Rigas, G. Economou, S. Fotopoulos, Biometric identification based on the eye movements and graph matching techniques. Pattern Recogn. Lett. **33**(6), 786–792 (2012)
35. V. Cantoni et al., GANT: Gaze analysis technique for human identification. Pattern Recogn. **48** (4), 1027–1038 (2015)
36. C.D. Holland, O.V. Komogortsev, Complex eye movement pattern biometrics: The effects of environment and stimulus. IEEE Trans. Inf. Forensic Secur. **8**(12), 2115–2126 (2013)
37. H.-J. Yoon, T.R. Carmichael, G. Tourassi, Gaze as a biometric, in *SPIE Medical Imaging*. International Society for Optics and Photonics, 2014

Chapter 16
Finger Knuckle-Based Multi-Biometric Authentication Systems

Aditya Nigam and Phalguni Gupta

1 Introduction

Human authentication is the prime concern in this current digital age due to the advancement in technology and availability of cheap hardware. It spans nearly all the domains required by humans for daily activities like banking [1], computer access, airport checking, and attendance management. Its objective is to secure the information from the malicious or unintended users. Personal authentication plays an important role in the society. It requires at least some level of security to assure the identity. Security can be realized through one of the three levels.

1. *Level 1[Possession]:* The user possesses something which is required to be produced at the time of authentication, for example, key of a car or a room.
2. *Level 2 [Knowledge]:* The user knows something which is used for authentication, for example, PIN (personal identification number), password, or CVV (card verification value) of a credit card.
3. *Level 3 [Biometrics]:* The user owns certain unique physiological and behavioral characteristics, known as biometric traits, which are used for authentication, for example, face, iris, fingerprint, signature, gait, etc.

However, there are drawbacks in Level 1 and Level 2 security. For example, key or smart cards may be lost or mishandled, while passwords or PIN may be forgotten or guessed. Since both possession and knowledge are not intrinsic user properties, they are difficult to be managed by the user. But this is not the case with Level

A. Nigam
Indian Institute of Technology, Mandi, India
e-mail: aditya@iitmandi.ac.in

P. Gupta (✉)
National Institute of Technical Teachers' Training & Research, Kolkata, India
e-mail: phalgunigupta@nitttrkol.ac.in

M. S. Obaidat et al. (eds.), *Biometric-Based Physical and Cybersecurity Systems*,
https://doi.org/10.1007/978-3-319-98734-7_16

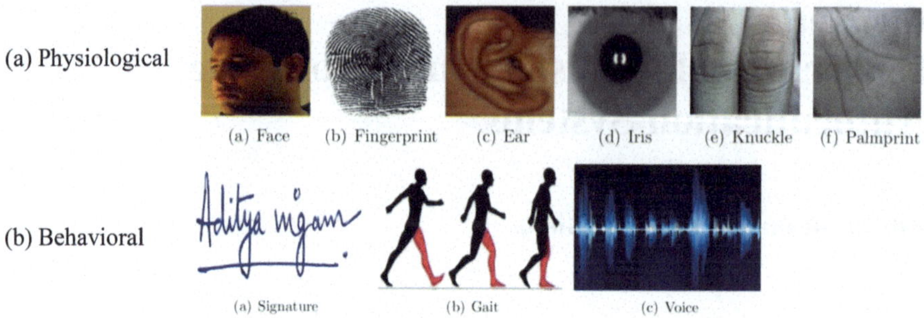

Fig. 16.1 Some common biometric traits. (**a**) Physiological. (**b**) Behavioral

3 security which is based on biometrics which can be considered as the science of personal authentication using the physiological (e.g., fingerprint, face, iris, etc.) and behavioral characteristics of human beings (e.g., signature, gait, voice, etc.). Examples of some well-known biometric traits are shown in Fig. 16.1.

A biometric-based authentication system is better than the traditional possession or knowledge-based system because of the following reasons:

- Biometric traits are intrinsically related to the user. They cannot be lost, forgotten, or misplaced; hence they are easy to manage.
- There is a need of physical presence of the trait for authentication.
- Features characteristics are unique.
- It is hard to spoof.

Any biometric trait can be used in the authentication system provided the trait makes the system more reliable, user-friendly, and cost-effective. Hence, the trait should possess the following characteristics in order to design an authentication system.

1. *Universality*: the biometric trait should be owned by everyone and should not be lost.
2. *Uniqueness*: characteristics associated with the biometric trait should be different for everyone.
3. *Performance*: it should instantaneously provide accurate matching results.
4. *Permanence*: it should be time invariant and can handle the environmental conditions.
5. *Collectability*: it can be easily and instantaneously acquired from anywhere and at any time.
6. *Acceptability*: it should be well accepted in the society, i.e., hygienic and free from social stigma.
7. *Measurability*: it should be acquired in a user-friendly manner.
8. *Circumvention*: it should be difficult to forge.

The traits along with their characteristics are depicted in Table 16.1. It can be observed that none can provide all the characteristics required in an authentication

Table 16.1 Properties of some well-known biometric traits

Biometric trait	University	Uniqueness	Performance	Permanence	Collectability	Acceptability	Measurability	Circumvention
Face	H	M	L	M	H	H	H	L
Fingerprints	M	H	H	H	M	M	H	M
Ear	M	M	M	M	M	H	M	M
Iris	M	H	H	H	L	L	L	H
Palmprint	M	H	H	M	M	M	M	M
Hand geometry	M	M	M	M	M	M	H	M
Retina	H	H	M	H	L	L	L	H
Hand vein pattern	H	H	H	H	M	M	H	M
DNA	H	H	H	H	L	L	L	M
Signature	L	M	M	M	M	M	L	H
Voice	M	M	M	L	H	H	H	M
Gait	M	L	L	L	H	H	M	L
Knuckleprint	H	H	H	H	M	H	H	L

system. A biometric-based personal authentication is a multistaged process. In the initial stage, the raw image is captured using an acquisition sensor. This is very critical and important stage because accuracy of any biometric system is highly dependent on the quality of images. In the second stage, the desired part from the image, termed as region of interest (ROI), is extracted from the acquired image. Third stage estimates the quality of the ROI. If the quality of the ROI is poor, then one may go for reacquisition of the image. In next stage, the ROI is preprocessed using some enhancement technique. Some transformations are also performed to get a robust ROI. Discriminative features from the enhanced ROI are extracted in the next stage. Finally, features of a query image have been matched against those of image(s) in the database to authenticate the claim.

Typically, any biometric authentication system performs the following three operations:

1. *Enrollment*: The user features are stored in the database corresponding to its unique identity. The quality of acquired trait is improved by mitigating the noise artifacts using preprocessing techniques. Several trait-specific features are extracted from region of interest (ROI). Enrollment process is shown in Fig. 16.2.
2. *Verification*: This operation is to determine whether the user has correctly revealed the identity. It verifies the user on the basis of the claimed identity and accordingly decides whether the user is genuine or impostor. The process of verification is shown in Fig. 16.3.
3. *Identification*: In case of identification, the aim is to reveal the correct identity of the user. It attempts to provide the best template match from the available database to reveal the user identity as shown in Fig. 16.4.

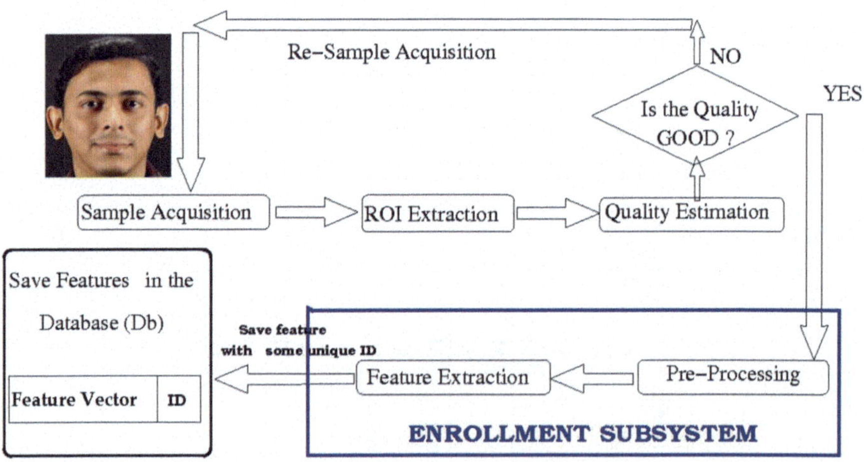

Fig. 16.2 Subject Enrollment

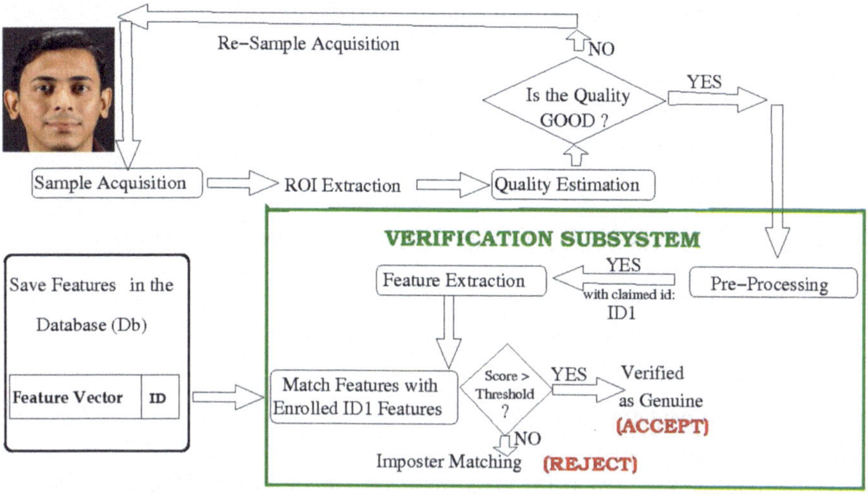

Fig. 16.3 Subject verification

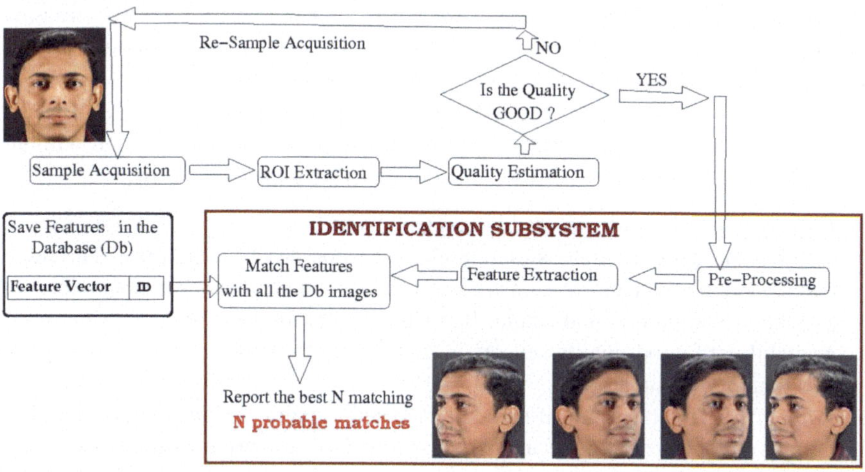

Fig. 16.4 Subject identification

1.1 Biometric Traits

There does not exist any biometric trait which satisfies all desired properties strictly. For example, facial features are not permanent throughout the life span, fingerprints are not visible for hardworking people, etc. However, there exist several well-known biometric traits which satisfy more or less all biometric properties. Biometric traits can be divided based on physiological and behavioral characteristics.

Table 16.2 Advantages and challenges of various biometric traits

Biometric trait	Advantages	Challenges
Face	Easy to capture	Pose variation, aging, and spoofing
Fingerprints	Unique and stable	Capturing good-quality prints
Ear	Easy to capture	Occlusion
Iris	Unique	Difficult to capture and spoofing
Palm print	Stable and unique	Illumination, translation, and rotation variations
Knuckle print	Uniqueness and easy to capture	Illumination, translation, and rotation variations

1.1.1 Physiological-Based Traits

Face, fingerprint, ear, iris, palm print, and knuckle print belong to the class of physiological biometric traits. However, each biometric modality has some advantages and challenges associated with it, as mentioned in Table 16.2.

Hand-Based Biometrics The major advantage of hand-based biometric traits is that they contain lots of discriminative information; they are noninvasive in nature and are quite stable as they do not vary with age. Fingerprints, palm prints, finger veins, and knuckle prints all come under the category of hand-based biometric traits. These traits can easily be captured by using low-cost sensing devices and without any much inconvenience to the user. Few of them are described below:

- *Fingerprint*: It is made up of ridges. There exist large amount of discriminative textures and patterns such as loop, arch, and whorl over a fingerprint. The ridge ending and the ridge bifurcation are known as minutia features. These features are assumed to be unique and stable. It is easy to capture these features because of availability of good-quality sensors in market, but the major challenge is to get a good quality of fingerprint.
- *Palm print*: The inner part of the hand is called as palm, and the extracted region of interest in between fingers and wrist is termed as palm print. Even monozygotic twins are found to have different palm print patterns. Huge amount of textures in the form of palm lines, ridges, and wrinkles is available over palm print. Prime advantage of palm print over fingerprint includes its higher social acceptance because it is never being associated with criminals.
- *Knuckle print*: The outer part of a finger is considered as finger knuckle. These patterns are presumed to be related with the anatomy of fingers that involves complex interaction with finger bones and skin. They can be acquired through a low-cost sensor.
- *Finger veins*: The vein pattern is also believed to be robust and invariant to aging. They are acquired in infrared lighting to get better light absorption. The major challenge is its variability; hence, it is considered as a soft biometric trait with limited application.

Fig. 16.5 Fingerprint Sensors

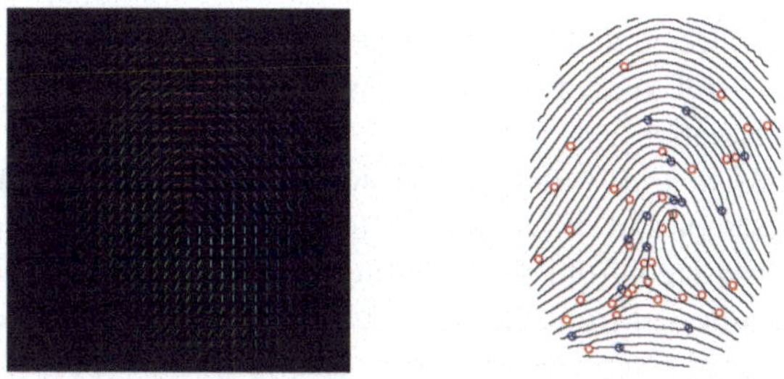

(a) Orientation Field Based Algorithm (b) Minutia Based Algorithm

Fig. 16.6 Two fingerprint matchers (multi-algorithm). (**a**) Orientation field-based algorithm. (**b**) Minutia-based algorithm

1.2 Multi-biometric System

The performance of any unimodal biometric system is often restricted due to variation and uncontrolled environmental condition, sensor precision, and reliability as well as several trait-specific challenges such as pose, expression, aging, etc. for the face. Moreover, it considers features to take decision on matching; hence, it is difficult to improve its accuracy. Hence, one can explore the possibility of fusing more than one biometric samples, traits, or algorithms. This is termed as multi-biometrics. There exist different types of multi-biometric system.

- *Multisensor system*: It considers images of the same biometric trait where images are captured with the help of multiple sensors. Figure 16.5 Shows three types of fingerprint scanners which can be used to build a multisensor biometric system. These sensors use different technologies to acquire data; hence, the quality and discriminative features of their samples are significantly different.
- *Multi-algorithm system*: It considers multiple matching algorithms to improve the performance of the system. Images of the selected trait are captured using single sensor. In Fig. 16.6, it is shown that one can use different algorithms applied over the same image. One algorithm may be using some global texture like orientation field features, while the other one may use minutia-based local features. Fusion of these matchers is expected to perform better than any of these two algorithms.
- *Multi-instance system*: It considers more than one image of the same trait per user. Multiple samples are collected. In Fig. 16.7, three samples of the same finger

Fig. 16.7 Samples of same fingerprint (multi-instance)

collected under controlled environment are shown. This redundant information is useful to address the issues related to local as well as environmental condition variations.

- *Multi-trait system*: It considers more than one biometric trait per user. Multiple samples of different traits (e.g., fingerprint, knuckle print, palm print) have been acquired from users during enrollment. These different types of information collected from the same subject enhance the collective discrimination of aggregated sample enormously. Individual trait-specific matching, followed by score level fusion, can improve the system performance significantly and can cater to external, local/global, or any other intra-class variability issues. Such systems also become more and more spoof resistant.

2 Performance Analysis

It is necessary to analyze the performance of any biometric system. There exist several performance measures to analyze any verification or identification system.

2.1 Verification Performance Parameters

Like any pattern recognition system, there are two types of errors, viz., false acceptance rate (FAR) and false rejection rate (FRR). When two feature vectors are matched, it generates a matching score. This score is either dissimilarity or similarity score. For a dissimilarity (similarity) score, if it is less (greater) than a predefined threshold, we assume these two feature vectors are matched. FAR is the probability of accepting wrongly an imposter as a genuine user. FAR is defined as follows:

$$\text{FAR} = \frac{M}{N} \times 100\% \tag{16.1}$$

where N is the number of distinct imposter matchings and M is the number of wrongly accepted genuine matchings.

Similarly, FRR is defined as the probability of rejecting a genuine user wrongly. That means, if we perform N distinct genuine matchings and M of them have been got rejected wrongly, then FRR is given by:

$$\text{FRR} = \frac{M}{N} \times 100\% \tag{16.2}$$

(a) *EER*: Equal error rate (EER) is defined as the value of FAR for which FAR and FRR are same. More clearly, we can say that it is the point of intersection of FAR and FRR curves.

(b) *Accuracy*: Accuracy is defined as:

$$\text{Accuracy} = \left(100 - \frac{A+R}{2}\right)\% \tag{16.3}$$

where A and R are FAR and FRR at threshold T.

(c) *Receiver operating characteristics (ROC) curve*: It is a graph plotting FAR against various FRRs. It helps to analyze the behavior of FAR against FRR. It computes the distinguishing strength of the system between genuine and imposter's score.

(d) *Decidability index (d')*: It measures separability between imposter and genuine matching scores and is defined by as:

$$d' = \frac{|\mu 1 - \mu 2|}{\frac{1}{2}\sqrt{V1 + V2}} \tag{16.4}$$

where $\mu 1$ and $\mu 2$ are the mean and $V1$ and $V2$ are the variance of the genuine and imposter scores, respectively. An example of the genuine and the imposter score distributions is shown in Fig. 16.8.

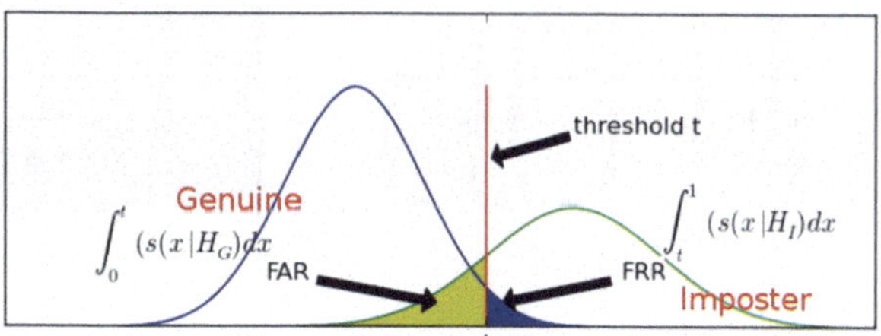

Fig. 16.8 Graph showing genuine and imposter similarity score distribution

2.2 *Identification Performance Parameters*

Similarly, for identification, performance analysis is done by calculating correct recognition rate (CRR) and genuine vs imposter best match graph.

(a) *CRR*: The correct recognition rate, CRR, is also known as the **Rank 1** accuracy. It is defined as the ratio of the number of correct topmost best matches and the total number of matching performed in the entire query set. If we have *N* images in the test set and out of that, M images have got the non-false top best match, then CRR is given as:

$$\text{CRR} = \frac{M}{N} \times 100\% \tag{16.5}$$

(b) *Genuine* vs *imposter best match graph (GvI graph)*: This graph shows the separation of genuine vs imposter best matching score plots. In Fig. 16.8, one such plot is shown from which one can observe that genuine matching scores are well separated from imposters and overlapping scores are errors.

3 Finger Knuckle Print-Based System Designing

Anatomical structure of a knuckle print is shown in Fig. 16.9a. In any knuckle print, there are lines like (i.e., knuckle lines) rich pattern structures in vertical as well as horizontal directions. These horizontal and vertical pattern formations are believed to be very discriminative. The main reason of using them is their unique anatomy; they are also noninvasive in nature and are quite stable which can be captured at low cost without any overhead of installing extra hardware device.

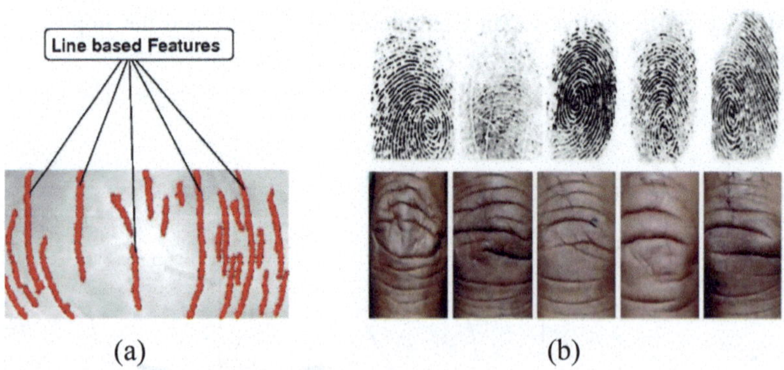

Fig. 16.9 (**a**) Knuckle anatomy. (**b**) Finger prints vs knuckle prints

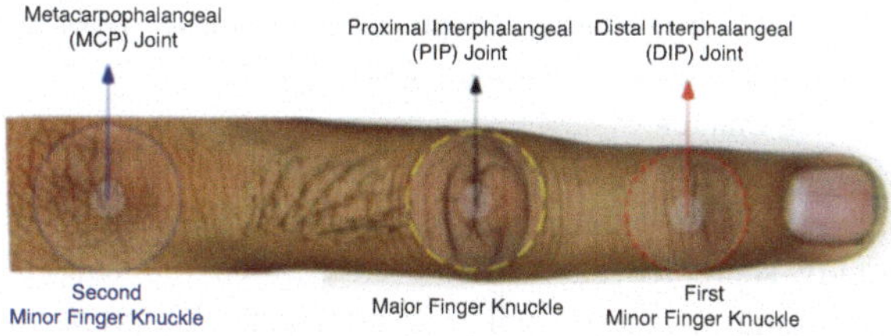

Fig. 16.10 Detailed finger knuckle anatomy. (Image taken from [9])

Agriculture is the main occupation of people living in South Asia. Here most of the persons are engaged in blue collared jobs. As a result, they use their hands in very rough and tough manner which potentially damages their course structure of hands including palm prints and fingerprints. In such cases quality of knuckle prints remains unchanged as they cannot be used for doing any hardship and thus they are less prone to any damage. Also, the minutia and singular points over fingerprints are observed to be faded for cultivators and laborers as shown in Fig. 16.9b.

In such scenarios finger dorsal knuckle print can serve as an effective biometric feature. The outer part of a finger is considered as finger knuckle. These knuckle prints are categorized into three categories; mainly they are major finger knuckle print, first minor finger knuckle print, and second minor knuckle print as shown in Fig. 16.10. These patterns are presumed to be unique. They can be acquired through any sensor. Many studies as presented in [9] have proved that combining first minor finger knuckle print with second minor and major finger knuckle print can give more accuracy than using them in isolation for personal identification. However, there exist big challenges in contact-less finger knuckle image (FKI) recognition such as efficiently identifying the middle knuckle line for image enrollment, exploring discriminative features, and matching nonrigid deformable knuckle regions. Any typical biometric system consists of sample acquisition, quality estimation, ROI extraction, preprocessing, feature extraction, matching, and finally fusion of multiple traits.

3.1 Quality Estimation of FKP Image

System performance is highly dependent on the quality of the input image. Quality of images acquired from different types of sensors plays an important role in getting good performance of a biometric system. Hence, quality assessment should be done as early as possible during data acquisition to achieve better results. Deficiency can also be revealed in any image by assessing its quality parameters.

Quality is defined as the amount of information observed or captured during the data acquisition phase. Designing an algorithm for assessing the quality of a biometric image is a very challenging task. An ideal quality system is the one that produces uniform results. In case of finger knuckle print-based biometric systems, quality assessment is quite cumbersome. Finger knuckle print images are generally of poor quality mainly because of reflection, poor uniformity, out of focus and camera reflection. Not much work has been done for quality assessment of finger knuckle images.

In [14], six quality assessment factors of finger knuckle images have been defined. These features are also quantified and fused together to obtain a single quality score for each image. Likelihood ratio-based fusion method is used for combining all these quality parameters. Comprehensive quality assessment of a finger knuckle print is a difficult problem because of its inherit complex structure. It has assessed the finger knuckle print quality by computing the amount of well-defined focus edges F, amount of clutter C, and distribution of focused edges in the image S and by observing block-wise entropy of focused edges E, reflection caused by light source and camera flash Re and the level of contrast Con.

3.1.1 Focus (F)

A focused image possesses a uniform spectrum where as a defocus image it usually contains dense spectrum toward lower frequencies. Well-focused pixels of an image are computed by convolving the proposed 6×6 kernel K, [14] with the input image where the kernel is given by:

$$K = \begin{pmatrix} 1 & 1 & 1 & 1 & 1 & 1 \\ 1 & 1 & 1 & 1 & 1 & 1 \\ 1 & 1 & -8 & -8 & 1 & 1 \\ 1 & 1 & -8 & -8 & 1 & 1 \\ 1 & 1 & 1 & 1 & 1 & 1 \\ 1 & 1 & 1 & 1 & 1 & 1. \end{pmatrix} \quad (16.6)$$

It selects only those pixels whose convolved value is greater than some empirically selected threshold value t_f constituting the set of pixels which are well focused, termed as *wf*. Let *vle* be the set of pixels lying on the vertically long edges of the image. The set of pixels F_{map} is obtained by computing the set intersection of pixels between *vle* (vertically long edges) and *wf* (well focused), as shown in Fig. 16.11c. It is defined as the most significant region constituting the focus quality parameter.

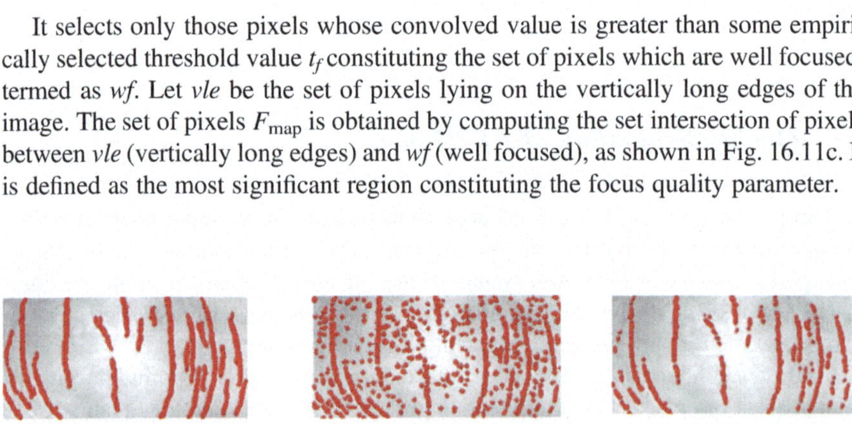

(a) *vle* (b) *wf* (c) F_{map} (anding *vle* and *wf*)

Fig. 16.11 Focused quality attribute F. (*vle* (vertically long edges) and *wf* (well focused)). (**a**) *vle*. (**b**) *wf*. (**c**) F_{map} (anding *vle* and *wf*)

3.1.2 Clutter *(C)*

The short vertical edge pixels which are well focused constitutes the quality parameter clutter. Clutter creates fallacious features that can degrade the performance of any recognition algorithm. It is defined as the ratio of short vertically aligned strong edge pixels to the longer edge pixels. Clutter is inversely proportional to the image quality.

3.1.3 Uniformity-Based Quality Attribute (*S*)

Uniform distribution of texture features throughout the image can constitute a good-quality biometric sample. *K*-Means clustering is used over pixel set F_{map} with $K = 2$. Finger knuckle images are somewhat symmetrical along *Y*-axis, due to which $K = 2$ has been chosen. Various statistical and geometrical parameters are used to determine the value of *S*. This method is discussed in detail in Algorithm 16.1.

3.1.4 Entropy-Based Quality Attribute (*E*)

Information content in any gray-scale image has been computed in terms of entropy which is defined as:

$$e = -\sum\nolimits_{i=0}^{255} \text{hist}[i] * \log(2 * \text{hist}[i]) \tag{16.7}$$

where hist*[i]* is the ith element of the 256 valued gray-level histogram, hist, of the input image *I*. Block-wise entropy is calculated using Eq. (16.7), and for computing it, each image is divided into blocks of size 5 × 5. Since all blocks do not have the same amount of importance, only blocks that are having well-defined focused long vertically aligned edge pixels (using F_{map}, more than a predefined experimentally selected threshold value $t_{f\ m}$) are considered as significant blocks. Entropy-based quality attribute is computed by summing up the entropy values of individual significant blocks.

3.1.5 Reflection-Based Quality Attribute (Re)

Whenever there is high reflection in the image, a very-high-intensity gray-level patch has been formed, mainly because of light source reflection or due to camera flash. In order to identify this reflection patch, several methods can be used; one such method is an adaptive thresholding methodology. The input image is repeatedly thresholded using different gray-level values. At the end of every step, number of pixels having value greater than thc threshold has been calculated. This thresholding procedure gets terminated when this count value becomes saturated. The full reflection patch is correctly recognized. The reflection-based quality attribute (Re) is defined as the fraction of pixels belonging to the reflection patch; and it is inversely proportional to the image quality.

3.1.6 Contrast-Based Quality Attribute (Con)

Lightning condition effect the finger knuckle print images to a large extent. Large illumination variation in biometric sample severely affects the discriminative line features and hence degrades the sample quality. Dynamic gray-level range of an image is depicted by its contrast. Therefore, we can use it to judge the uniformity in illumination of the entire image. In [14], the whole gray-level range is divided into three groups (0, 75), (76, 235), and (236, 255). The contrast-based quality attribute (Con) is defined as the fraction of pixels belonging to the middle gray-level range values (i.e., (76, 235)), because that indicates the moderated intensity range.

Algorithm 16.1 Uniformity-Based Quality Attribute (S) [14]
Require: The *vle* and *wf* pixel set for the input image (I) of size $m \times n$.
Ensure: Return the value S for the input image (I).

1. $F_{\text{map}} =$ and (wf, vle); [focus mask].
2. Let M_1 *and* M_2 be midpoint of left half $(\frac{n}{2}, \frac{n}{2})$ and right half $(\frac{m+n}{2}, \frac{n}{2})$ of the input image (I), respectively.
3. Apply 2-mean clustering over pixel set F_{map}.
4. Let C_1,C_2, nc_1, nc_2, std_1,and std_2 be mean loc., number of pixels, and standard dev. of left and right cluster, respectively.
5. Let d_1 and d_2 be Euclidean distance between point C_1 and M_1 and that of between C_2 and M_2, respectively.
6. $d = 0.7 * \max(d_1, d_2) + 0.3 * \min(d_1, d_2)$.
7. 7. $P_r = \frac{X}{Y}$ where $X = \max(\text{nc}_1, \text{nc}_2)$ and $Y = \min(\text{nc}_1, \text{nc}_2)$ [cluster point ratio].
8. 8. $\text{std}_r = \frac{A}{B}$ where $A = \max(\text{std}_1, \text{std}_2)$ and $B = \min(\text{std}_1, \text{std}_2)$ [cluster standard dev. ratio].
9. $\text{comb}_r = 0.8 * p_r + 0.2 * \text{std}_r$.
10. $D_{\text{std}} = 1 - \frac{d}{\sqrt{C}}$ where $C = \text{std}_1{}^2 + \text{std}_2{}^2$.
11. $D_{\text{nc}} = 1 - - \frac{d}{\sqrt{E}}$ where $E = \text{nc}_1{}^2 + \text{nc}_2{}^2$.
12. $S = 0.5{*}d + 0.2{*}\text{comb}_r + 0.15{*}D_{\text{std}} + 0.15{*}D_{\text{nc}}$.

3.1.7 Fusing Quality Attributes

All these mentioned six parameters are calculated and normalized using the max-min normalization technique. All six quality parameters should be fused in order to predict the overall quality of the input image. The likelihood ratio-based fusion has been used to fuse the obtained individual quality attributes for any input image I has been done as:

$$\mathrm{Quality}_{\mathrm{fused}}(I) = \prod\nolimits_{\forall q \in F, C, S, E, \mathrm{Re}, \mathrm{Con}} \frac{\mathrm{PD}(Iq|Gq)}{\mathrm{PD}(Iq|Bq)} \tag{16.8}$$

This *fusion* strategy classifies the input image into two classes (i.e., good and bad, binary classification) where PD(.) denotes the probability distribution of all the abovementioned quality parameters and Gq and Bq are the good and bad quality image samples with respect to the quality q, respectively.

3.2 *Finger Knuckle Print ROI Extraction*

The major task of any ROI extraction technique is to segment same region of interest correctly and consistently from all images. The central knuckle point as shown in Fig. 16.12b can be used to segment any knuckle print. Since knuckle print is aligned horizontally, there is no difficulty in extracting the central region of interest from any knuckle print that contains rich and distinctive texture as shown in Fig. 16.12c.

For major FKP ROI extraction, there are mainly two techniques, viz., convex direction coding [18] and curvature Gabor filter [15], which are discussed in the following subsections.

3.2.1 ROI Extraction Based on Convex Direction Coding

This method of ROI extraction is based on determining the local coordinate system. The following are the main steps involved in it.

- During data acquisition phase, all fingers are taken flatly, on the basal block; therefore, the bottom boundary of the FKP images is nearly consistent. This bottom boundary is extracted using the Canny edge detector. By fitting this

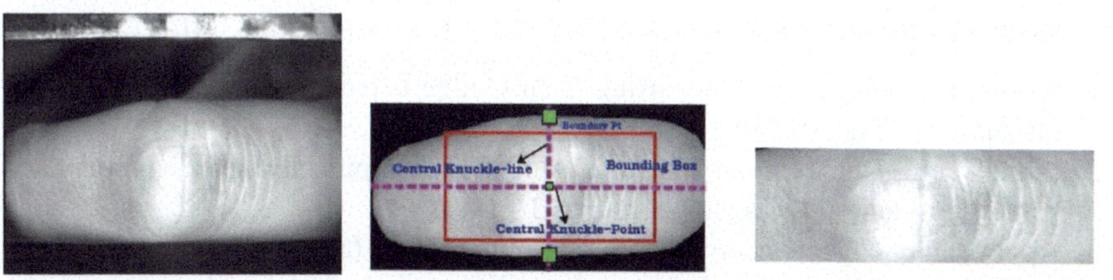

(a) Raw knuckleprint (b) Annotated Knuckleprint (c) Knuckleprint ROI

Fig. 16.12 Knuckle print ROI annotation. (**a**) Raw knuckle print. (**b**) Annotated knuckle print. (**c**) Knuckle print ROI

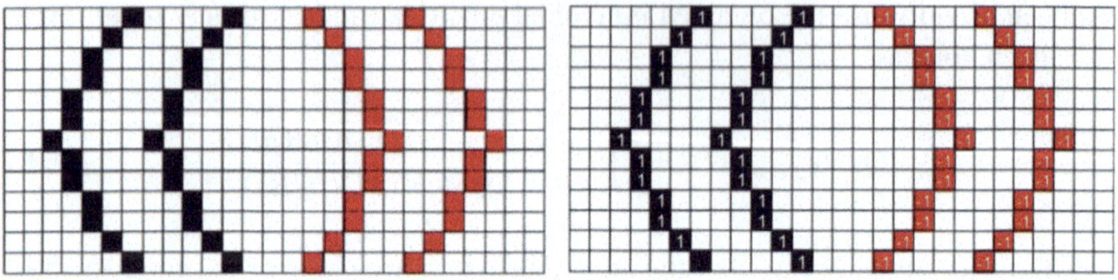

Fig. 16.13 FKP curve modeling. (Image is taken from [18])

boundary as a straight line, the *X*-axis of the local coordinate system can be determined.

- The top and bottom boundaries are estimated based on the actual size of the real fingers, obtained using Canny edge detector. The left and right boundaries are fixed and evaluated empirically, to obtain cropped image I_S as a sub-image.
- Canny edge detector is applied on I_S to obtain the edge map I_E.
- Convex direction coding is used to setup *Y*-axis, so as to get the most consistent line features from the finger knuckle print images. Every point on the image edge map can be given a code to identify its local direction. The curve is either convex leftward (+1) or convex rightward (−1), as shown in Fig. 16.13. Remaining pixels that are not on any curve are set to (0), as formulated below:

$$\mathrm{A_{code}} = \begin{cases} 0 & \mathrm{if\,(A = 0)[Not\,on\,curve]} \\ 0 & \mathrm{else\ if\,(}P7\&P8\mathrm{\ are\ Bifurcation\ point)} \\ +1 & \mathrm{else\ if\,(}P7\&(A\in \mathrm{UH})\|P5\&(A\in \mathrm{LH})\mathrm{[convest\ Left])} \\ -1 & \mathrm{else\ if\,(}P5\&(A\in \mathrm{UH})\|P7\&(A\in \mathrm{LH})\mathrm{[convext\ Right])} \end{cases} \tag{16.9}$$

In Eq. (16.9), UH and LH are considered as upper and lower half of the finger knuckle print image, respectively. The various steps involved in it are given in Algorithm 16.2 and are shown in Fig. 16.14.

Algorithm 16.2 Finger Knuckle Print ROI Extraction (Algorithm Taken from [18])
Require: The acquired finger knuckle print image I of size $m \times n$ (as shown in Fig. 16.14a)
Ensure: The cropped finger knuckle print ROI (as shown in Fig. 16.14h)

1. Extract the bottom boundary using Canny edge detection and assign as *X*-axis (as shown in Fig. 16.14b).
2. Initially crop the original image I using three empirically selected boundary values to obtain I_{cmp} (as shown in Fig. 16.14c).
3. Apply canny edge detector over I_{cmp} to obtain I_{canny} (as shown in Fig. 16.14d).
4. Apply convex direction coding over I_{canny} image to obtain I_{code} where every pixel is assigned a code representing its convex direction (as shown in Fig. 16.14e).

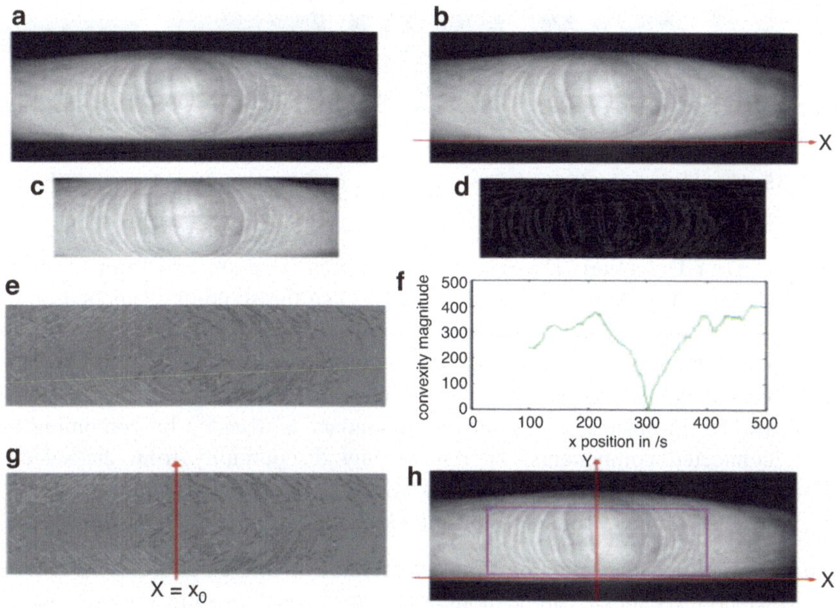

Fig. 16.14 Finger knuckle print image ROI extraction steps. (Images are taken from [18])

5. Obtain convexity direction which measures the strength of dominant direction locally as fined by conMag$(x) = |\sum_W I_{CD}|$ (as shown in Fig. 16.14f).
6. Since curves along small area around phalangeal joint do not have obvious/ dominant convex direction; hence, $x_0 = \text{argmin}_x(\text{conMag}(x))$ (as shown in Fig. 16.14g).
7. Set $X = X_0$ as the Y-axis to obtain the local coordinate system for the required *ROI* (as shown in Fig. 16.14h).

- The Y-axis of the coordinate system is determined. The characteristic of a FKP image has been shown in [18], suggesting that the convexity magnitude reaches a minimum around the center of the phalangeal joint and this position can be used to set the Y-axis of coordinate system.
- Since X-axis and Y-axis is fixed, the local coordinate system can be determined, and the ROI sub-image, I_{ROI}, can be extracted of a fixed size.

3.2.2 ROI Extraction Based on Curvature Gabor Filter

For extracting ROI, three main steps are performed under this technique: (a) detection of knuckle area, (b) central knuckle line, and finally (c) central knuckle point.

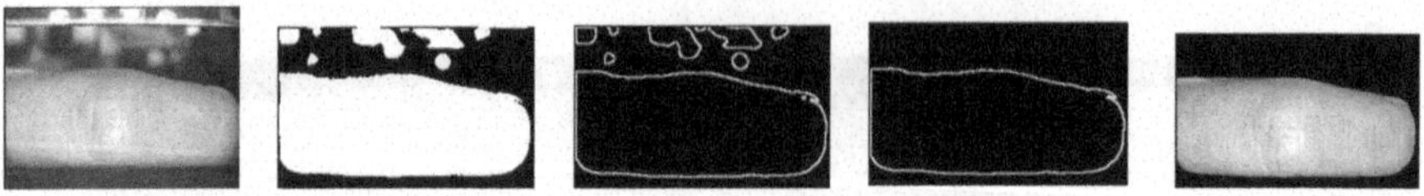

(a) Raw (b) Binary (c) Canny (d) Boundary (e) FKP Area

Fig. 16.15 Knuckle area detection. (**a**) Raw. (**b**) Binary. (**c**) Canny. (**d**) Boundary. (**e**) FKP Area

Knuckle Area Detection The whole knuckle area is segmented from the background to discard background region. Initially, Otsu thresholding is applied on the enhanced knuckle image to be binarized, as shown in Fig. 16.15b. It can be observed that the knuckle region may not be accurate because of sensor noise and background clutter. This image can be further be refined by using Canny edge detector as shown in Fig. 16.15c. From this image, knuckle boundary is detected by computing the largest connected components. To remove any discontinuity from the selected region, it is eroded as shown in Fig. 16.15d. All pixels that fall within the convex hull of detected knuckle boundary form the knuckle area as shown in Fig. 16.15e.

Central Knuckle Line Detection The central knuckle line is defined as the column w.r.t to which the knuckle can be thought of symmetrical, as shown in Fig. 16.12b. This line is used to extract the knuckle print ROI. Area around central knuckle line is quite rich in texture; thus it can be used for detection. To conserve such a specific texture, a knuckle filter is created by modifying the Gabor filter, as defined below.

(a) *Knuckle filter*: Traditionally Gabor filter is formed when a complex sinusoid is multiplied with a Gaussian envelope, as defined in Eq. (16.12), and is shown in Fig. 16.16c. If x and y are the spatial coordinates of the filter, X and Y are obtained by rotating x and y by an angle θ, using the following equations:

$$X = x* \cos\theta + y* \sin\theta \tag{16.10}$$

$$Y = -x* \sin\theta + y* \cos\theta \tag{16.11}$$

Then:

$$G(x, y, \gamma, \theta, \psi, \lambda, \sigma) = \text{Gaussian Envelope}^{*}\text{Complex Sinusoid} \tag{16.12}$$

where Gaussian Envelope $= e^{-\frac{K}{L}}$ for $K = X^2 + Y^2 - \gamma^2$ and $L = 2\,\sigma^2$ and complex sinusoid $= e^{i\left(\frac{2\pi x}{\lambda}+\psi\right)}$ (Fig. 16.16).

In order to preserve the curved convex knuckle lines, a knuckle filter (Fig. 16.17) is obtained by introducing curvature parameter in the traditional Gabor filter. Only X component has been modified as follows (Y remains to be same as in Eq. (16.11)):

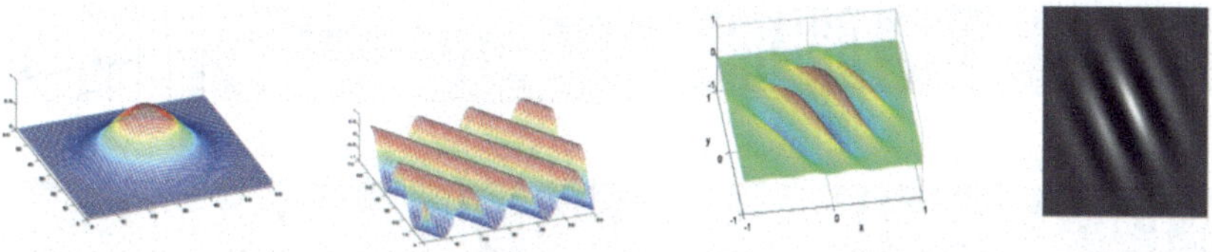

(a) 2D Gaussian (b) 2D Sinusoid (c) 2D Gabor (d) 2D Gabor Filter

Fig. 16.16 Conventional Gabor filter. (**a**) 2D Gaussian. (**b**) 2D Sinusoid. (**c**) 2D Gabor. (**d**) 2D Gabor filter

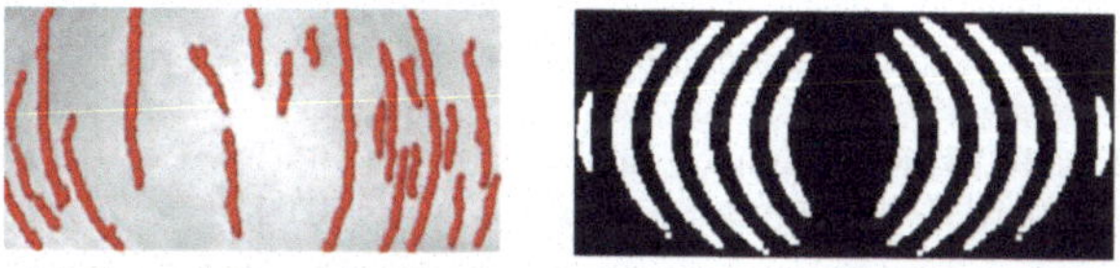

(a) Curved Knuckle Lines (b) Curvature Knuckle Filter

Fig. 16.17 Curvature knuckle filter. (**a**) Curved knuckle lines. (**b**) Curvature knuckle filter

$$X = x* \cos\theta + y* \sin\theta + c^*(-x* \sin\theta + y* \cos\theta)^2 \tag{16.13}$$

$$Y = -x* \sin\theta + y* \cos\theta \tag{16.14}$$

The curvature of the Gabor filter can be modulated by the curvature parameter. The value of curvature parameter is varied as shown in Fig. 16.18, and its optimal value is selected heuristically. The knuckle filter is computed by combining two such curved Gabor filters (f_1, f_2), in such a manner that the distance between them is d. The first filter (f_1) is obtained using the abovementioned parameters, while the second filter ($f_2 = f_2^{\text{flip}}$) is obtained by vertically flipping the first filter. In Fig. 16.18, several knuckle filters are shown with varying curvature and distance parameters.

(b) *Knuckle line extraction*: All pixels that belong to knuckle area are first convolved with the knuckle filter $F_{\text{kp}}^{0.01,30}$. As a result, all pixels that fall on central knuckle line have higher response. Then the filter response for each filter is calculated by using threshold as $f *$ max where max is the maximum knuckle filter response and $f \in 0$ to 1 is a fractional value. The binarized filter response is shown in Fig. 16.19a in which it is super-imposed over the knuckle area with a blue color. Finally, the central knuckle line has been extracted from that column which is having the maximum knuckle filter response as shown in Fig. 16.19b.

Central Knuckle Point Detection This is required to crop the knuckle print ROI that must lie over central knuckle line. It is computed by computing the midpoint of top and bottom points of central knuckle line as shown in Fig. 16.19b. The required knuckle print ROI is extracted as a region of window size $(2 * w + 1) \times (2 * h + 1)$

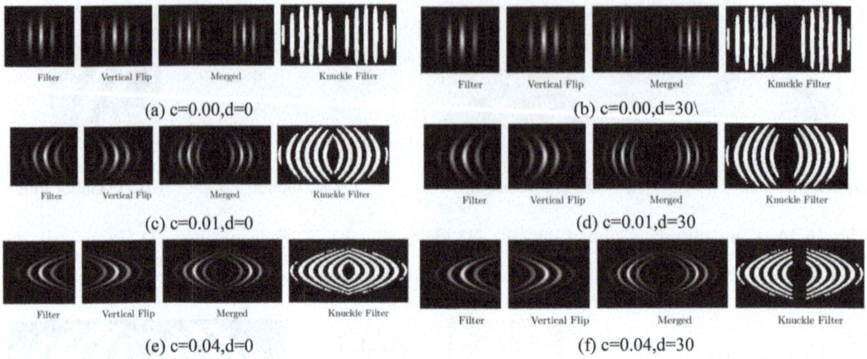

Fig. 16.18 Knuckle filter. (**a**) $c = 0.00$, $d = 0$. (**b**) $c = 0.00$, $d = 30\backslash$. (**c**) $c = 0.01$, $d = 0$. (**d**) $c = 0.01$, $d = 30$. (e) $c = 0.04$, $d = 0$. (**f**) c $= 0.04$, $d = 30$

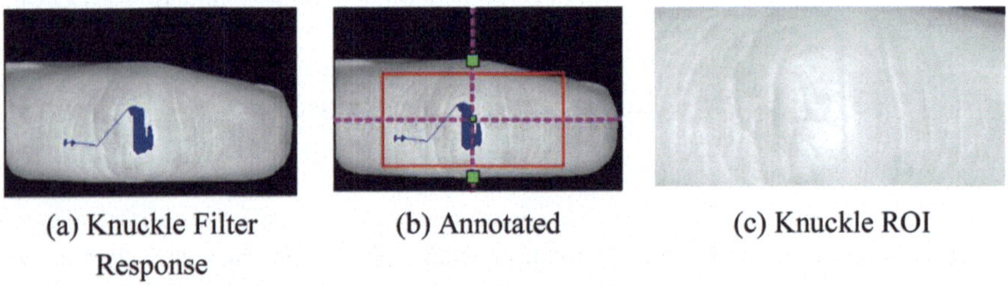

Fig. 16.19 Knuckle print ROI detection. (**a**) Knuckle filter response is superimposed over the knuckle area with blue color, (**b**) full annotated, and (**c**) FKP (FKP_{ROI})

considering central knuckle point as the center as shown in Fig. 16.19c where w and h have been considered as semi-width and semi-height, respectively, of the extracted knuckle print ROI.

3.2.3 Minor FKP Segmentation

Major steps involved in minor FKP segmentation have been illustrated in Fig. 16.20, for segmenting fixed-size minor finger knuckle print. Binarization of the acquired finger dorsal images is done by Otsu thresholding. This binarized finger shape is used to find the location of the fingertip using convex hull of images followed by background removal. The detailed segmentation strategy is mentioned in [8]. Some of the intermediate figures generated during segmentation are shown in Fig. 16.21. Major and minor segmented finger knuckle images are shown in Fig. 16.21c, e, respectively.

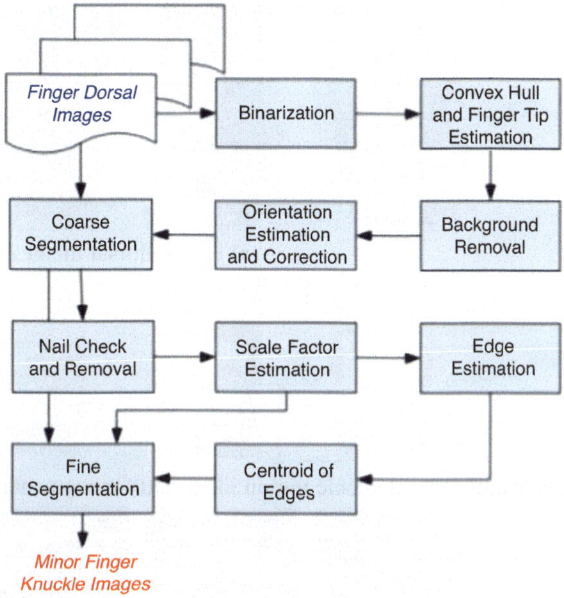

Fig. 16.20 Block diagram illustrating key steps performed for segmenting fixed-sized minor FRP ROI. (Images are taken from [8])

Algorithm 16.3 Knuckle Print ROI Detection (Algorithm Taken from [15])
Require: Raw knuckle print image I of size $m \times n$.
Ensure: The knuckle print ROI $\mathrm{FKP_{ROI}}$ of size $(2 * w + 1) \times (2 * h + 1)$.

1. Enhance the FKP image I to I_e using CLAHE.
2. Binarize I_e to I_b using Otsu thresholding.
3. Apply Canny edge detection over I_b to get I_{cedges}.
4. Extract the largest connected component in $\mathrm{I_{cedges}}$ as FKP raw boundary, \$\$$(\mathrm{FKP}^{\mathrm{raw}}_{\mathrm{Bound}})$.
5. Erode the detected boundary $\mathrm{FKP}^{\mathrm{raw}}_{\mathrm{Bound}}$ to obtain continuous and smooth FKP boundary, $\mathrm{FKP}^{\mathrm{smooth}}_{\mathrm{Bound}}$.
6. Extract the knuckle area K_a = All pixels in image $I \in$ the ConvexHull \$\$$(\mathrm{FKP}^{\mathrm{smooth}}_{\mathrm{Bound}})$.
7. Apply the knuckle filter $\mathrm{F}^{0.01,30}_{\mathrm{kp}}$ over all pixels $\in K_a$.
8. Binarize the filter response using $f *$ max as the threshold.
9. The central knuckle line (c_{kl}) is assigned as that column which is having the maximum knuckle filter response.
10. The midpoint of top and bottom boundary points over $c_{\mathrm{kl}} \in K_a$ is defined as the central knuckle point (c_{kp}).
11. The knuckle ROI $(\mathrm{FKP_{ROI}})$ is extracted as the region of size $(2 * w + 1) \times (2 * h + 1)$ from raw knuckle print image I, considering c_{kp} as its center point.

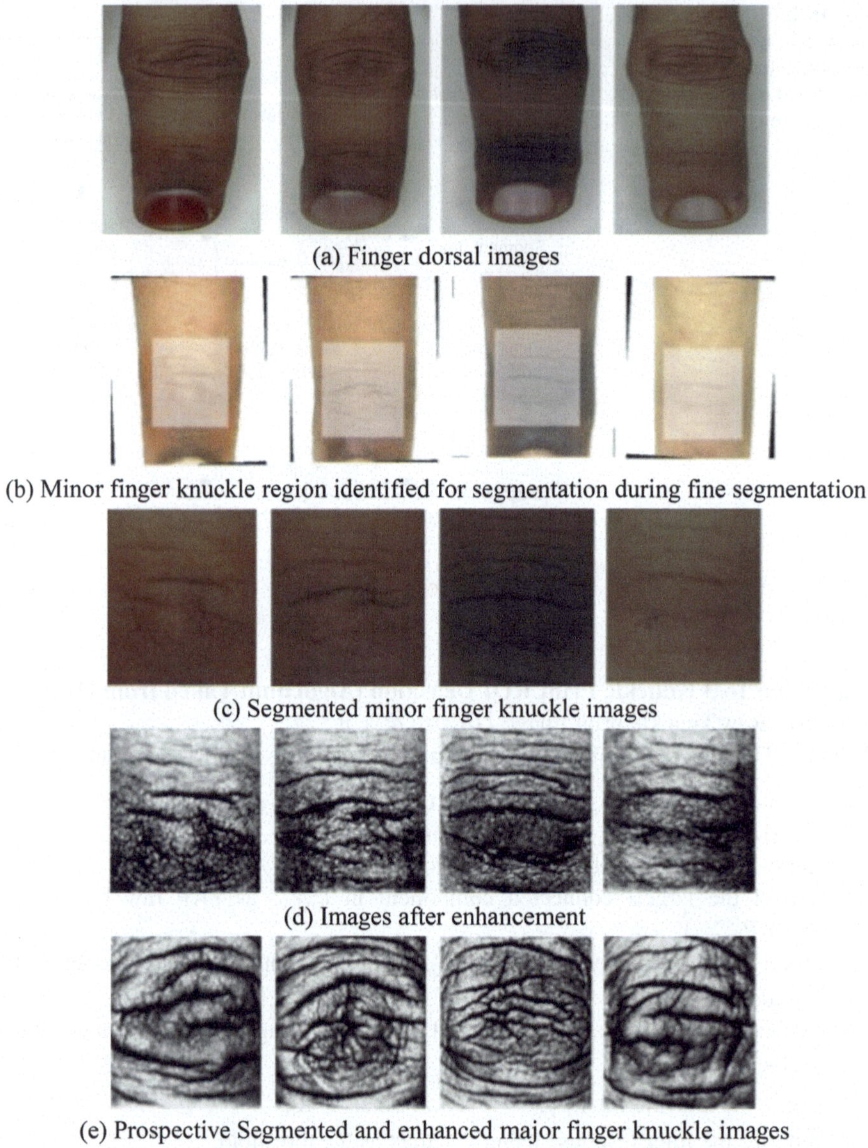

Fig. 16.21 Major and minor segmented finger knuckle images. (**a**) Finger dorsal images. (**b**) Minor finger knuckle region identified for segmentation during fine segmentation. (**c**) Segmented minor finger knuckle images. (**d**) Images after enhancement. (**e**) Prospective segmented and enhanced major finger knuckle images. (Images are from [8])

3.3 FKP Image Enhancement

The extracted region of interest (ROI) of knuckle print is generally of poor contrast. Hence several image enhancement algorithms can be applied over the ROI. Some of the FKP image enhancement techniques are discussed in the following subsection.

3.3.1 Histogram Equalization

In [2], FKP image is enhanced by dividing the FKP image into sub-blocks of 11×11 pixels. Mean of each block which estimates the reflection of the block is calculated. The estimated coarse reflection is expanded to the original size of the FKP image using bi-cubic interpolation. For the coarse estimate of reflection, if the block size is very small, the estimate is almost the same as the extracted FKP, and if the block size is high, the estimate becomes improper. The estimated reflection is subtracted from the original image to obtain a uniform brightness of the image. Histogram equalization is performed on blocks of 11×11 pixels to improve the contrast in the texture of FKP and then is subjected to perform filtering to smooth the boundaries between blocks.

3.3.2 CLAHE

In [5], ROI sample is partitioned into non-overlapping fixed size cells (8×8). The size is selected empirically which ensures that its mean closely indicates the coarse estimate of illumination. Each estimated cell is deducted from corresponding cell of original image and gets its uniformly brightened ROI sample. The resulting ROI is enriched using CLAHE which also lowers the blocking effect as shown in Fig. 16.22.

Weiner filtering is applied to smooth the boundaries and to minimize the additive noise. ROI sample is enhanced to preserve discriminative features. The enhanced knuckle print image possesses better-quality texture as shown in Fig. 16.22.

Fig. 16.22 Local CLAHE-based FKP enhancement. (**a**) Original. (**b**) Bg illum. (**c**) Uni. illum. (**d**) Enhanced. (**e**) Noise removal

(a) Original Knuckleprint (b) Knuckle *vcode* *(c) Knuckle hcode*

Fig. 16.23 Original and transformed (*vcode, hcode*) for knuckle print ROIs. (**a**) Original knuckle print. (**b**) Knuckle *vcode.* (**c**) Knuckle *hcode*

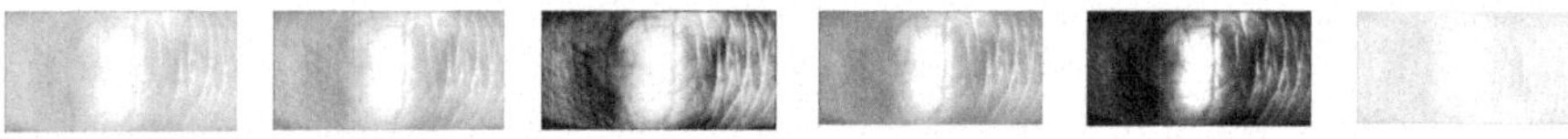

(a) FKP Original (b) FKP i1 (c) FKP i2 (d) FKP i3 (e) FKP i4 (f) FKP i5

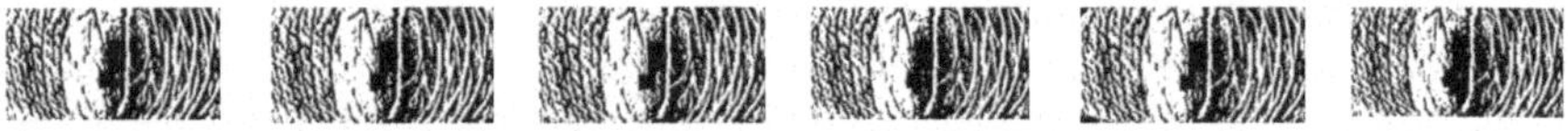

(g) vcode Orig *(h) vcode i1* *(i) vcode i2* *(j) vcode i3* *(k) vcode i4* *(l) vcode i5*

Fig. 16.24 Illumination in-variance of *LGBP* transformation. Here first row depicts same image under five varying illumination conditions. Second row depicts their corresponding *vcodes.* (**a**) FKP Original. (**b**) FKP i1. (**c**) FKP i2. (**d**) FKP i3. (**e**) FKP i4. (f) FKP i5. (**g**) *vcode Orig.* (**h**) *vcode i1.* (**i**) *vcode i2.* (**j**) *vcode i3.* (**k**) *vcode i4.* (**l**) *vcode i5*

3.3.3 Local Gradient Binary Pattern (LGBP)

The FKP images can also be enhanced by using robust representations (*vcode* and *hcode*) that can endure small change of illumination variation. Images are transformed using the LGBP transformation [16]. An original knuckle along with its *vcode* and *hcode* is shown in Fig. 16.23. In Fig. 16.24, one raw knuckle print image is considered under varying illumination and is shown along with the corresponding *vcode*. One can observe that the original knuckle print (as shown in Fig. 16.24a) has undergone very severe illumination variation (as shown in Fig. 16.24b–f). But the corresponding *vcodes* may not be varying much (as shown in Fig. 16.24g–l).

However, any of the abovementioned image enhancement technique can be used over ROI samples to preserve its discriminative features.

3.4 FKP Feature Extraction and Matching

Feature extraction and matching is one of the main steps involved in any biometric system. Features are extracted from preprocessed image which we get from the image enhancement step. Extracted features can be local features or it can be global features. Few FKP matching algorithms are discussed in the following subsections.

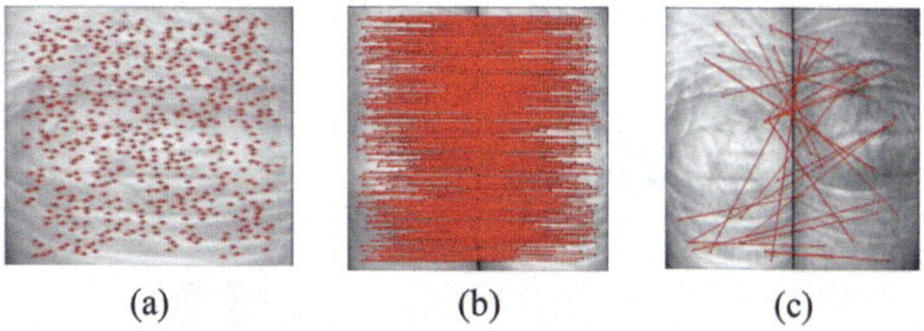

Fig. 16.25 (**a**) SIFT key points detected. (**b**) Genuine matching of SIFT key points. (**c**) Imposter matching of SIFT key points

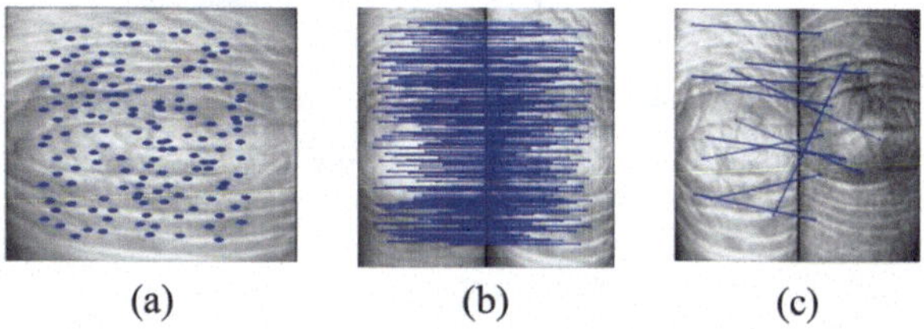

Fig. 16.26 (**a**) SURF key points detected. (**b**) Genuine matching of SURF key points. (**c**) Imposter matching of SURF key points

3.4.1 Handcrafted Features [2]

FKP is represented by two local feature vectors extracted by SIFT and SURF. In the recognition phase, the input FKP image is matched with the knuckle print features present in the database. Nearest neighbor ratio [11] has been used for to calculate the matching scores between corresponding feature vectors. When FKP images of the same user are matched, then it is known as the genuine matching; otherwise it is termed as an imposter matching. Let M_T and M_S be SIFT and SURF matching scores, respectively, between the query and an enrolled FKP. These SIFT and SURF matching scores are combined together to obtain the total matching score. An example of genuine matching and imposter matching using SIFT is shown in Fig. 16.25. Similarly, Fig. 16.26 shows an example of genuine matching and imposter matching using SURF. One of the major drawbacks of using SIFT is that SIFT fails to match nonrigid deformable regions, image patches with weak or repetitive textures.

3.4.2 FKP Matching Using Deep Matching [5]

Images with nonrigid deformation can be matched successfully using deep matching. The matching algorithm is based on a multistage architecture with *n* layers (depending on the image size) like a construction very similar to deep convolution nets but does not learn any NN model for feature representation. In this multilayered architecture, images are subdivided into patches as quad-tree, but quadrants in target image are not at fixed positions. These blocks can move locally to re-optimize their positions. This algorithm proceeds in a bottom-up and top-down fashion which performs convolution at every patch. The SIFT/HOG descriptor is a popular approach to match regions between images, with 4×4 spatial cells to generate a real vector (V) in 128-dimensional space, where $V \in \Re^{4 \times 4 \times 8}$. The SIFT

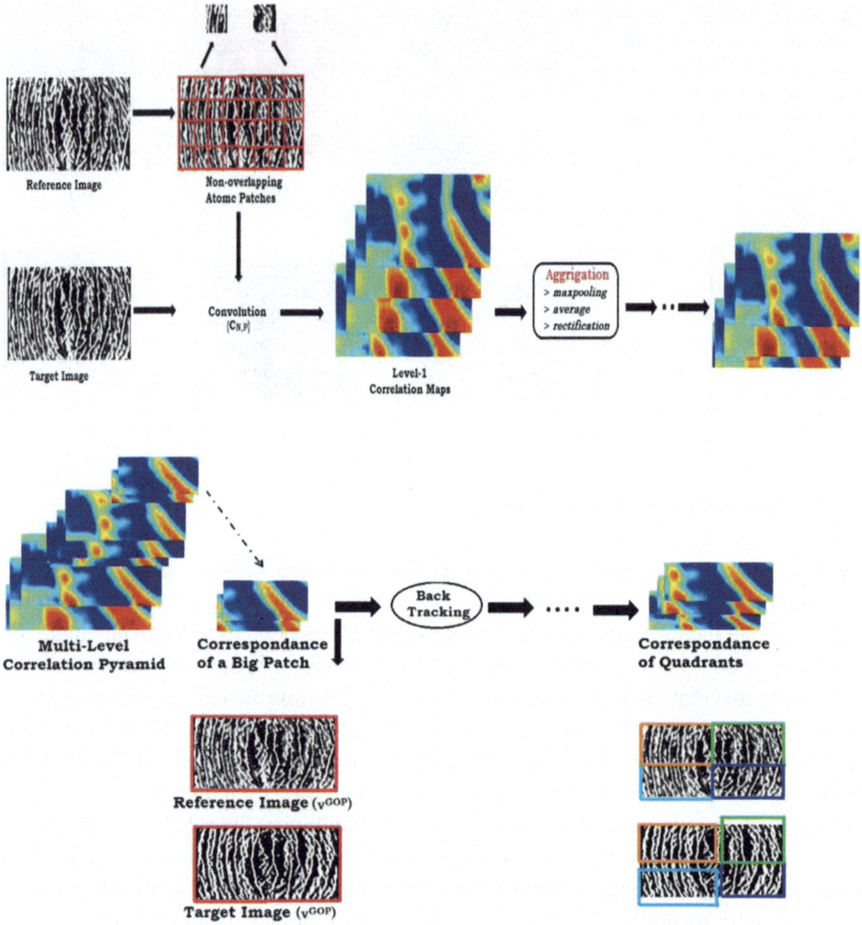

Fig. 16.27 Multi-scale correlation pyramid by bottom-up stage and top-down correspondence

patch can be split into four quadrants represented by $V = [V^1, V^2, V^3, V^4]$, with $V^n \in \Re^{2 \times 2 \times 8}$. There are two such descriptors, one for reference and other for target. In the target descriptor, the quadrants of 4 × 4 sized grids have not been kept fixed. Their positions can be optimized by maximizing Sim $(V, Q(p)) = \sum_{n=1}^{4} \max_{pn} (V_n^T Q(p_n))$, where $Q(p) \in \Re^{32}$ is the descriptor of a single quadrant extracted at position p. The similarity can be estimated efficiently with the assumption that each of these quadrants can move independently (up to some extent), which gives a coarse nonrigid matching. This method can perform reasonably well for nonrigid matching with explicit pixel-wise correspondences. The number of feature points matched correctly through deep matching (as shown in Fig. 16.27) has been used as a matching score.

3.4.3 Statistical Features [10]

In this paper, three appearance-based approaches (principal component analysis, linear discriminant analysis, and independent component analysis) are used for generating matching score from the knuckle/palm image. Euclidean distance metric is used to generate the distance between the projection coefficients obtained from training and test images. The final matching score is computed by fusing the individual subspace method scores (such as *PCA*, *LDA*, and *ICA*).

3.4.4 Correlational Features and Feature Ensemble [21]

Correlation can also be used for matching knuckle prints. It is computed by using band-limited phase-only correlation (*BLPOC*) [12]. Knuckle prints are presumed to have small range of frequencies; as a result, Fourier-based frequency band is limited, and band-limited phase only correlation (*POC*) [12] can be calculated. The *POC* value between two images can be used as similarity score and is obtained using the cross-spread spectrum of the Fourier transform of both images. Gabor filters can also be used for extracting patterns. They are applied over a pixel and its neighborhood to extract a resulting discriminating local pattern that can be used to represent that pixel. For achieving good results, fusion of local and global features has been done as shown in Fig. 16.28. For extracting global features, band-limited phase-only correlation (*BLPOC*) is used, whereas for extracting local features, Gabor filter bank has been used. For estimating the local orientation of a pixel, six Gabor filters at an angle of $\frac{\pi}{6}$ are used.

In another technique [19], three local features, viz., phase congruency, local orientation, and local phase (as shown in Fig. 16.29), are extracted. These features are calculated by fusing them together at score level, and computation is done using the quadrature pair filter. Finally, Gabor-based local features and *BLPOC*-based global features are also combined together to achieve better performance.

3.4.5 Compcode-Based Ordinal Features [20]

A bank of six Gabor filters at an angle of $\frac{\pi}{6}$ is applied to extract features [6], only from those pixels that are having varying Gabor responses. All pixels do not contribute to discrimination equally. Some of them may be at a similar background patch; hence, they may not possess any dominant orientation. Such pixels are ignored and an orientation code is computed. Similarly, magnitude code is obtained by the real and imaginary part of the Gabor filers as shown in Fig. 16.30. The orientation and magnitude information are fused to achieve better results.

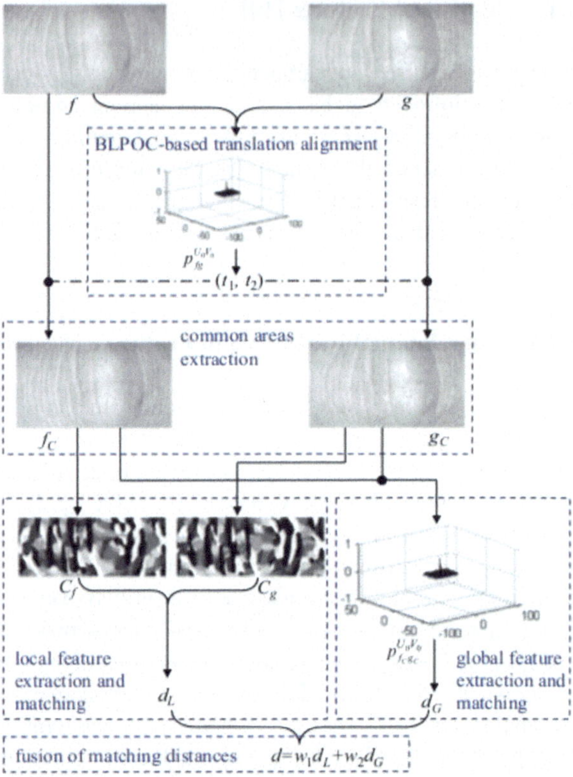

Fig. 16.28 Ensemble of local and global features [21]

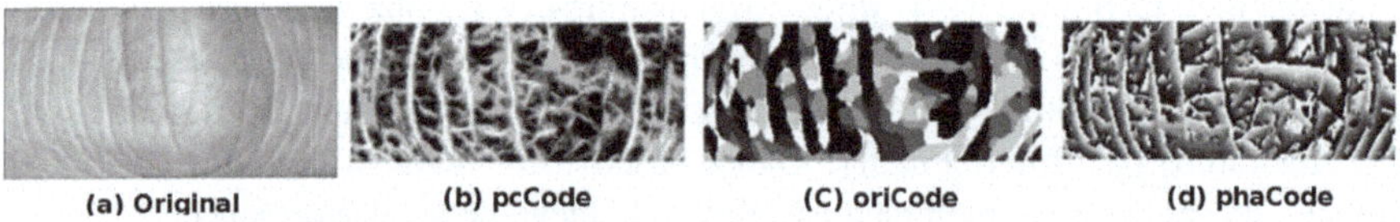

Fig. 16.29 Local feature extraction using Gabor filters. (a) Original. (b) pcCode. (c) oriCode. (d) phaCode

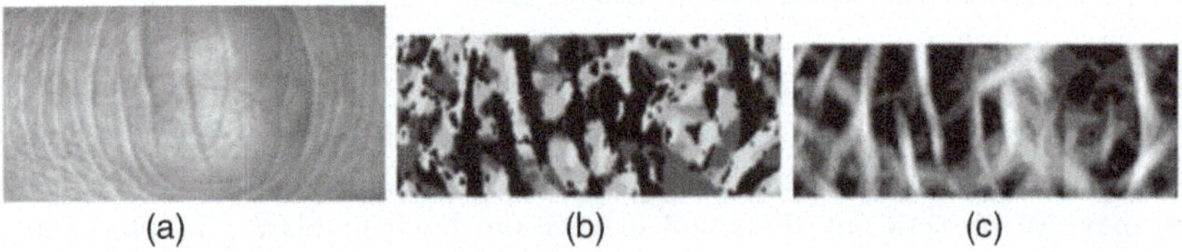

Fig. 16.30 (a) Original image. (b) Orientation code (ImCompCode). (c) Magnitude code (MagCode) images [20]

3.5 *Publicly Available Databases*

There exist two publicly available well-known finger knuckle print databases, namely, PolyU FKP database [17] and IITD FKP database [7]. The PolyU FKP database is the largest publicly available database for the knuckle prints. This database consists of 7920 *FKP* sample images obtained from 165 subjects in two sessions. On an average, time interval between the sessions has been 25 days. In each session, six images from four fingers (distinct index and middle fingers of both hands) are collected. Hence, a total of 660 distinct knuckle print data is collected. Out of 165, 143 subjects are belonging to an the 20 − 30 age group, and others are belonging to the 30–50 age group.

Another database has been acquired at the Indian Institute of Technology, Delhi (IIT Delhi), with the help of a digital camera. It consists of 790 sample images obtained from 158 subjects. All subjects in the database belong to the 16 − 55 age group. All images are in bitmap format.

3.6 *Comparative Performance Analysis*

Various matching algorithms that are discussed previously can be compared on different performance parameters, such as equal error rate (EER), correct recognition rate (CRR), error under ROC curve, and decidability index (DI), along with computation time. The testing strategy considers first six images as training and last six for testing. Based on matching score, any matching can be declared as genuine or imposter. There exist several different testing protocols defined so as to perform a fair comparison as discussed below.

3.6.1 Testing Protocol 1

In testing protocol 1, four categories of PolyU FKP database, right index (RI), right middle (RM), left index (LI), and left middle (LM), are considered independently. All FKP of 660 subjects and their all corresponding images are considered. Six images of first session are taken as training and remaining are considered as testing images. Therefore, a total of 23 and 760 and 15, 657, and 840 genuine and imposter matchings are performed as shown in Table 16.3.

3.6.2 Testing Protocol 2

There are 165 subjects, as mentioned in Table 16.3. The FKP samples are collected in two sessions with six images per session. Images of the first session are taken as training, and the remaining are taken as testing images. This results in 5 and 940 and

Table 16.3 Database specifications

Subject	Pose	Total	Training	Testing	Genuine matching	Imposter matching
PolyU (knuckle print)						
165 (660 knuckle print)	12	7920	First 6	Last 6	23,760	15,657,840
PolyU individual finger knuckle print (left index [LI], left middle [LM], right index [RI], right middle [RM])						
165 Finger knuckle print class	12	1980	First	Last 6	5940	9,74,160

Table 16.4 Comparative performance analysis for full PolyU finger knuckle print database [18]

Algorithm	Equal error rate	% Drop
Compcode [6]	1.386	35.64
BOCV [4]	1.833	51.33
ImCompcode and MagCode [13]	1.210	26.28
MoriCode [3]	1.201	25.72
MtexCode [3]	1.816	50.88
Moricode and MtexCode [3]	1.0481	14.89
$vcode^{GORP}$ [16]	2.9122	–
$hcode^{GORP}$ [16]	6.6023	–
$vcode^{SGORP}$ [16]	2.8067	–
$vcode^{SGORP}$ [16]	3.5276	–
Fusion [16]	0.8920	–

9, 74, and 160 genuine and imposter matchings as shown in Table 16.3. One such comparison is done on PolyU FKP dataset on the basis of EER rate and is shown in Table 16.4. Table 16.5 shows the comparative analysis over individual finger knuckle database.

A comparative analysis on the basis of false accept rate is done in [16] and is also shown in Fig. 16.31. It clearly indicates that local and global information combination (LGIC) scheme performs significantly better in terms of the verification accuracy in comparison to CompCode [6], BLPOC [20], and ImCompCode [13]. One such comparison is also done in [21] as shown in Fig. 16.32.

4 Conclusion

The soft biometric traits such as finger knuckle have been successfully applied over a wide range of subjects, exhibiting very good performance. Several state-of-the-art finger knuckle print-based authentication systems have been proposed so far in literature, in the last decade. All of them exploit the finger knuckle symmetric and enriched curvature features and performed well over publicly available benchmark datasets such as IITD and PolyU. All the presented literature suggests that finger

Table 16.5 Performance analysis over individual finger knuckle print databases

Description	DI	EER (%)	Accuracy (%)	EUC	CRR (%)
Left index finger knuckle print PolyU database					
vcode^{GORP}	1.6606	3.2972	97.2727	1.1332	99.79
hcode^{GORP}	1.3274	6.8694	94.7770	3.4261	97.87
vcode^{SGORP}	1.3671	4.0669	97.0546	2.1503	99.58
vcode^{SGORP}	1.7183	3.2988	97.5353	1.6264	99.09
Fusion	1.6357	1.07764	99.1548	0.3027	99.79
Left middle finger knuckle print PolyU database					
$vcode^{GORP}$	1.7326	3.0303	97.5978	1.0385	99.59
$hcode^{GORP}$	1.3326	6.3289	95.1156	3.1180	97.57
$vcode^{SGORP}$	1.3922	3.1132	97.6070	1.7210	98.58
$vcode^{SGORP}$	1.7839	2.2895	98.1309	1.0097	99.79
Fusion	1.6878	1.0099	99.1940	0.3562	100
Right index finger knuckle print PolyU database					
$vcode^{GORP}$	1.6559	2.8282	97.8285	0.9358	99.89
$hcode^{GORP}$	1.3115	7.1244	94.7995	3.6453	96.86
$vcode^{SGORP}$	1.3296	3.9551	97.0794	1.8999	97.87
$vcode^{SGORP}$	1.7161	3.0302	97.8734	1.5389	99.49
Fusion	1.6144	0.7407	99.3711	0.0860	100
Right middle finger knuckle print PolyU database					
$vcode^{GORP}$	1.7065	3.2496	97.6092	1.4161	99.59
$hcode^{GORP}$	1.3311	6.9324	94.8332	3.6938	97.97
$vcode^{SGORP}$	1.3560	3.7173	97.1706	2.0238	99.69
$vcode^{SGORP}$	1.8122	3.7052	97.4697	1.8090	99.09
Fusion	1.6693	1.0615	99.1268	0.2658	100

Taken from [16]

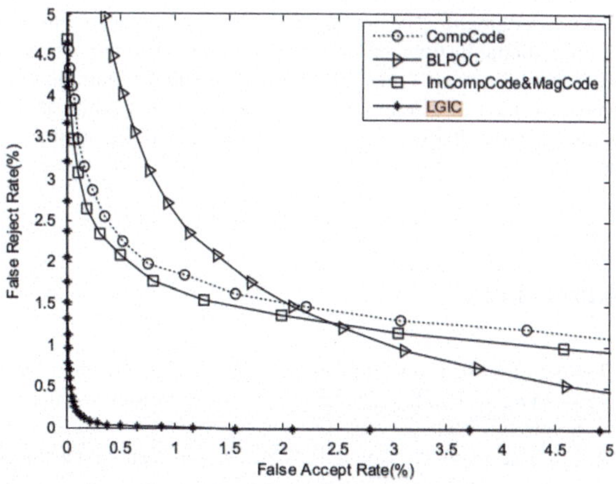

Fig. 16.31 DET curves for FKPs (images taken from [21])

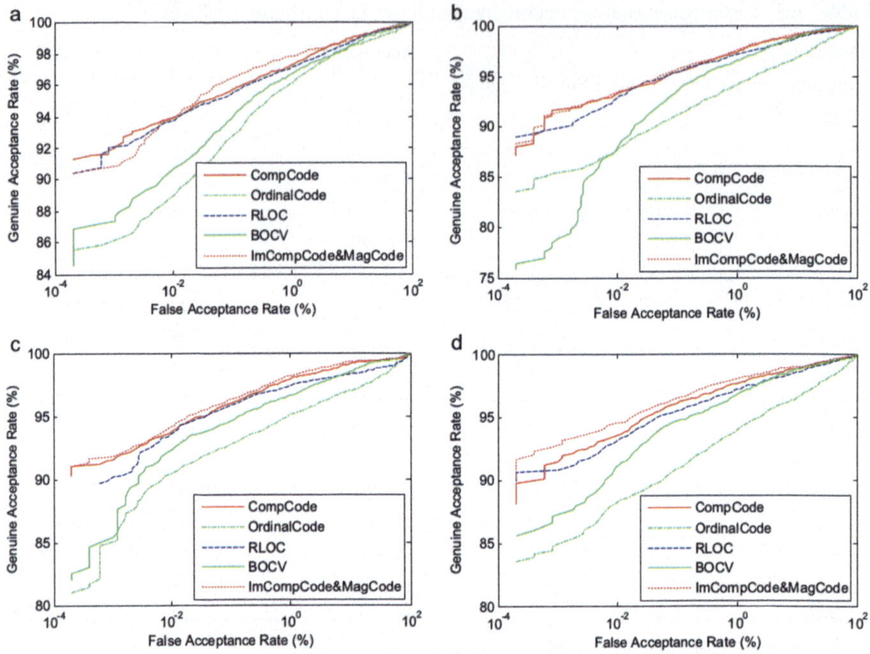

Fig. 16.32 DET curves for FKPs from (**a**) left index fingers, (**b**) left middle fingers, (**c**) right index fingers, and (**d**) right middle finger images considered in [20]

knuckle can be used as an acceptable biometric for personal authentication. There are also several challenges pointed out by them that have to be addressed in the future.

Acknowledgment The authors acknowledge the secretarial support provided by Mr. Subir Basak of the National Institute of Technical Teachers' Training and Research, Kolkata, India. Some of the work were reported in the Ph D Thesis entitled "Multimodal Biometric Recognition using Iris, Knuckleprint and Palmprint" of the first author submitted at Indian Institute of Technonlogy, Kanpur, India, 2015. Authors also acknowledge the support provided by the Institute to carry out the work.

References

1. Role of biometric technology in aadhaar authentication, authentication accuracy report, uidai, 27-March-2012. http://uidai.gov.in/images/role-of-biometric-technology-in-aadhaar-authentication-020412.pdf
2. G.S. Badrinath, A. Nigam, P. Gupta, An efficient finger-knuckle-print based recognition system fusing sift and surf matching scores, in *International Conference on Information and Communications Security*, (Springer, 2011), pp. 374–387
3. G. Gao, J. Yang, J. Qian, L. Zhang, Integration of multiple orientation and texture information for finger knuckle print verification. Neurocomputing **135**, 180–191 (2014)

4. Z. Guo, D. Zhang, L. Zhang, W. Zuo, Palmprint verification using binary orientation co-occurrence vector. Pattern Recogn. Lett. **30**, 1219–1227 (2009)
5. G. Jaswal, A. Nigam, R. Nath, Deep knuckle: Revealing the human identity. Multimedia Tool Appl. **76**(18), 1–30 (2017)
6. A.W.K. Kong, D. Zhang, Competitive coding scheme for palmprint verification. International Conference on Pattern Recognition (ICPR), **1**:520–523 (2004)
7. A. Kumar, The IIT delhi finger knuckle image database - 2006, (http://www4.comp.polyu.edu.hk /csajaykr/fn1.htm)
8. A. Kumar, Can we use minor finger knuckle images to identify humans? in *IEEE Fifth International Conference on Biometrics: Theory, Applications and Systems (BTAS)*, (IEEE, 2012), pp. 55–60
9. A. Kumar, Z. Xu, Personal identification using minor knuckle patterns from palm dorsal surface. IEEE Trans. Inf. Forensics Secur **11**(10), 2338–2348 (2016)
10. G. Lu, D. Zhang, K. Wang, Palmprint recognition using eigenpalms features. Pattern Recogn. Lett. **24**(9), 1463–1467 (2003)
11. K. Mikolajczyk, C. Schmid, A performance evaluation of local descriptors. IEEE Trans. Pattern Anal. Mach. Intell. **27**(10), 1615–1630 (2005)
12. K. Miyazawa, K. Ito, T. Aoki, K. Kobayashi, H. Nakajima, An effective approach for iris recognition using phase-based image matching. IEEE Trans. Pattern Anal. Mach. Intell. **20**(10), 1741–1756 (2008)
13. A. Morales, C. Travieso, M. Ferrer, J. Alonso, Improved finger-knuckle- print authentication based on orientation enhancement. Electron. Lett. **47**(6), 380–381 (2011)
14. A. Nigam, P. Gupta, Quality ·assessment of knuckleprint biometric images, in *20th IEEE International Conference on Image Processing (ICIP)*, (2013), pp. 4205–4209
15. A. Nigam, P. Gupta, Finger-knuckle-print ROI extraction using curvature gabor filter for human authentication, in *Proceedings of the 11th Joint Conference on Computer Vi-Sion, Imaging and Computer Graphics Theory and Applications (VISIGRAPP 2016)*, vol. 3, (VISAPP, Rome, February 27–29, 2016), pp. 366–373
16. A. Nigam, K. Tiwari, P. Gupta, Multiple texture information fusion for finger-knuckle-print authentication system. Neurocomputing **188**, 190–205 (2016)
17. Z. Lin, L. Zhang, D. Zhang. The polyu finger-knuckle-print database - 2009, (http://www4.comp.polyu.edu.hk/ biometrics/fkp.htm)
18. Z. Lin, L. Zhang, D. Zhang, Finger-knuckle-print: A new biometric identifier, in *16th IEEE International Conference on Image Processing (ICIP)*, (2009), pp. 1981–1984
19. Z. Lin, L. Zhang, D. Zhang, Z. Guo, Phase congruency induced local features for finger-knuckle-print recognition. Pattern Recogn. **45**(7), 2522–2531 (2012)
20. Z. Lin, L. Zhang, D. Zhang, H. Zhu, Online finger-knuckle-print verification for personal authentication. Pattern Recogn. **43**, 2560–2571 (2010)
21. Z. Lin, L. Zhang, D. Zhang, H. Zhu, Ensemble of local and global information for finger-knuckle-print recognition. Pattern Recogn. **44**(9), 1990–1998 (2011)

Part IV
Enabling Technologies

Chapter 17
Leveraging Cloud-Based Resources for Automated Biometric Identification

Wei Lu, John Hancock, Benjamin Osowiecki, Aram Taft, and Wei Li

1 Introduction to Cloud Computing

Cloud computing is one of the biggest subjects in the technology industry. In fact, it has become so widely known it has broken out of the tech world to become a household name. This should not come as much of a surprise, as cloud computing has become ubiquitous in many of our daily routines. However, if you ask someone what cloud computing actually is, there is a good chance they will not have a definitive answer. So what exactly is cloud computing?

The term cloud computing is often associated with the Internet giants of the mid-2000s, around the debut of Amazon's Elastic Compute Cloud, yet the term had come to be about 10 years earlier. The origin of the term, in the modern sense, has been traced back to a Compaq business plan dated 1996. Within is a prediction for the future, where "application software is no longer a feature of the hardware – but of the Internet" and would be known as "the cloud" [1]. Unfortunately, this still paints as very broad picture of what cloud computing actually is, simply defining it as software that runs remotely over the Internet, instead of on a local machine. Since this time the industry has exploded, but up until recently, there has been little guidance on what actually fits the definition of cloud computing. In the past few years, the industry has determined concrete guidelines as to what cloud computing is, going so far as to break it up into three distinct categories. Formally, cloud

W. Lu (✉) · B. Osowiecki · A. Taft
Department of Computer Science, Keene State College, USNH, Keene, NH, USA
e-mail: wlu@keene.edu

J. Hancock
Fidelity, Durham, NC, USA

W. Li
New York Institute of Technology, Vancouver, BC, Canada
e-mail: wlu@keene.edu

M. S. Obaidat et al. (eds.), *Biometric-Based Physical and Cybersecurity Systems*,
https://doi.org/10.1007/978-3-319-98734-7_17

Fig. 17.1 Five characteristics of cloud computing

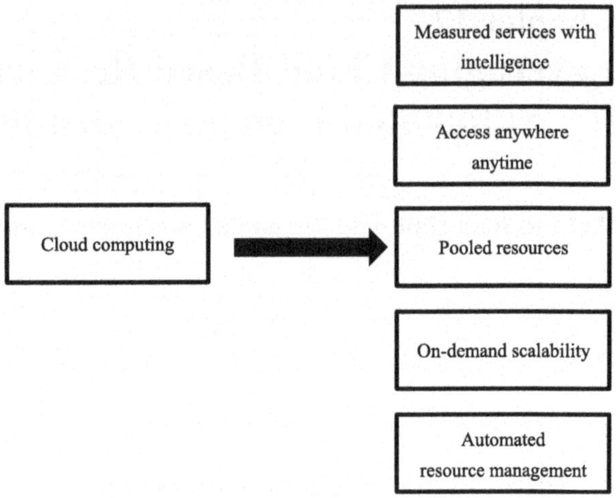

computing is defined as "a model for enabling ubiquitous, convenient, on-demand network access to a shared pool of configurable computing resources that can be rapidly provisioned and released with minimal management effort or to service provider interaction" [2]. This is to say that a cloud computing service is one where a user can log in from almost anywhere and be given the service offered, requiring the backend components to be allocated and de-allocated as needed.

As illustrated in Fig. 17.1, there are also five characteristics that all cloud computing services have in common. The first is that they allow users to use their services and access required resources (such as server time or network storage) without having to interact with a human. Secondly, cloud computing services should be designed to allow devices of all different form factors to connect and utilize the service, whether it is a mobile phone, laptop, or desktop workstation. Third, resources should be pooled in such a manner that multiple users can be served simultaneously and do so in a way which is transparent to the user. The fourth characteristic builds off the third; a cloud computing system should have the ability to scale as demand increases and decreases. Lastly, the system should have the ability to automatically track, control, and optimize resource use.

Despite these rather detailed guidelines, the world of cloud computing is still extremely diverse. This is where the three categories of Software as a Service come into play. The models illustrated in Fig. 17.2 describe three distinct forms of cloud computing service, with each consecutive tier building on the one prior. At the lowest level is Infrastructure as a Service (IaaS) [18], a step up is Platform as a Service (PaaS) [19], and then at the highest level is Software as a Service (SaaS) [12].

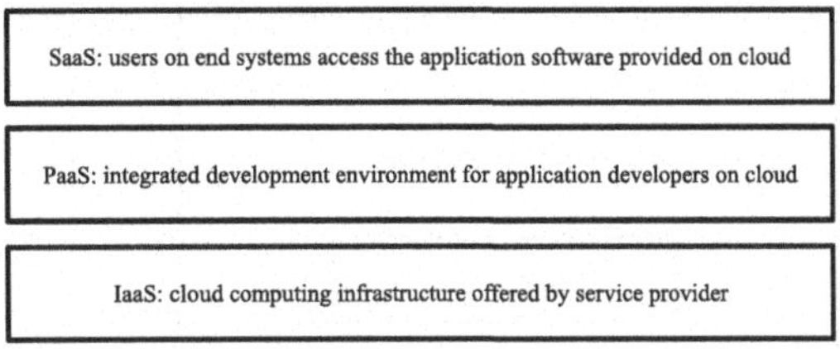

Fig. 17.2 Three typical cloud computing models

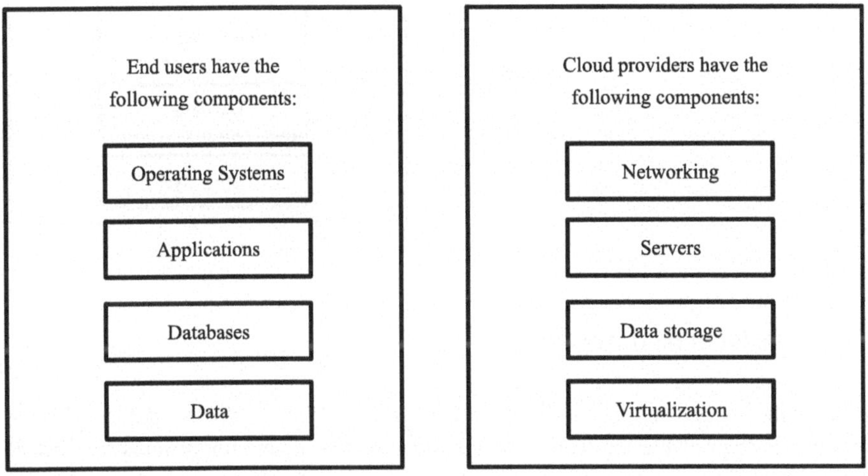

Fig. 17.3 Infrastructure as a Service

1.1 Infrastructure as a Service

As illustrated in Fig. 17.3, IaaS is the most "bare bones" of the three cloud computing models. In this model, the service provider provides only the computing infrastructure, including virtualization, servers, storage hardware, and networking capabilities. Often IaaS providers use this virtualization to hide the hardware from the customer, as it is irrelevant to them. The virtualization is accomplished through the use of a hypervisor, otherwise known as a virtual machine monitor. These systems allow for vast numbers of virtual machines, known as guest machines, to be hosted. These virtual machines are then provided to the customer for use with their software of choice. As such it is up to the user to supply the data, applications, databases, and operating system. Examples of such systems include Google Compute Engine [3] and Amazon EC2 [4]. IaaS systems are designed for those who have built (or are building) their application for either a specific or a custom platform and need complete control of what can be run, but do not want to have to acquire and maintain computing hardware.

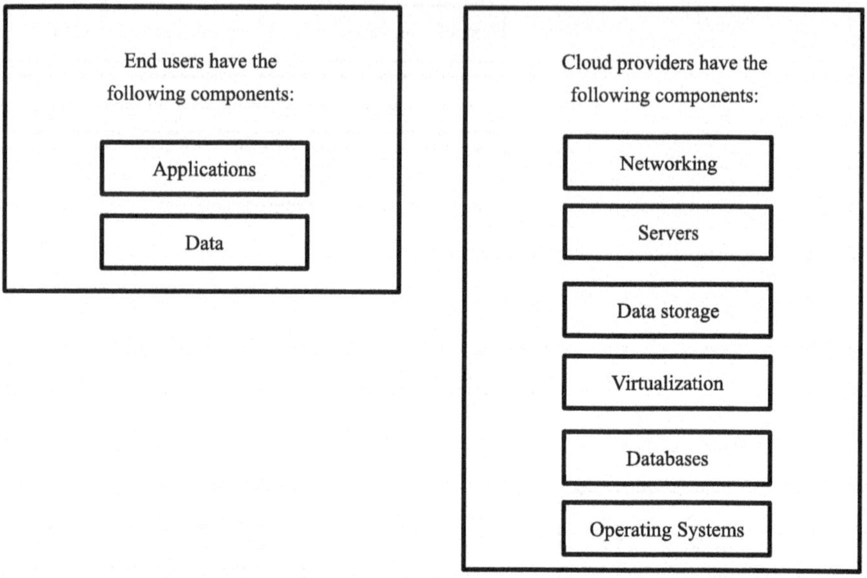

Fig. 17.4 Platform as a Service

1.2 *Platform as a Service*

PaaS is the second model that is illustrated in Fig. 17.4. In this model the service provider is responsible for everything related to hardware and for systems which will be required by a customer's applications (such as databases and the operating system). In other terms, Platform as a Service providers supply application developers with development environments on which they can develop and host their applications. Often times the service provider will create APIs, toolkits, and distribution systems for their respective platforms. Together, these allow application developers to develop apps without having to acquire, manage, and maintain the hardware, operating system, databases, or web server. Examples of PaaS services include Microsoft Azure [5] and Google App Engine [6].

1.3 *Software as a Service*

SaaS is the cloud computing model most consumers consider "the cloud." It is at this level that services such as Google Drive, Facebook, or Amazon Marketplace exist. As illustrated in Fig. 17.5 in this model, the users are often nontechnical people, as they are using already built application software. They do not have to deal with any hardware and OS or have to program the applications; they simply provide their data into the application. These platforms are otherwise referred to as "on-demand

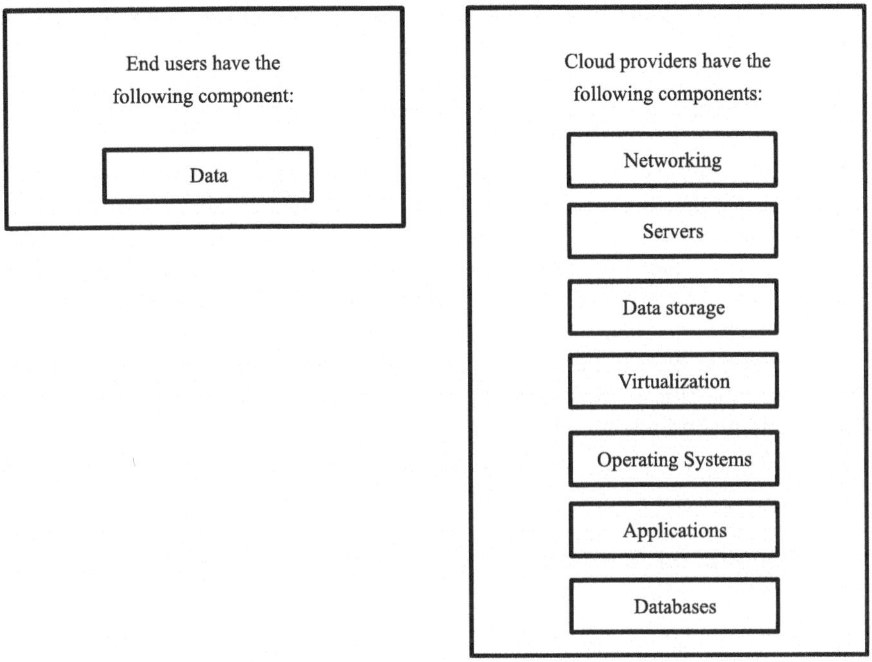

Fig. 17.5 Software as a Service

software" as they can be accessed on an as needed basis, from almost any device. This also alleviates the need for the user to install application software locally, as it is hosted on the provider's servers. To keep up with large demand this can cause, these services use load balancers to distribute work over sets of virtual machines; however, this process is invisible to the user. Often times these systems are priced on either a reoccurring subscription plan or on a pay as needed (often seen on cloud storage services).

2 Introduction to Biometrics for Authentication

Biometrics is defined as the "automated recognition of individuals based on their biological and behavioral characteristics" by the International Standardization Organization [7]. Biometrics is made to have an authentication which is not so easily spoofed by social engineering or hacking into a terminal. The complexity of the human body makes it rare for people to have the exact same fingerprint or vein pattern when trying to gain access to something with such biometric security protocols. Scanning these characteristics makes it available as a service to prove one's identity and gain access to things such as bank account, credit cards, and online profiles, to name a few. When you scan your fingerprint, iris, or palm, the reading from the scanner will be compared to what was originally stored when the account was set up, similar to passwords systems. Once checked it will get a result from the checking algorithm and either allow or deny access to what is trying to be accessed. The improved security comes with the human body characteristics that are similar in structure but unique to a majority of the population on earth as well as how many points of scanning the system does for authentication. Having multiple points will help with authentication given you have not a single point of failure where that one trait being scanned was altered or removed from you. An example can be if you were to cut your finger on the bottom and if the system scans one specific ridge in your print, that would be the point of failure. If the system was grabbing multiple points on your fingerprint, it could get a close enough match to still uniquely identify you and not someone else. In the following we will discuss some of the more popular options for biometrics including such as facial scanning, iris scanning, vein recognition, and fingerprint scanning.

2.1 *Fingerprint Scanning*

Fingerprint is a popular form of biometrics and is widely implementable in many security systems. There is a one in 64 billion chance that your fingerprint will match up exactly with someone else [8]. One of the problems of times now is coming up with ways to have a unique hash or password encryption. MD5 hashing algorithm is a great example and has been having hash collision for a while now, and you can even type it into google, and it will give you what was hashed back out. With so many possibilities in a fingerprint, it serves as a great authentication for a system or application.

Fingerprints have an assortment of friction ridges that form in the development process of human beings. As illustrated in Fig. 17.6, there are some of the various fingerprint styles, including such as arch, whorl, double loops, and simple loop. There are glands underneath the skin of your fingers are what make the distinct ridges visible on surfaces you touch such as glass or the fingerprint readers. By looking at the amount of ridges, direction, and pattern of the fingerprint, people were able to be identified uniquely from one another. This was used as early means of crime investigating where matching a set of print to an ink card someone had made

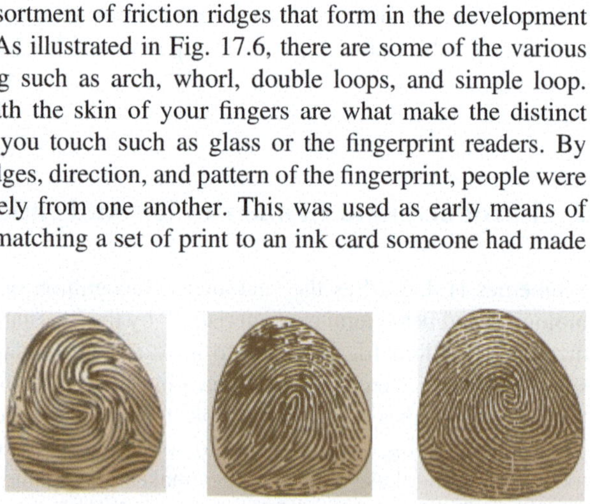

Fig. 17.6 Typical fingerprint styles

Fig. 17.7 An example of a scanned fingerprint

could tell if someone had left their fingerprints at a location. Now in the digital age, we have evolved to have algorithms to analyze the pattern and characteristics of the fingerprint and come up with a result of it is a match to the record they have on file.

Scanning a fingerprint can be done by using an optical scanner which involves a charged coupled device which is in term a light sensor, the same as found in digital cameras. The array of diodes underneath make up the many sensors for reporting how much light is hitting that sensor in the array. The device will emit its own light source when you start the scanning process and place your finger on the designated area, it will start recording the light that is be reflected back by the ridges of your fingerprint and then the software will invert the image to have the ridges of your finger as the darker portions instead of the areas between ridges; Fig. 17.7 shows what the inverted image looks like for the software to scan and compare to a stored set of data.

Lastly the software will check to see if there are sufficient characteristics between the print on record and the print you just scanned in. That way when you scan your print if the scanner does not pick up on every ridge or pattern, it will still grant you access because it still is proving to be a match to your print that it has stored. Another device to pick up fingerprint imaging is the capacitance scanner which uses electrical currents to have your salty skin complete the circuit and generate a map of the fingerprint. This device is made up of an array of cells; each cell is made up of two conductor plates for the positive and negatives within a capacitor circuit. These capacitors are very tiny so that they can be in between ridges of the fingerprint and get an accurate representation of the layout of the fingerprint. To get the signal out to the computing unit, the signal from the capacitors is so small that integrated into the device is an amplifier to step up the voltage and have the computing device an easily distinguishable reading, and signal is needed to travel the distance of the wiring connecting the sensor to the computing device, and we do not want the signal to degrade between the capacitor and the computing device. Before scanning begins all the capacitors need to be reset and drop any charge they may have then, when the fingerprint is applied that will start the charging of the capacitors again for imaging. Advantages of the capacitor scanning include having to use a real-time fingerprint

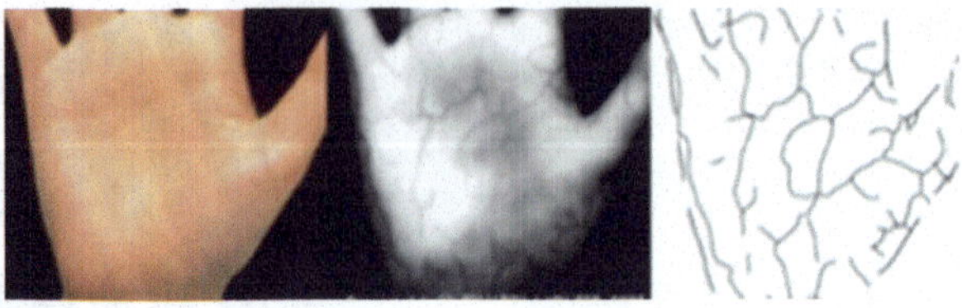

Fig. 17.8 An example of a scanned vein

reading which means without an actual finger you cannot so easily fool the scanner into thinking it is the right fingerprint on the sensor. An optical sensor takes an image, so having a very detailed fingerprint can trick the sensor into thinking you are someone else because it has no idea on the difference between image and a real finger.

2.2 *Vein Recognition*

Vein recognition is similar to fingerprint scanning, but instead of ridges formed in development, it is scanning the veins within your palm for distinct patterns. Veins offer another form of security with the matter of it is located beneath the skin where without surgery it is not an easy task to alter for the scanning device. Veins can be seen near the surface, but there are many veins underneath that you cannot see which also helps for creating the secure authentication and makes it harder for people to counterfeit a vein pattern. Veins also act as a viable use for authentication because they change shape and geometry very little as a person ages. The vein scanners work by using near-infrared light waves and projecting them onto your palm. Your red blood cells naturally absorb the light waves and appear darker because of the light being absorbed instead of reflected back; Fig. 17.8 gives an image of what the scanner is seeing when sending and receiving the wave lengths back. The scanner will then start mapping the darker lines to create a map of where the veins in your hand are located. After the geometry of your veins are scanned, they can be compared to the record stored and start checking points to which match and create an outcome within certainty that you are the person trying to authenticate with a specific user level of permission and access.

2.3 *Iris Scanning*

Iris scanning is another form of biometrics with complexity within its generation creating personal and unique characteristics for authentication making it another great resource of the biometrics industry. The iris can be seen in Fig. 17.9 for clarification on where the iris portion of the eye is located [9]. The iris is comprised of melanin pigments that form during development of the body and is still today not

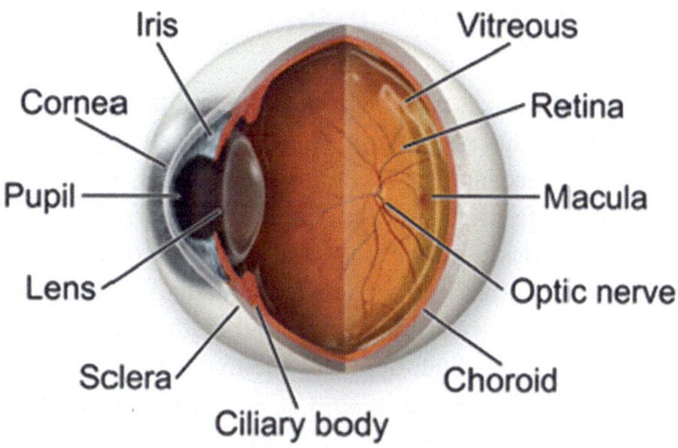

Fig. 17.9 Iris location in a normal eye anatomy

an exact science of how a specific color is produced, but we do understand how the colors come to be. Within the iris, eumelanin (appears as a brown/black pigments) and pheomelanin (appears as red or yellow pigments) combine in different strengths to create everyone's eye color. Other factors that play into the complexity of the iris are the fibers and blood vessels within the iris stroma. These randomly generated formations during the development process of the aforementioned factors make interesting geometric patterns within the iris and then through software and mathematics algorithms are created to scan your iris and create a data set for comparison.

Iris scanning makes for a more secure system because the iris can be scanned for 200 or more points for identification of an iris data set, whereas a fingerprint scan is normally 60–70 points for authenticating. The software for iris scanning is more complex in where the different regions of the eye as depicted in Fig. 17.9 shows that algorithms have to identify and map edges of the pupils, iris, sclera, and the eyelid itself. These regions are to focus in on the iris and see the patterns for authentication and ignore the regions you scanned but do not want to use for data analyzing. The iris also makes for a more reliable tool in biometrics as it is more of an internal organ with protection from the elements by the transparent but sensitive layer called the cornea. The protection of the iris means it is not going to break down or be worn down like fingerprints can be with manual labor and extensive use of the fingertips or hands.

Currently hardware for scanning iris are not as prevalent as fingerprint scanners but require a bit more specific hardware even though they contain the same sensor for collecting data on your iris. The scanners have been coming to market for a while now, but with more options becoming available and the pricing going down, there are more affordable options coming to market. Iris scanners can use the same technology as optical fingerprint scanners with the CCD sending out and scanning near-infrared light waves as well as visible light waves to produce a high-contrast image of your eye. The NIR (near-infrared) waves are good at getting the structure of the eye for comparison and make a more accurate picture because of the light wave being recorded. The ambient light is not affecting the imaging software and/or hardware.

2.4 Facial Scanning

Facial scanning is similar in number of points to fingerprint scanning with a bit higher 80 points of recognition to check between stored records and a current image scan. Facial scans can be done in either 2D or 3D depending on the machine and software suite running. 2D images are checked between images for similar patterns and geometrics with the image stored in the database. Facial images can be scanned for landmarks that make faces distinguishable from one another. Some of the notable landmarks that are scanned included distances between eyes, width of the nose, the shape of the jaw line, depth of the eye sockets, and shape of the cheekbones. The main drawback of 2D images is that the face needs to be 35 degrees toward the camera for calculating the similarities and differences between them. 3D imaging of facial features proves to be more accurate with more freedom of images whether its graphing is from a 2D image or from video to get a true 3D image. 3D imaging also allows for up to 90 degrees of rotation from the camera to be able to calculate facial features. The 3D imaging targets areas of the face including the chin, nose, rigid bone, and tissue areas as well as the curvature of the eye sockets. The 3D software allows for taking a 3D capture and calculates the distances on the face to create an accurate 2D image that can be turned to face the camera directly. Once generated, it can be compared with an older 2D image within an older database of facial scans.

Figure 17.10 shows the steps how a video feed can be converted from an image to a comparable image to search a database of faces. Another resource being developed for better recognition was to zoom in on the skin of the facial scan and verify the texture of the skin. The process named Surface Texture Analysis is breaking the skin into patches for scanning the texture, lines, and pores to even distinguish twins. It has been mentioned that combining the facial scanning with the texture analysis increases the match accuracy by 20% to 25%.

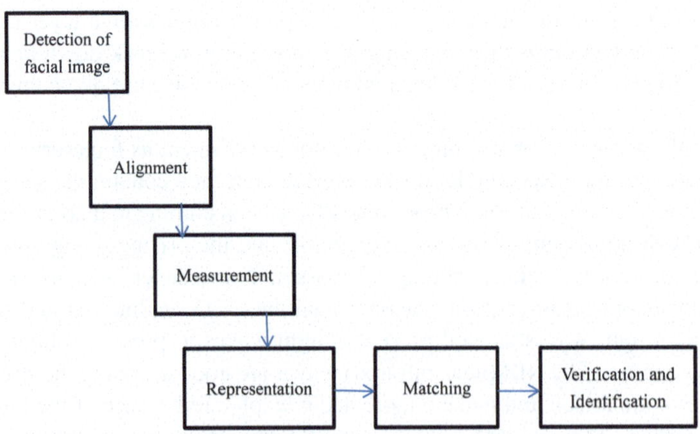

Fig. 17.10 Typical steps for facial recognition

3 Biometrics as a Service

Breaking down the term "Biometrics as a Service," biometrics – as defined by the International Standardization Organization – is the "automated recognition of individuals based on their biological and behavioral characteristics" [7]. Service is defined as "a facility supplying maintenance, repair and demand to the consumer" [10]. BaaS is a means of secure authentication using biometrics as layer of security. Biometrics since the 1990s has been slowly coming into the forefront as the leading method of securing data. In [11], Risch and Devlin stated that "individual DNA profiles are extremely uncommon, if not unique." It is much better in many ways regarding the fact that each human has unique biometric makeup. Also, forgetting your "password" is now a lot more difficult since people cannot forget to bring their hands, fingers, or eyes with them since they are attached to your person.

BaaS is considered a variant of Software as a Service. SaaS is defined as a software distribution model in which a third-party provider hosts applications and makes them available to customers over the Internet. All SaaS variants have these three things in common: they are licensed on a subscription basis, they are hosted by the service provider, and they are inseparable from cloud computing. Biometric authentication has been widely adopted by financial service industries, medical centers, law enforcement, and governments across the world.

As stated above, biometric services are inseparable from cloud computing technology. Typically, capabilities of cloud computing include (1) elasticity and agility, i.e., the ability to shift and deploy resources across disparate infrastructures; (2) network-centricity, i.e., the availability on any network and device; (3) on-demand domain, i.e., being able to be accessed anytime, anywhere in the world; and (4) data supremacy, i.e., data over the cloud ensures data supremacy [13]. Mobile strategies are being adopted by businesses, enterprises, and service providers around the world. Most companies today have a policy regarding the term bring your own device (BYOD). By incorporating mobile devices in everyday business affairs, it tremendously facilitates smooth cooperation with all parties involved. BYOD brings numerous benefits to businesses including job efficiency and flexibility, accelerated chain and business operations, and cost-saving regarding IT departments. Employees and employers are constantly using software to access sensitive personal information along with transactions related to payments. The cost that comes with efficiency from mobile devices is security.

There are two categories of biometric identifiers: behavioral and physiological attributes. Behavioral biometrics assess unique and measureable human traits, including patterns in attributes including voice, a person's gait (manner of walking), and typing rhythms. Physiological biometrics identifies a person's identity by using their fingerprints, facial recognition, palm veins, and retina and iris recognition. Biometric scanners use algorithms to take the data and inputs from the user to match said data points, which give authentication access for users.

This is where software and hardware from companies like ImageWare Systems and IriTech come into play. Solutions to cloud-based biometric authentication include topics ranging anywhere from mobile security and credentialing to healthcare and government that are being tackled by many different companies at many different angles. Leaders in the industry of BaaS currently are ImageWare Systems, Inc. who recently partnered with Fujitsu, BioID, IriTech, FingerCheck, and MorphoTrust. The topics covered in the portion will be directly referencing these five companies and their technologies.

3.1 ImageWare Systems, Inc.

ImageWare Systems has their hands in a variety of different markets. They offer an abundance of services regarding topics of access control to healthcare. On November 15, 2016 ImageWare Systems announced their Enterprise Suite for GoVerifyID. The product was created to provide full support for a Microsoft Ecosystem using end-to-end biometric authentication. GoVerifyID is a secure, flexible, and scalable biometric user authentication service. The way this software works is by having the servers pinging ImageWare Systems servers when an event – like a retail transaction – occurs. The user is asked for biometric verification which is then verified with anonymous templates on the server which will approve or deny the access or transaction.

GoVerifyID offers a demo of their product upon request via email. You are given a temporary account to test the product out. The way the software works is when you attempt to log in to a site that uses GoVerifyID, their server will push a request to your mobile device. This request opens the GoVerifyID application which then prompts you for a combination of your fingerprint, a voice passphrase, and facial recognition. Once you have successfully passed all the verification steps, the application sends a confirmation back to the website that enables you to log in.

3.2 BioID

BioID specializes in facial and voice recognition. BioID is designed to protect user information. Just like the other software services covered in this chapter, anonymity is a prevalent feature when verifying face and voices. The software uses a template that has no information about the user that is not necessary. Some biometric services usually do storage on a local, so nothing leaves the user's device. BioID – just like ImageWare Systems – uses secure data centers they are in control of due to the vulnerability of keeping data on a client's device.

BioID offers a trial version of their product which offers two ways to demonstrate their algorithms. The google play store has an application which offers a limited demo which features their facial recognition technology, which is displayed in Fig. 17.11. Their website does too but also offers voice recognition. You will have to sign up for an account as illustrated in Fig. 17.12. Then you must enroll your face so they can then recognize you for future authorization attempts.

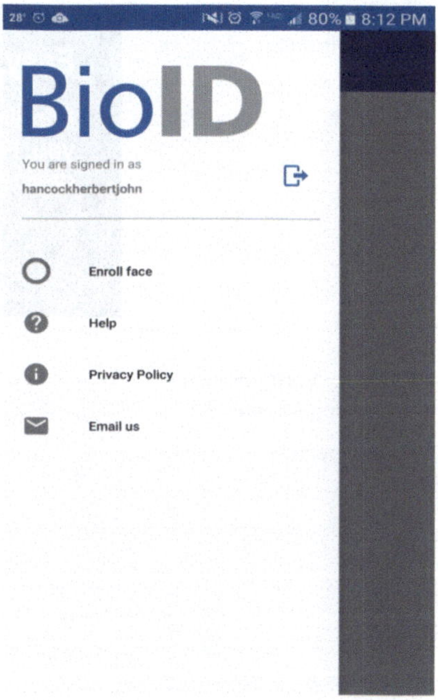

Fig. 17.11 Sign in GUI of BioID application

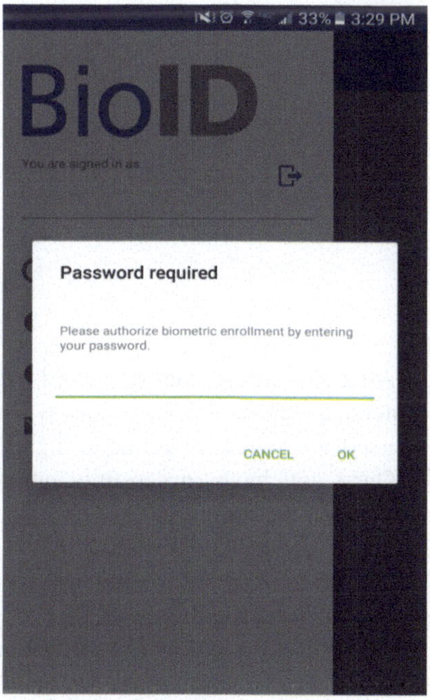

Fig. 17.12 User verification of BioID application

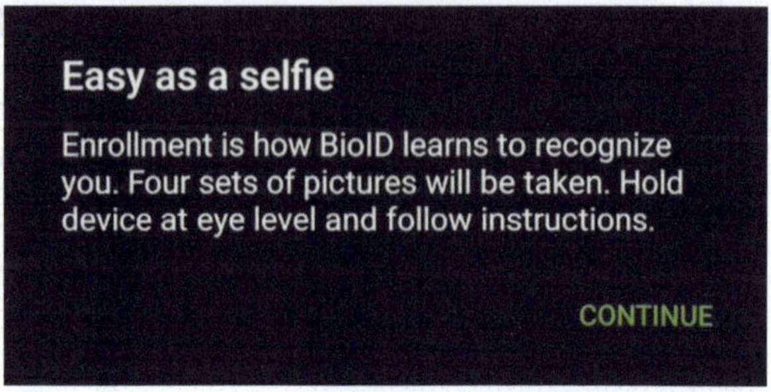

Fig. 17.13 Pictures taken in BioID application

Fig. 17.14 Verify prompt in BioID application

After submitting your password, you are then asked to position your face in the center at a reasonable distance away from the camera. It will ask you to then position yourself 90 degrees to get better lighting as illustrated in Fig. 17.13.

Once you have successfully enrolled your face, you can then start the trial verification by selecting verify prompt depicted below in Fig. 17.14.

This is the same procedure you will go through if you wanted to do this trial on a web browser via your computer or laptop. Once BioID has fully captured your bio template, you can then test out the accuracy by having a friend or colleague try to get their face to substitute for yours. You will most likely find that even if someone gets your password to initiate the bio-authorization, they will fail the facial recognition portion. Figure 17.15 shows the interface after completing the setup of BioID.

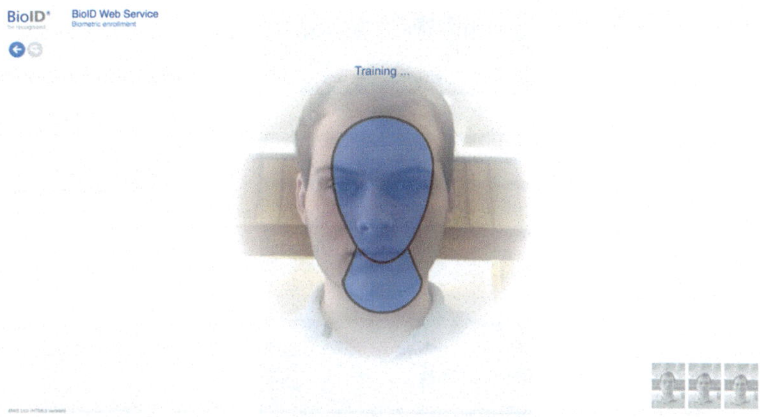

Fig. 17.15 Interface after completing the setup of BioID

3.3 IriTech

As you may perceive from IriTech's name, they specialize in iris scanning. The thing about IriTech is that their initial products were not only software but hardware. Obviously with scanning anything, you need a device to capture the data. Before smartphones were up-and-coming, there was not a mobile friendly way of securing your data via iris scanning. IriTech sells hardware and software as a package for a unique all-in-one solution to your biometric iris scanning needs. Now, they offer a cloud bases solution which can be used by just downloading an app on your smartphone. This software coupled with cloud-based servers is very secure and easily integrated into existing applications and services that support standard web service architecture.

Iritech's Iritracker was created as a time and attendance system which can help employers eliminate manual input, time theft, and identity card theft/exchanging. Their key features can be summarized as followed: The user interface is easy-to-use. Attendance data is generated from different locations. It has flexibility and manageability that offers administration create, replace, update, and delete functionality. The system's time reports are highly accurate and reliable. It supports multiple languages, capable of working in biometrically intolerable workplaces, and has automated message delivery with a bulletin board function.

Iritech – as discussed above – also offers a mobile option. The mobile application offers no option to test the functionality due to needing specific hardware from Iritech or having a mobile device that is supported by the application. The mobile application works by making a template of your iris using their underlying algorithm to use as a reference for authentication. More details on Iritech and its mobile application can see [14, 15].

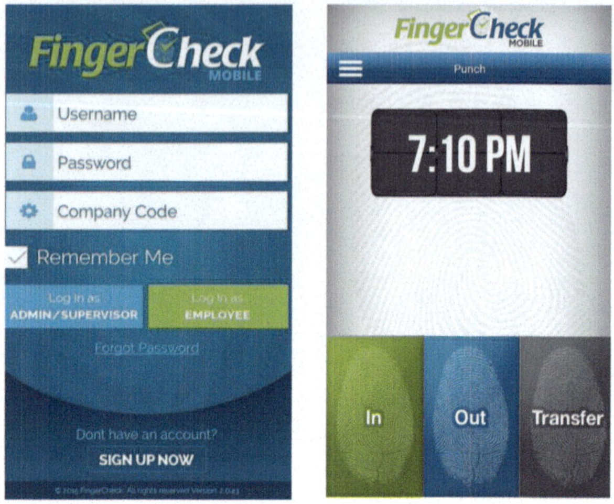

Fig. 17.16 An example of FingerCheck application

3.4 FingerCheck

FingerCheck has dedicated its software to combine time tracking and payroll into one solution. FingerCheck's software is reliable that uses fingerprint scanning via your phones scanner to keep track of worker's hours on the clock. They offer two ways to utilize their service. Download an application on a smartphone or tablet with fingerprint scanning capabilities coupled with a monthly paid service or hardware you can purchase that comes with the software and a fingerprint scanner. This software also uses cloud-based servers – specifically Amazon's redundant servers – to keep your data safe and secure. FingerCheck offers hardware and software solutions involving an all-in-one system and a mobile application. Covering the application, a login screen is the first thing you see as illustrated in Fig. 17.16 [16]. There are two options: one is for administrators and the other for employees. You will then be greeted by a selection to punch in and out or a transfer.

3.5 MorphoTrust

MorphoTrust specializes in identity protection and storage. Having physical identification can become cumbersome, so MorphoTrust has created a service that uses biometrics to safely store and authenticate identification cards like licenses for vehicles and aircraft, badges and IDs for law enforcement to hospital staff, and insurance cards. The main theme from these five companies is that they all use cloud services, license on a subscription basis, and are hosted by the provider and not the client. There are 42 states in the USA that use MorphoTrust's service. As illustrated in Fig. 17.17, MorphoTrust's way of processing identity includes three steps, i.e., enroll, verify, and issue [17].

Fig. 17.17 The MorphoTrust identity lifecycle

4 Conclusions

Biometrics technologies have already been widely adopted by government sectors such as law enforcement and border control. With growth of mobile computing everywhere, we can expect to see the growing demand for applying BaaS for identity authentication and online transactions in the field of high-risk industries such as health insurances, banks, and financial institutions because of the highly secure capability provided by the biometric identification and verification in managing risk and recruiting and retaining customers. It is our belief that leveraging cloud-based resources for automated biometric identification will be expanded across markets led by convergence trends, thus affecting biometrics technology providers, networking infrastructures, security and cloud storage industries, banking, healthcare, and retail as well as device manufacturers including smartphone makers, to name a few.

References

1. A. Regalado, Who coined 'cloud computing'? MIT Technology Review 1–10 (2011)
2. P. Mell, T. France, The NIST definition of cloud computing. In *NIST Special Publication* 800-145, Sept. 2011
3. D. Harris, What google compute engine means for cloud computing. GigaOM-Tech News, Analysis and Trends (2012)
4. E. Walker, Benchmarking Amazon EC2 for high-performance scientific computing. The Magazine of USENIX and SAGE **33**(5), 18–23 (2008)
5. D. Agarwal, S. Prasad, AzureBench: benchmarking the storage services of the azure cloud platform. In *Proceedings of the* 2012 *IEEE 26th International Parallel and Distributed Processing Symposium Workshops and PhD Forum* (IPDPSW '12). IEEE Computer Society, Washington, DC, pp. 1048–1057
6. R. Prodan, M. Sperk, S. Ostermann, Evaluating high-performance computing on Google app engine. Software, IEEE **29**(2), 52–58 (2012)
7. ISO/IEC 17788:2014 1st Information technology – Cloud computing – Overview and vocabulary JTC1/SC38
8. P. Pakutharivu, M.V. Srinath. A comprehensive survey on fingerprint recognition systems. Indian J. Sci. Technol. **8**(35) 1–7 (2015)

9. Z. Zhu, Q. Ji, Novel eye gaze tracking techniques under natural head movement. IEEE Trans. Biomed. Eng. **54**(12), 2246–2260 (2007)
10. Service https://www.merriam-webster.com/dictionary/service. Retrieved in Mar. 7 2017
11. N. Risch, B. Devlin, On the probability of matching DNA fingerprints. Science (New Series) **255**(5045), 717–720 (1992)
12. F. Liu, W.P. Guo, Z.Q. Zhao, W. Chou. SaaS integration for software cloud. In *Proceedings of 2010 I.E. 3rd International Conference on Cloud Computing*, 2010, pp. 402–409
13. Frost and Sullivan. Cloud-based Identity and Authentication: Biometrics-as-a-Service. Fujitsu. http://www.fujitsu.com/us/Images/Fujitsu-FrostSullivan_Cloud_WP_Biometrics-as-a-Service.pdf. Retrieved in Mar. 1 2017
14. Cloud-Based Iris Recognition Solution, http://www.iritech.com/products/solutions/cloud-based-iris-recognition-solution-0. Retrieved in Mar. 23 2017
15. IriShield Demo, https://play.google.com/store/apps/details?id=com.iritech.iddk.demo. Retrieved in Mar. 23 2017
16. FingerCheck, https://fingercheck.com/latest-updates-to-the-fingercheck-mobile-app/. Retrieved in Mar. 23 2017
17. MorphoTrust, http://www.morphotrust.com/Portals/0/MorphoTrust_Intro_Brief.pdf. Retrieved in Mar. 23 2017
18. A. Iosup, R. Prodan, D. Epema, Iaas cloud benchmarking: approaches, challenges, and experience, in *Cloud Computing for Data-Intensive Applications*, (Springer, New York, 2014), pp. 83–104
19. B. Butler, PaaS primer: what is platform as a service and why does it matter?. Network World, February 11, 1–5 2013

Chapter 18
Automated Biometric Authentication with Cloud Computing

Hisham Al-Assam, Waleed Hassan, and Sherali Zeadally

1 Introduction

Over the last few years, cloud computing has become one of the fastest-growing IT environments for providing services to individuals and businesses of all sizes. Cloud computing, as defined by the National Institute of Standards and Technology (NIST), is "a model for enabling ubiquitous, convenient, on-demand network access to a shared pool of configurable computing resources (e.g., networks, servers, storage, applications, and services) that can be rapidly provisioned and released with minimal management effort or service provider interaction" [1]. The so-called cloud service providers (CSPs) are the key players in cloud computing responsible for providing clients with a wide range of services that vary from applications such Microsoft Office 365 and Google Docs to a complete infrastructure such as Amazon Elastic Compute Cloud (EC2) [2]. This introductory section provides the reader with a brief background on four related topics: (1) the main characteristics of cloud computing, delivery models, and deployment models, (2) security challenges in cloud computing, (3) biometric-based recognition, and (4) the limitations of conventional biometric solutions for remote cloud authentication.

H. Al-Assam (✉) · W. Hassan
Applied Computing Department, University of Buckingham, Buckingham, UK
e-mail: hisham.al-assam@buckingham.ac.uk

S. Zeadally
College of Communication and Information, University of Kentucky, Lexington, KY, USA

M. S. Obaidat et al. (eds.), *Biometric-Based Physical and Cybersecurity Systems*,
https://doi.org/10.1007/978-3-319-98734-7_18

1.1 Cloud Computing

The convenience provided by cloud computing has led to an increasing trend of many business organizations, government agencies, and customers to migrate their services and data into cloud environments. The recent success of cloud computing in attracting such a great attention can be attributed to the following five characteristics [2]:

- *On-demand self-service*: A client can immediately get computing resources (e.g., CPU time, applications, network storage, etc.) without a need for human intervention at the CSP side.
- *Broad network access*: Cloud resources are network accessible from different clients' applications installed on different platforms such as smart phones, tablets, PCs, and laptops.
- *Resource pooling*: The CSPs aggregate their resources to meet clients' need by utilizing multi-tenant approaches based on physical as well as virtual resources which can be dynamically added or withdrawn based on clients' requirements. The *pooling* factor means that the clients do not need to know where the resources are coming from or where the data is physically stored.
- *Rapid elasticity*: The capabilities of cloud services should be flexible enough to rapidly shrink or expand to meet the requirements of different clients at different times.
- *Measured* service: CSPs have the ability to measure any resources used by each tenant (client) using charge-per-use mechanisms.

Cloud services are typically delivered to clients using pre-packaged combinations of IT resources provided by CSPs based on one of the following three common cloud service models [3]:

Software as a Service (SaaS) This model of delivery is also called "on-demand software." The software and associated data are centrally hosted on CSP's servers (i.e., instead of using a software installed on clients' machine, they can use it as a service where no maintenance or upgrades are required). In this model, clients have no control or management permission over the underlying cloud infrastructure. Common examples of SaaS include Google Docs, Dropbox, and Microsoft Office 365.

Platform as a Service (PaaS) This kind of service is typically used by application developers. This type of service provides access to computing platforms that include operating systems, programming languages, software tools, databases, web servers, etc. The clients have control only over their deployed applications. Some examples of PaaS include Google AppEngine, Microsoft Azure, and Apache Stratos.

Infrastructure as a Service (IaaS) This delivery model supplies clients with computing resources (physical or more often virtual) processors, storage, firewalls, load balancers, virtual local area networks, and so on. Therefore, the clients are not only able to deploy and execute various software, but they also have control over the operating systems, storage, processing power, and networking components. Amazon's EC2 is a very good example of IaaS.

The above three categories of services can be deployed in different environments. Deployment models define ownership and the size of cloud resources and most importantly define who can access them. Currently, four basic models of deployment have been identified by the cloud community [3]:

- *Private cloud computing*: The cloud infrastructure and services are offered exclusively to one enterprise, and it might be owned, managed, as well as operated by the enterprise, a third party, or a combination of both. This deployment model not only gets an optimal use of existing in-house resources, but it also provides better data security and privacy. It should be noted that the cloud environment in this model might be located in or outside of the premises of the enterprise.
- *Community cloud computing*: The cloud infrastructure is shared by a group of clients or organizations to provide shared policies, values, and security procedures. The ownership, management, and operation of this model are given to one or more members of the group.
- *Public cloud computing*: The cloud infrastructure is open for public use. The ownership and management are given to businesses, academic institutes, government bodies, and so on.
- *Hybrid cloud computing*: More than one deployment model can be combined to form a hybrid cloud environment to meet clients' needs.

It can be argued that each type of service and deployment model meets the demands of some business more than others. For example, while a large enterprise might benefit from the private cloud, smaller businesses will most likely opt for a public cloud for cost consideration. Figure 18.1 illustrates typical cloud computing service layers along with their cost and timeline impact.

1.2 Security Challenges in Cloud Computing

Although cloud computing offers great advantages over other traditional IT solution, it poses serious security concerns. In fact, security and privacy are essential factors for an enterprise when deciding on whether to migrate their data, applications, and other relevant services to cloud environments. Service agreements between clients and CSPs tend to include details on how to access and utilize cloud services, service duration, and data storage and management when the contract ends [1]. However, the main challenge is how to guarantee that the data is accessible by authorized users only. When data owners decide to use the cloud environment, they rely entirely on third parties to make decisions about their data. Therefore, it is very important for data owners to have the right technologies or methods to prevent CSPs from utilizing such data without their permission. Both technical and nontechnical methods have to provide effective means to fulfil this goal [4, 5]. A wide range of possible solutions have been proposed to implement different mechanisms to prevent unauthorized access to cloud data even by untrusted CSPs [6–11]. In general, to address clients'

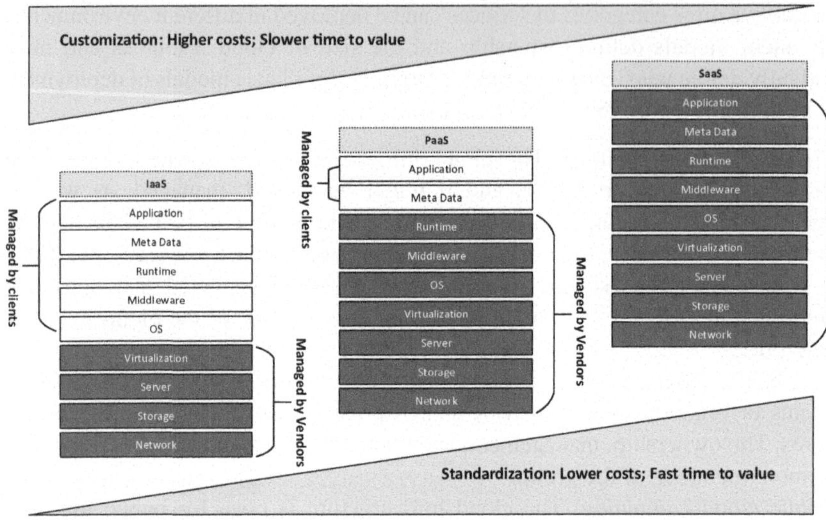

Fig. 18.1 Typical layers of cloud computing services. (Adapted from Sahai and Waters [2])

concerns about security and privacy of the cloud environment, the following three essential challenges must be addressed [4]:

- *Outsourcing*: In the traditional IT environment, clients can exercise full control over their data. However, they usually lose all means of physical control over the data once it is migrated to cloud environments, which is the key security concern. To overcome this problem, clients need to ensure that the cloud service providers are trustworthy and are capable of meeting the requirements related to secure data storage, correctness, and integrity of cloud data and computation at all times and maintaining clients' privacy.
- *Multi-tenancy*: Cloud environments can share their resources and services among multiple clients simultaneously. Both the virtual machines provided by CSPs and the cloud data of different clients are eventually located on a single physical machine based on particular resource allocation policy. Hence, a legitimate cloud client can potentially act as an adversary by exploiting some holes in the policies to gain unauthorized access to the data of other users.
- *Big data and intensive computation*: Cloud environment requires dealing with large volumes of data supported by powerful processing capabilities. Hence, traditional security techniques might be difficult to apply on such data because of the volume of high computation and communication overheads. For instance, to guarantee the integrity of remotely stored data, it is computationally infeasible to hash the whole data. Consequently, new strategies and protocols are needed to overcome such difficulties.

1.3 Biometric Authentication

In the cloud environment, a reliable identity management system is a key component to prevent identity theft and control access to different resources. Establishing the correct identity of a person is an essential task in any identity management system. Typically, there are three ways to authenticate an individual, each of which has its own advantages and limitations [12]:

- *Knowledge-based authentication*, or "something you know," that typically relies on a memorized password or PIN. A random 12-character password, for example, offers a strong security mechanism for user authentication. However, in practice, humans have difficulties in memorizing complex passwords, and passwords that they can easily remember are often short and therefore simple to guess or determined by a brute-force/dictionary attack.
- *Object-based authentication*, or "something you have," which relies on the physical possession of an object, such as a token. The main drawback of a physical token is that, when lost or stolen, an impostor can easily gain unauthorized access.
- *Identity-based authentication*, or "something you are," biometric-based authentication offers an advantage over other authentication factors in that a genuine user does not need to remember or carry anything. Moreover, biometric-based authentication is known to be more reliable than traditional authentication because it is directly linked with the identity of individuals. This is practically important for cloud environments as it associates data access with its ownership. However, biometric systems were not initially design for remote authentication in cloud environments. In fact, they can be subject to replay attack, and, unlike other credentials such as PINs, passwords, or smart cards, once biometric-related information is compromised, it is impossible to be changed again.

Biometric systems in general aim to identify or verify an individual's identity based on physical characteristics (e.g., face, iris, fingerprint, DNA, or hand geometry) and/or behavioral characteristics (e.g., speech, gait, or signature). A typical biometric system has two stages, enrolment and recognition. Figure 18.2 illustrates the process of the biometric enrolment stage, in which a user starts by presenting their biometric data to a biometric sensor (usually in a controlled environment). If the quality of the captured biometric sample is found to be adequate, the enrolment process proceeds to a preprocessing procedure to prepare the sample for the next step. A feature extraction technique is then used to extract a digital discriminating feature vector of the individual, called biometric template (BT), which will then be stored (often also called "enrolled") alongside the individual's identifier (ID) in a database.

At the recognition stage, biometric systems can function in two modes depending on the application context, namely, authentication or identification mode.

Biometric-based authentication (also known as verification) is a one-to-one comparison of a freshly captured biometric sample(s), known as query, against an

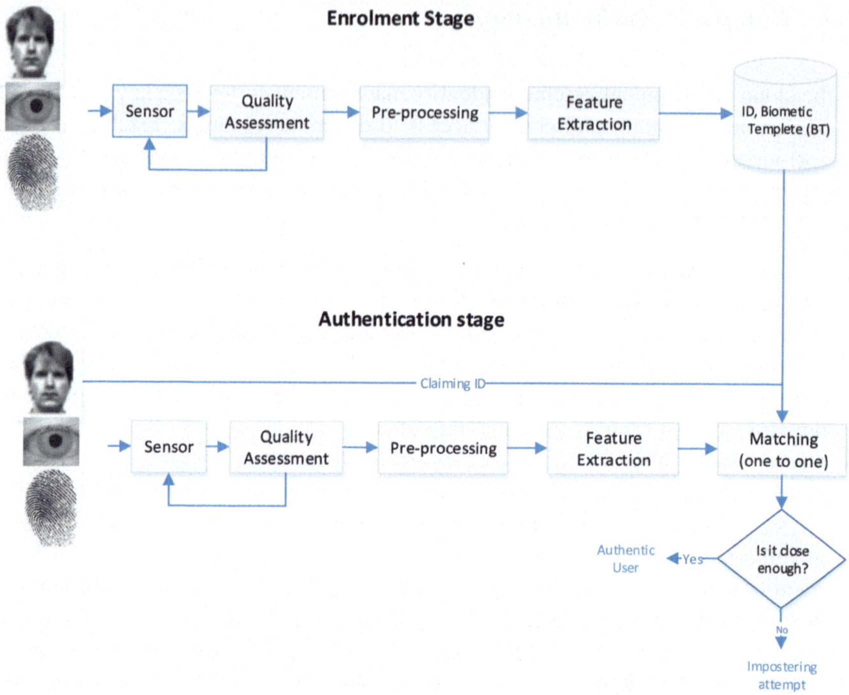

Fig. 18.2 A typical enrolment stage of a biometric system (the face image was used from the Extended Yale Face Database B [13])

enrolled BT as illustrated in Fig. 18.2. In this mode, a user claims an identity, and the biometric system verifies the authenticity of the claimed identity (e.g., the system answers the question: "Are you who you say you are?"). For example, authentication might be used when a user wants to access his/her bank account or computer. The matching process uses a distance or similarity function to calculate a score indicating the similarity between the stored BT and the fresh feature vector extracted from the query sample. If the matching score is high enough, i.e., close enough to the enrolled template, the biometric system grants access to the user. Otherwise the requested access is rejected. The term "high enough" is determined by the administrator depending on the level of tolerance necessary for the specific application. This allows the system administrator to adjust the rates of false acceptance (i.e., wrongly accepted imposters as genuine users) and false rejection (i.e., wrongly rejected genuine users) to the desired levels. Typically, there is a trade-off between the false acceptance rate (FAR) and the false rejection rate (FRR), in which the reduction of one rate causes an increase in the other. Most biometric systems are configured to be highly secure by maintaining a very low (e.g., 1 in 10,000) FAR and an acceptable FRR. It is generally less problematic to have a false rejection by asking the genuine user to re-scan their biometric, rather than a false acceptance in which an unauthorized individual will be granted access.

1.4 The Limitations of Conventional Biometric for Remote Authentication

As we have mentioned previously, many business organizations, government agencies, and customers are rapidly shifting many of their services and data onto the cloud which has necessitated the need for secure remote authentication schemes that are immune from fraud and identity theft. Although a biometric-based authentication system is known to be more reliable than traditional authentication schemes, biometric systems can be subject to failure due to the intrinsic factors mentioned earlier or adversary attacks. The security of biometric remote cloud-based authentication in particular can be undermined in a number of ways. For example, a biometric template can be replaced by an imposter's template in the cloud database, or it could be stolen and replayed [14]. As a result, the imposter will gain unauthorized access to a place or a system. Moreover, it has been shown that it is possible to create a physical spoof starting from biometric templates [15]. Adler et al. proposed a "hill climbing attack" [16] on a biometric system which, in a finite number of iterations, generates a good approximation of the target template. Another method has been also proposed in [17] to reconstruct fingerprint images from standard templates which might fool fingerprint recognition. Furthermore, biometric data on its own is not very secret. Individuals usually unintentionally leave (poor-quality) fingerprints everywhere, and a hidden camera can capture a picture of a face or an iris [18]. In fact, the level of secrecy varies among different biometric modalities (e.g., covertly collecting face images or voice samples is much easier compared to collecting retina and palm vein samples).

Remote biometric authentication in cloud environments is particularly vulnerable to eight possible points of attack highlighted by Ratha et al. [19]. As illustrated in Fig. 18.3, Attack 1 occurs when an attacker presents fake biometric sample at the sensor such as photo of a face, fake fingerprint, copy of a signature, and a recorder voice. Attacks 2 and 4 are replay attacks by resubmitting an old signal by bypassing the sensor or the feature extractor. Attacks 3 and 5 are Trojan horses that produce feature set or matching score chosen by the attacker. Attack 6 is to target the enrolled

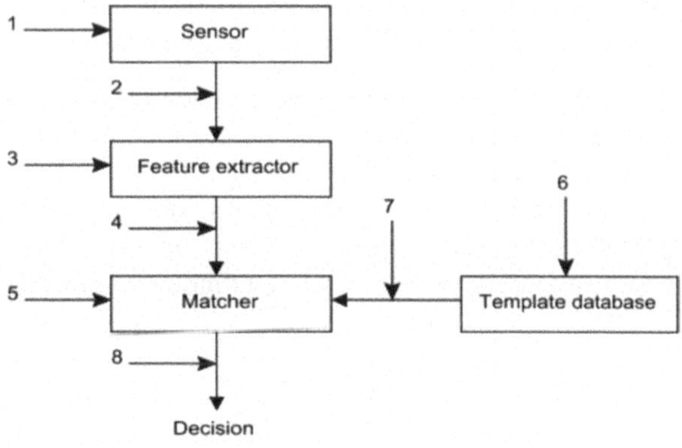

Fig. 18.3 Eight points of attacks in a biometric authentication system. (Adapted from Georghiades et al. [19])

templates database stored on the cloud and tamper with the template values. Attack 7 is on the channel between the template database and the matcher where an attacker might tamper with the template before it reaches the matcher. Finally, Attack 8 is to override the decision by the attacker.

Most of the eight points of attacks explained above can be exploited in unattended remote authentication when accessing cloud resource. Therefore, the deployment of such systems for remote authentication is still very limited. Solutions such as cancellable biometric systems and biometric cryptosystems are inherently not immune to replay, man-in-the-middle, and other remote attacks. This necessitates the need for new innovative solutions that help in fraud detection and identity theft prevention in cloud environment as highlighted in Sect. 3.

2 Authentication Challenges for the Cloud Computing Environment

With more individuals and organizations opting to use cloud storage service, remote access control to data and cloud resources is becoming a key requirement. This is particularly important as cloud clients do not usually know where their data is physically stored (i.e., the overall security control has shifted form data owners to the hand of service providers [3]). Therefore, the question of how to limit data access to the data owner and authorized user has been one of the biggest challenges in cloud environments. Combination of ID cards and password-/PIN-based authentication is the most widely used form of remote user authentication [5]. As such authentication factors are not linked with the identity of data owner (see Sect. 1.3), biometric authentication seems the ideal option for access control. However, employing biometric-based solutions for cloud authentication is far from a straightforward task due to security issues highlighted in Sect. 1.4. Although biometric authentication is perceived to be more reliable than traditional authentication schemes, the open nature of unattended remote authentication makes biometric systems vulnerable to replay and other remote fraudulent attacks. Generally speaking, an effective remote authentication in cloud environments has to be a component of a security package that meets the following requirements [3]:

- The original data should be intractable to restore by cloud providers.
- Offering effective means to distinguish between legitimate users and impostors and prevent the latter from gaining unauthorized access.
- Preventing any intruder from altering original messages.
- Minimizing response time to clients' requests, which is a vital requirement for security mechanisms that rely on timestamp synchronization.

2.1 Traditional Biometric Solutions for Cloud Authentication

The concept of revocable or cancellable biometric templates generates biometric templates that are not fixed over time. In this case, such templates can be revoked in the same way as lost PINs or passwords [20]. Biometric cryptosystems, on the other hand, aim to generate biometric keys and bio-hashes that are used as a proof of identity instead of biometric templates. In general, the key approaches to protect biometric templates rely on the use of a non-invertible secret transformation on the biometric feature vectors [21].

An interesting implementation of revocable biometrics schemes is the multifactor biometric authentication (MFBA) which makes use of user-based transformations (UBTs) on biometric feature vectors. These MFBA schemes have been proposed to improve security and privacy of biometric systems. Figure 18.4 shows the key steps of a MFBA system based on the UBT approach at the enrolment and authentication stages. UBTs tend to use transformation keys produced from passwords/PINs or stored on a token.

It can be argued that revocable biometric systems and biometric cryptosystems enhance the overall security and user's privacy, but at the same time they are inherently not robust against replay, man-in-the-middle, and other remote attacks in cloud environments.

Another traditional approach to remote biometric authentication relies on combining MFBA with challenge-response approach between clients and cloud authentication service [22]. As illustrated in Fig. 18.5, such an approach employs a blinding random vector as an essential ingredient of a one-time representation of multifactor biometric that prevents replay attacks and enhances the security of the MFBA. The

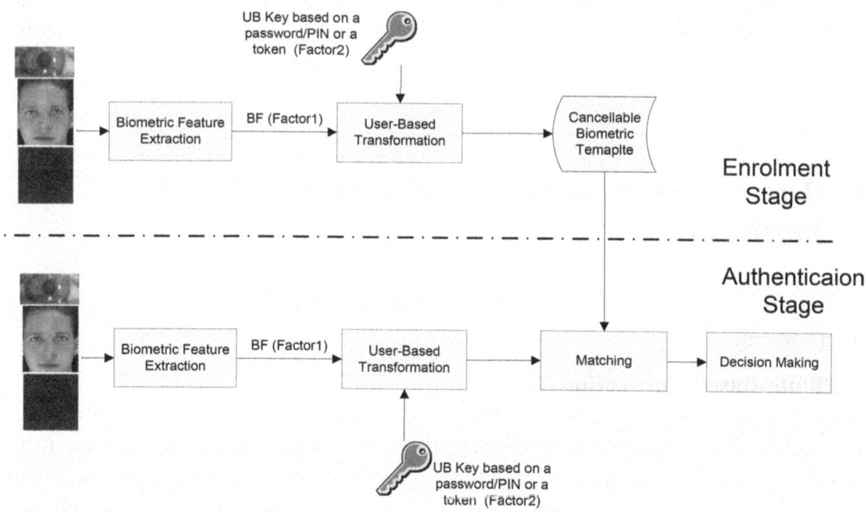

Fig. 18.4 Key steps of a MFBA system based on applying UBTs at the enrolment and authentication stages

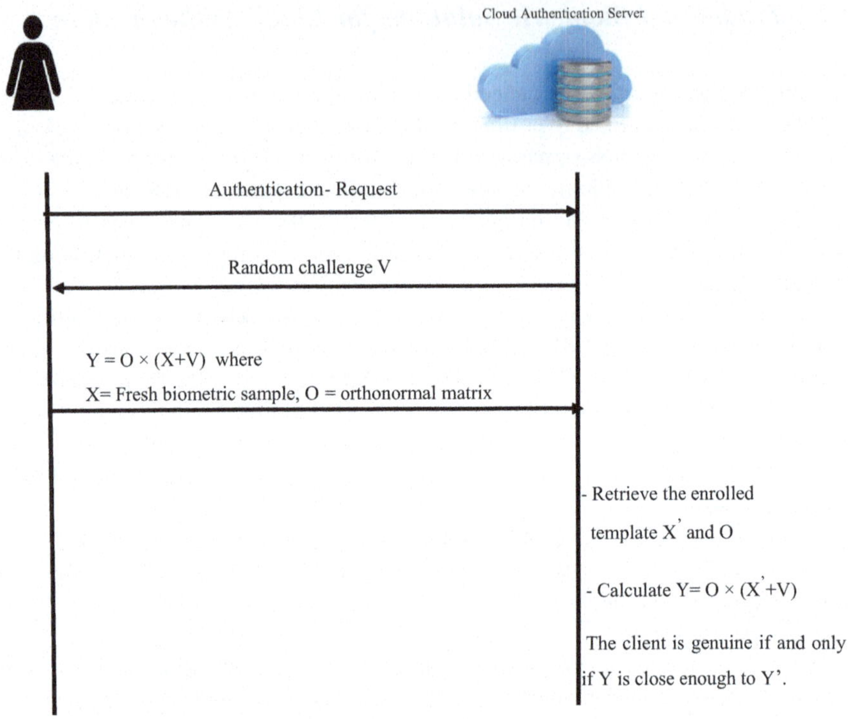

Fig. 18.5 Multifactor one-time biometric representation based on challenge/response for remote cloud authentication. (Adapted from Adler [22])

production of one-time, fresh revocable biometric template is based on transformations that are generated fresh with contribution from both the client and the cloud authentication server to provide the non-repudiation feature.

3 Recent Solutions to Address the Security in Cloud Environment

Next we discuss some recently proposed solutions that provide security in the cloud environment.

Attribute-Based Encryption (ABE)
Data sharing between different parties is a key feature of cloud computing. As discussed earlier, data should be encrypted before being migrated to cloud environments to ensure the security of cloud storage. The main challenge associated with using symmetric/asymmetric key encryption is how to securely store and exchange the keys between different parties in an open environment such as cloud computing. The public-key infrastructure (PKI) has been providing a practical solution for

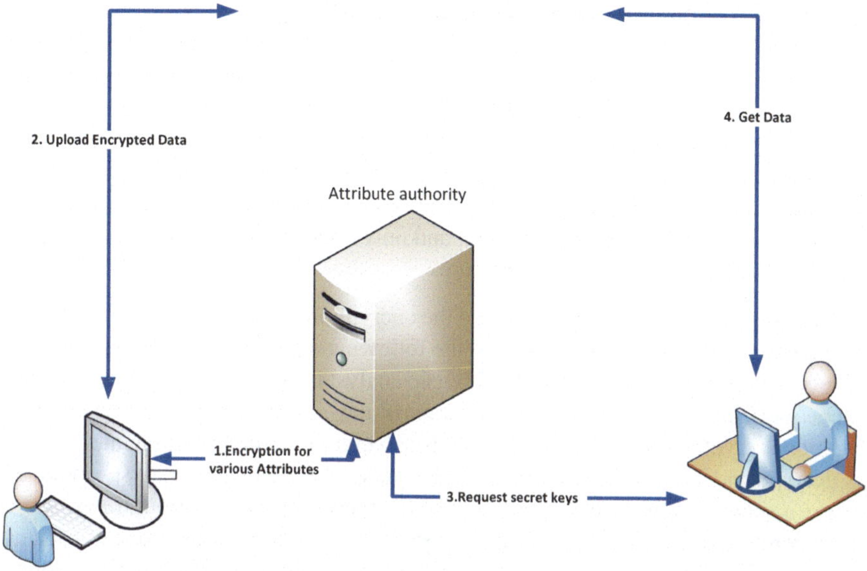

Fig. 18.6 General attribute-based encryption architecture. (Adapted from Hao et al. [24])

session key exchange for many web services. The key limitation of the PKI solution is not only the need for a trusted third partly (e.g., certificate authority) but also the missing link between data owner and the encryption keys.

To link encryption keys with users' identities in cloud environments, attribute-based encryption (ABE) has been recently proposed as a new formulation of public/private key infrastructure in which the encryption keys are directly derived from a user's identity [23]. In traditional public-key encryption, a message is encrypted for a specific receiver using the receiver's public key, whereas ABE revolutionizes this idea by linking the public key with the identity (e.g., email address) and/or the attribute (e.g., roles) of the receiver. Thus, the key feature of ABE is to enable data owners to share encrypted data with a set of individuals who have a matching set of attributes.

For example, ABE enables the head of a computing department to encrypt a document and share it with members of staff who have attributes {lecturers, admission committee, exam committee}. A threshold to specify the minimum number of required attributes can be used to offer better level of flexibility on who can access the data. In the above example, if the threshold is set to 2, then any staff member with at least two of the three attributes would be able to decrypt the message. The encryption and decryption keys are generated by a trusted third parity based on a set of descriptive attributes as illustrated in Fig. 18.6.

The following subsections further explain the two main extensions of ABE approaches proposed, namely, the key-policy ABE (KP-ABE) and the ciphertext-policy ABE (CP-ABE).

Key-Policy Attribute-Based Encryption (KP-ABE)

The KP-ABE was initially proposed by Goyal et al. [24] to provide a development version of ABE. In KP-APE, data owners generate a master key to encrypt the data in such a way that the corresponding ciphertext is labelled with a set of attributes. The decryption key (private key) given to the user is associated with an access policy (i.e., a tree-like access structure that specifies which ciphertext the key can decrypt). The leaves of the tree-like access policy are associated with attributes of the users. As a result, a user can decrypt a ciphertext if and only if the attributes associated with the ciphertext satisfy the key access structure.

One application of KP-APE could be the encryption of Audit Log Entries of a big organization. Suppose the entries have the following structure {user name, date and time of action, type of action} and a forensic analyst is assigned the task of carrying out a particular investigation on the log. If the entries are encrypted using a traditional cryptography, the analyst needs a secret key which will enable him/her to decrypt and access ALL entries. However, in KP-APE, the analyst would be issued a secret key associated with a specific access structure, through which the corresponding decryption key enables a particular kind of encrypted search such as accessing log entries whose attributes satisfied the conditions {"username = John" OR (access date between 01/01/2017 and 01/06/2017)}. The KP-APE also makes it unfeasible for multiple analysts to access unauthorized entries from the audit log even if they collaborate and share their keys [24]. Another simple example of KP-ABE is illustrated in Fig. 18.7.

Figure 18.7 shows a simple access structure which dictates who can retrieve the decryption keys. In this case, Alice would be able to access the decryption key and unlock the ciphertext or part of it if and only if his/her attributes satisfy the corresponding access structure (i.e. she has to be a {Dean OR (a member of

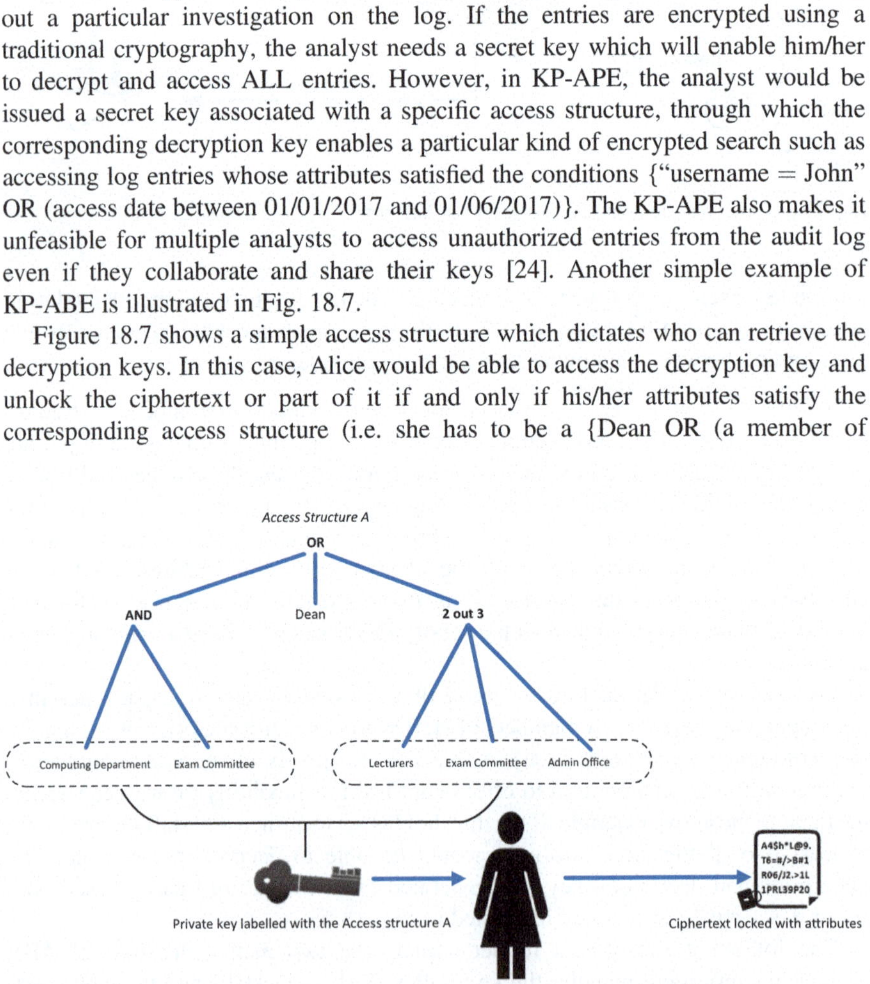

Fig. 18.7 Key-policy attribute-based encryption model. (Adapted from Hao et al. [24])

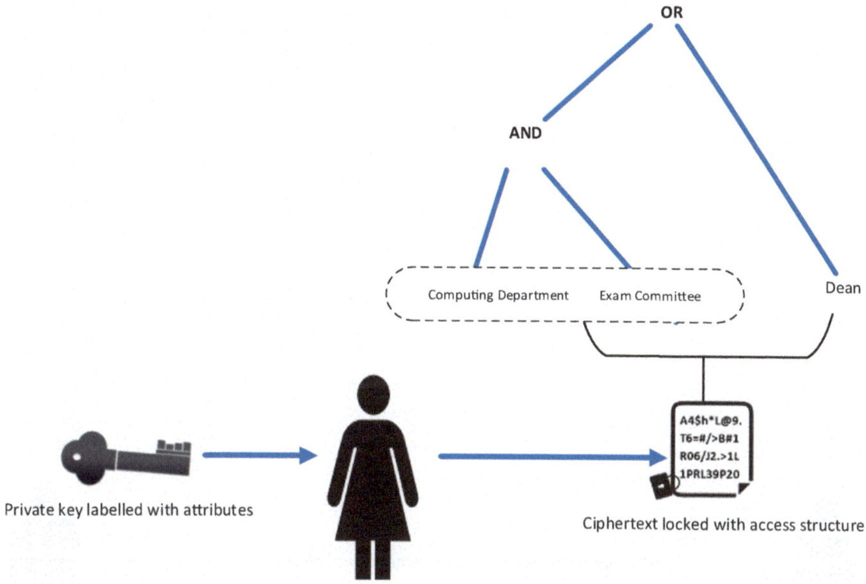

Fig. 18.8 Ciphertext attribute-based encryption model. (Adapted from Boneh and Boyen [9])

Computing Department AND Exam committee) OR (she belongs to two of the three: Lectures, Exam Committee, Admin Office)}).

Ciphertext-Policy Attribute-Based Encryption (KP-ABE)

The main limitation of the KP-ABE is that data owners have no control over who can access the encrypted messages because the access policy which is typically managed by a third-party public-key generator (PKG) is not attached with the ciphertext (i.e., the access policy controls the access to the decryption keys instead of controlling the access to ciphertext). On the other hand, the ciphertext-policy attribute-based encryption (CP-ABE) shifts the focus to the ciphertext by giving data owners the power of locking their encrypted data with different access policies (i.e., for each message they can decide on who can decrypt that particular message as illustrated in Fig. 18.8).

The example given in Fig. 18.8 demonstrates the flexibility provided to by CP-ABE in locking different messages with different access structures (i.e., data owners are able to choose access policies based on the sensitivity and the security of their encrypted messages). For example, the figure shows that the encrypted message in this scenario can be only decrypted by the Dean OR (a member of both Computing and Examination staff).

Standard Identity-Based Encryption (IBE)

The first proposal of identity-based encryption (IBE) was presented by [25] as an alternative approach to traditional public-key encryption in which the public key of a user is some unique information about the identity of the user such as email address.

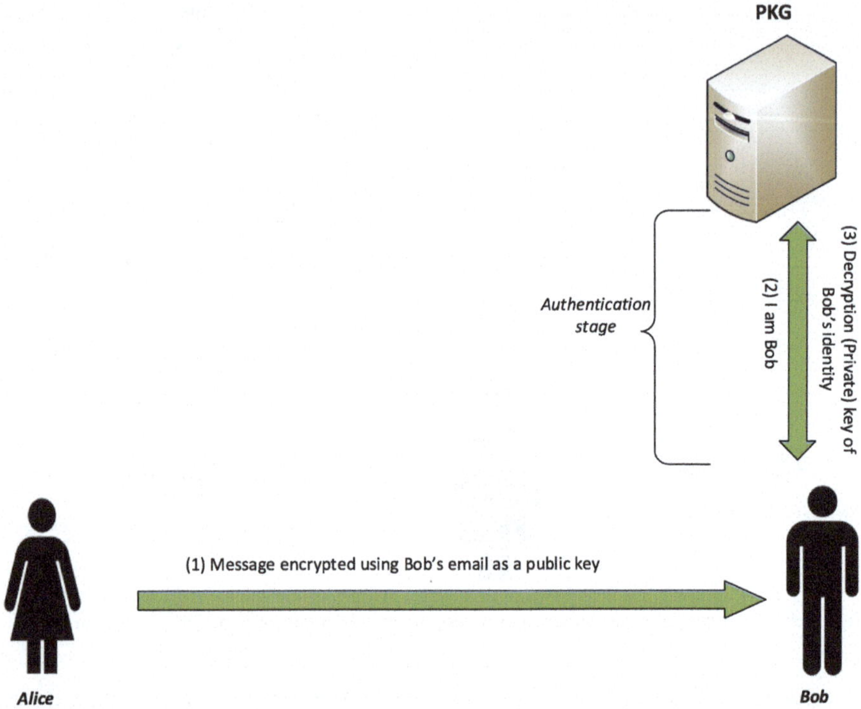

Fig. 18.9 Example of identity-based encryption architecture. (Adapted from Ratha et al. [25])

If Alice wants to send a secure message to Bob, she can use the text value of his email address as an encryption key using the public parameters of the system stored on the PKG. The basic concept of IBE can be illustrated as in Fig. 18.9.

The figure shows that Alice sends Bob a message encrypted using Bob's identity such as email, phone number, and so on even if Bob has no public-key certificate. The private decryption key is then retrieved from the PKG which holds public parameters as well as the master secret key msk. IBE scheme can offer more power to Alice because she can apply more restrictions on her encrypted message by adding more roles to Bob's ID such as {Bob's email, Access time/data, Role}.

4 Biometric Solutions for Cloud Computing

4.1 Fuzzy Identity-Based Encryption

The initial idea of fuzzy identity-based encryption (F-IBE) was presented in [23] where the identity was modelled as a set of descriptive attributes. The key feature of F-IBE is that the private key of an identity x has an ability to decrypt a ciphertext that has been encrypted with another identity *y* if and only if the distance between *x and y* is less than or equal to a certain threshold value. The F-IBE plays a key role in utilizing biometric data such as a fingerprints or face images as identity. F-IBE is a promising solution that bridges the gap between the exactness of encryption/decryption keys and the fuzziness of biometric data (i.e., the enrolled biometric samples and the freshly captured ones are never the same). This feature enables a private key of biometric identity to decrypt a message that was encrypted with a public key of a slightly different biometric identity.

It can be argued that the weakness associated with the use of traditional IBE is that the identity such as a "name" or "email" needs to be authenticated first before retrieving the corresponding decryption key in [6, 7, 26]. Therefore, the user might need to provide additional "supplementary documents" to link the name and/or the email with her identity. In contrast, the biometric-based F-IBE offers a natural way of authentication by providing biometric data, which is part of user's identity, to retrieve the decryption private key. It has been proved that F-IBE can withstand collusion attacks (i.e., a group of users cannot integrate their keys in a manner that enables them to decrypt messages without individual permission).

In F-IBE, the user's private key is a set of **n** private components or features that are linked within the identity of the user. Shamir's secret sharing [25] is typically employed to distribute the master secret key over the components of the private key by using a polynomial of degree **(d-1)** where **d** < =**n** is the minimum number of private components that the user needs to present to retrieve the decryption (private) key. Figure 18.10 shows an example of two identities X and Y where the number of overlapped features is 16 out of 20 features. If the threshold **d** is set to be 16 or less, then the two identities will be deemed to be the same.

4.2 *Recently Proposed Biometric Solutions*

BIOmetric identity-based encryption (BIO-IBE) is an interesting application derived from F_IBE, in which the identity x is a feature vector $(f_1, f_2, \ldots, f_n)$ of size **n** that is extracted from biometric data using a particular feature extractor technique. On the other hand, the identity **x**′ represents a feature vector $(f'_1, f'_2, \ldots, f'_n)$ extracted from a

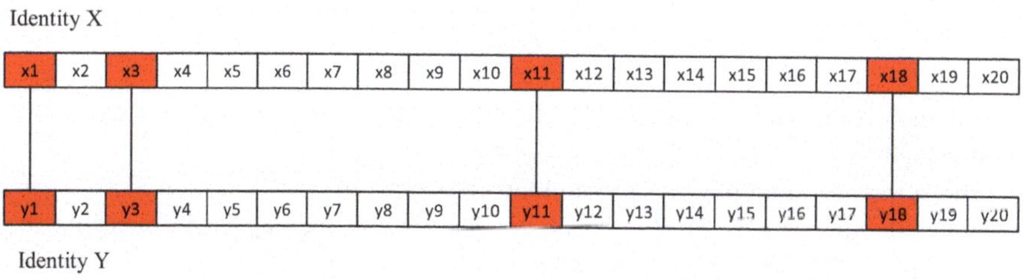

Fig. 18.10 Two identities X and Y with 16 out of 20 overlap

fresh biometric query. A polynomial of degree $(\boldsymbol{d}-\mathbf{1})$ is used to distribute and reconstruct the secret value over set of overlapping feature between **x** and **x**′ in such away $|\boldsymbol{x} \cap \boldsymbol{x}'| \geq \boldsymbol{d}$. Interpolating the subset of *d*-features would be enough to reconstruct the secret value and decrypt the ciphertext (encrypted by public key of identity **x**) using the private key components of identity **x**′.

The second application of the F-IBE is ABE, in which a set descriptive attributes are used to encrypt/decrypt a message. For example, the following attributes {Applied computing department, Staff member, age > =40, exam committee member} can be used during encryption. At the decryption stage, anyone has who d-attributes (e.g., 3 out of 4 attributes) should be able to decrypt the message. If $d=3$, then a person with {Dept = Applied computing, staff member, age = 42} will be able to decrypt the message.

In [27], IBE was developed to build a new scheme of digital signature based on the so-called identity-based signature in which the sender uses the recipient's biometric template to encrypt a message. The recipient then authenticate himself/herself to the PKG by presenting a fresh biometric template. The recipient then will be able to decrypt the message if the distance between the two templates is less than a pre-define threshold. In [28], a new protocol for key exchange using biometric identity-based encryption was proposed to enable parties to securely exchange cryptographic keys even when an adversary is monitoring the communication channel between the parties. The protocol combines biometrics with IBE in order to provide a secure way to access symmetric keys based on the identity of the users in the unsecure environment.

The message is first encrypted by the data owner using a traditional symmetric key before migrating it to a cloud storage. The symmetric key is then encrypted using public biometrics of the users selected by the data owner to decrypt the message based on fuzzy identity-based encryption. Only the selected users will be able to decrypt the message by providing a fresh sample of their biometric data. Such a solution eliminates the needs for a key distribution center in traditional cryptography and gives the data owner the power of fine-grained sharing of encrypted data by controlling who can access his/her data.

As illustrated in Fig. 18.11, the key stages of the biometric-based IBE framework to exchange keys for sharing encrypted data in the cloud environment [28] are:

- Alice encrypts her data using traditional encryption (symmetric/ asymmetric) techniques such as AES or RSA:

$$\mathcal{E}_M \longleftarrow Enc(sk, M)$$

  where *sk* is the encryption key, *M* is the original message, and $\mathcal{E}_M$ is the encrypted message.
- She stores the encrypted data in a cloud environment.
- Now, if Alice wants to allow Bob to decrypt the message, she encrypts the encryption key *sk* using a public key of Bob's unique identity *w'* (i.e., Bob's biometric such as a photo of his face) to produce $\mathcal{E}_{sk}$:

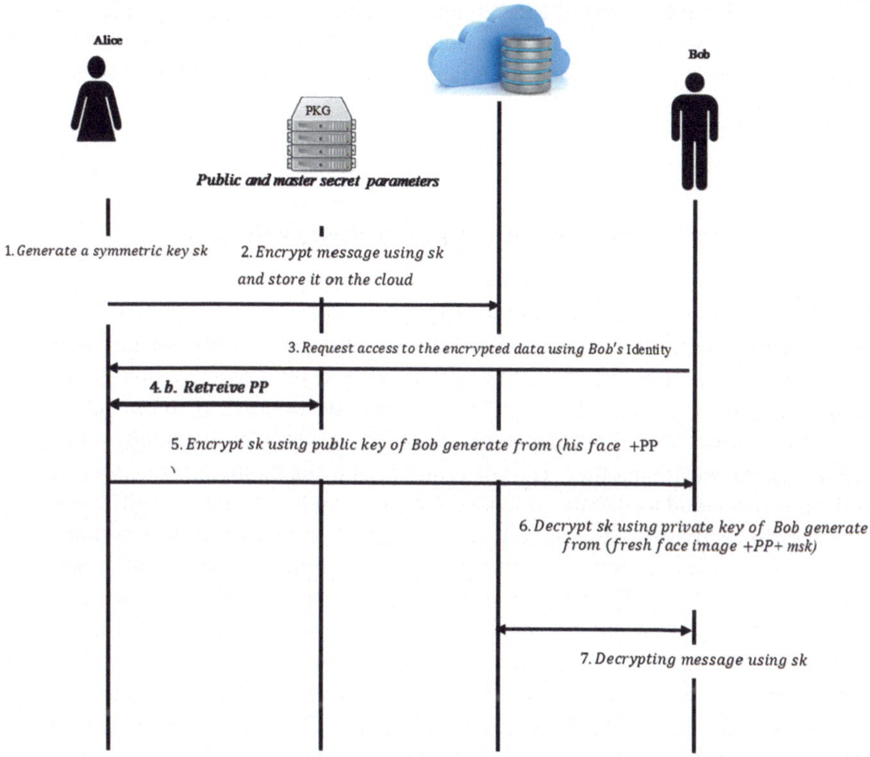

Fig. 18.11 Biometric-based IBE framework to exchange keys for sharing encrypted data in cloud environment. (Adapted from Waters [29])

$$\mathcal{E}_{sk} \longleftarrow Enc(\, pk_{id}, sk)$$

where pk_{id} is the public key of Bob's identity, and $\mathcal{E}_{sk}$ is the encrypted secret key.

- Alice sends the output $\mathcal{E}_{sk}$ to Bob.
- To get the *sk*, Bob needs to provide a fresh biometric sample *w*.
- If and only if the overlap between *w* and *w'* is greater than a threshold value, Bob will retrieve the corresponding private key of his identity and decrypt the ciphertext to get the *sk*.

$$sk \longleftarrow Dec(\, sk_{id}, sk)$$

- Bob brings the encrypted data stored in the cloud environment to his local device and uses *sk* to retrieve the original message/data.

$$M \longleftarrow Dec(\, sk, \mathcal{E}_M)$$

It can be argued that since the face biometric data, for example, is public between parties who know each other, it can be obtained from many resources such as social media resources (e.g., Facebook, Instagram, etc.). Hence, face recognition is an ideal biometric trait for our proposal.

5 Challenges, Solutions, and Future Research Directions

While the use of biometric data provides a great level of convenience by eliminating the need to remember complex passwords or carrying security tokens, it raises serious questions as to whether, and to what extent, the privacy of the biometric owners can be breached. The fact that we are surrounded by more and more biometric sensors that can capture our biometric traits may eventually limit our privacy in one way or another. The convenience of using biometrics for cloud-based authentication could lead to the loss of privacy as a result of being tracked in many of our daily life activities. Moreover, recent research into biometrics shows that more and more personal information can be revealed from biometric raw data and templates such as gender, age, ethnicity, and even some critical health problems including diabetes, vision problems, Alzheimer's disease, and so on [30]. Such personal and sometimes confidential information might be used to discriminate against individuals when it comes to insurance, jobs, border entry enforcement, and many other issues. Therefore, the challenges in biometric research activities have been expanding to include the maintenance of user's privacy in biometric-based systems in addition to the traditional work on enhancing accuracy, scalability, and usability.

In cloud computing, the challenge of incorporating biometrics for robust authentication is not only related to designing a system that is highly accurate and convenient to use, but it should also provide an acceptable level of user privacy. The big question here is: "Do users understand the level of invasion into their privacy as a direct result of improving the security of their data or applications?" The answer is rather tricky because the acceptable level of privacy invasion is not yet clearly defined in the trade-off between security and privacy. Academic efforts have not stopped over the last few years to come up with a common understanding of the security-privacy trade-off at different levels, which aims to propose best practices and guidelines to national and international policymakers. For example, the SurPRISE (surveillance, privacy, and security) project (2012–2015) [31] was an EU-funded project that aimed to examine the trade-off between security and individual privacy and addresses the difficult questions of "Does more security justify less privacy?" and "What is the balance between these two?" in which different security-related technologies and measures were evaluated.

5.1 *Privacy-Preserving Biometric for Cloud Authentication*

Combining challenge/response approach with revocable biometrics generated from UBTs could lead to developing an effective privacy-preserving cloud-based biometric authentication [22].

At the enrolment stage and to address the privacy concerns highlighted earlier, only a revocable biometric features X_{AC} and a hash of a PIN used to generate user-based transformation are stored in the cloud authenticator's database as illustrated in Fig. 18.12. As an example, a client captures an image of his/her face using his/her mobile's camera and key in a four-digit PIN to produce a UBT. It has been shown that such a transformation improves privacy without compromising on accuracy [22].

During the authentication stage, the client captures a fresh biometric sample and applies the same UBT to produce a revocable feature vector X_C, which is then

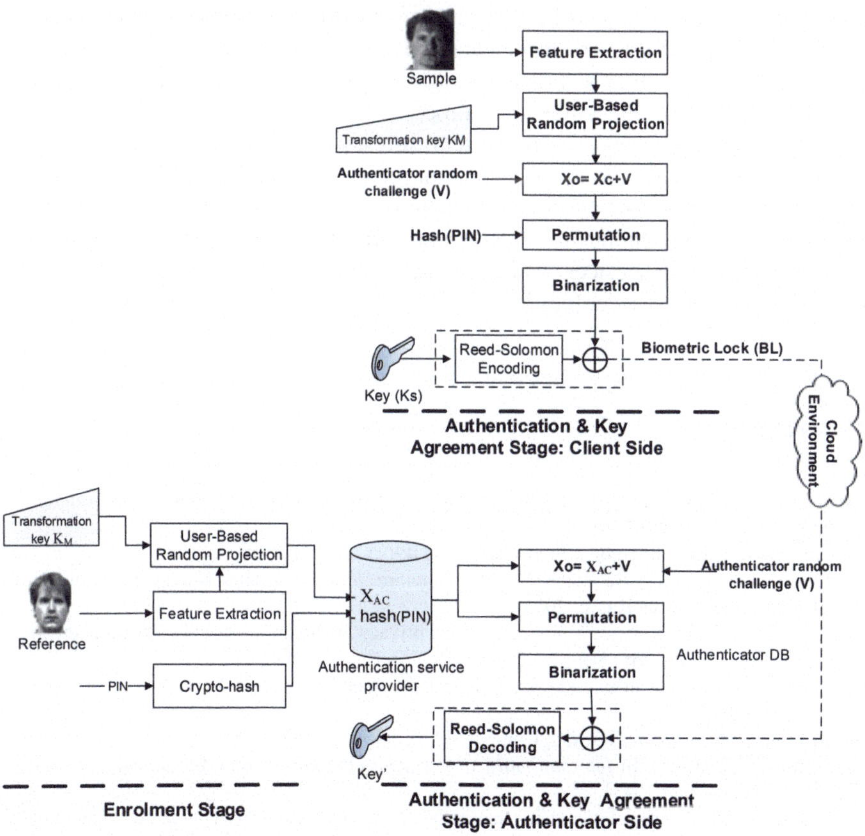

Fig. 18.12 Authentication stage of the privacy-aware authentication scheme for cloud service. (Adapted from Adler [22])

combined with a one-time random vector V generated by the cloud authentication server. After that, the output is shuffled using a permutation key generated from the PIN. Due to the variation between the freshly captured sample and the enrolled template, error-correcting codes such Reed-Solomon can be used. At the cloud authenticator server, if the error-correcting codes were successful, the cloud service produces a key K' that matches the original key if the distance between the biometric template (stored at the cloud authenticator database) and the fresh biometric sample is less than the agreed threshold (i.e., they both belong to one client).

5.2 Conclusion and Future Research Direction

While biometric-based cloud authentication can provide a convenient way to manage access control to cloud resources, biometric data can be misused to track individuals and leak confidential information related to health, gender, ethnicity, and so on. It can be argued that privacy of cloud-based biometric authentication cannot be solely addressed by technical means. Therefore, there is an urgent need for new legislation to enforce privacy-aware measures on cloud service providers related to biometric collection, data processing, and template storage. Although some types of regulation related to users' privacy and data protection do exist in many countries, many of these regulations related to managing the privacy and security of biometric data are either not there yet or insufficient. In the end, technical solutions have to complement legal frameworks to enforce certain measures on cloud authentication services, which would eventually lead to wider public acceptance of biometric-based solutions.

References

1. W. Jansen, T. Grance, et al., Guidelines on security and privacy in public cloud computing. NIST Spec. Publ. **800**(144), 10–11 (2011)
2. S. Dustdar, IEEE computer. Cloud Computing **49**(2), 12–13 (2016)
3. J.W. Rittinghouse, J.F. Ransome, *Cloud Computing: Implementation, Management, and Security* (CRC Press, Boca Raton, 2016)
4. H. Takabi, J.B. Joshi, G.-J. Ahn, Security and privacy challenges in cloud computing environments. IEEE Secur. Priv. **8**(6), 24–31 (2010)
5. Z. Xiao, Y. Xiao, Security and privacy in cloud computing. Comm. Surv. Tutor. IEEE **15**(2), 843–859 (2013)
6. D. Boneh, M. Franklin, SIAM J. Comput. **32**(3), 586–615 (2003)
7. D. Boneh, X. Boyen, In *International Conference on the Theory and Applications of Cryptographic Techniques*, 2004
8. X. Boyen, B. Waters, Anonymous hierarchical identity-based encryption (without random oracles). In *Advances in Cryptology-CRYPTO 2006*, Springer, 2006, pp. 290–307
9. J. Bethencourt, A. Sahai, B. Waters, Ciphertext-policy attribute-based encryption. In *Security and Privacy*, 2007. *SP'07. IEEE Symposium on*, 2007

10. M. Chase, Multi-authority attribute based encryption, in *Theory of Cryptography*, (Springer, Berlin, Heidelberg, 2007), pp. 515–534
11. N. Cao, C. Wang, M. Li, K. Ren, W. Lou, Privacy-preserving multi-keyword ranked search over encrypted cloud data. IEEE Trans. Parallel Distrib. Syst. **25**(1), 222–233 (2014)
12. A. Jain, A.A. Ross, K. Nandakumar, *Introduction to Biometrics* (Springer, Boston, 2011), pp. 1–49
13. A.S. Georghiades, P.N. Belhumeur, D.J. Kriegman, From few to many: generative models for recognition under variable pose and illumination. IEEE Trans. Pattern Anal. Mach. Intell. **23**(6), 643–660 (2001)
14. A.K. Jain, K. Nandakumar, A. Nagar, Biometric template security. EURASIP J. Adv. Signal Process. **113** (2008)
15. K. Nandakumar, Multibiometric systems: fusion strategies and template security, PhD thesis, Michigan State University, 2008
16. A. Adler, Vulnerabilities in biometric encryption systems. In *Proc. of the 5th Int Conference on Audio and Video-Based Biometric Person Authentication*, 2005
17. R. Cappelli, A. Lumini, D. Maio, D. Maltoni, Fingerprint image reconstruction from standard templates. IEEE Trans. Pattern Anal. Mach. Intell. **29**(9) (2007)
18. F. Hao, R. Anderson, J. Daugman, Combining cryptography with biometrics effectively. IEEE Trans. Comput., 1081–1088 (2006)
19. N.K. Ratha, J.H. Connell, R.M. Bolle, An analysis of minutiae matching strength. In *Proc. of Third International Conference on Audio- and Video-Based Biometric Person Authentication*, 2001, pp. 223–228
20. H.S.S.J. Hisham Al-Assam, A lightweight approach for biometric template protection, 2009, pp. 73510P-73510P-12
21. S.J. Hisham Al-Assam, Security evaluation of biometric keys. Comput. Secur. **31**(2), 151–163 (2012)
22. S.J. Hisham Al-Assam, Robust biometric based key agreement and remote mutual authentication. In *The 11th International Conference on Trust, Security and Privacy in Computing and Communications (TrustCom)*, Liverpool, UK, 2012
23. A. Sahai, B. Waters, Fuzzy identity-based encryption. In *Advances in Cryptology–EUROCRYPT 2005*, Springer, 2005, pp. 457–473
24. F. Li, Context-aware attribute-based techniques for data security and access control in mobile cloud environment, 2015
25. A. Shamir, Identity-based cryptosystems and signature schemes. In *Advances in Cryptology*, 1984
26. R. Canetti, S. Halevi, J. Katz, A forward-secure public-key encryption scheme. In *Advances in Cryptology—Eurocrypt 2003*, Springer, 2003, pp. 255–271
27. B. Waters, Efficient identity-based encryption without random oracles. In *Annual International Conference on the Theory and Applications of Cryptographic Techniques*, 2005
28. H.A.-A. Waleed Hassan, Key exchange using biometric identify based encryption for sharing encrypted data in cloud environment. In *SPIE Defense, Security, and Sensing*, 2017
29. W.K. Hassan, H. Al-Assam, Key exchange using biometric identity based encryption for sharing encrypted data in cloud environment. In *Proc. SPIE 10221, Mobile Multimedia/Image Processing, Security, and Applications*, 2017
30. K.E. Forbes, M.F. Shanks, A. Venneri, The evolution of dysgraphia in Alzheimer's disease. Brain Res. Bull. **63**(1), 19–24 (2004)
31. [Online]. Available: http://www.surprise-project.eu/

Chapter 19 Biometric Security and Internet of Things (IoT)

Mohammad S. Obaidat, Soumya Prakash Rana, Tanmoy Maitra, Debasis Giri, and Subrata Dutta

1 Introduction

The Internet of Things (IoT) is the interconnection of different devices of our daily life like cars, refrigerators, mobile phones, smart doors, devices for patient monitoring, or any other monitoring devices. These devices are attached with smart sensor RFID tag, actuator, and internetwork connectivity, which enable the devices to exchange or collect and send data to a server. This type of technology is called Internet of Things. IoT is basically the combination of different fundamental types of technology and has different layer of communication level (see Fig. 19.1). Different level demands require different degrees of security arrangements. For the level

M. S. Obaidat
ECE Department, Nazarbayev University, Astana, Kazakhstan

King Abdullah II School of Information Technology (KASIT), University of Jordan, Amman, Jordan

University of Science and Technology Beijing (USTB), Beijing, China

Fordham University, New York City, NY, USA

S. P. Rana
Division of Electrical and Electronic Engineering, London South Bank University, 103 Borough Rd, London SE1 0AA, United Kingdom

T. Maitra
School of Computer Engineering, KIIT University, Bhubaneswar 751024, Odisha, India

D. Giri (✉)
Department of Information Technology, Maulana Abul Kalam Azad University of Technology, Nadia 741249, West Bengal, India

S. Dutta
Department of Computer Science and Engineering, National Institute of Technology Jamshedpur, Jamshdpur 831014, Jharkhand, India

M. S. Obaidat et al. (eds.), *Biometric-Based Physical and Cybersecurity Systems*,
https://doi.org/10.1007/978-3-319-98734-7_19

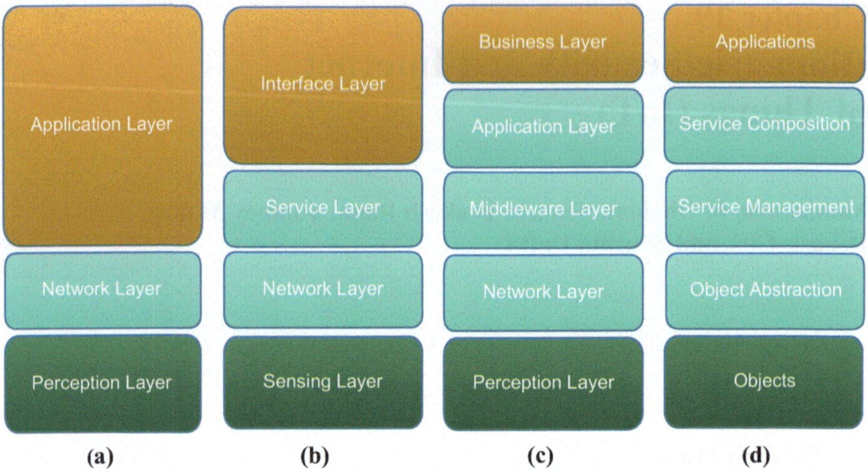

Fig. 19.1 IoT architecture: (**a**) 3-layer, (**b**) 4-layer, (**c**) 5-layer, and (**d**) SOA-based [5]

where the direct human access is needed, biometric security arrangements are highly recommended by researchers. Biometric security arrangements ensure a scalable solution [1–4] for IoT to work against unauthorized access, ID swapping, and manual badge checks.

Biometrics deals with recognition of individuals based on their behavioral or biological characteristics. The human organs, which might be considered as biometric means, should have the following major desirable properties: universality, permanence, uniqueness, performance, collectability, acceptability, and circumvention [1–6].

Applications of IoT

There are several domains where IoT is being successfully implemented. The potentialities of IoT can still be exploited to develop new applications for the benefit of society. It can boost the role of information and communications technology (ICT) so that the quality of our lives can be improved. In the application environments of IoT, smart objects can communicate with each other and represent a context perceived from the environment. The potentiality of IoT can be exploited in many domains like healthcare, transportation systems, environmental monitoring, personal and social, smart city, industrial control, and many more. In this section, we discuss few promising application domains and pointed out their shortcomings.

Smart environment (homes, buildings, office, and plant): Sensors and actuators deployed or attached with household equipment like refrigerator, lighting, and air conditioners can monitor the environment inside a house, plant, or office. The lighting system of a house can change according to the time of the day, like in the evening most of the lights will be on, while they will be off late at night. Based on the reading of a temperature or a smoke detector sensor, a fire alarm can be set off automatically. Such type of application is very helpful for elderly people staying alone at home. Based on the movement of occupants in home, some appliances like doors in room can be opened, lights can be turned on at current room, and water taps/faucets will be open at kitchen. Air conditioners, refrigerators, and washing machines will now be IoT-enabled and controlled over the Internet to save energy. In the near future, a smart malfunctioning refrigerator will send a message to a service man automatically without user's intervention. Industrial automation is improved by deploying RFID tags with products. Production process is controlled to ensure quality of product by getting different parameter values from sensors [1, 2].

IBM has launched Smart Home solution [7], better known as "Stratecast," to provide services to users allowing seamless communication among various smart devices in the house, like medical devices, computers, mobiles, TVs, lighting, security, and sound system. IBM is collaborating with Verizon as a communication service provider (CSP) and Philips as a device vendor to implement the architecture. Siemens, Cisco, Xerox, Microsoft, MIT, and many others are working in this domain. They have set nearly 20 home labs using more than 30 home appliances, 5 network protocols, and 3 artificial intelligence (AI) techniques [8]. The Intel smart home platform supports recognition of family members by voice or face and personalizes the home. Intel provides IoT solutions for smarter building to support personalization by controlling over the office/living environment, mobility by enabling managers to monitor property remotely, and sustainability and efficiency in terms of saving energy, water, and other building resources.

Healthcare: Another impact area is healthcare [9, 10], where IoT can be used in tracking of people (i.e., patients and staff) or objects, recognition and verification of people, and automatic data collection [11]. Real-time tracking of person or objects in motion, in the case of patient-flow monitoring, helps to manage staffing and improve the workflow in hospitals. Identification and authentication of staff secure the patients and help to avoid mistakes like wrong drug/dose/time. Automatic collection of clinical data must be enabled to enrich medical inventory. Real-time monitoring of patients is possible by attaching different sensors to measure body temperature, blood pressure, and heart response. These IoT-enabled sensors can detect abnormality locally and, in no time, send this information to the physicist. A community health service [12] has already been proposed with three-layer architecture to monitor the healthcare remotely from hospital. This application can prevent hospitals from overcrowding. Analyzing patient's data doctors can send advice through text/video/voice/email means. The limitation of such application is that there is no security and privacy of patient's data. Timely delivery of accurate response to critical patient is another challenge in this domain. The IoT-based application powered by artificial intelligence, called ambient assisted living (AAL) [13, 14], can assist elderly individual in his or her residence in a convenient and safe manner. In [15], a modular architecture, security, control, and communication are included. The m-IoT promotes a mobile computing platform consisting of sensors and communication technologies. In [16], authors describe eHealth and IoT policies and regulations in accessing healthcare services. The potentiality of m-IoT has been tested in noninvasive sensing of glucose level [17], body temperature monitoring [18], and blood pressure monitoring [13] for ambient assisted living. Pulse oximetry is noninvasive nonstop monitoring of blood oxygen saturation. IoT-based pulse oximetry is an integrated approach used in CoAP-based healthcare services [19] and remote patient monitoring [20]. However, implementing IoT technology in this domain has many challenges [17]. Context awareness and m-IoT ecosystem are two prominent challenges in this area [17]. IoT is applied in electrocardiogram (ECG) monitoring to provide maximum information in real time [21]. Research efforts have been done toward full automation of smart wheelchair based on IoT [22] for disabled people. Intel developed connected wheelchair which is connected with data-driven machines [23]. Awareness around children's health [24] in monitoring emotional, behavioral, and mental health is exercised by a specialized IoT service called Children Health Information (CHI) where an interactive totem may be positioned in a pediatric ward offering CHI services for educating, entertaining, and empowering hospital kids. The work in [25] proposes an IoT-based m-health service that can be used to develop a healthy nutritional habit among children. The potentiality of medical semantics and ontologies in providing huge medical information and knowledge influences IoT-based healthcare applications. A medical monitoring service called Semantic Medical Access (SMA) is proposed in [26] that is based on IoT sensors. To analyze large medical data stored in cloud medical rule engines may be employed in healthcare.

Despite its immense benefit to wellness of people, it throws many challenges. As health data is highly sensitive, if misused can deteriorate relation and destroy reputation among individuals. Data centers must be capable of handling of streaming and digitizing large volume health data. Without proper security of such data, there is a risk of cyber-attack. But implementing security measures in healthcare is constrained by the computational, energy, and memory limitation. As healthcare devices are mobile in nature, developing a mobility-compliant security algorithm is a big challenge. Moreover, standard rules and regulations for compatible interfaces and protocols are not followed by devices made by different vendors. It raises the interoperability issue in IoT. Immediate effort is needed for the standardization of IoT-based healthcare services. In [16], authors describe eHealth and IoT policies with regulations in accessing healthcare services. Many countries like Japan, France, China, and India have already announced eHealth policies.

Smart cities: Smart city is a cyber-physical ecosystem emerging by deploying advanced communication facility along with the novel services over a city [1, 27, 28]. Use of IoT in smart city optimizes usage of physical city infrastructure such as power grid, road networks, parking space, etc. and improves the quality of life of its citizens in cities [2, 29, 30] like Glasgow, Barcelona, Masdar, etc. It can be used to monitor the traffic in cities or highways and divert traffic accordingly in order to avoid congestion. Smart parking facility is made available to smart cars through RFID and sensors technology to find out currently available parking space nearby in city [31]. Using IoT sensors can send air pollution data such as amount of carbon dioxide, etc. to an agency. A Smart City Platform is developed to smart waste management in a European city [32]. Sensor data can be used to detect violator of traffic rules and to analyze the accident scenarios.

Another important application of IoT that can be provided to citizens is water network monitoring and quality assurance of drinking water. To ensure high quality of water, sensors measuring critical water parameters are placed at important locations around the city. This technique eventually detects accidental contamination of drinking water, rain water, and sewage disposal. Smart grid and smart metering are being implemented around the world to ensure efficient energy consumption [33]. Using smart meters, energy consumption is monitored at every electricity point in the house and the information used to modify the way the electricity is consumed. Similarly, monitoring citywide electricity consumption pattern helps to balance load within the grid, thereby ensuring good quality of service.

The sensors used in smart cities are vulnerable to physical as well as cyber-attacks. What will happen if a smart meter is tampered with? Associating authentication to a smart meter to avoid anonymous meter readings is difficult now. The data communication in smart city must be secured against such attacks.

Transportation and logistics: IoT is used heavily in intelligent transport systems and autonomous vehicles. Trains, buses, and cars can now be equipped with sensors, actuators, and processing power and can provide necessary information to passengers, drivers, or agencies for better monitoring and management in order to safely navigate passengers. Transported goods attached with RFID tags can be monitored and the status of delivery could be enquired through IoT. RFID and NFC provide wide market opportunity in the domain of supply chain management, transportation, inventory management, and mobile ticketing. Intel in collaboration with BMW group and computer vision leader, Mobileye, is working toward realizing the next-generation platform for automated driving [34]. Cars are embedded with sensors, which can generate 360 degree of data. Intel processors transform the data into actionable insight to assist and/or automated driving. By 2020, self-driving is expected to have traffic signal recognition, emergency braking, pedestrian detection, park assistance, cross traffic alert, and many more. Providing security for the connected cars and instant driving assistance is essential for safer, more productive and enjoyable travel. An intelligent monitoring system for refrigerator trucks is proposed in [35] to monitor temperature and humidity inside it using RFID, sensors, and wireless communication technology.

Personal and social: *Social Internet of Things* (SIoT), introduced by Atzori et al. [36], describes a world where things around the human can intelligently sense and can form a network. SIoT helps individual to interact with other and maintain social relationship [37]. Twitter and Facebook are web portals through which people remain in touch with friends by getting and posting real-time updates [38]. Existing privacy and protection techniques in social networks may be imposed in IoT to improve the security of IoT. Tracking of IoT-enabled objects against losses or thefts is possible by developing applications. For example, smart objects such as laptops and mobiles will send SMS automatically to their owners on reaching a new location or unauthorized access. Many companies have implemented SIoT for their products for collecting data from users. These data are communicated over the Internet to social networks of people and devices who can respond to a problem, deliver a service, or sell a solution [39]. In the current scenario, implementing human-to-thing interactions is a challenge to achieve the complete vision of SIoT. Again, SIoT must be intelligent enough to starting, updating, and terminating the objects' relationships in SIoT. In SIoT, research must address issues like interoperability, device management, security and privacy, fault tolerance, and heterogeneity.

Agriculture: Agriculture is one of the emerging fields for implementing IoT [40–42]. If IoT can be implemented in third-world countries like India, Bangladesh, and Brazil where agriculture is the main profession, then human effort can be used optimally. In [42], authors proposed a framework called AgriTech toward agricultural automation. Deploying sensors like humidity and nutrients and getting data from the field of agriculture farmers can save their time. The excess human effort can be used for the industrialization in these countries. The potential impact of IoT may be exploited in agricultural automation to optimize utilization of water, fertilizers, and insecticides.

However, implementing AgriTech in third-world countries is challenging due to initial setup cost. Sensors deployed in the field are vulnerable to physical attack. Again, improper deployment of sensors may result in unwanted information from field not belonging to the farmer.

Pharmaceutical industry: Safety and security of pharmaceutical products are of utmost important. In this view, smart labels are attached to drugs for monitoring their status while being transported and stored. Patients are directly benefitted from smart labels on drugs by knowing expiry, authentication, and dosages of medicines. Counterfeiting [43] in this area can be stopped by IoT. Smart medicine cabinet also helps patients to track timely delivery of medicines. In [44], a pharmaceutical intelligent system built on IoT has been employed to inspect drugs in order to perceive drug reaction, injurious effects of excipients, problems associated with liver and renal defect. It helps and assists physicians toward clinical decisions and prescription. Here, near-field communication (NFC) and barcode verification techniques are combined in different gadgets. Along with this drug, identity is matched with intelligent information system to detect the suitability of a drug for a patient.

In any IoT system, huge amount of data transactions will take place. Some decisions and control signals or suggestions will be sent to the application. For ensuring security of those data, messages of suggestion and control signal, a strong and efficient authentication protocol will be needed.

The organization of this chapter as follows: Sect. 19.2 describes the traditional different types of biometric used in IoT applications. Biometric security system and its benefits to use in different IoT applications are discussed in Sect. 19.3. Different techniques to extract features from biometrics are given in Sect. 19.4. Section 19.5 provides the brief description of some biometric-based security protocols for IoT application. Finally, we conclude this chapter in Sect. 19.6.

2 Types of IoT Security

IoT security is the domain, which worries researchers and users due to the vulnerable attack on "things" or connected devices and network. The maximum connection in IoT is derived from the devices, embedded sensor systems (ESS) employed in industrial communication, building computerization system, vehicle communication, and wearable gadgets. Therefore, devices, which are connected in this giant network, also raise the scope of potential attack for hackers and other cyber criminals. There are five types of attack that occur in IoT internetworking systems [45–47]:

- Botnet is a network of systems, which take control remotely. Command-and-control servers (C&C server) is used to regulate the system and used by criminals for stealing private information, exploiting online-banking data, and phishing emails.
- Man-in-the-middle attack is a notion where the invader or hacker interrupts and breaks communication link between two discrete systems. The invader covertly interrupts and sends fake messages, but the sender and receiver believe that they are communicating via authentic message with each other.
- Data and identity theft occurs in case of careless handling of Internet-connected devices such as mobile phones, kindles, smart watches, etc. The goal of identity theft is the accumulation of data, which can say a lot about a person. The information accessible on the Internet including social media and smart devices gives an overall clue of personal identity.
- Social engineering is an action of influencing people, so they hand over confidential information like device or email passwords or bank details. It also includes the installation of mischievous software that can open the door to access personal data. This type of threat can be done by phishing emails or redirecting to websites like banking or shopping sites that look genuine but always ask or influence to enter secret information [3].
- Denial-of-service (DoS) attack occurs when a service that usually works is unobtainable. A large number of systems maliciously attack on a particular target in case of distributed denial-of-service (DDoS) attack. This is done through botnet, where many devices are programmed to demand a service at the same time. This type of attack does not try to steal information or leads to security loss which affects reputation of a company that can cost a lot of time, money, and reputation.
- A report by Hewlett Packard shows that 70% of the normally used IoT components are serious vulnerabilities [48]. These components have weaknesses due to absence of transport encryption, insecure website interface, poor software protection, and inadequate authorization. On average, each component contains almost 25 risks of compromising the home network. The properties, such as confidentiality, integrity, authentication, authorization, non-repudiation, availability, and privacy, must be guaranteed [49, 50] to ensure the security of IoT components. There are mainly four types of security requirements like secure bootstrapping of objects and transmission of data, secure authentication and identification, security on IoT data, and secure access to data by authentic users [1, 3].

The probable solution to certify security of things is identification technology (IT), which bids the mapping of unique identifier or UID, to an entity for making it unambiguous. UIDs may be made as sole measure such that the combination of their values is exclusive. In IoT, the "things" have a digital identity (demonstrated by a unique identifier) that is identified with a digital name, and the relationships among "things" can be specified in the digital field [51]. There are two categories of IoT security technique to assign a UID to an entity or things, traditional and biometric-based techniques. Traditional IoT security is either knowledge based (like password, PIN, or any type of personal information which can be used for point-to-point protocol (PPP) to authenticate a user) or object based (like smart cards are implemented to deliver user verification by data storage). Smart cards offer a robust security confirmation for single sign-on (SSO) within big establishments, where biometric IoT security states the measurement associated with human characteristics. IoT biometric validation or real-time verification is employed to recognize entities in groups that are under observation. As this chapter concentrates on biometric security in IoT, therefore it is explained in depth in the rest of the section.

2.1 *Biometric Security in IoT*

Biometric recognition or biometric security considers two grounds about human body characteristics, distinctiveness and permanence [3, 52]. Some of the most popular physiological traits, which are used in IoT biometric security, are fingerprint, face, iris, hand geometry, gait, DNA (deoxyribonucleic acid), etc. The selection of a biometric generally depends upon the necessities of the authentication application. For example, voice biometrics is suitable in mobile security matters because the device which senses vocal sound is previously embedded in the mobile phone, and the finest part of the IoT biometric authentication is that it can identify the person who is not registered in the system, but still trying to get the access. There are two types of biometric modalities, physiological and behavioral [3], and brief description is made in the next sections.

2.2 *Physiological*

Physiological features are based on direct measurements of a part of the human body, e.g., face, fingerprint, iris, and hand geometry recognition schemes belong to this category (see Fig. 19.2).

1. *Face*: Face recognition algorithms generally use relationship among the locations of facial features such as eyes, nose, lips, chin, and the global appearance of a face. State of art on face recognition technology can be found in [53, 54]. There are some significant factors like brightness, motion, makeover, obstruction, and

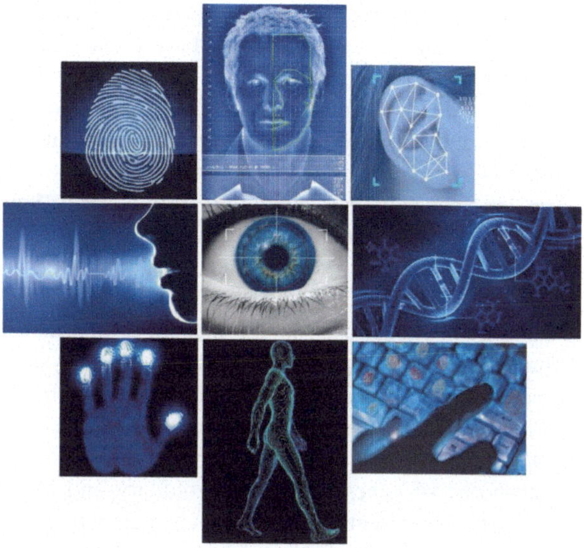

Fig. 19.2 Set of popular physiological and behavioral IoT biometrics

posture disparities, which disturb the execution of a face identification algorithm. Face is a widely accepted biometric and has a good level of accuracy in different environments.

2. *Fingerprint*: Fingerprint based recognition is the most successful and widespread method for person authentication in IoT. It contains a unique texture pattern, which is made of ridges and valleys for a person. These ridges are categorized using some points, known as "minutiae." Therefore, the spatial distributions are proved to be unique for a person [55]. These "minutiae" points are used to match two different person's fingerprints. This biometric has received larger consideration since forensic departments of many countries employed Automated Fingerprint Identification Systems (AFIS) for their purpose. Beside this many civil and commercial application also use fingerprint for authentication.
3. *Iris*: Iris is a colored loop around the pupil holds complex pattern in human eye, and it is scientifically proved that it contains unique characteristics for every human being. It has some individual characteristics like stripes, pits, and furrows, which are being considered for proof of identity. The work of Daugman [56] showed the working principle of iris recognition, matching speed, and accuracy (is very high) of this biometric consideration. The pattern of the texture is steady and idiosyncratic here [56]. Some large-scale systems integrate iris authentication as a part of their identification procedures. It is also noticed that lack of legacy for iris pattern databases may be a challenge for several government applications.
4. *Palm Print*: Palm print is another popular and well-accepted biometric which is used in IoT security systems. It contains distinct ridges and flexion creases like fingerprint [57, 58]. Mainly, forensics are using this scheme, and research shows 30% of palm print samples are being collected from criminal cases (like knifes,

guns, etc.). Minutiae and creases are considered for matching and finding out the required palm print or person. Many palm print samples are available in the databases, which are gathered by forensics. Generally, the resolutions of these samples are low around (75 dpi). From the research point of view, it is a challenge to extract the texture features for feeding into intelligent expert systems.

5. *Hand Geometry*: It is demanded that identification of persons can be made based on the shape of their hands [59, 60]. Identity verification by hand geometry employs low-resolution hand pattern images to extract geometrical features such as finger length, width, thickness, perimeter, and finger area. The accuracy of person authentication by hand geometry is quite limited. Due to this reason hand geometry is used in 1:1 matching for low safety access mechanism and measurement of attendance like areas. The physical size of the hand geometry measurement systems is large; therefore, it is hard to embed those systems in existing security systems for IoT biometric-based security schemes.
6. *DNA (Deoxyribonucleic Acid)*: DNA is present in each cell of a human body and composed of genes. This is a hereditary and highly stable material, which is utilized to represent physiological characteristic and identify person [61]. Usually DNA is unique for each person except identical twins. Patterns of DNA are developed from different body parts like the hair, finger nails, saliva, and blood. Forensics and law enforcement agencies first make a unique profile of a DNA using some intermediate steps, and then the matching process is conducted. This process is very expensive and time-consuming. Also, DNA profiles may get contaminated if they are not done in an ideal environment. Therefore, it is challenge to make an automated IoT security using DNA, and not appropriate for outsized biometric implementation for public usage.
7. *Hand Veins*: The configuration of blood vessels concealed under the skin is distinct in persons, even among the identical twins and constant over long period of time. The veins present in hands (e.g., palm, finger, and palm dorsal surface) are acquired by near-infrared illumination and employed for person authentication [62]. The pattern of veins is steady for adult age but changes after that because of bone and muscle strength and also sometimes due to diseases. Till now there is no known large-scale vascular biometric system. Therefore, it is challenging to make hand vein recognition-based IoT security because of cost of the system and absence of large-scale studies on vein uniqueness and steadiness. Beside this, these systems are touchless, which repeatedly pleas to the user.

2.3 Behavioral

Behavioral features are one kind of indirect human characteristics measurement through the feature extraction and machine learning. Some popular behavioral biometrics are signature, key stroke dynamics, gait, and voice [63] which are elaborated later (shown Fig. 19.2).

1. *Gait*: Human walking pattern is considered as the gait, which is almost unique for each person and has the potential to demonstrate a person based on his or her gait biometric. Wearable and non-wearable sensors are used to capture the gait data, and later statistical features are extracted to explain the complex gait dynamics for a person [64]. Then a machine learning model is trained to recognize a person using these features. There are several important parameters like velocity, linear distance between two successive placements of the same foot, linear distance between the placements of both feet, number of steps per time unit, linear distance between two equivalent points of both feet, and direction of the foot during the step, which are used to describe a gait pattern. Generally, gait biometric is used for medical and healthcare applications, but these days it is being implemented for IoT biometric applications also.
2. *Signature*: This is another behavioral biometric, which is used every day for business transactions. Academics and industry have made several attempts for concrete signature recognition model which is not successful yet. These systems capture the characteristics of signature by measuring the pressure-sensitive pen pad. The shape, speed, acceleration, and speed of strokes are captured from the real-time signing [65]. These features are learned using machine learning algorithms to improve the signature recognition and verification performance along with circumvent signature forgeries. However, very few automatic signature verification systems have been deployed.
3. *Voice*: Speech or voice recognition schemes find the persons based on their vocalized words [66]. Human voice includes a mixture of behavioral and physiological characteristics. The physiological factor of voice relies on the shape and size of vocal tracts, nasal cavities, lips, and mouth. The behavioral elements are established by the movements of the jaws, lips, tongue, larynx, and velum and can change with the person's age. Another side, duration, intensity, pitch information, and quality are the spectral information to train a machine learning system, which would be able to verify a person's identity by the vocal sound. Voice biometrics is mainly used for verification purpose.
4. *Keystroke*: It states the skills are established for automatic person authentication based on the learning of typing patterns. These technologies present numerous problems related to demonstrating and matching dynamic orders with high intraclass variability (e.g., samples from the same user show huge differences) and variable performance (e.g., human behavior is strongly user-dependent and varies significantly between subjects) [1, 2, 63, 67].

3 Biometric Security and Internet of Things (IoT)

This section discusses the biometric security system and its benefits in different IoT applications.

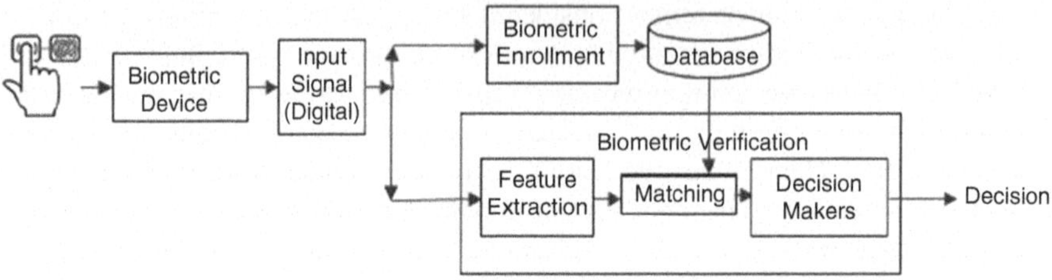

Fig. 19.3 Working procedure of biometric security system: a part of IoT system

Table 19.1 Comparison of the popular biometric technologies [68]

Biometric type	Accuracy	Ease of use	Acceptability of users	Implementation viewpoint	Cost
Fingerprint	⇑	⍰	⇓	⇑	⍰
Hand geometry	⍰	⇑	⍰	⍰	⇑
Voice	⍰	⇑	⇑	⇑	⇓
Retina	⇑	⇓	⇓	⇓	⍰
Iris	⍰	⍰	⍰	⍰	⇑
Hand writing	⍰	⍰	⇑	⇓	⍰
Face	⇓	⇑	⇑	⍰	⇓

⇑, high; ⇓, low; and ⍰, medium

3.1 *Biometric Security System*

Matching problem of the identity can be classified into two different types of problems with different complexity: (a) authentication and (b) recognition or identification. Verification proves or disproves a claim of a person, whereas identification process identifies. Figure 19.3 depicts the scenario of biometrics used in the security system.

Table 19.1 shows the performance comparison of different biometric organ of human being. Table 19.1 is given to compare different types of biometrics like fingerprint, hand geometry, voice, retina, iris, signature, face, etc. However, though the accuracy and ease of implementation for fingerprint are higher than the signature of people, still signature (biometric type) is more popular than fingerprint.

3.2 *Comparison of Biometric Security Over Other Conventional Security System*

Biometric-based security is far better than password-based/PIN-based security. Biometric-based security has many advantages over conventional password-based security system which are discussed below:

1. *Accurate information*: Biometric-based security can get accurate more secure information than PIN-/password-based security systems. No need to remember the password/PIN by the user or no evil person can break the security by duplicating, guessing, or hacking the password from the server of the system.
2. *Easy and safe for use*: Biometrics is very easy and safe to use. Hackers cannot possess the biometric information of the legitimate users.
3. *Accountability*: A person who is using a system using biometrics cannot deny the activity, which has been done by him in the future. Therefore, it can be said that the accountability of a system will also increase from the user end for any kinds of malfunction and misuse of the system.
4. *Security of biometric information*: Biometrics cannot be guessed or stolen. Therefore, it may provide a long-term security solution. However, in the password systems, a sequence of numbers, symbols, and letters are used to build a password, which is very tough to remember. Furthermore, in the token-based system, token can be stolen easily or lost. Therefore, both the password- and token-based systems have a high risk to use as the secret information is being shared or disclosed. In such case, verifier may not be sure about the legal user. But, these would not be the case with biometric characteristics, and thus biometric-based systems are free from the problem of fraud, sharing, and duplication.
5. *Scalability*: Biometric-based systems provide flexibility and scalability from the users' viewpoints. Users can use higher versions of sensors and security systems based on their needs. User can use their characteristics, which are not very discriminative. However, to get the higher level of security in a large-scale database of a user with higher identification accuracy, this kind of systems can be used with more discriminable features. This is because chances of collision of hash value of biometrics are lower than in conventional security systems.
6. *Time-saving*: Biometric identification is very fast to execute, which is another advantage over other traditional security techniques. A person can be verified (i.e., rejected or accepted) in a fraction of seconds. Moreover, the use of this technology can only be beneficial to the office revenue by increasing productivity and reducing costs by removing fraud and wastage of time in the verification process.
7. *User-friendly systems*: Users can install the biometric systems easily into their e-devices, and then, they can do their job uniformly, quickly, and reliably. To understand the operational functionality of the biometric system, minimum amount of training is needed for the users, and there is no need for expensive password administrators. New problems are arising with the aging of the population due to increased life expectancy and declining birth rate. Nowadays, there are 600 million aged people in the world. The number will be doubled in 2025 and will reach $\approx$ 2000 million in 2050. Therefore, user friendly system is very much mandatory for the aged people who will use IoT system as end user.
8. *Convenience*: It is a convenient security mechanism because people do not need to remember passwords, as well as do not need to carry secret documents or identity cards for verification.

9. *Versatility*: Nowadays, different types of biometric scanners are available in the market, and they are suitable to use in various applications. Many organizations and companies use the biometric devices at the security check points like doorways, entrances, and exits. Beside this, users can build the most out of the biometric mechanism to obtain the knowledge about accessibility on the systems. Companies also use such biometric scanners to monitor the entry or exit time of an employee and their attendance. In case of the remote patient monitoring systems, the biometric identity can be used to send an emergency message to a rescue team of remote patient monitoring system. A soldier can ask help to get rescue from danger situation by pressing a button, which is basically a transmitter along with biometric identification system. Thus, it can be said that the biometric security systems have versatile applications in different IoT environments.
10. *Return on investment*: This is definitely high, because human can avoid fraud including "buddy punching," besides lowering payroll costs, accurate computation of work hours, and reduced management time. While the security is enhanced, users can also easily apply consistent policies and procedures at the same time. More thinking is required about the initial cost of the biometric system. Industries can benefit from biometric systems to a great extent. Rather than remembering the password, biometric systems offer unique biometric information for individual so that users can get their accessibility to obtain services after verification. Therefore, from business point of view, biometric technology is very much profitable.
11. (Table 19.2) demonstates the advantages and disadvantages of most common authentication procedures.

3.3 Benefits in Using Biometric-Based Security Schemes for IoT Applications

Biometric-based security can be used in every application area of IoT. Biometrics is used in the application level where man and machine interaction is required. Figure 19.4 describes the security system needed in different application areas. Security systems can be usable in smart home systems. A person can use smart lock system in the door using biometric locking system. An old person can report his/her health condition by logging in to the IoT healthcare system using biometric identification/verification system. In the IoT of transportation systems, the system can verify the identity of an end user while parking a car or paying a traffic fine, etc. Traffic police can verify whether a car belongs to a driver or not using a biometric verification means. Before using the IoT applications, which are related to national project or application area like smart grid [68] or defense applications, an end user needs to verify his/her identity using biometric security to increase the reliability of the system. Biometric security can be useful in case of IoT healthcare systems [69]. All the medical persons need to pass biometric verification before prescribing

Table 19.2 Comparison of three main authentication approaches [69]

Authentication procedure	Advantages	Drawbacks
Handheld tokens (card, ID, and passport)	A new one can be generated People can get same facility to the different country by accessing the tokens as it seems to be standard	It can be stolen A fake or duplicate can be generated It can be shared A user can register him-/herself with several identities
Knowledge based (password and PIN)	It is a manageable and has low cost to fabricate For any problem, it can be replaced with new one easily	It can be guessed or cracked Long or complicated password is hard to remember It can be distributed A user can be registered with different identities
Biometrics	It cannot be guessed, lost, forgotten, stolen, and shared One person with multiple identities can be verified easily It provides a greater degree of security than others	For any problem (oily skin for fingerprint), replacement is not possible It is impossible to replace if biometric data of a person is stolen

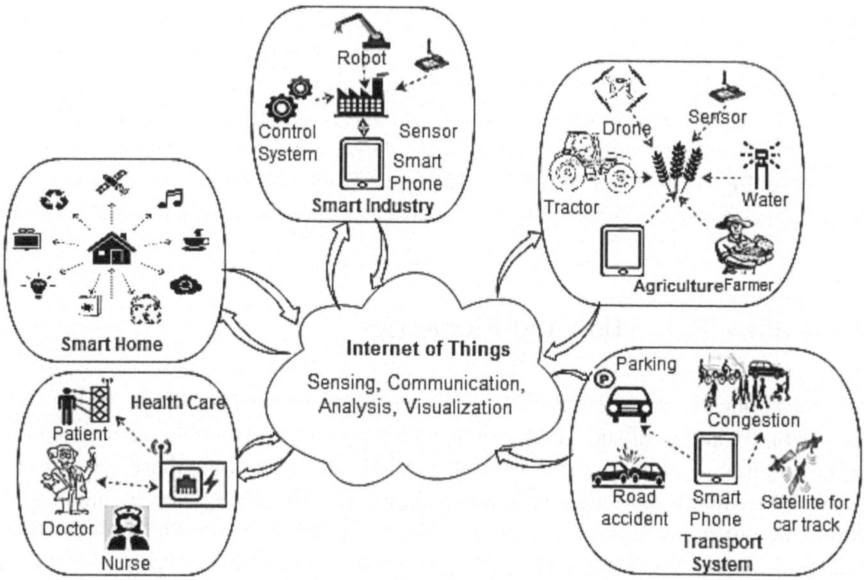

Fig. 19.4 End users and application areas based on data in IoT

or accessing the data of a patient. If any person faces any accident then biometric information is able to help identity and get his medical history. Presently, researchers are trying to introduce the IoT in agricultural systems [70]. Third-world countries like India, Indonesia, Bangladesh, etc. are the largest producers of food grain, but farmers in these countries have very low literacy rate. Therefore, it is mandatory that the IoT in agricultural systems should be easy to use. Here, biometric security systems are very much easier to use. Therefore, biometric-based systems are very useful in agricultural IoT systems.

The IoT enables the objects so that they can participate in daily activities. However, in complex system, controlling such objects along with their security is a very challenging task. Generally, IoT is a scope where persons relate to the technology built on smart devices. Relations of four IoT components, i.e., individual, intelligent devices, high-tech ecosystem, and method, provide a complete and perceptive characteristic for IoT security. In this regard, secrecy on human information is required when they interact with the smart technological ecosystem. Similarly, during the communication with the control processes, safety on data should be guaranteed. Methods should safeguard their dependability and preserve the aims. Beside this, due to the increasing autonomy of objects, IoT security based on cognitive and systemic techniques is going toward a greater autonomy so that the different security threats can be protected. Here, we will discuss the role of each actor, i.e., person, intelligent and smart devices, technological ecosystem, and process, in IoT security to highlight the research issues.

According to [71], the major component of IoT can be categorized in four different nodes. There are four numbers of fundamental node, and they are *process*, *person*, *technological ecosystem*, and *intelligent object*. In the imaginary, those nodes form a tetrahedron shape due to the nonlinear relationship among them. Four planes of that tetrahedron represent different scenarios of security management system of IoT like safety, security, access, and cyber security (see Fig. 19.5).

Figure 19.6 is the representation of each plane of Fig. 19.5 in two-dimensional area. From Fig. 19.5, it can be said that safety, security, and access plane consist of a node named as person. According to [72], the edge of each side is named as tension.

4 Feature Extraction and Biometrics

Several applications are employed to ensure the identity of a “thing” in the IoT paradigm. Examples of such applications include secure and reliable access to smart buildings, mobile phones, cloud databases, ATMs, etc. Therefore, these systems are vulnerable without a robust verification algorithm. The emergence of biometric-based identification or verification systems has pointed out to the problem that affects traditional verification methods by using physiological or behavioral features related to the person under consideration. IoT biometric systems use hand geometry, fingerprints, retina, iris, face, signature, and voice, among others, to validate a person’s uniqueness. A simple IoT biometric system contains four components: (a) sensing or acquisition, (b) feature extraction, (c) pattern matching, and (c) decision-making [73]. A typical biometric authentication-enabled IoT architecture is shown in Fig. 19.7, where fingerprint biometric-based verification is considered to get access control over the “things.”

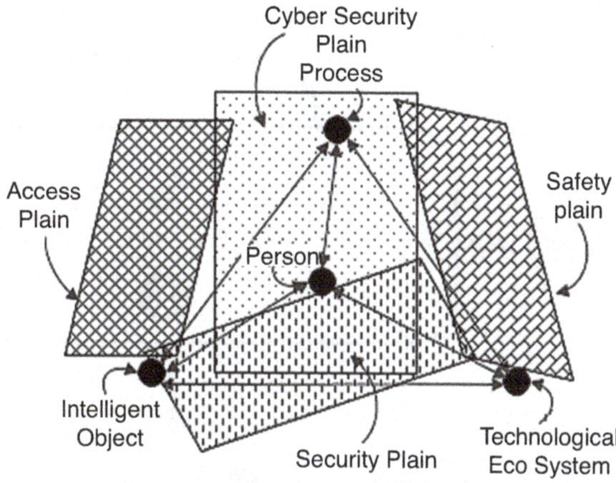

Fig. 19.5 Tetrahedron shape of security system in IoT [72]

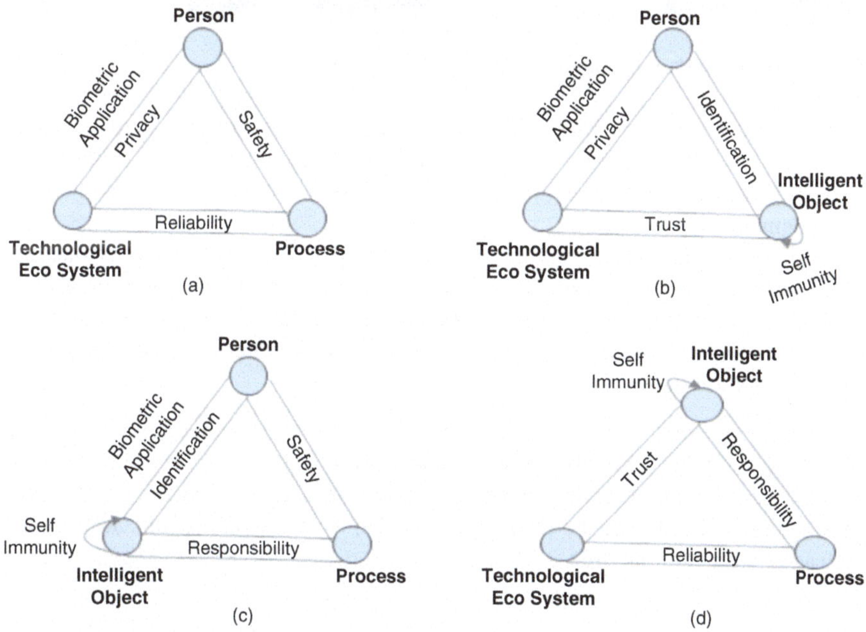

Fig. 19.6 Projections of the 3D pyramid on each of the planes: (**a**) safety plane, (**b**) security, (**c**) access, and (**d**) cyber security [72]

According to Fig. 19.3, a brief description on the components is included here, and feature extraction unit is described later.

- *Sensor or acquisition module*: It obtains the biometrics of an individual. As an example, fingerprint sensor captures the fingerprint impressions of a user (see Fig. 19.3).

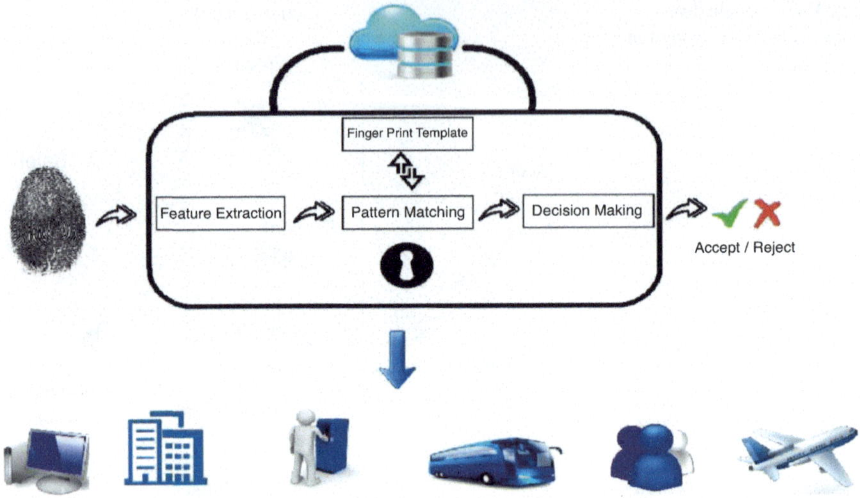

Fig. 19.7 Framework of IoT biometric-based security system

- *Feature extraction module*: The assimilated data is managed to extract feature values. As an example, the positional and orientation-related features are extracted from a fingerprint image in the feature extraction module of a fingerprint recognition system.
- *Matching module*: The feature values are compared and calculated against the template by making matching. An example, here the matching score is calculated between query and template in this module, which helps in the next section.
- *Decision-making module*: The requested identity is either rejected or accepted here based on the similarity score.

The effectiveness of a biometric scheme is assessed by taking account the true positive rate (TPR) and false positive rate (FPR). These two measurements are graphed together in receiver operating characteristic (ROC) curve, which plots TPR against FPR. As per the interest of this chapter, the next sections are focused on feature extraction module.

4.1 Different Techniques to Extract Biometric Features

The framework of IoT biometric security shows the fingerprint template is fed into a feature extractor module, which transforms the template image into a set of features or feature vector. These features contain distinct properties of the input patterns that assist in discriminating between the classes of input patterns. The idea of feature extraction and the selection of features in a computer vision problem are highly reliant on the specific problem at hand. There are two types of feature extraction methods in computer vision: low-level and high-level feature extractors. Low-level feature extractor transforms the visual content of a biometric template by associating features such as color, gradient orientation, texture, and shape with the content of the input template. For example, link an extracted color such as blue with the sea or sky, white with a car or dress, red with an apple, and so on, whereas high-level algorithms are typically associated with the machine learning field. These procedures are concerned with the transformation or classification of a scene. Multiple methodologies are presented for each of these visual features, and each of them symbolizes the feature from a different perception. Some popular feature extraction techniques are described in the next section.

4.1.1 Fourier Transform (FT)

Fourier transform is a concept, which states that any function can be expressed as the integral of sines and/or cosines multiplied by a weighting function. The function, stated in a FT, can be reconstructed completely via an inverse procedure. These significant properties of FT allow working in the "frequency domain" and then reappearance to the original form without missing any information. The Fourier transform, *FT*(*s*) for a single variable, and *f*(*a*) continuous function, is defined below:

$$FT(s) = \int_{\infty}^{\infty} f(a)e^{-j2\pi sa}da$$

where $j = \sqrt{-1}$, and the inverse Fourier transform (IFT) is given by:

$$f(a) = \int_{\infty}^{\infty} FT(s)e^{j2\pi sa}ds$$

These two forms indicate no loss of information in forward and inverse FT. These mathematical expressions are easily mapped for two variables, s and t:

$$F(s,t) = \int_{\infty}^{\infty}\int_{\infty}^{\infty} f(a,b)e^{-j2\pi(sa+tb)}dadb$$

Also, inverse transform is:

$$f(a,b) = \int_{\infty}^{\infty}\int_{\infty}^{\infty} F(s,t)e^{j2\pi(sa+tb)}dsdt$$

FT is complex-valued function of frequency, where absolute part indicates that the frequency exists in the original function and complex part indicates phase offset of the frequency. These frequency components are used as feature value for further classification or learning algorithms. FT is a useful feature extraction and filtering means for face, gait, voice, speech, and heart beat recognition [74, 75].

4.1.2 Local Binary Pattern (LBP)

LBP is an efficient texture operator. It labels the pixels of an image by thresholding the neighborhood of each pixel and contemplates the consequences as binary numbers or patterns. This was first proposed in 1990 [76]. LBP is popular for biometric applications [77, 78] in machine learning domain because of its high discriminative power and computational simplicity. This is usually a divergent statistical and structural model of texture feature extraction and robust to monotonic gray-scale changes caused by illumination variations. LBP divides the input window template into cells (e.g., 16 × 16 pixels for each cell). The objective patterns are usually extracted in a circularly symmetric neighborhood by comparing each image pixel with its neighborhood, which is expressed by:

$$LBP(P,R) = \sum_{i=0}^{P-1} u(g_{\mathrm{i}} - g_{\mathrm{c}})2^{i}$$

where P is the number of neighboring samples and R is the radius of neighborhood, g_{i} denotes the intensity value of neighboring pixel $I(i = 0,\ldots,P - 1)$, g_{c} is the intensity value of the center pixel, and $u(x)$ is a step function with:

$$u(x) = \begin{cases} 1, & \text{for } x \geq 0 \\ 0, & \text{otherwise} \end{cases}$$

The intensities of neighboring pixel values, which do not fall exactly on the image grid, are obtained by bilinear interpolation. Then, after getting the interpolated template from the function *LBP*(*P*, *R*), we find a histogram h_i which is the total number of observations, the total number of bins (b), and the histogram m_i that counts the number of observations that fall into the specified bins defined as:

$$h_i = \sum_{i=1}^{b} m_i$$

This histogram is used as a feature vector later on for the input of a machine learning algorithm. An example to generate local binary pattern is included in Fig. 19.8, where a center pixel value 70 is surrounded by 8 neighboring pixel intensity values. It shows the differentiation among the pixels and thresholding to make a binary pattern. After differentiation and thresholding, the binary pattern will be 11110000. Then the final decimal value 15 is substituted instead of 70.

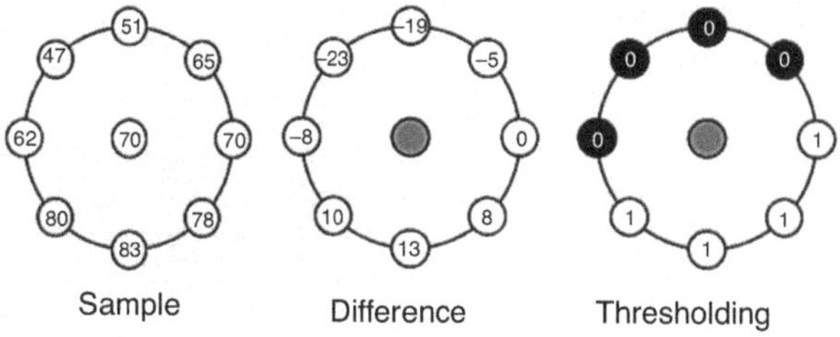

Fig. 19.8 An example of local binary pattern generation

4.1.3 Gabor Filtering (GF)

Gabor filtering is popular feature extraction technique in computer vision domain. It is widely used for biometric authentication also. GF is an efficient texture feature descriptor for palm print, face, fingerprint, and iris image template [79, 80]. Gabor elementary functions are Gaussians modulated by sinusoidal functions. It is shown that the functional form of GFs conforms closely to the receptive profiles of simple cortical cells, and GF is an effective scheme for image representation A two-dimensional (2D) even Gabor filter can be represented by the following equation in the spatial domain:

$$G(x, y; \theta, f) = e^{-\frac{1}{2}\left[\frac{x'^2}{\delta_{x'}^2} + \frac{y'^2}{\delta_{y'}^2}\right]} \cos(2\pi f x')$$

$$x' = x \cos\theta + y \sin\theta$$

$$y' = y \cos\theta - x \sin\theta$$

where f is the frequency of the sinusoidal plane wave along the direction θ from the x-axis and $\delta_{x'}$ and $\delta_{y'}$ are the space constants of the Gaussian envelope along x' and y' axes, respectively. Therefore, GFs over different frequency and directions are considered as feature values for a biometric pattern. Figure 19.9 shows an input template of fingerprint and its enhanced version using the GF, where the enhanced version Fig. 19.9b contains smooth edges than the input Fig. 19.9a.

4.1.4 Radial Zernike Polynomials (RZP)

RZP is first proposed by Fritz Zernike in 1934 [81]. It is useful in describing the wave front data since these are of the same form as the categories Zernike polynomials of aberrations are often experimented in optical tests. This is a well-known shape feature descriptor in computer vision domain. The complex Zernike moments of order n with repetition l are defined as:

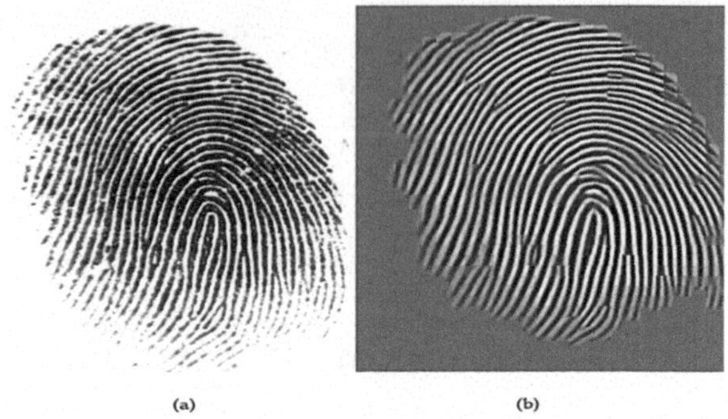

Fig. 19.9 Fingerprint: (a) input template, and (b) enhancement of input using GF

$$A_{nl} = \frac{n+1}{\pi} \int_0^{2\pi} \int_0^{\infty} [V_{nl}(r,\theta)]^* f(r\cos\theta, r\sin\theta) r dr d\theta$$

where $n = 0, 1, 2, \ldots, \infty$ and l take on positive and negative integer values subject to the conditions:

$$n - |l| = \text{even}, |l| \leq n$$

The symbol "*" denotes complex conjugate. The Zernike polynomials:

$$V_{nl}(x,y) = V_{nl}(r\cos\theta, r\sin\theta) = R_{nl}(r)e^{il\theta}$$

are a complete set of complex-valued functions orthogonal on the unit disk $x^2 + y^2 \leq 1$,

$$\int_0^{2\pi} \int_0^{\infty} [V_{nl}(r,\theta)] * V_{mk}(r,\theta) r dr d\theta = \frac{\pi}{n+1} \delta_{mn}\delta_{kl}$$

The real-valued radial polynomials $\{R_{nl}(r)\}$ satisfy the relation:

$$\int_0^1 R_{nl}(r) R_{ml}(r) r dr = \frac{1}{2(n+1)} \delta_{mn}$$

and are defined as:

$$R_{nl}(r) = \sum_{s=0}^{(n-|l|)/2} (-1)^s \frac{(n-s)!}{s!\left(\frac{n+|l|}{2} - s\right)!\left(\frac{n-|l|}{2} - s\right)!} = \sum_{k=|l|,\, n-k=\text{even}}^{n} B_{n|l|k^{r^k}}$$

The function $f(x, y)$ can be expanded in terms of the Zernike polynomials over the unit disk as:

$$f(x,y) = \sum_{n=0}^{\infty} \sum_{\substack{l=-\infty \\ n-|l|=\text{even} \\ |l|\le n}}^{\infty} A_{nl}V_{nl}(x,y)$$

where the Zernike moments A_{nl} are calculated over the unit disk. If the series expansion is truncated at a finite order N, then the truncated expansion is the optimum approximation to $f(x, y)$:

$$f(x,y) \approx \sum_{n=0}^{N} \sum_{\substack{n-|l|=\text{even} \\ |l|\le n}} A_{nl}V_{nl}(x,y)$$

Therefore, if maximum value of radial degree $n = 5$, and azimuthal degree $m = 5$ are considered, then 21 radial Zernike polynomials are generated from unit disk for each gait pattern; this means 21 features can be generated or the feature vector length is 21. Zernike polynomials are very popular in iris and face recognition research [82, 83].

4.1.5 Scale-Invariant Feature Transform (SIFT)

SIFT is a procedure in computer vision to identify and define local features in a template. This is a high-level feature extraction algorithm in machine learning domain. The method is patented in the United States by the University of British Columbia and first published by David Lowe [84]. It takes the original input template, produces gradually blurred images, and converts original image to its half of size. Mathematically, "blurring" is known as the convolution of the Gaussian operator and the image template. It has an operator that is applied to each intensity value and that results a blurred image.

$$L(x,y,\sigma) = G(x,y,\sigma) * I(x,y)$$

where the symbol L is a blurred image, G is the Gaussian blur operator, I is an image, x and y are the location coordinates, and σ is the scale parameter; this means that the greater the value is, the greater the amount of blur, and * is the convolution operator in x and y, which uses Gaussian blur G over the image I. The mathematical expression of Gaussian blur operator is given by:

$$G(x,y,\sigma) = \frac{1}{2\pi\sigma^2} e^{-\frac{(x^2+y^2)}{2\sigma^2}}$$

Now these blurred images are used with another set of images, which is the *difference of Gaussians* (DoG). Those are then used to calculate *Laplacian of Gaussian* (LoG) approximations that are scale invariant. There are two parts to find out key points: (a) locate maxima/minima in DoG images and (b) find out subpixel maxima/minima. This is done by the Taylor expansion of the image around the approximate key point. Mathematical expression can be written as:

$$D(x) = D + \frac{\partial D^T}{\partial x}x + \frac{1}{2}x^T\frac{\partial^2 D}{\partial x^2}x$$

It is easy to find out the extreme points of this equation which increase the chances of matching and stability of the algorithm. After that, the extreme points are needed to be free from low-contrast key points. If the magnitude of the intensity at the current pixel in the DoG image (i.e., being checked for minima/maxima) is less than a certain value, it is rejected. Now gradient directions and magnitudes around each key point are collected, and we can figure out the most prominent orientation (s) in that region. Then these orientations are assigned to the key points. Gradient magnitudes and orientations are computed using these formulae:

$$m(x,y) = \sqrt{(L(x+1,y) - L(x-1,y))^2 + (L(x,y+1) - L(x,y-1))^2}$$

$$\theta(x,y) = \tan^{-1}((L(x,y+1) - L(x,y-1))/(L(x+1,y) - L(x-1,y)))$$

Finally, a histogram is created for 360° of orientation and are broken depending upon the problem statement. A typical example of the outcome of SIFT algorithm is included in Fig. 19.10 where (a) are the input images of face and then LoG is used to find out the key points on each face image and resultant images are shown in (b). There are many researchers who used SIFT as a high-level feature extractor [85, 86].

As per the focus of the chapter, some famous, efficient, and robust biometric feature extraction techniques are discussed here for IoT. But, the evolution of deep learning and its high capacity for learning the different categories are slightly changing the choice of the people. This giant deep learning algorithm itself is a perfect package for preprocessing, feature extraction, and classification steps, and it is going to change the way researchers think today. It will open huge research scopes for IoT, Big Data, and machine learning.

5 Secure Biometric-Based IoT Systems: A Review

Section 19.5.1 demonstrates several applications of IoT like eHealth systems. Based on the applications viewpoint, in this section, we discuss some application-based biometric security protocol in details.

(a)

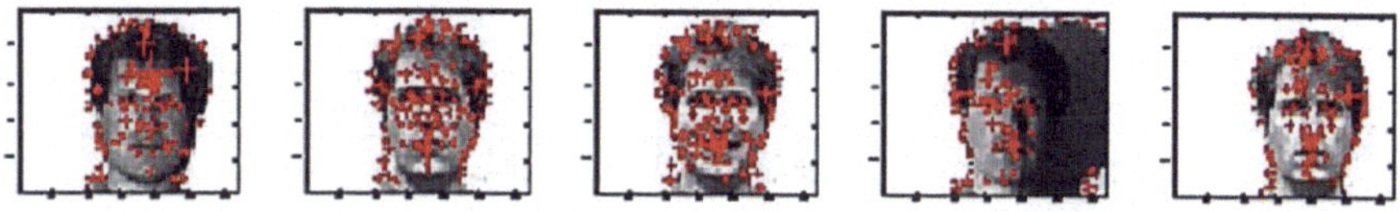

(b)

Fig. 19.10 (**a**) Input set of facial expression image, (**b**) detection of scale invariant key points from the input template

5.1 *Biometric-Based eHealth System: An Authentication and Key Agreement Approach*

Maitra and Giri [87] proposed a biometric-based authentication and key agreement scheme for secure communication in medical systems. In such systems, a patient can get his/her treatment from home by accessing his/her biometric smart card using a mobile device through the Internet. For this purpose, patients do their registration to an authentication server (*AS*) to get their biometric smart card, and then they can get several medical services from medical servers (*MS*) by punching their smart card. Figure 19.11 shows the network structure of the scheme of Maitra and Giri [87].

5.1.1 Description of Maitra and Giri's Scheme

There are four phases of Maitra and Giri's scheme [87]: (a) setup phase, (b) registration phase, (c) log-in phase, and (d) authentication and session key agreement phase.

Setup Phase

The authentication server (*AS*) makes two large prime numbers x and y such that $x = 2y + 1$. *AS* picks a secret key $d \in_R Z_y^*$, where Z_y^* is the set of integer numbers (except 0) over modulo y. *AS* also selects a cryptographic one-way hash function $f(\cdot)$: $\{0, 1\}^* \rightarrow Z_y^*$. Finally, *AS* declares y and $f(\cdot)$ publicly and keeps d as secret. Note that, cryptographic one-way hash function is collision resistant as defined in [2, 88].

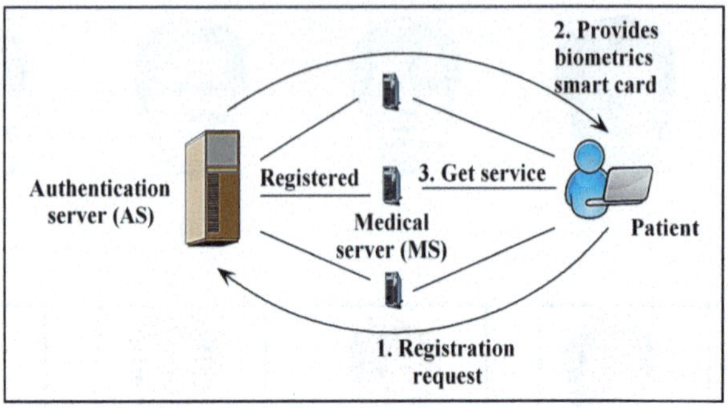

Fig. 19.11 Network structure of Maitra and Giri's scheme [87]

Registration Phase

All the medical servers (*MS*) and the patients do their registration under the authentication server (*AS*) in this phase.

- *Registration phase of medical server*

By performing the following steps, a medical server MS_i performs its registration under *AS* by executing the following steps:

Step 1: MS_i picks its identity MID_i and transmits it to *AS* using secure channel. As the registration is performed only one time in offline mode, thus the communication pattern can be termed as secure channel.

Step 2: After getting MID_i from MS_i, *AS* picks a number a_i randomly from Z_y^* and calculates the unique secret key S_{key}^i of MS_i as $f(a_i \| d)$. Then, *AS* supplies S_{key}^i to MS_i secretly.

- *Registration phase of patient*

A patient P_i does his/her registration under *AS* by executing the following steps:

Step 1: Sensor of mobile device scans the biometrics of patient P_i and extracts the biometric feature b_i, by following a suitable technique as discussed in Sect. 19.4. P_i selects a unique identity PID_i and password pw_i through mobile application. Then the mobile device of P_i computes $bpw_i = f(pw_i \| b_i)$ and sends $\{PID_i, bpw_i\}$ to *AS* via secure channel.

Step 2: Upon getting $\{PID_i, bpw_i\}$, *AS* calculates $A_i = f(PID_i \| d)$, $C_i = bpw_i \oplus A_i$, and $D_i = f(A_i \| PID_i \| bpw_i)$. AS further computes p number of key-plus-id combinations $\{\text{key}_p, MID_p | 1 \leq p \leq i + n\}$, where key_p is calculated as $ENC_{A_i}\left[f\left(\underbrace{f(a_p \| d)}_{S_{\text{key}}^p} \| PID_i\right)\right]$, ENC_{A_i} is the symmetric key encryption (i.e., ASE-128 [89]) using a key A_i, and p is the number of medical servers present in the e-medical system to provide medical services.

Step 3: AS finally burns $\{PID_i, C_i, D_i, \{\text{key}_p, MID_p | 1 \leq p \leq i + n\}, f(\cdot), y\}$ into the memory of biometric smart card and issues the card for patient P_i.

Log-In Phase

To reach several medical facilities by accessing the medical servers (MS), the patient P_i punches the biometric smart card into his/her mobile device, and also the sensor of the mobile device scans the biometrics of P_i and extracts the feature b_i' of P_i. Then the patient P_i supplies his/her password pw_i to the mobile device. After that the mobile device of P_i executes the following operations:

Step 1: Mobile device calculates $bpw_i' = f\left(pw_i \parallel b_i'\right)$, $A_i' = C_i \oplus bpw_i'$, and $D_i' = f\left(A_i' \parallel PID_i \parallel bpw_i'\right)$. Then, it checks $D_i = ?D_i'$. For the correct value, the device executes the next step; otherwise, it stops the current session.

Step 2: Mobile device provides the permission to select an identity of medical server from which P_i wants to get service. P_i submits an identity MID_i of MS_i.

Step 3: The mobile device retrieves key_i corresponding to MID_i. It then extracts $f(f(a_i \parallel d) \parallel PID_i)$ by decrypting key_i using the key A_i' as $f(f(a_i \parallel d) \parallel PID_i) = DEC_{A_i}[\text{key}_i]$.

Step 4: The device picks a number r_i randomly from Z_y^* and calculates $L_i = ENC_{f(f(a_i \parallel d) \parallel PID_i)}[r_i \parallel f(b_i \parallel r_i \parallel T_1)]$ and $R_i = f(r_i \parallel T_1 \parallel MID_i \parallel PID_i \parallel f(b_i \parallel r_i \parallel T_1))$, where T_1 is the current timestamp. Then the mobile device transmits a log-in message $\{PID_i, L_i, R_i, MID_i, T_1\}$ to the medical server MS_i via the Internet.

Authentication and Session Key Agreement Phase

Upon getting the log-in message $\{PID_i, L_i, R_i, MID_i, T_1\}$ at timestamp T_2, the medical server MS_i checks the validity of timestamp as $(T_2 - T_1) \leq \Delta T$, where ΔT is the threshold value of time span. For the correct result, MS_i executes the following steps:

Step 1: MS_i calculates $f\left(S_{\text{key}}^i \parallel PID_i\right)$, retrieves $\left[r_i^* \parallel (f(b_i \parallel r_i \parallel T_1))^*\right]$ by decrypting L_i as $DEC_{f\left(S_{\text{key}}^i \parallel PID_i\right)}[L_i]$, and calculates $R_i^* = f\left(r_i^* \parallel T_1 \parallel MID_i \parallel PID_i \parallel (f(b_i \parallel r_i \parallel T_1))^*\right)$. Then MS_i checks $R_i = ?R_i^*$. For the false result, MS_i rejects the current session; otherwise, it executes the next step.

Step 2: MS_i picks a number e_i randomly from Z_y^* and calculates $re_i = r_i^* \oplus e_i$, $SK_i = f\left(r_i^* \parallel e_i \parallel T_1 \parallel T_2 \parallel (f(b_i \parallel r_i \parallel T_1))^*\right)$, and $K_i = f(SK_i \parallel e_i)$. Then, MS_i transmits a reply message $\{re_i, K_i, T_2\}$ to the mobile device of P_i.

Step 3: After getting $\{re_i, K_i, T_2\}$ at timestamp T_3, the mobile device verifies the validity of sent timestamp. For the correct result, it computes $e_i' = r_i \oplus re_i$, $SK_i' = f\left(r_i \parallel e_i' \parallel T_1 \parallel T_2 \parallel f(b_i \parallel r_i \parallel T_1)\right)$, and $K_i' = f\left(SK_i' \parallel e_i'\right)$. Then the device checks $K_i = ?K_i'$. For the correct equality, both the patient P_i and medical server MS_i agree upon a session key SK_i for secure data communication into the same session.

6 Conclusion and Future Scope

Biometric technology has been around for decades, but it is mainly popular for identity authentication or verification in highly secure environment. Biometric-based security systems are becoming popular day by day. Figure 19.12 discussed the changes of revenue of total biometric revenue market with respect to year. The change is monotonically and exponentially increasing in nature. Therefore, it can be said that day by day, biometric-based security is becoming more and more popular. Still there are some challenges being faced by biometric technology. The first challenge is the cost of biometric technology. There are some reasons for increasing cost of biometric technology like hardware maintenance, processing power for databases, experimental infrastructure, real-time implementation, salary for employees, training for the employees, marketing cost, exception handling, productivity loss, and system maintenance, among others.

Figure 19.13 shows the possible threat of biometric security system. At the sensor level where end user puts his/her biometric information, there are different attacks that can occur like collective attack, spoofing attacks, and mimicry attacks. However, at the time of biometric features extraction, different attacks might take place like eavesdropping, man-in-the-middle attack, replay attack, and brute force attack. The storage data can be attacked by reading template, replacing template attack, and change-binding attack (ID biometric). Those attacks influence the error of matching during the matching process of biometric information inserted by the end user with the stored but tempered biometric information. If we can consider that stored biometric information is correct, then also attacks can occur at the time of matching phase. The type of attack occurred at matching phase is insertion of imposter data, component replacement, hill climbing, and manipulation of score guessing attack.

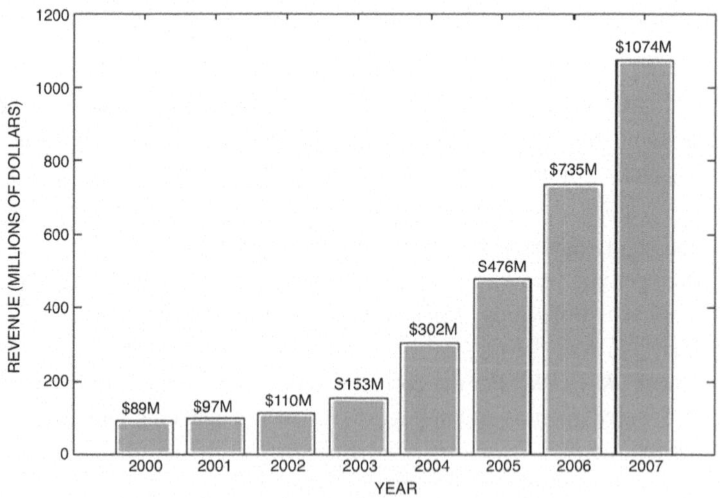

Fig. 19.12 Total biometric revenue market: 2000–2007 [68]

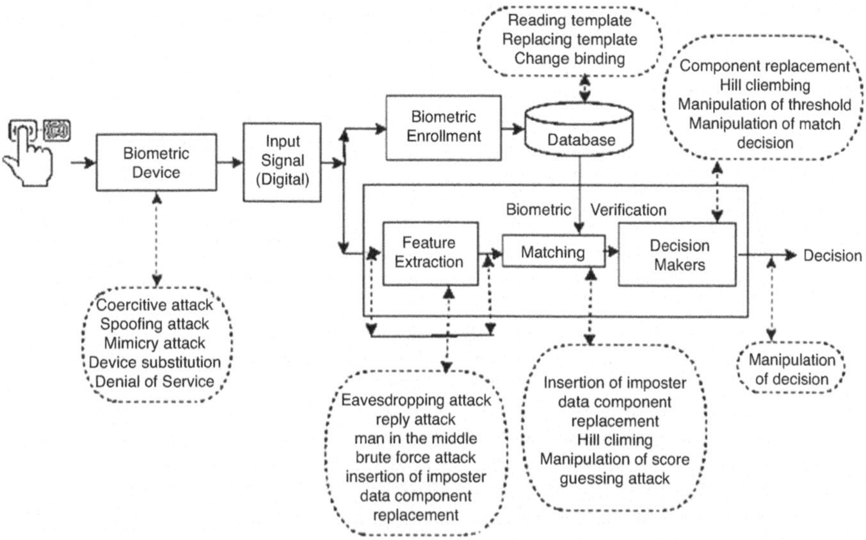

Fig. 19.13 List of security threat in different steps of biometric security system

Eavesdropping attack, reply attack, and man-in-the-middle attack can occur during the transmission of biometric data from data storage to the matching block. After matching process is done, the message regarding percentage of matching information is to transfer to the decision block. During transfer process, the message might be tempered with by hill climbing and manipulation of match score process. Even hackers can change the data at the decision level. Moreover, attacks can occur after transmitting the decision, and such attacks are component replacement, hill climbing, and manipulation of decision, among others.

References

1. M.S. Obaidat, P. Nicopolitidis, *Smart Cites and Homes: Key Enabling Technologies* (Elsevier, 2016)
2. M.S. Obaidat, S. Misra, *Principles of Wireless Sensor Networks* (Cambridge University Press, 2014)
3. M.S. Obaidat, N. Boudriga, *Security of e-Systems and Computer Networks* (Cambridge University Press, 2007)
4. G.I. Davida, Y. Frankel, B.J. Matt, On enabling secure applications through off-line biometric identification, in *Proceedings of IEEE Symposium on Security and Privacy (Cat. No.98CB36186)*, 1998, pp. 148–157
5. A. Al-Fuqaha, M. Guizani, M. Mohammadi, M. Aledhari, M. Ayyash, Internet of things: A survey on enabling technologies, protocols, and applications. IEEE Commun. Surv. Tutorials **17**(4), 2347–2376 (2015)
6. M. Hansen, P. Berlich, J. Camenisch, S. Clauß, A. Pfitzmann, M. Waidner, Privacy-enhancing identity management. Inf. Secur. Tech. Rep. **9**(1), 35–44 (2004)

7. M. Jude, IBM: Working towards a smarter connected home (2014), http://docs.caba.org/documents/IBM-Smart-Cloud-Home-SPIE2012.pdf
8. L. Jiang, D.-Y. Liu, B. Yang, Smart home research, in *Proceedings of 2004 International Conference on Machine Learning and Cybernetics (IEEE Cat. No.04EX826)*, vol. 2, 2004, pp. 659–663
9. N. Bui, M. Zorzi, Health care applications: A solution based on the internet of things, in *Proceedings of the 4th International Symposium on Applied Sciences in Biomedical and Communication Technologies*, 2011, pp. 131:1–131:5
10. A.-M. Rahmani, N.K. Thanigaivelan, T.N. Gia, J. Granados, B. Negash, P. Liljeberg, H. Tenhunen, Smart e-health gateway: Bringing intelligence to internet-of-things based ubiquitous healthcare systems, in *Consumer Communications and Networking Conference (CCNC), 2015 12th Annual IEEE*, 2015, pp. 826–834
11. A.M. Vilamovska, E. Hattziandreu, R. Schindler, C. Van Oranje, H. De Vries, J. Krapelse, RFID application in healthcare–scoping and identifying areas for RFID deployment in healthcare delivery. *RAND Europe*, February 2009
12. L. Lei-hong, H. Yue-shan, W. Xiao-ming, A community health service architecture based on the internet of things on health-care, in *World Congress on Medical Physics and Biomedical Engineering*, May 26–31, 2012, Beijing, China, 2013, pp. 1317–1320
13. A. Dohr, R. Modre-Opsrian, M. Drobics, D. Hayn, G. Schreier, The internet of things for ambient assisted living, in *Information Technology: New Generations (ITNG), 2010 Seventh International Conference on*, 2010, pp. 804–809
14. G. Acampora, D.J. Cook, P. Rashidi, A.V. Vasilakos, A survey on ambient intelligence in healthcare. Proc. IEEE **101**(12), 2470–2494 (2013)
15. M.S. Shahamabadi, B.B.M. Ali, P. Varahram, A.J. Jara, A network mobility solution based on 6LoWPAN hospital wireless sensor network (NEMO-HWSN), in *Innovative Mobile and Internet Services in Ubiquitous Computing (IMIS), 2013 Seventh International Conference on*, 2013, pp. 433–438
16. S.M.R. Islam, D. Kwak, M.D.H. Kabir, M. Hossain, K.-S. Kwak, The internet of things for health care: A comprehensive survey. IEEE Access **3**, 678–708 (2015)
17. R.S.H. Istepanian, S. Hu, N.Y. Philip, A. Sungoor, The potential of Internet of m-health things "m-IoT" for non-invasive glucose level sensing, in *Engineering in Medicine and Biology Society, EMBC, 2011 Annual International Conference of the IEEE*, 2011, pp. 5264–5266
18. Z.L. In, Patient body temperature monitoring system and device based on internet of things. Chin. Pat. **103**, 577–688 (2014)
19. H.A. Khattak, M. Ruta, E. Di Sciascio, CoAP-based healthcare sensor networks: A survey, in *Applied Sciences and Technology (IBCAST), 2014 11th International Bhurban Conference on*, 2014, pp. 499–503
20. E.C. Larson, M. Goel, G. Boriello, S. Heltshe, M. Rosenfeld, S.N. Patel, SpiroSmart: Using a microphone to measure lung function on a mobile phone, in *Proceedings of the 2012 ACM Conference on Ubiquitous Computing*, 2012, pp. 280–289
21. P.K. Dash, Electrocardiogram monitoring. Indian J. Anaesth **46**(4), 251–260 (2002)
22. L. Yang, Y. Ge, W. Li, W. Rao, W. Shen, A home mobile healthcare system for wheelchair users, in *Computer Supported Cooperative Work in Design (CSCWD), Proceedings of the 2014 I.E. 18th International Conference on*, 2014, pp. 609–614
23. Dr. Hawking's connected wheelchair project (2014), http://smartcitiescouncil.com/resources/stephen-hawking-and-intel-connected-wheelchair-project
24. Awareness Day 2014 Activities by Program Type, National Children's Mental Health Awareness Day (May 2014), https://www.samhsa.gov/sites/default/files/children-awareness-day-activities-by-program-2014.pdf
25. M. Vazquez-Briseno, C. Navarro-Cota, J.I. Nieto-Hipolito, E. Jimenez-Garcia, J.D. Sanchez-Lopez, A proposal for using the internet of things concept to increase children's health awareness, in *Electrical Communications and Computers (CONIELECOMP), 2012 22nd International Conference on*, 2012, pp. 168–172

26. G. Zhang, C. Li, Y. Zhang, C. Xing, J. Yang, SemanMedical: a kind of semantic medical monitoring system model based on the IoT sensors, in *e-Health Networking, Applications and Services (Healthcom), 2012 I.E. 14th International Conference on*, 2012, pp. 238–243
27. D. Miorandi, S. Sicari, F. De Pellegrini, I. Chlamtac, Internet of things: Vision, applications and research challenges. Ad Hoc Netw. **10**(7), 1497–1516 (2012)
28. A. Zanella, N. Bui, A. Castellani, L. Vangelista, M. Zorzi, Internet of things for smart cities. IEEE Internet Things J. **1**(1), 22–32 (2014)
29. T. Nam, T.A. Pardo, Conceptualizing smart city with dimensions of technology, people, and institutions, in *Proceedings of the 12th Annual International Digital Government Research Conference: Digital Government Innovation in Challenging Times*, 2011, pp. 282–291
30. T. Bakıcı, E. Almirall, J. Wareham, A smart city initiative: The case of Barcelona. J. Knowl. Econ. **4**(2), 135–148 (2013)
31. H. Wang, W. He, A reservation-based smart parking system, in *Computer Communications Workshops (INFOCOM WKSHPS), 2011 I.E. Conference on*, 2011, pp. 690–695
32. D. Bonino, M.T.D. Alizo, A. Alapetite, T. Gilbert, M. Axling, H. Udsen, J.A.C. Soto, M. Spirito, Almanac: internet of things for smart cities, in *Future Internet of Things and Cloud (FiCloud), 2015 3rd International Conference on*, 2015, pp. 309–316
33. M. Yun, B. Yuxin, Research on the architecture and key technology of internet of things (IoT) applied on smart grid, in *Advances in Energy Engineering (ICAEE), 2010 International Conference on*, 2010, pp. 69–72
34. E. Qin, Y. Long, C. Zhang, L. Huang, Cloud computing and the internet of things: Technology innovation in automobile service, in *International Conference on Human Interface and the Management of Information*, 2013, pp. 173–180
35. Y. Zhang, B. Chen, X. Lu, Intelligent monitoring system on refrigerator trucks based on the internet of things, in *International Conference on Wireless Communications and Applications*, 2011, pp. 201–206
36. L. Atzori, A. Iera, G. Morabito, SIoT: Giving a social structure to the internet of things. IEEE Commun. Lett. **15**(11), 1193–1195 (2011)
37. L. Atzori, A. Iera, G. Morabito, M. Nitti, The social internet of things (siot)–when social networks meet the internet of things: Concept, architecture and network characterization. Comput. Netw. **56**(16), 3594–3608 (2012)
38. J. Kleinberg, The convergence of social and technological networks. Commun. ACM **51**(11), 66–72 (2008)
39. P. Semmelhack, *Social Machines: How to Develop Connected Products that Change Customers' Lives* (Wiley, 2013)
40. Y. Bo, H. Wang, The application of cloud computing and the internet of things in agriculture and forestry, in *Service Sciences (IJCSS), 2011 International Joint Conference on*, 2011, pp. 168–172
41. J. Zhao, J. Zhang, Y. Feng, J. Guo, The study and application of the IOT technology in agriculture, in *Computer Science and Information Technology (ICCSIT), 2010 3rd IEEE International Conference on*, vol. 2, 2010, pp. 462–465
42. A. Giri, S. Dutta, S. Neogy, Enabling agricultural automation to optimize utilization of water, fertilizer and insecticides by implementing internet of things (IoT), in *2016 International Conference on Information Technology (InCITe) – The Next Generation IT Summit on the Theme – Internet of Things: Connect Your Worlds*, 2016, pp. 125–131
43. T. Kelesidis, I. Kelesidis, P.I. Rafailidis, M.E. Falagas, Counterfeit or substandard antimicrobial drugs: A review of the scientific evidence. J. Antimicrob. Chemother. **60**(2), 214–236 (2007)
44. A.J. Jara, A.F. Alcolea, M.A. Zamora, A.F.G. Skarmeta, M. Alsaedy, Drugs interaction checker based on IoT, in *2010 Internet of Things (IOT)*, 2010, pp. 1–8
45. L. Toms, 5 Common cyber attacks in the IoT-threat alert on a grand scale, in *Global Sign* (GMO Internet Group, 2016)
46. N.K. Ratha, J.H. Connell, R.M. Bolle, Enhancing security and privacy in biometrics-based authentication systems. IBM Syst. J. **40**(3), 614–634 (2001)

47. M. Abomhara, Cyber security and the internet of things: Vulnerabilities, threats, intruders and attacks. J. Cyber Secur. Mobility **4**(1), 65–88 (2015)
48. Hewlett Packard, HP study reveals 70 percent of internet of things devices vulnerable to attack, July 2014
49. T. Heer, O. Garcia-Morchon, R. Hummen, S.L. Keoh, S.S. Kumar, K. Wehrle, Security challenges in the IP-based internet of things. Wirel. Pers. Commun. **61**(3), 527–542 (2011)
50. S. Cirani, G. Ferrari, L. Veltri, Enforcing security mechanisms in the IP-based internet of things: An algorithmic overview. Algorithms **6**(2), 197–226 (2013)
51. H. Sundmaeker, P. Guillemin, P. Friess, S. Woelfflé, Vision and challenges for realizing the internet of things. CERP-IoT, Eur. Comm. **3**(3), 34–36 (2010)
52. A.K. Jain, S.Z. Li, *Handbook of Face Recognition* (Springer, New York, 2011)
53. C. Ding, C. Xu, D. Tao, Multi-task pose-invariant face recognition. *IEEE Trans. Image Process.* **24**(3), 980–993 (2015)
54. L.B. Rowden, A.K. Jain, Longitudinal study of automatic face recognition. *IEEE Trans. Pattern Anal. Mach. Intell.* **40**(1), 148–162 (2018)
55. R. Teixeira, N. Leite, A new framework for quality assessment of high-resolution fingerprint images. IEEE Trans. Pattern Anal. Mach. Intell. **39**(10), 1905–1917 (2017)
56. J. Daugman, How iris recognition works. IEEE Trans. Circuits Syst. Video Technol. **14**(1), 21–30 (2004)
57. L. Zhang, Y. Shen, H. Li, J. Lu, 3D palm print identification using block-wise features and collaborative representation. IEEE Trans. Pattern Anal. Mach. Intell. **37**(8), 1730–1736 (2015)
58. L. Zhang, L. Li, A. Yang, Y. Shen, M. Yang, Towards contactless palm print recognition: A novel device, a new benchmark, and a collaborative representation based identification approach. Pattern Recogn. **69**, 199–212 (2017)
59. A. Tkach, M. Pauly, A. Tagliasacchi, Sphere-meshes for real-time hand modeling and tracking. *ACM Trans. Graph.* **35**(6), 222 (2016)
60. S. Sharma, S.R. Dubey, S.K. Singh, R. Saxena, R.K. Singh, Identity verification using shape and geometry of human hands. Expert Syst. Appl. **42**(2), 821–832 (2015)
61. J.G. Rodriguez, P. Rose, D. Ramos, D.T. Toledano, J.O. Garcia, Emulating DNA: Rigorous quantification of evidential weight in transparent and testable forensic speaker recognition. IEEE Trans. Audio Speech Lang. Process. **15**(7), 2104–2115 (2007)
62. A. Kumar, K.V. Prathyusha, Personal authentication using hand vein triangulation and knuckle shape. IEEE Trans. Image Process. **18**(9), 2127–2136 (2009)
63. M.S. Obaidat, B. Sadoun, Verification of computer users using keystroke dynamics. IEEE Trans. Syst. Man Cybern. B **27**(2), 261–269 (1997)
64. A.M.D.L. Herran, B.G. Zapirain, A.M. Zorrilla, Gait analysis methods: An overview of wearable and non-wearable systems, highlighting clinical applications. Sensors **14**, 3362–3394 (2014)
65. A. Fischer, R. Plamondon, Signature verification based on the kinematic theory of rapid human movements. IEEE Trans. Hum. Mach. Syst. **47**(2), 169–180 (2017)
66. D.P. Jarrett, E.A.P. Habets, P.A. Naylor, *Theory and Applications of Spherical Microphone Array Processing* (Springer, 2017), pp. 1–10
67. K. Ali, A.X. Liu, W. Wang, M. Shahzad, Recognizing keystrokes using Wi-Fi devices. IEEE J. Sel. Areas Commun. **35**(5), 1175–1190 (2017)
68. R. de Luis-Garcıa, C. Alberola-López, O. Aghzout, J. Ruiz-Alzola, Biometric identification systems. Signal Process. **83**(12), 2539–2557 (2003)
69. M. Faundez-Zanuy, Biometric security technology, in *Encyclopedia of Artificial Intelligence* (IGI Global, 2009), pp. 262–269
70. Y. Zhen, X. Li, Q. Ou, L. Zeng, Internet of things and smart grid. Digit. Commun. **39**(5) (2012)
71. A. Kulkarni, S. Sathe, Healthcare applications of the internet of things: A review. Int. J. Comput. Sci. Inf. Technol. **5**(5), 6229–6232 (2014)

72. A. Riahi, E. Natalizio, Y. Challal, N. Mitton, A. Iera, A systemic and cognitive approach for IoT security, in *Computing, Networking and Communications (ICNC), 2014 International Conference on*, 2014, pp. 183–188
73. A.K. Jain, A. Ross, S. Prabhakar, An introduction to biometric recognition. IEEE Trans. Circuits Syst. Video Technol. **14**(1), 4–20 (2004)
74. W. Hwang, H. Wang, H. Kim, S.C. Kee, J. Kim, Face recognition system using multiple face model of hybrid Fourier feature under uncontrolled illumination variation. IEEE Trans. Image Process. **20**(4), 1152–1165 (2011)
75. S.D. Choudhury, T. Tjahjadi, Silhouette-based gait recognition using procrustes shape analysis and elliptic Fourier descriptors. Pattern Recogn. **45**(9), 3414–3426 (2012)
76. D.C. He, L. Wang, Texture unit, texture spectrum, and texture analysis. IEEE Trans. Geosci. Remote Sens. **28**, 509–512 (1990)
77. U. Park, R.R. Jillela, A. Ross, A.K. Jain, Periocular biometrics in the visible spectrum. IEEE Trans. Inf. Forensics Secur. **6**(1), 96–106 (2011)
78. Z. Guo, L. Zhang, D. Zhang, X. Mou, Hierarchical multi scale LBP for face and palmprint recognition, in *Proceedings of 17th IEEE International Conference on Image Processing (ICIP)*, September 2010, pp. 4521–4524
79. C. Gottschlich, Curved-region-based ridge frequency estimation and curved Gabor filters for fingerprint image enhancement. IEEE Trans. Image Process. **21**(4), 2220–2227 (2012)
80. S. Xie, S. Shan, X. Chen, J. Chen, Fusing local patterns of Gabor magnitude and phase for face recognition. IEEE Trans. Image Process. **19**(5), 1349–1361 (2010)
81. V.F. Zernike, Diffraction theory of the cutting process and its improved form, the phase contrast method. Physica **1**(7–12), 689–704 (1934)
82. C.W. Tan, A. Kumar, Unified framework for automated iris segmentation using distantly acquired face images. IEEE Trans. Image Process. **21**(9), 4068–4079 (2012)
83. C.W. Tan, A. Kumar, Accurate iris recognition at a distance using stabilized iris encoding and Zernike moments phase features. IEEE Trans. Image Process. **23**(9), 3962–3974 (2014)
84. D.G. Lowe, Object recognition from local scale-invariant features, in *Proceedings of the IEEE International Conference on Computer Vision*, vol. 2, Washington, DC, 1999, pp. 1150–1157
85. X. Wu, Y. Tang, W. Bu, Offline text-independent writer identification based on scale invariant feature transform. IEEE Trans. Inf. Forensics Secur. **9**(3), 526–536 (2014)
86. D. Smeets, J. Keustermans, D. Vandermeulen, P. Suetens, meshSIFT: Local surface features for 3D face recognition under expression variations and partial data. Comput. Vis. Image Underst. **117**(2), 158–169 (2013)
87. T. Maitra, D. Giri, An efficient biometric and password-based remote user authentication using smart card for telecare medical information systems in multi-server environment. *J. Med. Syst.* **38**(12), 142 (2014)
88. T. Maitra, M.S. Obaidat, R. Amin, S.K. Hafizul Islam, S.A. Chaudhry, D. Giri, A robust ElGamal-based password-authentication protocol using smart card for client-server communication. *Int. J. Commun. Syst.* **30**(11), e3242 (2017)
89. Advanced Encryption Standard (AES), Federal Information Processing Standards Publication 197 (2001). https://www.nist.gov/publications/advanced-encryption-standard-aes; http://nvlpubs.nist.gov/nistpubs/FIPS/NIST.FIPS.197.pdf

Part V
Technology and Society

Chapter 20
Protecting the Integrity of Elections Using Biometrics

Mohammad S. Obaidat, Tanmoy Maitra, and Debasis Giri

1 Introduction

From an extensive understanding on conventional voting systems, it can be said that voting systems should be computerized so that (a) the counting time of votes is reduced, (b) frauds on the system are minimized, (c) provided votes are accounted correctly, and (d) people with special needs/requirements are accommodated. e-Voting has introduced new opportunities to both government and public through enabling voters to cast their votes from anywhere. It provides different voting channels, broadening access to the disable people, conveying voting results quickly and reliably, decreasing costs, and enlarging turnout. Eventually, e-voting replaces the conventional paper-based systems with Internet-based e-services. But in e-voting systems, there are several security issues and challenges, given that democratic principles depend on integrity of the electoral process. It is not a trivial task to secure the voting systems. Apart from the traditional features of security like confidentiality, availability, and integrity, some other properties are also required as well, for example, authenticity of voters, anonymity of votes, proper proof of vote

M. S. Obaidat
ECE Department, Nazarbayev University, Astana, Kazakhstan

King Abdullah II School of Information Technology (KASIT), University of Jordan, Amman, Jordan

University of Science and Technology Beijing (USTB), Beijing, China

Fordham University, New York City, NY, USA

T. Maitra
School of Computer Engineering, KIIT University, Bhubaneswar 751024, Odisha, India

D. Giri (✉)
Department of Information Technology, Maulana Abul Kalam Azad University of Technology, Nadia 741249, West Bengal, India

M. S. Obaidat et al. (eds.), *Biometric-Based Physical and Cybersecurity Systems*,
https://doi.org/10.1007/978-3-319-98734-7_20

counting mechanism of the voters, allowing voters to cast their vote via the Internet after protecting voters' coercion, preserving the uniqueness of the vote in decentralized manner, and software and hardware compatibility of the system, among others. However, to ensure the security requirements in e-voting systems, several cryptographic techniques can be adopted, such as visual cryptography to get receipts on submitted vote, a shared secret key establishment between voter and vote to decrypt a vote using homomorphic encryption [1, 2], and mix networks to make anonymous channels, which can protect the anonymity of the voters and their votes.

On the other hand, biometric systems have started to be vastly adapted in e-voting system in order to authenticate the identity of the voters as well as to reduce fraud. However, biometrics can be described as a process of personal identification based on independent characteristics [3]. Biometric technologies calculate a variety of anatomical, physiological, and behavioral features and utilize the features to differentiate the people from each other. Generally, many biometric features are documented by means of specialized devices such as hand geometry, iris, EEG, fingerprint, handwritten signature, and voice. By taking the advantages of biometrics of each voter, election authority can produce digital voter ID card for each voter; the voter ID card is basically a smart card. During the voting procedure, election authority verifies the legitimacy of the voters through that biometrics of the voters.

The organization of this chapter will be as follows. Section 20.2 describes the traditional electoral system. Generic architecture of e-election process is discussed in Sect. 20.3. Biometrics and its operational process are presented in Sect. 20.4. Section 20.5 discusses the properties to build the secure e-election system. The secure and biometric-based e-voting systems are demonstrated in Sect. 20.6. Section 20.7 provides a literature review of the secure e-voting systems. Different possible attacks on e-voting system are discussed in Sect. 20.8. Finally, conclusion of this chapter comes in Sect. 20.9.

2 Traditional Electoral System

Most traditional election systems are not faultless. Such systems depend on a number of reliable parties who have the potential to make secret plans jointly to commit an unlawful or harmful act like altering the results of the election or enforcing voters to vote in a particular way. However, this kind of systems is assumed to be secure because most of the reliable parties are either trustworthy, and therefore no collusion can take place.

In traditional voting systems, there is possibility for votes to be altered, lost, or wrongly counted at the time of counting process. As improper tallies may be the outcome of dishonesty, the recorded incorrectness in computerized voting tally may not show the misuse of the voting equipments and software. In an election of school board in Carroll County (1984), the election authority unfortunately runs the incorrect utility program for counting the votes; therefore, computerized voting tally system produced the improper tally [4]. The utilization of e-voting system provides the mobility property in national elections by allowing voters to submit their votes from anywhere through the Internet. However, e-voting systems tend to minimize privacy and maximize the possibility for vote to be altered or lost. Despite these procedural shortcomings, it would be hard for a national election to be thrown away because of the large number of enclosure and the variety of voting systems used. In addition, vote buying is a crucial issue because it is not possible for a voter to prove how he/she voted after leaving the polling booth. However, vote buying can happen easily when e-voting systems are used.

Huge number of experts, social, and interested organizations prefer to use mail-in voting systems in their electoral process. Such systems provide permission to the voters to submit their votes from remote location, but it may compromise the accuracy and privacy. However, these systems are usually used by several organizations in case of noncontroversial. Furthermore, the organizations employ an unbiased party to run their election process. Many states also use mail-in voting system for some elections, especially in small precincts. Normally, voters are asked to supply their vote in double envelopes to protect their privacy. Probably, Teamsters is the largest organization which used mail-in voting system.

Moreover, the traditional e-voting systems can be checked either by the agents of a party or by reliable third parties. Generally, for the voters, it is impossible to check if each vote has been counted properly or not. Beside this, when the identification process recognizes the strategic problems and inconsistency between the final tally and the number of voters who submitted the vote, then it normally cannot correct the results.

3 Generic Architecture of e-Election Process

Figure 20.1 shows an overview of a generic architecture of an election process [5]. There are the following entities: user/voter, a registration server, a voting console (mobile API), a voting server, an electronic ballot box, and an electronic election bulletin board for final tally. Electoral authority controls the registration server, voting server, and electronic ballot box.

(a) *User or voter*: End user submits his/her vote by applying mobile application after getting permission or getting ticket.

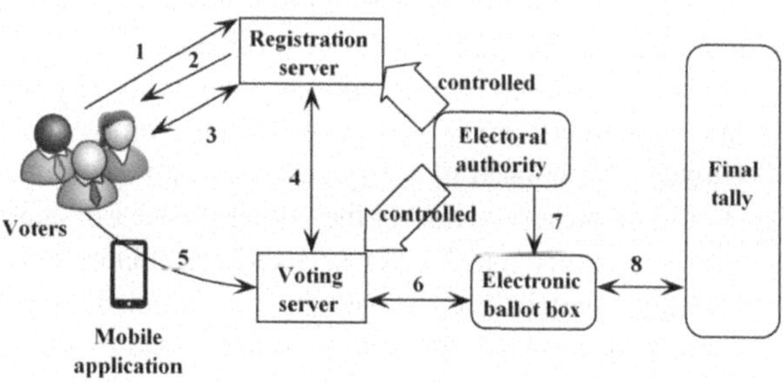

Fig. 20.1 Generic architecture of election process

(b) *Registration server*: Registration server is responsible to provide legal information to a user during registration process. During the logged-in into the system process, the server verifies the candidature of the user and provides permission to the user to submit his/her vote.
(c) *Voting server*: It helps to store the vote of each user securely.
(d) *Electronic ballot box*: It is a database where votes submitted by each voter are stored.
(e) *Final tally*: In the vote counting process, it counts the legal votes and declares the result of the election.

During the registration process, a voter sends request to registration server (RS) (event 1 in Fig. 20.1). RS issues a smart card by embedding the candidature of the voter (event 2). During log-in session, voters have to send a permission request message to RS to get a permission that allows them to cast their votes (event 3). RS communicates with the voting server to get the corresponding identity of ballot (event 4) and utilize them to make a permission message that is returned to the voters (event 3). After getting permission from RS, the voters use the voting console (mobile device) to cast their vote (event 5). Then the voting server reserves the submitted votes into the electronic ballot box (event 6). When the election process is finished, the electoral authority starts the counting phase (event 7). The electoral authority counts the votes and publishes the receipts through the election bulletin board i.e., known as final tally (event 8).

3.1 Secure Techniques for e-Voting System

In the e-voting system, several security mechanisms are used which are as briefed below [1, 6]:

- *Digital certificate*: Sender of a message digitally signs the message with his/her private key so that the receiver of the message can verify the received message by the sender's public key. Moreover, it is used to build authentication mechanism.
- *Public key infrastructure* (PKI) *mechanism*: Public key infrastructure is used to carry the utilization of digital certificate where each voter produces private and public key so that they can use these keys during the e-voting activity. Furthermore, PKI can be used to preserve secrecy on voters' information when it generates the unique identity of the voters.
- *Hashing mechanism*: It is a one-way technique by which plaintexts are converted into ciphertexts known as hashed value, but reverse procedure is not possible. However, it is used to verify accuracy and non-alteration of message. Accuracy should be preserved during the voting phase. For example, the hashed value of the voter's identity and his/her selected candidate is concatenated to the encrypted message, which is encrypted by the authority's public key. After getting the encrypted message, the authority decapsulates the encrypted message by using

its private key and compares its hashed value to the received hashed value in order to get assurance that no alteration of the original message has taken place during the vote submission process.

- *Blind signature mechanism*: By using blind signature mechanism, voters can sign their unique identity to make digital signature so that only proper recipients can verify the signature. However, public key signing algorithms like RSA-based blind signature scheme [7] are commonly used to create a blind signature mechanism in e-voting system. For better understanding, an RSA-based blind signature scheme is discussed here.

  (i) The sender S of a message m computes the blind message $\widehat{m}$ on m as $\widehat{m} = (a)^e m \bmod n$, where a is a random number, e is the receiver's public key, and n is the receiver's public modulus (1024 bits). Here, a must be chosen so that a and n are relatively prime. S then transmits the blind message $\widehat{m}$ to the receiver R.

  (ii) After obtaining $\widehat{m}$, R signs the blind message $\widehat{m}$ by computing the blind signature BS: $BS = (\widehat{m})^d \bmod n$, where d is the secret key of R. Then R transmits the blind signature BS to S.

  (iii) S then eliminates the blind factor from BS and calculates an RSA-based normal signature SG:

$$
\begin{aligned}
SG &= BS \cdot (a)^{-1} \bmod n \\
&= (\widehat{m})^d (a)^{-1} \bmod n, \quad \text{as } BS = (\widehat{m})^d \bmod n \\
&= ((a)^e m)^d (a)^{-1} \bmod n, \quad \text{as } \widehat{m} = (a)^e m \bmod n \\
&= (a)^{ed} (m)^d (a)^{-1} \bmod n \\
&= a\,(a)^{-1} (m)^d \bmod n, \quad \text{as } e \cdot d = 1 \bmod n \\
&= (m)^d \bmod n
\end{aligned}
$$

  Note that, only S can do the aforementioned computations as it knows the value of a. After that S sends the signature SG to the verifier (a third party who can verify the signature SG as well as has need of message m).

  (iv) The verifier can verify the signature and gets the message m by doing normal RSA signatures. It will satisfy requirement as $m = (SG)^e \bmod n = (m)^{ed} \bmod n$, since $e \cdot d = 1 \mod n$.

- *Biometric smart token*: Generally, fingerprint or iris is used as biometric template to recognize the voters in elections. With the biometric match using smart card technology, the provided biometrics is checked with the stored biometric template into the smart card in a secure manner. In such a scenario, the biometric template is stored into the smart card in such a way that cannot be extracted. However, signing on a document by using biometric key is the only one application where the biometric authentication through smart card can be used.

4 Biometrics

As discussed in Section 20.1, biometrics of the voters is a vital feature to identify the candidature or to verify the legitimacy of the voter during voting session. Basically, in the registration process, registration server (see Fig. 20.1) collects the biometrics of the voters during registration process and issues a digital voter ID card so that by using the digital ID card, voters can submit their votes remotely by accessing e-voting application from mobile phones, laptops, tablets, computers, or any other electronic devices via the Internet. According to Fig. 20.1, biometrics is needed in event 1, event 2, and event 3, where events 1 and 2 are under registration procedure which is performed before election process and event 3 is required during voting procedure to check the validity of the voter.

4.1 Operational Process

At time of registration phase, the registration server is responsible for voters' admission and certification. Figure 20.2 depicts the registration process for biometric-based e-voting system. During the registration process, voters input their biometrics like retina, finger print, iris scan, and hand writing to the sensor. The sensor extracts the unique feature from the provided biometrics for each voter and supplies it to the authentication server. The authentication server generates a unique voter ID card (smart card or photo ID card) by embedding the biometric features into the smart card and storing the biometric template into its database for each voter in order to verify the candidature of the voter during voting process. Later, during the election process, the voter can submit his/her vote by using the digital ID card if and only if the authentication server gives permission to submit vote after proper verification of the voter.

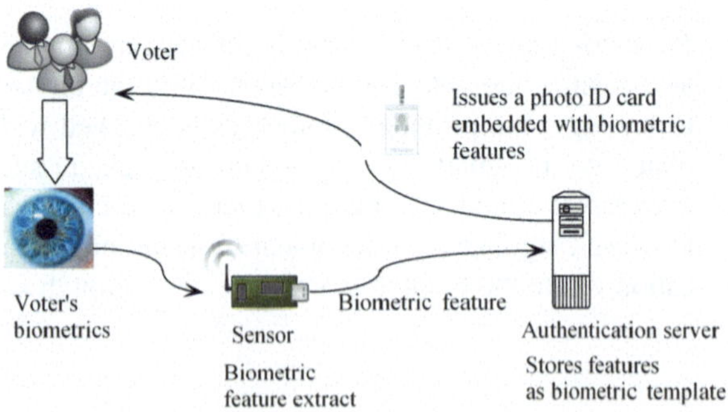

Fig. 20.2 Registration process of voters using their biometrics in e-voting system

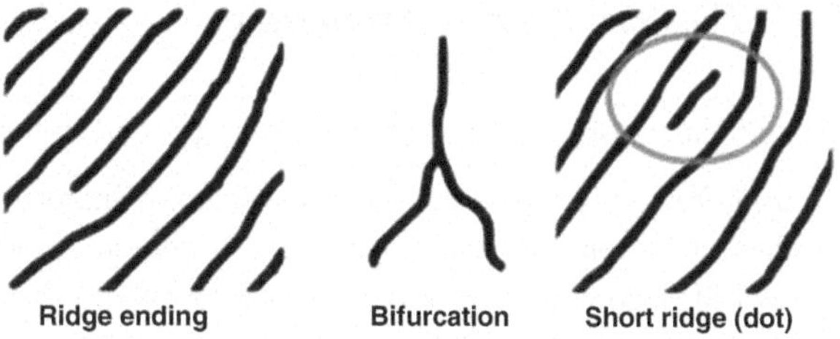

Fig. 20.3 Pattern of the fingerprint

4.2 *Biometric Technologies*

It is a quite difficult task to extract unique biometric feature. For this purpose, several biometric extractors are used, but fuzzy-based extractors [8] are commonly used in biometric feature extraction as they are simple and provide accuracy to extract feature from biometrics for different time instances. Furthermore, fuzzy extractor can tolerate some noise in different biometric time instances. Here, we provide the simple working flow of fuzzy extractor for fingerprint biometrics.

1. When the voter puts his finger on the fingerprint sensor, the sensor captures image.
2. The extractor finds the patterns of the image. Basically, there are three basic decorations of fingerprint ridges: (a) arch, (b) loop, and (c) whorl.

   Arch: The ridges start from one side of the finger. They rise in the middle of finger and form arches. Then they end to the other side of the finger.
   Loop: The ridges start from one side of the finger. They make curve and then end on that same side of finger.
   Whorl: Ridges form circularly around a middle point on the finger.

3. Based on the pattern of the fingerprint, we can identify ridge ending, bifurcation, and short ridge as shown in Fig. 20.3.
4. The scheme measures the distance (may use Euclidian distance) from each point to others and takes average distance for the image of the finger. The average distance should be unique for each voter as the patterns on fingerprint are not identical for any two persons.

5 Requirements and Proprieties for the Election Process

This section discusses the features, which should be preserved to build a secure e-election system.

5.1 Election Process Properties

To build an e-voting system, the following properties have to be maintained [5, 6, 9].

- *Soundness*: Any party or authority cannot interrupt or show its power to the election committee and final tally. Soundness should be maintained so that integrity on voting results is achieved. However, there are several ways to corrupt the e-voting system. For example, the authorities of the system may defraud by allowing invalid voters to register or a cheated voter may register him-/herself under another candidature. In this scenario, vote counting mechanism may be compromised.
- *Eligibility*: Only valid voters (ability to cast his/her vote) can join in the election process. To ensure the eligibility, each voter should register him-/herself before going to cast his/her vote.
- *Privacy*: It is the main concern in e-voting system. Voters' information like social security number, address, and delivered vote should be kept secret during the election process as well as at the end of election process for a long time.
- *Uncoercibility*: Any party and even authority should not get the information regarding the votes and the corresponding voters. Furthermore, no authority should force a voter to deposit his/her vote in a specific way provided by the fraud authority. Voting systems should provide a facility by which voters can deliver their vote freely.
- *Fairness*: No partial announcement regarding tally of votes is disclosed before the finishing of voting period in order to provide fair treatment to the all candidates. Even no authority should be able to have any knowledge about the voting results.

5.2 Election Security Requirements

- *Confidentiality*: Confidentiality guarantees that information in communication messages remains secret so that any outside observers of the system cannot learn anything about the transmitted messages. In case of e-voting systems, any information like identity of voter and his/her corresponding vote should be kept secret until the counting is over or until voting result is published.
- *Integrity*: Integrity of information ensures that information of a message is full and not corrupted. When the information is exposed to damage, corruption, demolition, or other disruptions of its authentic state, then the information integrity can be threatened. However, fraud can be mounted, while information is being saved or transmitted. At the end of election process in e-election systems, ultimate vote counting must exactly represent the total number of valid voters and total number of unique votes of each voter; the latter should be equal.
- *Authenticity*: Authentication of each entity during communication is highly required in e-voting system. The users i.e., voters and electoral authority may not be trusted all the time. Therefore, during communication both way

authentications are needed so that each entity gets knowledge and can verify the candidature of the received messages. However, in the e-voting systems, legitimacy of voters must be checked at two different phases: (a) at the time of voters' registration so that voters can enter the system and (b) before the submission of vote.

- *Non-repudiation*: Non-repudiation ensures that the communicators cannot oppose the authenticity of their signature on a document or message that they have produced. Here, the election authority should not deny that the voter is not authentic after generating a ticket for a voter (by which the voter can put his vote). This means, if the election authority gives permission to a voter to deliver his/her vote, the authority cannot say that the voter is fraud during the counting procedure. On the other hand, after delivering a vote, a voter cannot deny that he/she did not deliver that vote during the counting procedure.
- *Availability*: Availability of information in voting services should be provided by an e-voting system, and security requirements during the entire election process should also be preserved. Furthermore, all the services or options like name of the candidate list should be available to the voters so that they can select the candidate of his/her choice.

6 Biometric Voting Schemes

In this section, some biometric-based secure voting schemes [1, 6, 10] are demonstrated which are as follows.

6.1 Ahmed and Aborizka's Scheme

Before going to present Ahmed and Aborizka's scheme [6], it is important to know the used e-voting system architecture in this study. Figure 20.4 shows the pictorial view of the architecture used in their scheme [6].

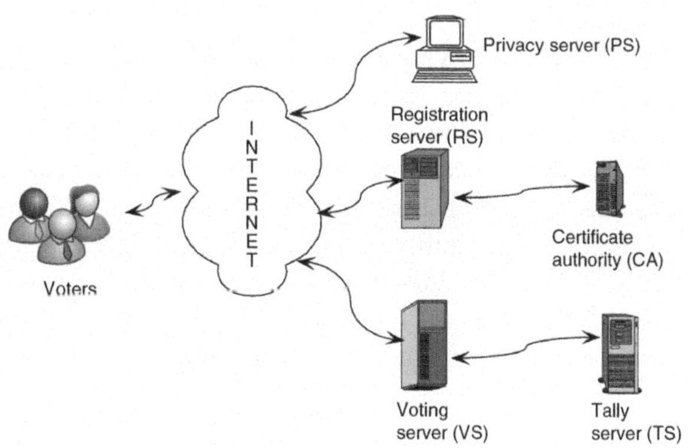

Fig. 20.4 Architecture for e-voting system used in Ahmed and Aborizka's scheme [6]

- Registration Server (RS): RS monitors and controls the registration procedure of the e-voting system. It provides registration certificate to the legal voters so that they can join in the elections by considering a set of rules declared by the election authority.
- Privacy Server (PS): PS maintains anonymity of the voter's identity so that the voter's specification like name and address cannot be traced from his/her vote. For this purpose, PS first verifies the voter's certificate, then checks the submitted anonymous identity generated by the voter is valid or not.
- Voting Server (VS): VS receives the votes from the voters and stores the votes in an e-ballot box. After verification of voting certificate of a voter, it stores the vote and sends it to the vote counting server.
- Tallying Server (TS): TS counts the votes at the end of the election process and publishes the result publicly.
- Certificate Authority (CA): CA is responsible for confirming the voter's identification data received by RS in the registration and verification phase. Moreover, it provides individual information about the voter where the registration server can utilize such information to determine whether the voter is eligible to cast his/her vote or not.

Each voter has to submit his/her personal smart token which contains the national digital certificate (NDC) and his/her biometrics to the election authorities (i.e., CA and RS) to get his/her voting rights. RS stores biometrics corresponding to NDC for each voter in its database. The voters supply their smart token and submit their identity, their password, and their current biometrics like fingerprint and retina scan to make their smart token active. The biometric scanner maps the biometrics into template (i.e., binary form of biometrics). Ahmed and Aborizka's scheme [6] contains three phases: (a) registration phase, (b) voting phase, and (c) tally phase. Next, we will discuss each phase in details.

6.1.1 Registration Phase

Step 1: A voter sends his/her biometric template along with his/her public components to the registration server (RS). The public components (NDC) are already stored into smart token.

Step 2: After receiving the request, RS checks the validation of received public components and biometric template by searching into its database. For the successful result on search, RS gives permission him/her to join election process by sending election certificate (VCD) and stops the registration session for the voter. For the unsuccessful result on search, RS sends a request to CA to make sure of the eligibility of the voter as well as to get credentials for the voter.

Step 3: CA replies with voter's credentials for the valid voter.

Step 4: RS verifies the uniqueness of the public components and biometric template of the voter received from CA. RS then makes a voting digital certificate for the voter and updates its database by adding certificate, biometric template, and public component of the voter securely.

RS then transmits the election certificate (VCD) to the voter which contains serial number of issued certificate, unique pair of public/private keys of VCD, digital stamp of current election, and the public key of election authority.

Step 5: After getting election certificate (VCD), the voter verifies the digital stamp. If it is true, the voter stores the certificate in the smart token securely; otherwise, it rejects the certificate.

6.1.2 Voting Phase

At the end of registration phase, the registered voters cast their votes in this phase. As each voter has been issued a VDC by RS, thus, he/she has permission to join in the current election process. Furthermore, in this scheme, voter can choose his/her pseudo-identity while submitting his/her vote, which can ensure that no one can trace the vote corresponding to the voter. Beside this, the system is not centralized where several authorities are involved to make success on e-voting system and every authority has its own task to fulfill which can be verified by other authorities in a distributed manner. This kind of systems may ensure that only eligible voters can cast their votes by obeying the rules and settings of the current election process. The following steps are executed to complete the voting phase of a voter:

Step 1: The voter computes a unique number *PID* for identification using the stored information into the smart token. *PID* can be calculated as $PID = h(ENC[stamp]_b)$, where *stamp* is the digital stamp of current election, *b* is the biometric template of voter as key, *ENC*[.] is the secret key encryption, and *h*(.) is the one-way hashing operation.

The voter blinds the *PID* using blind signature scheme like RSA [7] and signs the resulted value using the stored secret key of VDC to produce *Sign*.

The voter then encrypts *Sign* using public key of election authority to produce *ESign* and then transmits *ESign* to the anonymous identity server (AIS).

Step 2: After getting *ESign*, AIS decrypts *ESign* by using the private key of election authority to get *Sign*. AIS then computes a blind signature on *Sign* using the secret key of election authority.

After that, AIS encrypts the computed blind signature using the public key of VDC of the voter to produce *BS*.

Finally, AIS transmits *BS* to the smart token and stores the voter as valid so that the voter cannot send any other request by computing the blind signature again.

Step 3: After getting BS, the smart token decrypts BS using the secret key of voter and unbinds the blind signature on *Sign*. Then it checks the validity of Sign. If true, the smart token stores *Sign* and *PID* securely; otherwise, it terminates the current session of voting phase.

The voter sends a request message to VS to connect to cast his/her vote and also to establish him-/herself as authenticate or valid voter using PID and VCD.

Step 4: After getting request, VS checks the eligibility of the voter. For the valid voter, VS gives permission to the voter to deliver his/her voting choice(s); otherwise, it rejects the voter.

Step 5: Upon getting permission, the voter sends *Sign* and information about his/her voting choice to VS.

Step 6: VS checks the uniqueness of *Sign* by searching into its database. If it is not found, VS believes that *Sign* is unique and stores the information about voting choice of the voter and corresponding *Sign* into its database. Otherwise, VS rejects the vote.

Upon successful voting, VS sends a receipt to inform the voter for his/her successful vote.

6.1.3 Tally Phase

At the finishing of voting phase, tally server (TS) collects the voting database from VS and counts the votes. After ending of counting, TS publishes the voting result publicly.

6.2 Alrodhan et al.'s Scheme

Before going to present Alrodhan et al.'s scheme [10], it is important to know the used e-voting system architecture in this study. Figure 20.5 shows the pictorial view of the architecture used in their scheme [10].

The authentication certificate (AC) issuer verifies the validation of voters from remote places via the Internet so that it can be decided that the voter is eligible to cast his/her vote or not. Furthermore, AC, being a part of trusted authority, digitally signs the votes to make confident that no fake votes are casted by invalid voters. For this purpose, AC performs: (a) voter registration, (b) voter verification, and (c) ticket generation by signing the vote. Beside this, Ballot Center (BC) collects the votes

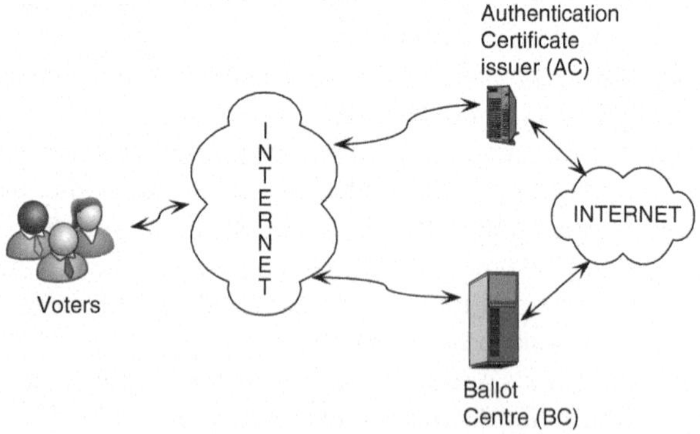

Fig. 20.5 The e-voting system architecture followed by Alrodhan et al.'s scheme [10]

from eligible voters after being verified by AC. For a valid vote, BC generates a corresponding receipt to the voter and, lastly, publishes the final result after the ending of election process.

Alrodhan et al.'s scheme [10] was based on RSA blind signature [7]. Alrodhan et al.'s scheme [10] contains four phases, namely, (a) registration phase, (b) authentication phase, (c) voting phase, and (d) result publishing phase. Before going to join in election process, every voter must do his/her registration to AC. k_{sec} is the shared secret symmetric key between AC and the BC.

6.2.1 Registration Phase

All the voters physically present to AC so that they can join in the system. For this purpose, AC checks voter's national ID card of each voter to verify the eligibility of voter to vote. If the process is successfully executed, AC collects the biometrics such as fingerprint of each voter and stores the biometric template for future use in voting process and issues a voter ID so that the voter can cast his/her vote. Two fingerprint samples from same hand or two fingerprint samples from two hands can be collected by AC from each voter. The voter notifies AC which fingerprint will be used to cast his/her vote, and according to the notification, AC stores the biometric sample into its database.

6.2.2 Verification Phase

The details of verification phase are discussed in the following:

Step 1: To enter into the voting process, the voter opens the e-voting terminal (i.e., application) on his/her smartphone. The voter submits his/her voter ID and also provides his/her registered biometrics to the terminal through sensor attached in smartphone. Then the terminal transmits this information to AC.

Step 2: AC checks that the voter ID and biometrics are registered and stored in its database or not. For the truth value of checking, the voter is legal and can submit his/her vote in voting phase. Otherwise, AC sends an error message to the voter and denies voting permission to the voter.

6.2.3 Voting Phase

After verification, AC provides a list of local candidates to the terminal on the basis of location of the voter in case of regional voting scheme; otherwise, AC provides the global candidate list to the terminal. However, the steps of the voting phase are as below.

Step 1: A voter selects a candidate from the list provided by AC through the e-voting terminal on his/her smartphone. The e-voting terminal sends a request to BC to get a nonce (i.e., a random number). However, the nonce is used to protect replay attack, and it is also used to denote a temporary pseudonym nickname of the voter. After getting the request, BC generates a nonce, i.e., $nonce_1$, and sends it to the e-voting terminal.

Step 2: Upon getting $nonce_1$ from BC, the e-voting terminal calculates a blind message $\widehat{m}$, signed by the AC's public key as $\widehat{m} = (a)^{e_{AC}} \overbrace{(c \parallel nonce_1)}^{n}$, where $\parallel$ is concatenation operation, e_{AC} is the public key of AC, a is the random number generated by terminal, and c is the reference information corresponding to the selected candidate. Then the e-voting terminal transmits the blind message $\widehat{m}$ to AC.

Step 3: Upon receiving $\widehat{m}$, AC signs the blind message $\widehat{m}$ as $BS = (\widehat{m})^{d_{AC}} \bmod n$, where d_{AC} is the private key of AC. Note that, AC does not know and get the value of c i.e., reference information corresponding to the selected candidate.

AC further computes a message λ as $\lambda = \left\{ENC[flag \parallel nonce_2]_{k_{sec}} \parallel nonce_2\right\}$, where *flag* is a status flag 0 means a fake vote or 1 indicates valid vote. $nonce_2$ is a nonce picked by AC to make fresh message λ in order to eliminate replay attack. Note that, *flag* and $nonce_2$ are encrypted using the shared secret key k_{sec} between AC and BC. Then AC transmits the blind signature BS and λ to the e-voting terminal.

Step 4: After receiving $\{BS, \lambda\}$, the e-voting application computes a signature S by eliminating the blind factor from the received blind signature BS: $S = BS \cdot (a)^{-1} \bmod n$. The e-voting terminal transmits the message $m = (c \parallel nonce_1)$, S, and λ to BC.

Step 5: Upon receiving $\{m, S, \lambda\}$, BC checks whether nonce is fresh or not as well it also verifies the signature S. For the truth value of both checks, BC encrypts the vote c and stores it in its database. Then BC executes the next step. For the negative result on both checks, BC terminates the voting session and transmits an error message to the terminal.

BC decrypts the message λ by using the shared secret k_{sec} and earns *flag* value after verifying the freshness of $nonce_2$. If the extracted *flag* value is 0, BC marks the vote as fake; otherwise, it marks the vote as valid and then saves the decision in its database. However, the votes with fake marked will not be accounted for counting in final result. Thereafter, BC transmits a new nonce (i.e., $nonce_3$) to the e-voting terminal.

Step 6: After getting $nonce_3$, the e-voting terminal builds another new blind message $\bar{m} = (\bar{a})^{e_{AC}} \cdot nonce_3 \bmod n$ where $\bar{a}$ is the random number generated by terminal and passes it $\bar{m}$ to AC.

Step 7: After getting $\bar{m}$, AC believes that the voter already casts his/her vote and marks as "done" against that voter and stores it in its database. However, this step is crucial by which AC will not verify the same voter in different session, because, in a voting mechanism, only one vote can be allowed by each voter electronically. AC then signs on $\bar{m}$ to produce $\overline{BS} = (\bar{m})^{d_{AC}} \bmod n$ and sends the blind signature $\overline{BS}$ to the e-voting terminal.

Step 8: After receiving $\overline{BS}$, the e-voting terminal computes a signature $\bar{S}$ by eliminating the blind factor from $\overline{BS}$ as $\bar{S} = \overline{BS} \cdot (\bar{a})^{-1} \mod n$. Then the e-voting terminal transmits $\bar{S}$ and $nonce_3$ to BC.

Step 9: Upon getting $\bar{S}$, BC verifies the signature $\bar{S}$. If the verification shows the true value, BC provides a receipt number to the e-voting terminal to ensure the success on casting vote and BC also stores the receipt number into its record. For the false value of the verification, an error message is generated by BC to inform the voter of the unsuccessful voting session and removes the vote from its database.

6.2.4 The Result Announcement Phase

After the ending of election process, this phase is executed to announce the result of the election in e-voting system. In the following, we discuss the steps executed in this phase.

Step 1: BC collects the number of submitted votes from its database and decrypts all the votes to get reference number of the selected candidate c.

Step 2: BC counts reference number of the selected candidate c. Only valid votes, i.e., flag value is 1 of the submitted votes, are considered for counting.

Step 3: BC publishes the total number of votes earned by each candidate publicly.

A schematic view of voting phase of Alrodhan et al.'s scheme [10] is given in Figs. 20.6 and 20.7.

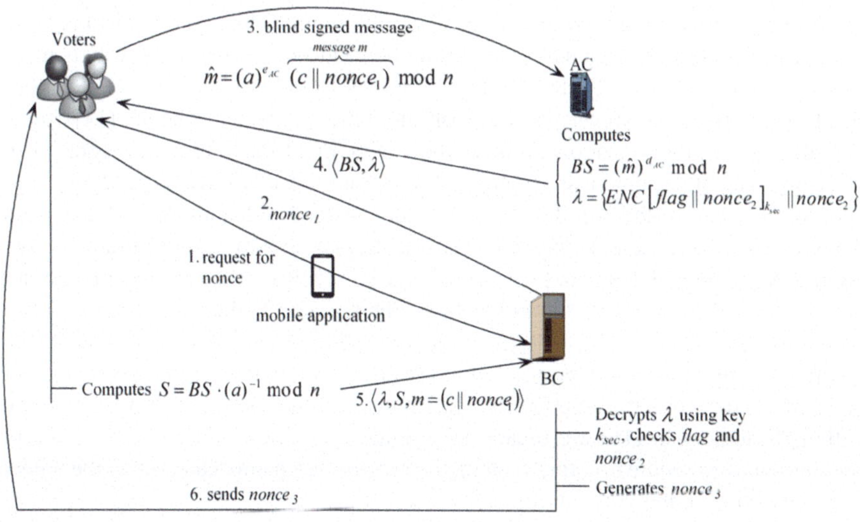

Fig. 20.6 Schematic view of voting phase (steps 1–6) of Alrodhan et al.'s scheme [10]

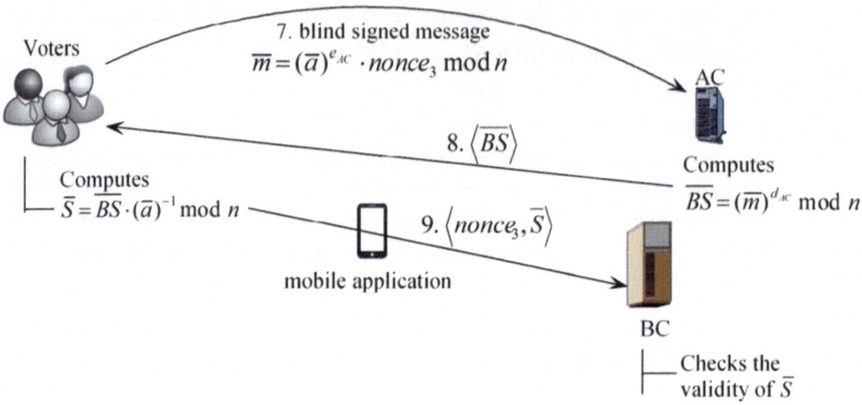

Fig. 20.7 Schematic view of voting phase (steps 7–9) of Alrodhan et al.'s scheme [10]

7 Security Protocols in Electoral System: A Review

Chaum [11] proposed the first cryptographic voting protocol on anonymous e-mail using digital pseudonyms. Chaum used public key cryptography and pseudo identities of voters. But, the protocol [11] does not provide the anonymity of voter's identity because it can be traced from outside of the system. Electoral protocol can be damaged by a single citizen and Chaum's protocol [11] can identify such damages, but it cannot resolve this problem without restarting the whole election process [12].

Cohen and Fischer [13] presented a secure election system. In the scheme reported in [13], the dishonest voters are unable to interrupt the election process. But, the protocol cannot provide the privacy of individuals like voters' identity from the election committee. Cohen [14] further introduced a scheme by which more privacy can be achieved by distributing the power of government into some authorized committees. However, because of the scheme's (i.e., scheme [14]) huge complexity in communications, it takes the voters long time to cast their votes [15].

Nowadays, some practical cryptographic techniques that did not need any connection between voters or did not need any specialized equipment have been devised. However, the existing schemes (mentioned above) cannot eliminate vote buying. Benaloh and Tuinstra [16] published two secret voting schemes that do not permit the voters to prove the contents of their votes. Unlike the other security protocols (i.e., schemes [13, 14]), the protocols [16] require the voters to vote inside a voting booth. However, the protocol in [16] does not guarantee that voters cannot be coerced, until one or more trusted election authorities are present in the system. Although authors in [16] are unable to provide a practical solution in real-world Internet-enabled secure e-voting system, the receipt-free nature can protect the voters from involving in the vote buying.

According to architecture of generic e-voting system (see Fig. 20.1), two election authorities a validator (i.e., registration server) and a tallier (tally server) are required in the simplistic schemes [6, 10, 17]. In such schemes [6, 10, 17], a voters encrypts his/her vote by the public key of the tallier, then signs it, and passes them to the validator. The validator strips off the voters' signatures, checks to make sure that the vote is provided by the valid and fresh voter or not. For the correct validation, validator passes the vote to the tallier. Then it decrypts the votes and stores the votes. Moreover, this kind of scheme can stop the illegal voters from voting as well as the legal voters from numerous voting.

In Nurmi et al.'s scheme [17], two agencies, i.e., registration server and voting server, have the responsibility to validate the registered voters and to declare the results after storing the votes securely. In [17], the validator generates a long prime identification tag for each voter and transmits it to m number of voters who want to join in election process. The validator also sends the list of identification tag to the tallier. After getting tag, a voter V sends a pair $\langle E_V, h(E_V, vote_V)\rangle$ to the tallier, where E_V denotes the voter's tag, $h(\cdot)$ is a one-way hash function, and $vote_V$ denotes the casting vote. After getting the pair, the tallier declares $h_V(E_V, vote_V)$. During the counting process, V sends another pair $\langle E_V, (h_V(E_V, vote_V))^{-1}\rangle$ to the tallier so that it can extract the vote $vote_V$. After ending of election process, the tallier announces the list of each vote $vote_V$ and its corresponding $h_V(E_V, vote_V)$. By this procedure, a voter can get conformation that his/her vote is counted successfully.

When Chaum first proposed the blind signature-based e-voting scheme [11], it has been argued that blind signatures can be used for securing the vote submitted by the voters. After some years, Fujioka et al. designed an e-voting scheme [18] in which, blind signatures have been used to resolve the collusion problem of Nurmi et al.'s scheme [17]. The authors in [18], argued that the proposed solution is able to reduce the overall time complexity in communications.

Similar to Fujioka et al.'s scheme [18], another protocol named as Sensus [19] has been proposed. However in [19], a voter encrypts his/her vote and transmits the encrypted form of vote to the tallier. Then the tallier sends an acknowledgement to the voter as a receipt of the submitted vote. After getting receipt, the voter supplies the decryption key to the tallier instantly, and election process will be ended in one session. But, in the Fujioka et al.'s scheme [18], a voter cannot supply his/her decryption key until the election is finished. Thus, publishing the vote cannot be done in a single session.

On the other hand, Sensus [19] is unable to correct following issues: (1) The dishonest election authorities (i.e., registration server or voting server) can count and publish votes for abstaining voters. The illegal votes can be identified by the abstaining voters only or by government itself after verifying the validation of signatures on all the submitted requests. Moreover, there is no such way to verify the illegal votes so that they are deleted from final tally. (2) Sensus [19] takes large time in communications when a voter submits his/her vote.

8 Security Attacks on e-Voting System

There are several security attacks that exist in e-voting systems. An adversary may try to mount various attacks on this system in off-line mode as well as online mode. As the communications are done through insecure channel like the Internet in e-voting system, the adversary can capture and alter the original message or try to establish himself/herself as a legal voter. Thus, the main aim of designing any biometric-based security system for e-voting is to eliminate the security loopholes as listed below:

Insider attack: In e-voting systems, employees under voting server (see Fig. 20.1) and many other entities associated with e-voting systems (except registration server as it is directly handled by government or trusted party) may try to extract the information regarding vote and identity of the voter. Thus it is very crucial attack in such system. However, proper authentication in both communications ways may eliminate insider attack.

Voter masquerading attack: In e-voting systems, after capturing the communication messages between a valid voter V_i and the other servers like registration server and voting server, an adversary may try to impersonate as V_i. For this purpose, the adversary may try to extract secret information like voters' biometrics and identity after capturing the communication messages. Thus a secure e-voting system should ensure that no third party from outside of the system can extract any secret information from captured messages. Furthermore, a valid voter V_j may also try to impersonate as another valid voter V_i. If V_j gets success to do so, then V_j can easily deliver V_i's vote. As a result, V_i will not able to produce his/her own choice to vote. Therefore, secure e-voting systems should remove the voter masquerading attack.

Server masquerading attack: As mentioned earlier the registration server may be trusted in e-voting systems as it is controlled by either government or trusted party, but all the other entities like voting servers may not be trusted. Thus, an employee under voting server may act as an adversary to mount server masquerading attack. In such an attack, a server S_j may act as another server S_i. As a result, the valid voters may receive voting receipt from S_j on behave of S_i and in the latter when actual receipt is sent by S_i, the voter will discard the actual receipt. However, proper mutual authentication along with no leak of secret information from communication messages is required to eliminate this attack.

Software module attack: The executable program at a module can be altered in such a way so that it always produces the results desired by the attacker. Trojan-horse attack is the common attack in this scenario. To overcome this attack, specialized hardware or secure code execution practices can be used to impose secure execution of the software. Algorithmic integrity is another solution, which also imposes the component of software integrity. Algorithmic integrity can be defined as software should manage any provided input in a desirable way. As an example of algorithmic drawback, assume a matching module in which a particular input B_0 is not managed properly and when B_0 is supplied to the matcher, it always produces the acceptance (yes) decision. However, this loophole may not affect the biometric-based functioning system. This is because the probability of generating such B_0 from a real-biometric data can be negligible. But, an attacker may utilize this drawback to easily break the security without being detected.

Attacks on the template database: An adversary from inside the system may try to alter or share the biometrics template with the other untrusted parties; therefore, this attack produces very much damage for the biometric-based security system. Attacks on the template can lead to the following two security threats: (a) a template can be substituted by a masquerader's template to achieve unauthorized access and (b) the stolen template may be replayed to the matcher to obtain uncertified entrance. For example, a biometric template stolen from a voter's database can be used to search criminals' biometric records or cross-link to person's banking records.

Replay attack: An adversary may repeat the same communication messages between the valid voter V_i and other servers associated with the voting system for different time instance. If the adversary gets success in this attack, the adversary may alter the vote submitted by V_i. To eliminate the replay attack, proper mutual authentication scheme along with freshness of each and every communication messages are highly desirable.

Man-in-the-middle attack: In this attack, an adversary stealthily relays and modifies the communication messages between two communicators who think that direct communication is going on. Moreover, in active eavesdropping, the adversary builds self-governing connections with the victims and relays messages between the victims to ensure them that they are communicating directly to each other through a private connection. But, the entire communications are in fact run by the adversary. The adversary intercepts all the relevant messages passing between the two victims.

Denial-of-service (DoS) attack: An adversary may flood the system resources to the point where the access of legal voters will be denied. Eventually, an adversary may send flooded messages on behalf of valid voters. For this kind of scenario, the valid voter will not be able to enter the e-voting system. However, to prevent this attack, proper verification of each communicator is required.

9 Conclusion

Many countries worldwide are going to build e-government system. Developing countries such as India provide such digital voter ID card for each person in which biometric feature of that human is incorporated. In the future, this smart card will be incorporated with the election authority so that people can submit their vote from remote place using their mobile application through the Internet. Beside this, the USA and the UK have already started to develop e-voting system. However, practical implementation of such e-voting system has some security issues. This is because, each entity associated with voting system cannot be considered as trusted. Furthermore, preserving the voting information of each voter along with his/her identity is a basic challenge in the development of such systems. During the communication, authentication of the communicators is highly desirable in each step. However, biometric authentication with the cryptographic technique explores the mutual authentication between the voters and other entities. However, it is bounded in research world and has no proper implementation. Researchers are trying to advance their work in e-voting systems so that privacy, eligibility, uniqueness, uncoercibility, fairness, accuracy, robustness, individual verifiability, and universal verifiability can be achieved under one umbrella.

References

1. M.S. Obaidat, N. Boudriga, *Security of e-Systems and Computer Networks* (Cambridge University Press, Cambridge/New York, 2007)
2. M. Naor, A. Shamir, Visual cryptography, in *Advances in Cryptology, Eurocrypt 94*, ed. by A. De Santis (Ed), vol. 950, (Springer, Berlin, Heidelberg, 1995), pp. 1–12
3. A.K. Jain, P. Flynn, A.A. Ross, *Handbook of Biometrics* (Springer, New York, 2007)
4. R.G. Saltman, Accuracy, integrity and security in computerized vote-tallying. Commun. ACM **31**(10), 1184–1191 (1988)
5. A.O. Santin, R.G. Costa, C.A. Maziero, A three-ballot-based secure electronic voting system. IEEE Secur. Priv. **6**(3), 14–21 (2008)
6. T.K. Ahmed, M. Aborizka, Secure biometric E-voting scheme, in *Proceedings of Intelligent Computing and Information Science*, vol. 134, (Springer, Berlin, Heidelberg, 2011)
7. S. Goldwasser, M. Bellare, *Lecture notes on cryptography*. Summer course on cryptography, MIT, 1996–2008 (2008), http://cseweb.ucsd.edu/~mihir/papers/gb.html
8. D. Giri, R.S. Sherratt, T. Maitra, A novel and efficient session spanning biometric and password based three-factor authentication protocol for consumer USB Mass Storage Devices. IEEE Trans. Consum. Electron. **62**(3), 283–291 (2016)
9. M. Rezvani, S.M.H. Hamidi, MIZAN: A secure E-voting schema with vote changeability, in *Proceedings of International Conference on Information Society*, London, 2010, pp. 548–552
10. W.A. Alrodhan, A. Alturbaq, S. Aldahlawi, A mobile biometric-based e-voting scheme, in *Proceedings of World Symposium on Computer Applications & Research (WSCAR)*, Sousse, 2014, pp. 1–6
11. D. Chaum, Untraceable electronic mail, return addresses, and digital pseudonyms. Commun. ACM **24**(2), 84–88 (1981)
12. K.R. Iversen, A cryptographic scheme for computerized general elections, in *Advances in Cyptology-CRYPTO'91*, vol. 576, (Springer-Verlag, Berlin, 1992), pp. 405–419
13. J.D. Cohen, M.J. Fischer, *A robust and verifiable cryptographically secure election scheme*. Technical Report YALEU/DCS/TR-454, Yale University, 1985
14. J.D. Cohen, *Improving privacy in cryptographic elections*. Technical Report YALEU/DCS/TR-454, Yale University, 1986
15. K. Sako, J. Kilian, Secure voting using partially compatible homomorphisms, in *Advances in Cryptology, Crypto'94*, ed. by Y. G. Desmedt (Ed), (Springer-Verlag, Berlin, 1994)
16. J. Benaloh, D. Tuinstra, Receipt-free secret-ballot elections, in *Proceedings of the Twenty-Sixth Annual ACM Symposium on the Theory of Computing*, 1994, pp. 544–553
17. H. Nurmi, A. Salomaa, L. Santean, Secret ballot elections in computer networks. Comput. Secur. **36**(10), 553–560 (1991)

18. A. Fujioka, T. Okamoto, K. Ohta, A practical secret voting scheme for large scale elections, in *Proceedings of Cyptology-AUSCRYPT'92*, vol. 718, (Springer-Verlag, Berlin, 1993), pp. 244–251
19. L.F. Cranor, R.K. Cytron, Sensus: A security-conscious electronic polling system for the Internet, in *Proceedings of 30th Hawaii International Conference on System Sciences*, Wailea, HI, vol. 3, 1997, pp. 561–570

Chapter 21
Ethical, Legal, and Social Implications of Biometric Technologies

Sudeep Tanwar, Sudhanshu Tyagi, Neeraj Kumar, and Mohammad S. Obaidat

1 Introduction

Personal identity of a human being is important due to several reasons. From a group point of view, as we narrow down the regions from a country to village routing, cities, and districts, individual communities desire to have their independent identities [1], whereas from an individual point of view, as the population is increasing, an individual tries to locate himself in the stereotype locations. This type of desire indicates his individual identity. However, another level of identity is also existing, which includes the validation of the identity. For any situation, this is very important in almost all situations like mobility domain, stationary domain, hospitals, residential areas, industrial areas, educational areas, and political, financial, and legal areas [2].

S. Tanwar (✉)
Department of Computer Science and Engineering, Institute of Technology, Nirma University, Ahmedabad, Gujarat, India

S. Tyagi
Department of Electronics & Communication Engineering, Thapar Institute of Engineering and Technology Deemed to be University, Patiala, Punjab, India

N. Kumar
Department of Computer Science and Engineering, Thapar Institute of Engineering and Technology Deemed to be University, Patiala, Punjab, India

M. S. Obaidat (✉)
ECE Department, Nazarbayev University, Astana, Kazakhstan

King Abdullah II School of Information Technology (KASIT), University of Jordan, Amman, Jordan

University of Science and Technology Beijing (USTB), Beijing, China

Fordham University, New York City, NY, USA
e-mail: m.s.obaidat@ieee.org

M. S. Obaidat et al. (eds.), *Biometric-Based Physical and Cybersecurity Systems*,
https://doi.org/10.1007/978-3-319-98734-7_21

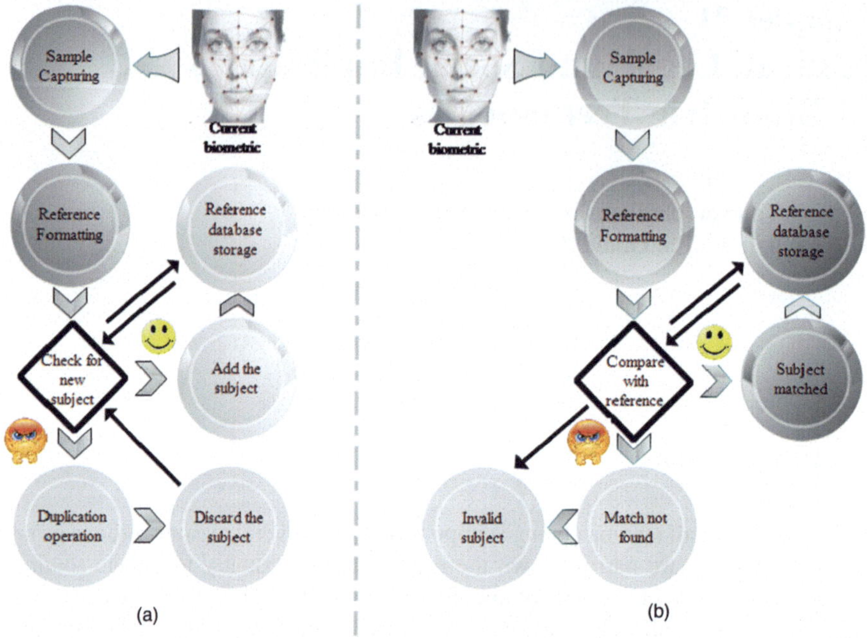

Fig. 21.1 The use of biometric technology. (**a**) Enrollment process. (**b**) Matching process

One of the interesting aspects in biometric schemes is how to execute the entire process. A detailed execution is illustrated in Fig. 21.1. Two segments have been shown, where Fig. 21.1a shows the performing of enrollment process and Fig. 21.1b shows the matching process. Enrollment process maintains a biometric data record and stores the same in a reference database. In this process, biometric sensors scan the current biometric and do the reference formatting. Now task manager checks to find out if the current subject is new or already exists. For existing subject it activates the duplication action that is discarded by the task manager, whereas new subject is stored in the reference biometric database. On the other hand, for matching process and after biometric scanning, the system performs capturing and reference formatting [3]. Now, verification compares captured/formatted biometric samples against previously stored biometric samples. Identification compares on one-to-many basis. If there is a match, this means that we have positively acknowledged and, for the situation where match is not found, this means a negatively acknowledged situation signaled to the task manager [4] on the basis of which appropriate action can be taken. Sometimes mismatch may occur due to poor performance of biometric sensors; therefore advanced check may be applied before taking the final decision [5, 6]. This process extracts a unique biometric identifier like face capturing from the raw biometric data. Different biometric identifiers are also available and are represented in Fig. 21.1.

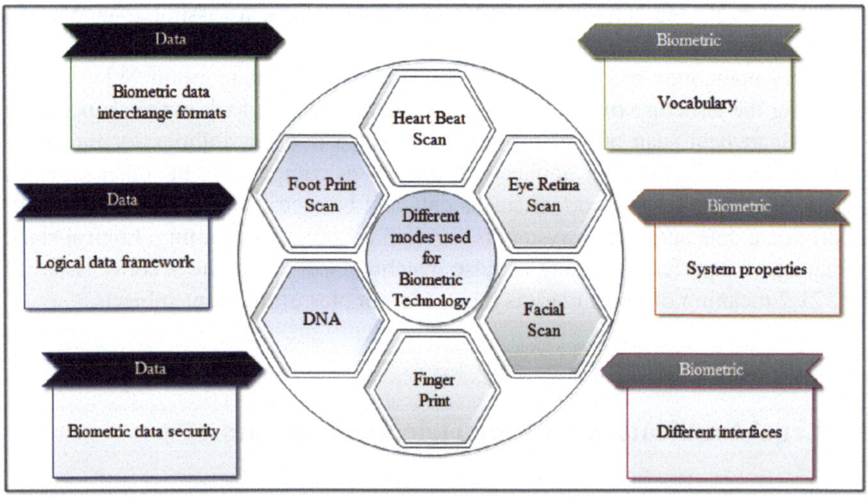

Fig. 21.2 Different modes used for biometric technology and essential stack

Fingerprint scanning, facial scanning, DNA, and eye retina scanning are frequently used biometric scanning schemes. Figure 21.2 shows different biometric scanning schemes. Fingerprint and facial scanning are most popular among the smartphone users, as using special characters and remembering lengthy password is a tedious task. The smartphones of Samsung Galaxy, Moto series of Motorola, latest version of iPhones, Oppo, and some other popular smartphones have the security option of fingerprint and facial scanning. Smartphone fingerprint readers are usually placed at the backside of the handheld device so that single-handed biometric authentication can easily be done [7]. A front-end camera has been used for the facial scanning. Out of these two, fingerprint scanning as a biometric authentication is a very easy and secured option and is considered an effective replacement to traditional password, whereas on computers and laptops, this is not a new concept; in fact banks are using this technique very commonly these days [8].

The iris of the human eye is a very interesting component of the biometric scanning technology. In medical science the circular region between the sclera and pupil is known as the iris. The iris is one of the stable biometric techniques [9] as it is stable over the long life of a person starting from childhood. The iris has very high pattern changeability between the twins and even two eyes of an individual. Due to this feature, nowadays, smartphones are also using this biometric technique for authentication access. However, this facility is available to the smartphones having high quality of front camera because iris recognition system has also a lot of challenges.

DNA has the structure, which is responsible for the body, cells, tissues and organs, and body components. Every individual has a unique composition, and verification sample can be taken from any part of body, like hair, blood cell, or nail, for the DNA testing. Definite areas of DNA contained chain that repeats it on regular interval, and interestingly different persons have a dissimilar number of repetitive sections. The least building block of DNA is called the nucleotide [10], and every nucleotide has a dioxide ribose base and phosphate group. When one is analyzing the structure of DNA for identification, then we look at the sequence of bases. Heart beat scan and footprint scan are also two possibilities for biometric scanning methods, but their applications are less as compared to the other schemes discussed above. Two important stacks data and biometric are shown in Fig. 21.2. Appropriate data acquisition system is required for proper formatting. Logical stack is required for perfect matching as false matching will create the security issue. In Fig. 21.2 meaning of vocabulary is biometric samples of different subjects.

2 Legal Implications of Biometric Technologies

In order to prevent the fraud, now governments are using biometric technologies like fingerprint, face recognition, and iris, which can be used in government official documents such as LPG gas services and Passport Seva Kendras. The latter is an Indian service that aims at delivering passport services to citizens in a timely, transparent, more accessible, and reliable manner and in a comfortable environment.

As time passes corporations in city are also using biometric technologies for their services in order to provide better services to the citizens of country in a timely and uninterrupted manner. Business organizations can also use the biometric technologies in their workplace to provide secure access control. But there exist the legal challenges while using the biometric technologies by both public institutions (government, corporations, etc.) and businesses. In Europe, with the use of biometric technologies, the most important legal challenges are in the extent of conformance to privacy and data protection mechanisms. To overcome it, the European Bio Sec consortium is working to develop a legal framework that ensures use of biometric technologies with full compliance as per European regulations to protect data.

Legal issues include the lack of precedence, ambiguous processes, imprecise definitions, and logistics of proofs and defense, as shown in Fig. 21.3. Main problems here are the type of information collected about individuals and how this can be used. Moreover, the ability to access information and reparation inaccuracies,

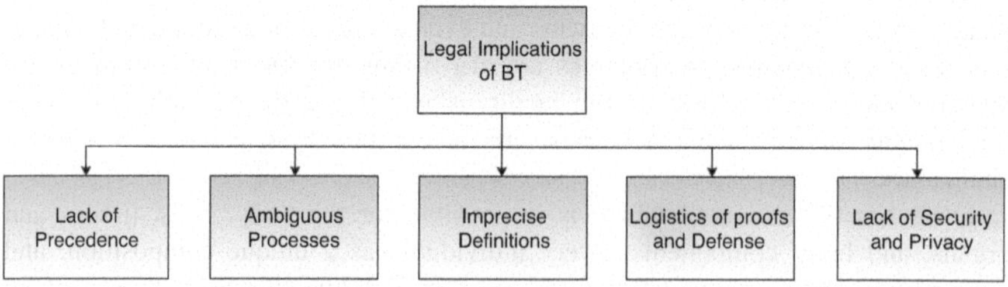

Fig. 21.3 Legal implications of biometric technologies

and providing secure mechanisms so that information related to the users cannot be compromised, especially while enrollment process of biometric schemes going on and throughout data transmission over public networks. Approaches used for storing biometric templates remain a critical issue, because RFID chips are used by users to process their information on smart card. Some challenges such as emergency procedures used during the time of failure in automatic technology processing need to be taken care.

Even though many biometric legal proceedings have been showed at global level, so far there are a few large-scale schemes operating in full swing. In some of the cases that have been conducted, they actually have been found to be illegal and in breach of users' fundamental rights. For example, if biometric project given to any data protection agency for use of biometrics stored on physical devices and for controlling physical access to the secured areas in some predefined applications (like airport, hospitals, etc.) were in proportion with the use for which the data was collected, it may be feasible that the collected data can be used for time and attendance control in that application was in breach of its data protection laws. In another example that highlights the legal challenges in the use of biometrics in ID cards for overseas residents, here one can claim that the technology targets a specific group. Another legal challenge associated with use of biometric technology is ensuring that rights can be given to citizens to correct information held about them, because biometrics contains mechanisms for identification and authentication purposes. Individual identity theft is a rising problem, and, even though the bulk of such theft is still done using offline methods, the probability of online identity theft is increasing rapidly. Without providing appropriate solution, the owner of the data has fear that his/her data is not safe and he could not get back the control of his identity and feels that the biometric data fall into the wrong hands. Further details regarding the legal issues in the use of biometrics can be found at [3].

Each citizen of the country has the right to decide in principle which personal information related to his/her can be posted on the social sites. The processing of data needs to be based on legislations. It is always advisable to use the personal collected data for specified purposes; data must not be processed to any third party without taking the permission from the user; otherwise it clearly violates the rule of personal privacy [47]. Moreover, there should be significant restrictions for the use of "personal/sensitive" data. For example, there exist some standard legal challenges, which can be followed by each country of the European Union. All security applications cannot be just based upon the collected data because preventive police measure will also play an important role here along with or without notice period. Keeping above points into consideration, there is an urgent need to fix up the legal boundary over the processing of biometric data. Like in Germany [11], it is very difficult to predict the usage of existing police system, and the data protection laws allow the ease use of biometrics in structured manner. In contrast, this may be feasible during the investigation of criminal case using fingerprints of persons individually.

In defensive situations, such as the scanning of baggage at the airports and personal monitoring as well as tracking of visitors through smart CCTV systems, however, no final outcome about any criminal person is established before processing his personal data. It has been noticed from the literature that upgrading in biometric systems emerged to new challenges for personal privacy and protection of data. These new upgrades highly depend on the new used technical design. Biometric technology helps the enforcement officers to take appropriate decision on the accused, based on automated collected biometric data. Enforcement officers also proactively collect the information about the unspecified group of persons, whose activities are not normal. Hence, the complete society will be under trouble without proper legal and technical safeguards. Further, data collection centers may require proper attention to the data so that user can decide when and at what situation he/she behaves freely without worrying about the interpretation of that behavior at the other end of the surveillance system.

2.1 *Legal Perspective of Biometrics with Reference to Enforcement Officer*

Error rates can be taken into consideration by the users of the biometric technologies for law enforcement purposes. From this it has been clear that police measures on the basis of biometric can not only play final role to put any person under suspicion, but police legislator along with the view of executing officer plays final role to put any person under suspicion.

The function of the law – particularly the fundamental rights to privacy and informational self-determination, as well as the respective data protection acts – needs to be the protection of the personal rights of the data subjects. To this end, specific provisions for certain application scenarios may be necessary in the future. Furthermore, a legally compliant technology design is able to significantly reduce privacy and data protection risks.

2.2 *Main Legal Issues Associated with the Use of Biometrics*

The main concern associated with the use of biometrics is the violation of privacy. Citizens of the country who had undergone to the biometric scan feel that such procedures are disturbing because they involve scanning of body parts. Another major concern is the matter of information security. Collected biometric data are usually stored in databases/clouds that can be accessed by the employer of the company, if the data is related to the organization, and can also be accessed by the government officials. But the question that comes here is what is the secrecy of this data, as this contains private information such as medical history of patients and bank data. If any person raised any legal issue regarding process of security of biometrics, then it is usually resolved using a traditional balancing test. Outcome of this test is the privacy rights. A court will consider the person's belief of privacy versus the public need to obtain such information. For example, consider a court that wants to analyze any person's handwriting skills for national security reasons. As the public concern for national security has high priority, then a judge might conclude that it is essential to assess the handwriting skills of that person in order to protect the security of the nation.

2.3 *Futuristic Vision of Biometrics*

Biometrics is rapidly changing from fiction to fact. Consider that Fast Identity Online (FIDO) Alliance, which has members like Google, Microsoft, Visa, and Mastercard, recently came with a final draft regarding its biometric standards; final note of this draft is that "today, we celebrate an achievement that will define the point at which the old world order of passwords and PINs started to wither and die." Progressively, Microsoft's new Windows operating system – Windows 10 – may drain passwords altogether and rely on biometric technologies, like face, iris, and fingerprint recognition, that are used to not only log you into your machine but also to "authenticate apps and websites without sending a password at all" [12].

For example, consider any person holding a glass of water for your biometric data. We can say that your body parts are your password; demanding a glass of water (user voice recorded), picking up the glass of water (user fingerprint captured), drinking the glass of water (the DNA in user saliva saved), and then walking back to your four-wheeler parking (user face captured by camera) are all possibly near to handing out user password on a business card to everyone that passes by.

The other legal issue associated with the biometric technology, consider that a hacker during the press conference has taken the photos of Cabinet Minister of any country from 8 feet way with high-resolution camera and was perfectly able to reproduce the fingerprints of Cabinet Minister, or consider a criminal who was sentenced on DNA that was taken from the armrest of the chair where he was sitting during the period of interrogation. It will be crucial for the law officer to keep the collected biometric data to be properly secured because large segment of the people wants to keep the DNA profile private. People want to conduct their biometric impressions in a secure environment. In a society, your biometric collected data and potentially your identity are at the danger of being exposed. For the abovementioned reasons, biometric data needs to be properly secured.

2.4 *Category of Biometric Security*

According to legal perspective, two categories of issues for biometric security can be explored. First, level of extent to which user wants to secure his biometric data? Consider an example, if the employer of an organization provides me the access of local cloud drive that contains some confidential documents and also secures that account with a password. If I pass this password to my department colleague in the organization, then after sometimes several security issues could arise. But, at the same time, if employer secures the account with my fingerprints or iris, in this situation employer is thinking that his system is safe until unless I do not misuse the system. But at the same time from my prospective, I have to be more focused about duplication of my fingerprints and picture of iris. Therefore, in order to maintain the security of the organization, I may have to wear the gloves and required the goggles all the time in my office. Second, at what extent can the users be granted to secure their captured biometric data? Consider an example, when you are in market your picture can be taken. But at the same time, can the freedom available to you force the person to stop taking your picture or force them to destroy all already taken pictures based upon your choice for protecting your collected biometric data?

2.5 Biometric Data Protection

It is up to end user to decide the mechanism of how to protect the biometric data of end users. It is not easy to handle the legal issues associated with the biometrics, but the actual fact is that law is not alone adequate to address all the legal issues as mentioned in previous section. Companies need to set up their own policies and producers for preventing the collected biometric data from being misused. Possibly, the most critical question to be asked is: how can I provide security to the biometric data that give appropriate solution to my problems instead of creating new ones? For proper guidance regarding current or potential legal issues, you may contact respective attorney of your city as soon as possible to obtain proper advice, direction, and illustration.

3 Biometric Technology in Government Use

Nowadays, there is a rapid increment in the use of biometric technology by government agencies worldwide. Several components of biometric technologies like posture identification through facial recognition are very much popular and effective to identify the criminals in huge public places [13, 14]. With the utilization of this method, governments can upgrade their database by identifying people in different situations. In the USA after September 11 attack, security has become a primary concern as well as a challenging issue. At that time US Transportation Security Administration (TSA) has proposed a complete body scanning mechanism to be implemented on all international visitors. A detailed list of abbreviation and their meanings is given in Table 21.1. But this process violates the civil liberties. This process required the nude images of an individual and that should be stored somewhere to maintain the biometric database, and there is no guarantee that the

Table 21.1 List of abbreviations and their meanings

Abbreviation	Description
DNA	Deoxyribonucleic acid
UIDAI	Unique Identity Authority of India
LPG	Liquid petroleum gas
RFID	Radio-frequency identification
ID	Identity
CCTV	Closed-circuit television
BT	Biometric technology
FIDO	Fast Identify Online
TSA	Transportation Security Administration
BSSCs	Biometric social security cards
US	United States
EU	European Union
VIS	Visa Information System
EUROSUR	European Border Surveillance System
SIS	Scheme Information System
SIENA	Secure Information Exchange Network Application
INS	Immigration and Naturalization Service
BFA	Basic Facility of Aliens
DAA	Dutch Alien Act
DPDPA	Dutch Personal Data Protection Act
DPDA	Dutch Policy Data Act
CJCN	Criminal justice chain number
CJCD	Criminal justice chain database
BN	Biometric numbers
HF	Human factor
AU	Authorization unique
TBSP	Total Biometric System Performance
FC	Function creep
FRS	Facial recognition system
NBC	National Biometrics Challenge report
OECD	Organization for Economic Co-operation and Development
NSTC	National Science and Technology Council
IPTS	Institute for Prospective Technological Studies
UID	Unique identification
CIDR	Central Identities Data Registry
RoI	Resident of India
KYC	Know your customer
ABIS	Automated Biometric Identity Subsystems
BSP	Biometric service provider
RDD	Rural Development Department
CSs	Civil Supplies
CAD	Consumer Affair Department

(continued)

Table 21.1 (continued)

Abbreviation	Description
MOUs	Memorandum of understandings
CIDR	Central Identity Repository
CSF	Critical success factor
ATV	Ability to verify
3CFA	Credit Contracts and Consumer Finance Act
CGA	Consumer Guarantee Act
PA	Privacy Act
FTA	Fair Trading Act
CECMA	Contract Enforcement and Contractual Mistakes Act
Bio-Sec	Biometric-Security Project

images are always secured. Another possible solution in this area was to use the biometric social security cards (BSSCs) to prevent prohibited visitors to enter the USA. In case of fake BSSC, the culprit has to be punished like if visitor have the visa for his job, then job could be discontinued, and visitor has to leave the USA. In addition to this, visitors and US resident and citizens biometric database cannot be placed together as there is always a chance to misuse the same.

3.1 The Social Implications of Biometrics

Let us consider the spectrum of all available security technologies. This comprises everything ranging from both perspective of hardware and software. For example, it includes hardware devices such as network connecting devices, switches, routers, smart cards, etc. Regarding software, it includes small patches, auto-upgrades, and antivirus mechanisms, among others. When both hardware and software items are positioned and installed, there is no query about their effect on the end user [15]. It is presumed that the end users will perform their related tasks and at the same time prevent the system from cyberattacks to occur. But, nowadays, after the inclusion of biometric technology in these systems, it often gets questioned, not from the angle of its capability to strengthen the lines of protection of a business, but also its effects to the end user. One of the popular and interesting questions can be asked at this step is what is the reason behind its popularity? What are the reasons of more concern of the individuals and what will be the next step after their fingerprint is scanned? The primary reason behind asking these questions has to do with our physiological or behavioral effects, which are being captured. We do not have any control over this, and actually, the individual/citizen does not know how this information is being processed by the biometric devices.

In real sense, it looks like a black box, where no knowledge about the process is happening in between, which results in hindrance in the acceptance of biometric technology, especially in the United States of America, where 9/11 attacks took place. However, there are other countries in the world where biometric technologies are extensively adopted, especially in the lesser developed nations. The next section will cover the comparative study of acceptance rate of this technology in the USA and other parts of the world.

3.1.1 The Acceptance Rate of Biometrics in the USA Versus the Underdeveloped World

If any individual wants to know the trends of various biometric technologies on worldwide basis, then outcome is very clear: The rate of adoption of biometric technologies tends to be much lesser in the USA than other parts of the world, especially for developing nations. This segment can be better explained by considering the fundamental rights of an individual. As the citizens of the USA, fundamental rights of the individuals are secured by the constitution of country. Meaning, each and every citizen of the country will be considered as distinctive individuals in the eyes of the Federal Government [16]. If for any reason citizens are not considered as distinctive individuals, then at least in principle, there exist certain types of legal recourse that citizens can take. Due to existence of this, if there is something that goes against rights of citizens, then they can easily assert that there is a violation of constitutional fundamental privacy rights and civil liberties. This is the main issue regarding the social acceptance of biometric technologies. In general, the claims of constitutional fundamental privacy rights and civil liberty violations fall into three general categories as briefly explained below.

3.1.1.1 Anonymity

Citizens of a particular country have the conviction that when they register themselves into a biometric system by following enrollment and verification process like Aadhar registration process in Indian subcontinent, they lose their total sense of anonymity. However, this may not be true always as strong security systems can maintain it. On the other side, in absence of appropriate security systems, there is a chance of hacking the ID number, which is associated with the said biometric template [17]. As an example when the local citizens in the USA experience the discomfort for a situation, then they will claim their right to remain unidentified. However, governments may say that this is not possible technology for the security purposes.

3.1.1.2 Tracking and Surveillance

Another category of privacy right is tracking and surveillance, which is disliked very much by most people despite efforts to justify them by governments. In the USA, much of this terror looks like just watching the "Big Brother." This tracking and surveillance is also hint on in the book wrote by George Orwell, titled *1984*. The primary substance of this fear comes when the Federal Government is misusing the stored biometric information and data.

3.1.1.3 Profiling

One of the biggest fears of the American citizens is using any form of biometric technology. In similar manner to that of "Big Brother" watching, the anguish-provoking approach is known as the "mark of the beast." Here, the main focus is on recording and analysis of a person's behavioral and psychological characteristics or to help in identifying classes of people. However, the citizens of most developing nations of the world, like Africa and Asia, hardly even have a democratic constitution (or other similar documents) in which their rights are protected. This results in that the individual is not treated as unique person by the government. But, after using the biometric technologies, citizens of these countries have shown their existence. As a result of deployment of these biometric technologies, the governments have to consider this fact and also consider these people as unique citizens of the country. Therefore, the acceptance rate of biometric technologies is much higher, because it helps these people by giving new hope to them in having freedoms and rights.

3.1.2 Uses of Biometric Technology by the European Union

Lot of migrants and refugees are facing the safety issues in parts of African countries where the European Union (EU) is facing number of complicated challenges like how to determine entry or fight for the fraud identity. Therefore, biometric digital systems are suited significantly to locating the culprit migrants who crossed the territory illegally. In addition to this, the utilization of biometric-based digital systems can also manage effectively the movement of migrants. This can also aid in monitoring their identification as well as aid risk assessment systems for decision-making.

3.1.2.1 Digitization of Borders

Authorities of border control and law enforcement agencies can use biometric identification technologies to differentiate the local citizens from immigrants and refugees. The Netherlands is one of the active members of the European Border Surveillance System (EUROSUR) (EC/1052/2013). EU law (2013/1052/EC) has managed the flow of population in matters like the number of refugees and immigrants who have crossed the border in order to ensure security and minimize illegal migration and international crime and terrorism.

3.1.2.2 Identification

Proper and accurate identification is vital to border security agencies in any country [18]. Authorities managing the immigration task can either permit or deny the entry of travelers. Biometric-based recognition and identification systems play a vital role to verify the accurate identity of travelers. Recently, identity-based management technologies are working to form the instruments which are expected to confirm that the ID holder is having a valid ID and his entry can either be granted or refused.

3.1.2.3 Profiling

Profiling is basically used to prevent the situations like illegal migration, human trafficking, fraud identification, and other international crimes. Sometimes profiling is also used to find new criminals. Profiling works on the group rather than the individual travelers. Profiles are construed based on the historical characteristics of particular persons, like who is earlier found guilty under criminal record illegal entry to the country or who has previous record of identity fraud [19]. Based on the above historical characteristics, profiles can be applied on each traveler at border crossings like airports, shipyards, and a line of control. The profile characteristics of a person must be regularly monitored and updated to find the risk flag of that person. This is the reason that EU law has been created to check the risk profiling of every traveler at border control and immigration. According to Directive 2004/82/EC, [22], an EU law, authorities of airlines are required to provide the set of passenger data to border control authorities prior to their arrival.

3.1.3 EU System's Policy Domain of Migration

This system includes a wide range of networked databases and biometric digital circuits of extensive shared dataset. The standards used in EU networks in broader landscape are Scheme Information Systems (SIS I and SIS II), EU's Visa Information System (VIS) (European Commission, 2015), and Eurodac (Council Regulation (EC) No. 2725/2000). Eurodac is a centrally controlled biometric dataset of fingerprints of registered refugees. Secure Information Exchange Network Application (SIENA) (European Commission, 2015) is used by the Netherlands as the primary channel for law enforcement information sharing and exchanging of the surveillance information about illegal migration and cross-border crime at the EU's land, sea, and air borders. The following are the policy domains of migration used in EU systems.

3.1.3.1 Immigration and Naturalization Service (INS) Console

Residence permit of the citizen is formed by encrypting the registered data after enrollment of third-country national. Here the meaning of third-country national is a

Table 21.2 Standard decisions used in PROGIS console

Decision	Use of	Working
Decision-007	Hand geometry	Work premises control
Decision-008	Fingerprinting	Professional premises control
Decision-009	Hand geometry	School and restaurant premises control
Decision-019	Vein pattern recognition	Professional premises control
Decision-027	Fingerprinting	Used in professional laptops as the security concern

AU authorization unique

citizen not having Dutch, EU, Norwegian, Icelandic, or Swiss nationality. Basic Facility of Aliens (BFA) is a centrally controlled government agency, which includes the photographs and the fingerprints as the identity of all immigrants [20, 21]. This Dutch Personal Data Protection Act (DPDPA) applies to all collected data by INS. Dutch Alien Act (DAA) was also formed to update the identity and set a limit for 10 years. In order to use the INS console, all third-country persons should be enrolled under INS console. However, this is not practically possible.

3.1.3.2 PROGIS Console

PROGIS is defined in Dutch language as *Programma Informatievoorziening Strafrechtsketen.* This standard was used for law enforcement agencies. PROGIS was formed under Dutch Police Data Act (DPDA). In this standard each PROGIS ID contains two numbers, criminal justice chain number (CJCN) and biometric numbers (BN). These two numbers are stored centrally in a specific database, which is called the criminal justice chain database (CJCD). BN and CJCN work together here; BN signifies State ID and CJCN indicate the criminal fellow [22]. PROGIS console is popular and reliable as it performs the identification of an immigrant before a policeman. A list of standard decisions used in PROGIS console is summarized in Table 21.2. In addition to this, Table 21.3 represents the deliberation review of biometric system for 10 years CNIL's report.

4 Perception of Biometrics

The major factor that affects the social recognition of biometric technology is the overall perception that how an exact modality behaves on the first impression. This segment is best suited for a scientific study known as "human factors" (HFs). In the market of biometrics industry, huge pressure is created on the vendors to develop the fastest and powerful algorithms. Ultimate goal of this is to attract the persons and provide them the platform so that they can use the biometric systems without any difficulty. Therefore, biometric suppliers are more focused about the theoretical and practical features of the modality, which is developed by them. Whereas, less

Table 21.3 10-year CNIL's deliberation review with respect to biometric system

Deliberation	Duration	Objective(s) of the deliberation
n^{0}57	16 November 2000	Controlling of the employees' working time
n^{0}23	17 February 2005	Control accessing of staff to sensitive areas
n^{0}031	17 February 2005	Working time management purpose
n^{0}034		
n^{0}035		
n^{0}036		
n^{0}037		
n^{0}113	7 June 2005	Controlling the area of ID cards and passport collection
n^{0}135	14 June 2005	Working time controlling of hospital staff
n^{0}101	27 April 2006	To control the working time management of staff
n^{0}051	21 March 2007	Accessing of sensitive areas of chemical plant
n^{0}080	25 April 2007	Control accessing of operation rooms of hospital
n^{0}146	21 June 2007	Control access to the specific casino
n^{0}254	13 September 2007	Ecureuil lease society-based biometric system
n^{0}256	13 September 2007	To control access of restricted areas
n^{0}038	7 February 2008	To control the presence of mentally disabled persons at workplace
n^{0}056	6 March 2008	Specific use
n^{0}084	27 March 2008	For the experimentation purpose
n^{0}178	26 June 2008	To control access of establishment
n^{0}324	11 September 2008	Access control on accommodation center
n^{0}328	11 September 2008	Specific use
n^{0}492	11 December 2008	To control access of community home of small age workers
n^{0}360	18 June 2009	Control access in examination rooms
n^{0}526	24 September 2009	Control access to the hotel
n^{0}033	11 February 2010	Simple biometric identification system
n^{0}131	20 May 2010	Control access to the specific casino
n^{0}464	9 December 2010	Control access to the satellite control posts
n^{0}147	19 May 2011	Control access to the catering
n^{0}185	23 June 2011	Specific use
n^{0}223	21 July 2011	To control access of restricted areas
n^{0}257	21 September 2011	To control the data processing center GROUPE MIT
n^{0}280	21 September 2011	To control access to the specific site
n^{0}282	21 September 2011	Specific use

(continued)

Table 21.3 (continued)

Deliberation	Duration	Objective(s) of the deliberation
$n^0$388	1 December 2011	To control access to the luggage storage
$n^0$423	15 December 2011	Automatic identification of a speaker or an author of a text
$n^0$236	12 July 2012	For the experimentation purpose
$n^0$322	20 Sept 2012	Upgraded version
$n^0$039	2 December 2012	For the experimentation purpose
$n^0$375	25 September 2014	For the experimentation of a system "talk to pay"

CNIL-The French data protection authority

attention is being paid on the comfort to have complete enrollment and verification processes. In fact, there is not much reported literature on the misconceptions of the biometric device by users as well as the tools and effects of external environment available to the individuals.

Hence, biometrics industry must ensure while developing a new biometric modality that time should be a crucial factor during enrollment and verification. Moreover, it should provide the equal priority as given to the HF variables. This approach has the type two-pronged approach and is known as "Total Biometric System Performance" (TBSP). In short it is defined as the incorporation of biometrics into wider range and must stand intelligently while taking the growth of such type of applications.

4.1 Function Creep

The main objective of biometric technology is to offer sources to verify the identity of an individual; for that any biometric scanning processes, as discussed above, can be used. This process is complicated and passes through enrollment and matching procedure, and finally it should be stored in the safe custody server. Most of the time, individuals are worried about the security of personal information, as it may be used either intentionally or unintentionally for any security reasons. This theory has been taken into consideration by US citizens who are worried as they do not know if their respective personal information is used or will be used for other purpose and even if their prior permission was not taken before using personal information for purposes other than what it is intended for. This phenomenon was known as the "function creep" (FC). In order to save time and money, if a facial recognition system (FRS) was used at the one entry point of a shopping mall, the same can be used for same individuals at another point of entry.

4.2 *Fraud or Identity Theft*

One of the major perceptions with the biometric technology is the belief that it is foolproof technology; actually it is not always 100% possible. Due to this weakness, there is always a risk of someone imitating an individual and stealing his ID by capturing some database. Expectation in terms of accuracy from the biometric database is huge, as database has the unique qualities of an individual [23]. At the same time, there is no guarantee from the biometric technology to prove the innocence of an individual. Considering the situation of changing the passwords in common security systems, it is impractical to replace biometric readings with another one from the same person. However, a crime-based situation may occur when a person will cut off the finger of another person to access the security system, laptop, tablet, or vehicle of that person.

4.3 *False Scanning by the Sensor*

Depending on the performance of scanning sensor, there is a possibility to have false-positive and false-negative readings while comparing current scan data and pre-existing biometric database. Having such a situation will break down the system, as it provides the false reading, which is nowhere required. This situation is very much complicated as a valid individual may be denied access through the system, whereas the access is given to someone who would not be allowed to do so. Hence, privacy issues of an individual with biometric technology are not safe, and important information could be known to others; examples include marital status, religion, age, gender, salary, social security number, or employment situation.

5 Ethical Issues

Biometrics is now progressively more used for the identification and verification of a person in several applications like in insurance sector, visa preparation, border controls, health systems, and attendance system in offices and educational institutions. Biometric technology is very popular among the customers as accessing mechanisms of this technology is very easy and practical. Further, many applications are easily compatible and safe with the involvement of this technique, low maintenance, and the decrement in price of the biometric equipments also attracting the customers.

The biometric system used for distinctive proof of identity of any individual has been available in literature. Some latest examples that evidently emerge are the Frenchman Alphonse Bertillon who had invented "anthropometrics" in 1888 and the Englishman Edward Henry who introduced the concept of fingerprint technology first time in the world in 1900 in order to trace the fingerprints of criminals. However, due to the increase in terrorism all over the world, biometric technology has been gaining popularity as there is a need for better security systems. There are many benefits in using biometric technologies such as their distinctive and unique features and ability to satisfy the necessity to provide accurate verification and authentication. Some applications where biometric technologies are used include identity cards, e-passports, visa service, e-borders for protection of borders, airport security, and police investigation [24–26]. From commercial point of view, the leading players that use biometric technologies are the tour and travel sector, action parks, transportation systems, banks, and financial institutions.

At the same time, using biometric technologies is leading to many serious ethical implications. Some of the issues are the disclosure of personal identity, the conflict with one's principles and morals, and use of his/her personal biometric data for any other purpose. The civil liberty organizations claimed that the biometric-related technologies reduce the human rights related to privacy. It is unpleasant and has the ability to make serious impact on personal freedom and democratic rights. The technology is always prone to failure and is not false proof as it can be deceived. Nevertheless, many issues and threats around the security world exist, such as risk of terrorism, stealing of personal identity and fraud, security, and entry of illegal immigrants. It has now become important to have the ability to check the person's identity for later identification and verification [27]. After what happened in 9/11, organizations and governments worldwide have increased the use of biometric schemes for identification, verification, and authentication. Nowadays, hardware required for installing biometric technologies has better quality in design and correctness. The prices have also decreased, which moved biometric technology to the mainstream, both by individuals and organizations.

The abovementioned issues cannot really be ignored. There is an urgent convincing need to find practical solutions, which can be deployed easily without any difficulty or hindrance. Academics play an important role through research and development, discussions, seminars, awareness, training, and education.

5.1 Ethics and Biometrics

Relevance of the ethics under biometric technology mainly focused around the security, in the very initial phase of this growing technology. Many issues have been related to the rights of an individual. These include safety of the individuals and their dependents, security of personal and official database, privacy, and autonomy. One of the challenging issues is to determine the relationship between the person and his cooperative rights. Biometric technology was initially planned to help out the individuals in terms of their security, but at the same time, this technology has enhanced their roots not only in personal activity, but in social and collective identity as well [28]. Subjects related to sociology are very basic and common at higher school levels and early age of college levels. But these subjects are not having the

Table 21.4 6-year project report issued by official bodies on ethical and social aspects of biometrics

Year	Report	Wide area of application	Remark
2001 [28]	RAND	US defense biometric	Used for sociocultural aspect
2003 [29]	–	European Commission biometric	Working party data protection
2004 [30]	BIOVISION	Integration between client and system	Deployed successfully
2004 [31]	OECD	–	Biometric technology
2005 [32]	IPTS-EC	Joint research	To check the effect on society
2006 [33]	NBC	NSTC	US based

Fig. 21.4 RAND report

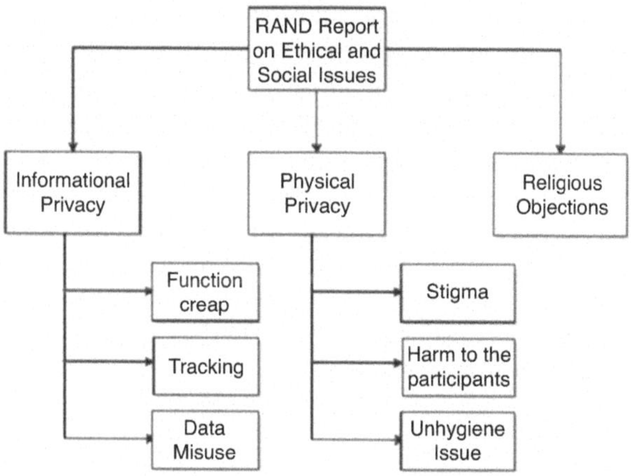

significant knowledge of the ethical issues of biometrics at international level, as this technology is very popular at global level. Now, encyclopedia of bioethics has been revised, for example, the word "biometrics" does not feature either as an entry or in the analytical index [31]. In fact, the academic press [32] says it the encyclopedia of human biology and few others encyclopedia of science and technology do the same [33]. Till 2006 few reports have been issued by official bodies on ethical and wider social implications of biometrics; details are given in Table 21.4.

5.2 *RAND Report*

Detailed structure of RAND report submitted by the RAND institutions on the request of US army is shown in Fig. 21.4. This structure was prepared to explore the legal, ethical, and sociological concerns raised by biometrics.

RAND report on ethical and social issues of biometrics use was basically prepared to focus on informational privacy, physical privacy, and the factors related to the religious concern. Citizens of any country are having the fear of insecurity of their personal information that was taken by biometric scanners for the specific application. Function creep, tracking, and data misuse are the important factors in this domain; out of these three, function creep has already been covered in Sect. 4.1. Tracking was similar to the function creep; it can perform the real-time monitoring of the actions performed by an individual. Misuse of data is a major concern as personal information of an individual can be either intentionally or unintentionally used to the situation that was not in the knowledge of that individual. In most of the reported literature, this type of factor has reported as unavoidable risk of the biometric database. The major concern of biometric technology was to enhance the security of any system, but as for appropriate knowledge concern, identification from biometric technique is based on the statistic rather than a full proof mathematic. Meaning is that the probability of mismatching of different templates of biometrics of a primary individual with secondary individual signifies the valid identification of primary. Under physical privacy concerns, report has suggested three kinds of risk: (i) the stigma linked with biometrics, (ii) chance of genuine injury to the person taking part in the process of enrollment either by technology or by surrounding, and lastly (iii) hygienic confirmation of the biometric devices being used for the process may not be available. The last two issues are directly related to the public in terms of direct harm; hence, extra precautions have been recommended in the report for these two issues. The stigma may be an important factor when biometrics is essential for a particular application. Finally, the RAND report addressed some religious objections to the biometric technology.

5.3 Business Ethics and Biometric Security

A list of ethical concerns with biometric identification techniques [34] have been investigated by users:

(a) Retina scans is one of the biometric identification techniques that is fairly invasive.
(b) Number of persons is having a thought that collecting the fingerprints is associated with a record keeping to the criminal behavior of that person.
(c) In general persons may feel the loss of their personal dignity and security while giving their detailed biometric information to a central platform.
(d) Sometimes matching sensor may have scanning problem that may cause the embarrassing feeling among the people if there is a matching error due to malfunctioning of the matching sensor device.
(e) Personal security of an individual may be affected during the automated face recognition taking at public places without prior information to that person.

(f) Most of the time, persons may ask many questions like, how this data will be stored and used? What are the security measures taken while storing the personal information? Who will be responsible for maintaining the safety of electronic biometric information and database?
(g) As we know every individual is worried about his own and his loving one's security. Therefore, the use of biometric scanning techniques in business and government can offer a one-step enhancement on the security of organizations and individuals. However, it may create some issues such as privacy of individual as this is largely affected and any misuse of the same can be harmful. The issues discussed above are some of the major concerns.

6 Case Study

Since the last few decades, biometric technologies are frequently used and very successful in their respective field. In addition to this, the technology is very reliable and provides secure access control. Several successful deployments on medium- to large-scale environments like airports, railway stations, educational institutions, and country borders prove their feasibility. However, applications involving biometric systems for security and privacy concerns, facing lot of challenges to satisfy the customers and end users. In order to find the difficulties for the implementation and possible solutions, a couple of case studies have been considered and discussed in details in the forthcoming sections of this chapter.

6.1 Case Study of Megaproject UIDAI of India

In order to maintain the security of the country, government takes several actions with the help of biometric technologies; this task has rectified several issues of security. Identity of the resident of that country can be taken through biometric scanners and stored to the secure server to maintain the security. One of the Asian countries, India, has started one program for their citizens to provide them with an authorized unique identification (UID) [35]. This program is the part of megaproject Unique Identity Authority of India (UIDAI). The actual purpose of UID is not only to provide the authorized identification but also to provide better services and support to the resident of India (RoI). This program is successfully executing in the all states of the country and is helping the government and RoI. This section is including the development of UID program in details like involvement of different process, execution, and applications. UID is a program of the Indian government that leverages emerging biometric scanning-based technologies to help various government and commercial agencies to identify and maintain the database of RoI. Importantly UID is not an identity card, but a number of 12 digits, which is stored on the cloud server. Cross verification of any UID can be done by comparing the number to biometrics that are collected and stored on a Central Identities Data Registry (CIDR). This number is unique. The UID program is similar to another program of the government of India, known as know your customer (KYC).

6.1.1 System Architecture of UID

At the organization level, biometric data processing was taken very carefully. CIDR collects three Automated Biometric Identity Subsystems (ABIS) that run simultaneously. Several organizations are using the single biometric scanning mechanism like offices, educational institutes, banks, insurance agencies, and shopping malls. Therefore, keeping in the view of Indian population (1.32 billion as of today, which is the second highest in the world), utilization of three ABIS enhanced the accuracy and minimizes the mismatch rate. The single ABIS enrollment can easily be tracked and misused by multiple systems that ultimately put a question mark against the security of an individual, it also decreased the dependence on single vendor and gave the UIDAI an ability to test and introduce the new solutions. A detailed system architecture of UID is given in Fig. 21.5. The three ABIS are operated by outsourced

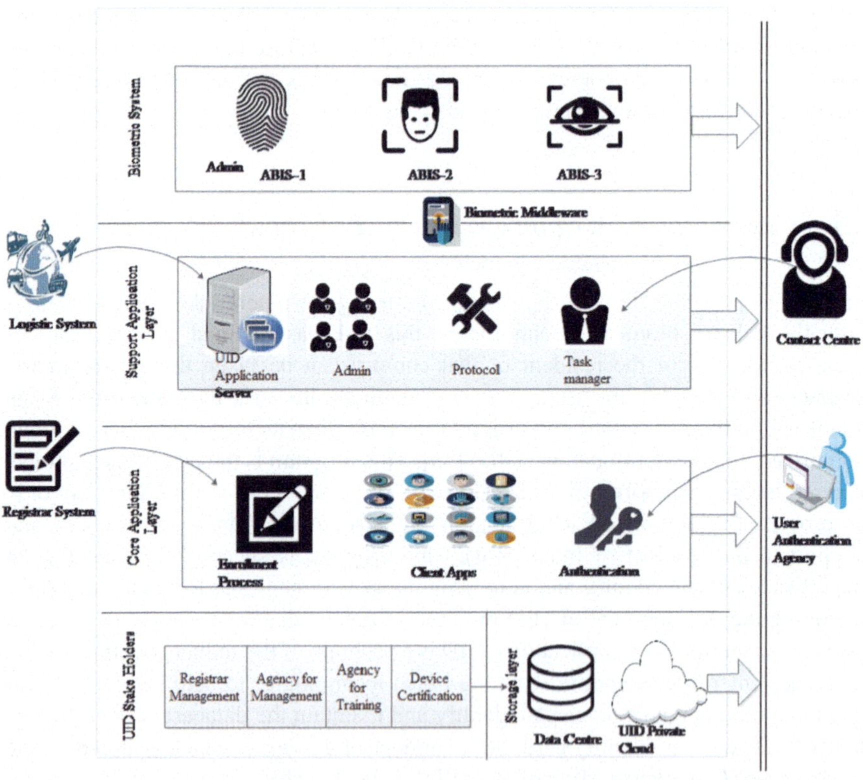

Fig. 21.5 System architecture of UID

biometric service providers (BSPs). They maintain the proprietary fingerprint and iris image templates in their respective database server. Every ABIS autonomously de-duplicates across the complete range of enrolled data and stored all enrollments safely. Quality of every ABIS is that it is independent of the vendor and flexible in terms of modification. The performance of every ABIS is examined over every ten million de-duplications, and as per ABIS's performance, data is reallocated. The collected information from individual vendor is template based and stored centrally at UIDAI level, and ABIS records have been maintained in the form of proprietary templates.

A possible situation may occur when a vendor does not want to continue the services; then a new vendor will take over the same in the form of copy of enrollment records. Further, it convert the same and store for de-duplication in order to help the new vendor demographic de-duplication can be provided to decrease the errors. The continuous accuracy monitoring protocol, the UID Middleware, has been inserted between the three ABIS and main application [36]. As a regular practice, the server cross-examines the entered data with the existing data and provides a scaling count known as fusion count that indicates the similarity of current data with the existing data. Here, lower fusion count represents the lowest similarity, whereas higher fusion count indicates the larger similarity. Main part of the system architecture is CIDR, which includes Tri-ABIS and UID Middleware, enrollment, authentication services, and supporting applications like administration, report preparation, and fraud detection. Supporting applications are interfaced with the logistics provider and the portal to control the internal administrative and information access. An intermediate administrative application is also used for proper user management, access control, status reporting, and business automation purpose.

To improve the data integrity and enhance the data quality, proper execution of algorithm mentioned in Table 21.5 is essential. Symbols used in Table 21.5 are defined as F^r, fingerprint scanning data; I^S, iris scanning data; F^C, facial scanning data; Q^t, biometric data using standard biometrics algorithm; B^d, biometric database;

Table 21.5 Algorithm for UID enrollment process

Input: F^r, I^S, F^C, Q^t, B^d & P^d
Output: Resident of the Country Enrolled on Aadhar Biometric System

1.	Q^t of biometric data measured using standard biometric algorithms
2.	If ($Q^t == T^h$) **then**
3.	B^d is in its required form
4.	Else
5.	Checks performed by client software to avoid any fraud
6.	B^d checked against the stored data base available with operator
7.	If (User $==$ P^d) **then**
8.	Additional photograph of hands and face taken
9.	Else
10.	Go to step 2
11.	Operator overrides of the policies set in software
12.	Further investigate the captured process
13.	All captured images sent to the central server

P^d, personal database; and T^h, threshold. It may be possible that a vendor can intentionally introduce the fake identity to the database. But the system has been designed in the beautiful way that the culprit vendor will be identified when the fake identity is identified, because a vendor can submit the respective database by providing his personal biometric as the identity of his database.

6.1.1.1 Enrollment Process

One of the main functions of UIDAI is to set up the standards for enrollment so that the governments of States and the Union Territories can identify the registrars and provide appropriate resources necessary to fulfill the enrollments of UIDs. The State departments, like the Rural Development Department (RDD), Civil Supplies (CSs), and Consumer Affairs Department (CAD), form an association with the UIDAI and sign memorandum of understandings (MOUs) as mentioned in [37] that asserts that the States are committed to "enhance efficiency in delivery of government benefits and services through accurate identification of beneficiaries and to have uniform standards and processes for verification and identification of beneficiaries." As per the agreement, a hierarchy has been formed as shown in Fig. 21.6 in order to collect the biometric data of RoI. Several enrolling agencies, either fixed or dynamic as per the requirement of data collection, have been outsourced to complete the enrollment

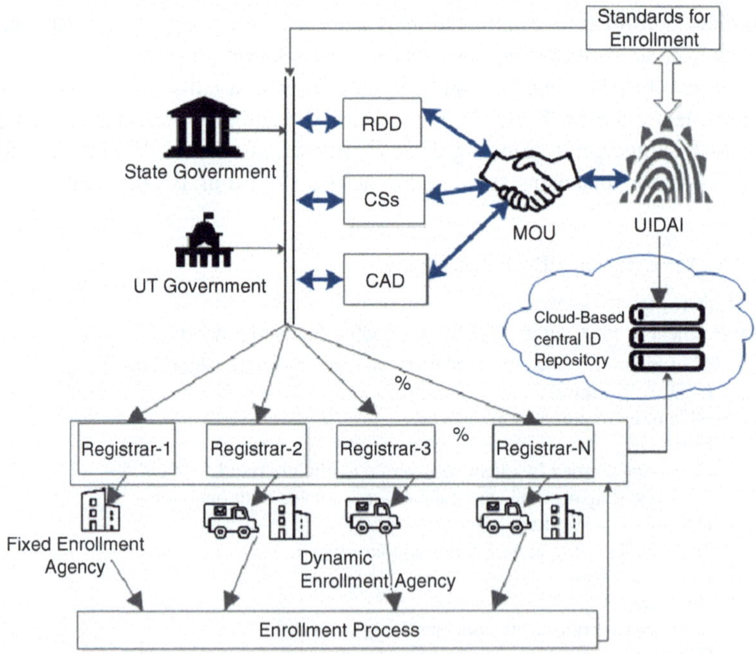

Fig. 21.6 Enrollment process

functions. Enrollments finally should be saving at UIDAI Central ID Repository after passing through registrars. Selection of enrolling agencies is based on meeting the requirements of standards of enrollment functions prepared by UIDAI. After satisfaction, they receive the respective certificates from UIDAI.

6.1.1.2 Updating and Maintenance of Database

Strong and secure cloud is a must at CIDR end because after enrollment process is over, unique identification numbers have to be assigned to the individuals. Further, this process is not fixed as the regular processing and modification on the database is the general practice. Databases stored in the CIDR cloud are of two types, viz., demographic and biometric; data may depend upon the time. RoI can relocate himself/herself as per convenience, but respective RoI has to timely inform the task manager of CIDR cloud. This information can be transferred to the CIDR via multimode which is easily accessible, like mobile phone, to the RoI. Updating process can also be maintained at the registrar end by arranging an online portal for the RoIs. This online portal is managed by the UIDAI and can be easily accessible to the smart RoIs. A systematic execution of updating process is shown in Fig. 21.7. Due to accident if appearance of the face of an RoI is changed, then this high sensitive information could not be processed either online or by mobile phone. In this situation either dynamic enrollment agency has to approach the respective RoI or RoI has to report to the nearest enrollment agency. As this is not the routine process hence, enrollment agency may charge for the updating. As per the current

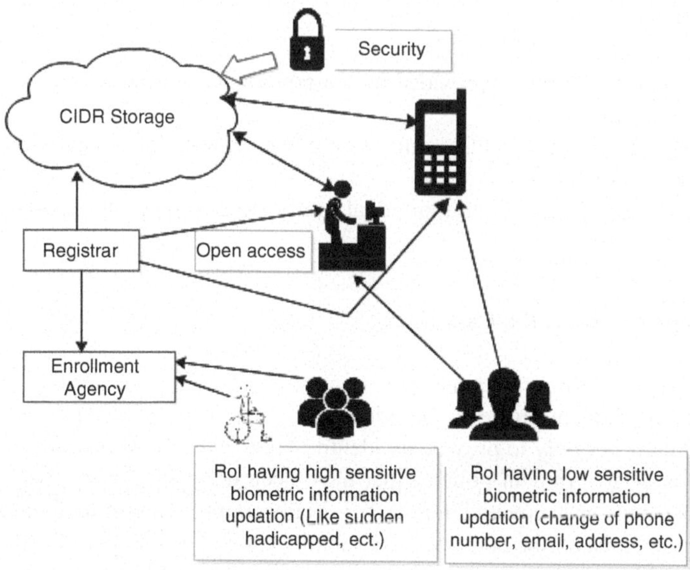

Fig. 21.7 Updating and maintenance process

knowledge, low sensitive information, like change of address, change of mobile phone, change of e-mail address, any correction in the date of birth, etc., are updated through online portal or by mobile phone. CIDR is not charging for the updating of low sensitive information.

6.1.2 Impact on Other Countries

The UIDAI project of India has received massive popularity around the globe. As per media report [37], few nations like, Mexico, Brazil, Indonesia, and Argentina are planning to change their national ID program based on the UIDAI. The India's UIDAI project is popular not only for its size and scope, but it is one of the first national ID projects to have been designed in a way that has the potential to touch every aspect of a society.

6.1.3 Applications of Megaproject UIDAI

The 12-digit UID of an individual along with his personal information in the form of a card, known as *Aadhaar card*, is provided to that individual. This *Aadhaar card* is used in several services; some of them are listed below:

(a) For the opening of a bank account
(b) For registration of new cellular connection
(c) To receive the subsidy directly to the bank account
(d) Passport making
(e) Insurance sector
(f) Health sector
(g) Ladies pension schemes provided by the government of India
(h) Income tax deduction
(i) Liquid petroleum gas (LPG) connection under Ujjwala Yojana provided by the government of India
(j) To verify the identity of an individual in the election by polling party to avoid fake polling

6.1.4 Outcome from the Case Study

We have seen in earlier sections that to increase the level of security, multiple biometrics components have to be added on a single platform. The purpose of UID in India is not only to provide an identity code of 12 digits but also to provide the benefits of government policies that have been designed for the RoIs. Several applications have been covered in Sect. 6.3. Many countries are using fingerprints as the only biometrics for the preparation of identity cards. The purpose of adding facial scanning and iris of the individual is to enhance the security level. A good number of Indian population are also involved in the labor job, so there may be a chance that quality of their fingerprint will never come to the appropriate level. In order to provide the UID irrespective of the job of an individual, UIDAI has involved tri-biometrics approach in their system. However, the project UIDAI is not fully completed in India but several RoIs are taking the benefits from this project. Hence, this is one of the successful projects of India in terms of the huge database of UID management of their citizens.

6.2 Case Study on Biometric-Based Security in Banking Sector

This case study includes identification and discussion on various issues and factors related to implementing biometrics in banking sector, specifically for the user authentication and control [38].

6.2.1 Objectives

The main objectives of this case study are shown below:

- To perform the analysis of security issues in the banking sector in New Zealand
- To check the biometric solutions as a key for security in banking sector
- To track different security strategies for banking sector

6.2.2 Background

This case study includes the survey of current banking system in New Zealand. In this study, analysis on different security models has been performed. This study also includes the involvement of biometric technology in New Zealand banking sector. We have taken this case study from [39]; the database has been collected from research journals, Internet, textbooks, and social websites. Authors in [39] have prepared the questionnaire that includes qualitative and quantitative questions for collection of information related to the New Zealand banking sectors. The questionnaire was intentionally prepared in such a fashion that gathers maximum-security issues in banking sector. Key areas of the investigation [39] for banking sector are given below:

- To know the current IT security processes
- To know the current IT security policies
- To know the current IT security infrastructure
- To know how to control the IT security
- To know how biometric technology affects the current banking sector

- To know how much knowledge of biometric-enabled security is known to the staff
- To classify the challenging issues and concerns

6.2.3 Analysis of Information Security Models Used in Banking Systems

Nowadays, computer-based secure environment is a major requirement of any banking system. As a policy of security in banking sectors, the following issues are essential along with proper auditing, confidentiality, accountability, availability, integrity, and non-repudiation [40, 41]. A systematic comparison of different models has been given in [39]. This task is executed on the physical layer of the protocol. Accessing of the banking account can be provided after cross-checking of several parameters. Several security models have been compared with those that are promptly used in the banking sectors.

6.2.4 A Detailed Discussion on the Case Study

In this study authors of [39] collected the information in the banking sector of New Zealand and performed the analysis to check the security issues in the subject. Here, the integration of biometric-based security technology with current system should not ignore the client privacy fears and account holder's tolerance levels, changing to the banking system and legal issues. A new paradigm in the case study was investigated that will support biometric-based security systems in banking sectors. This paradigm was known as critical success factor (CSFs). The CSFs have been broadly classified into four categories, viz., technology factors, monetary factors, management factors, and legal and ethical factors. Figure 21.8 represents a summary of CSFs along with related factors and respective solutions. CSFs have been analyzed by the authors of [39] according to the questioner's database. A detailed classification of individual factor of CSFs has been presented in Fig. 21.8. The next section includes the detailed discussion on every factor.

6.2.4.1 Technology Factors

(a) *Accuracy*: Accuracy is one of the measuring parameters of a biometric system; usually this parameter is used to check the involvement of technology, which is used in the current system. A number of measuring terms have been covered in the literature to check the accuracy of biometric system on banking sectors; some of them are rejection of false measurement, reduction in false matching rate, failure to false enrollment, and ability to verify (ATV) the correct measurement [28]. In the banking environment, the ATV has given the most priority in the biometric security-based banking system because it can maintain better accountability and also can give the attractive performance in terms of the fraud

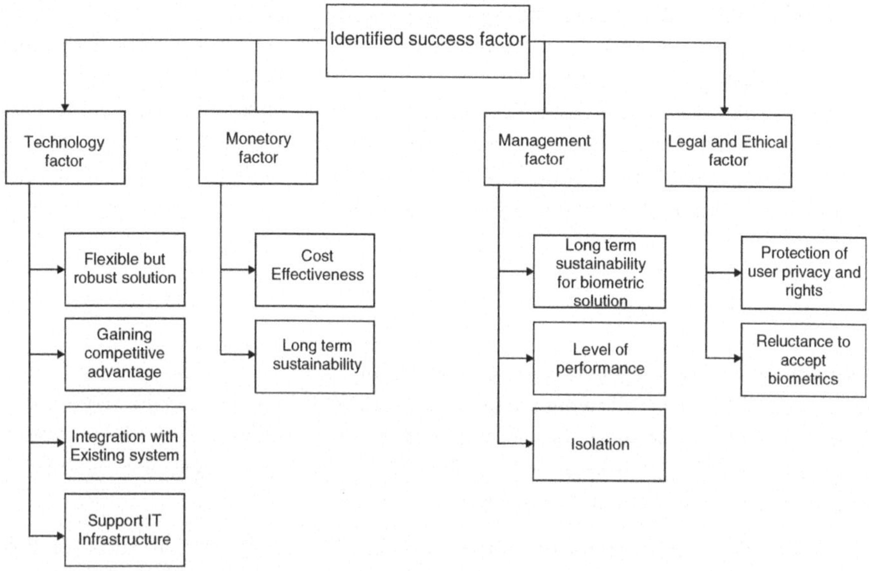

Fig. 21.8 Four main categories of success factors

detection and rejection. Therefore, on the basis of technology factor, CSF of the banking sector could be increased.

(b) *Flexibility*: Biometric systems stored the sensed database to a secure server or to the cloud. During the scanning process, some threshold value of the parameter is required. As time increases, the nature of the body of a human will also change, like wrinkles on the face and palm and rubbing of fingerprint due to hard work. Therefore biometric system must be fault-tolerant considering the abovementioned issues. During the cross verification of a person's biometrics for a low-level security system, hundred percent matching of his biometrics may not be required; hence, considering the abovementioned issues, the threshold should be decided accordingly. In banking sectors usually two levels of security are preferred. First could be biometric-based fingerprint, iris, or facial scanning and secondly either swiping of cards or password based. This mechanism involves some sort of flexibility to the client side, as user can change/update the password on regular basis. This parameter enhances the belief of client to the banking sector as security can be changed partially by himself. Hence, flexibility in the technology along with appropriate tolerance levels in biometric-based security system for the banking sector may increase the success of adoption of this system.

(c) *Privacy and confidentiality issues*: Security and privacy are always the prime issues to be considered in the system, especially for financial transaction-based system. Any system could collapse if biometric database is either stolen or intentionally misused by the service provider. Therefore, level of security must be defined earlier while enrolling the database for biometric system. Three levels of security, low, average, and high [42], can be planned for banking sectors. Different services have been offered by the bank; administrative authority of the bank, usually bank manager, can choose application-based security level. Decision for the selection of level of security is crucial as selection of the appropriate biometrics totally depends on this decision. In general low to average security could be verified from behavioral biometrics. On the other hand, physical biometrics is required for high-level security [43].

6.2.4.2 Monetary Factors

Biometric-based security systems are considerably more costly than ordinary security systems [44]. The following are the significant costs that play a major role for the success factors of case study. Integration to the existing system along with the testing costs and, secondly, high-level skilled trainers and maintenance costs are required to manage the updating in existing system.

(a) *Intangibles*: The authors of [39] survey have investigated that some of the intangible remuneration in banks can have competitive advantage and improve productivity and prosperity. By doing this, there will be a reduction in security risks which increases the confidence level of account holders of the banking system.
(b) *Long-term stability and sustainability issues*: In order to maintain the internal security system of bank, few banking systems have provided the facility of smart card to their clients. Smart clients are also in a habit to use these smart cards in their regular routine. Therefore, to maintain the long-term stability and sustainability, banking sectors have to arrange proper counseling and training modules to the existing clients to communicate the additional benefits of biometric system. To maintain the long-term relationship to existing clients, these training programs should be provided free of cost. However, a significant charge can be taken from the new clients, or it may depend upon the bank or respective government.

6.2.4.3 Management Factors

Database of employee segment collected by authors of [39] indicates that management also maintains a clear impact on the success of a biometric-enabled security in the banks. Support to the biometric innovation should be enthusiastically taken by the branch manager. Enthusiasm of the upper management staff of bank plays a very significant role on the adoption of biometric-based security to their bank. Branch manager is responsible and should arrange the minimum amount to be used in this security system. Sudden change in the working culture and the new policy may create the cause among the staff members of the bank. Hence, appreciation to the staff should be given on existing environment, and at the same time, staff should be motivated to the change in policy. By doing this activity of staff can be managed by themselves without any dispute to the possible change.

(a) *Availability of resources*: Availability of appropriate resources plays a vital role for making the biometric-based security system a great success. Some of the resources are as follows: training and orientation programs need to be arranged in a timely manner for the banking staff, especially those who are directly involved in this program. Selection of skilled and trained staff member for program is essential, infrastructure can't be tolerable for the program, sufficient capital should be arranged prior to start the program, and additional staff members and infrastructure will improve the stability of the program in case of any emergency situation.

6.2.4.4 Legal and Ethical Factors

New Zealand banking sectors are required to fulfill the government programs for the banking security like the Credit Contracts and Consumer Finance Act (2003) (3CFA), Consumer Guarantee Act (CGA), Privacy Act (PA), Fair Trading Act (FTA), and Contract Enforcement and Contractual Mistakes Act (CECMA) [39]. These factors are essential and will play a very important role for the biometric-based security system implementation process for the banking sectors.

(a) *Social and psychological issues*: The impact of biometric technology on social and psychological concern is divulging the bodies of an individual to the biometric scanning system [45, 46]. Few vendors supplying biometric equipments have been claimed that their products are of good quality and have less effect on the human biological system. However, in relation to the biometric system, this type of claims is simply myths and must be rejected.

After combining the factors as discussed above, altogether, it has been observed that in biometric-based security systems for banking sectors, legal and ethical factors could not be ignored to achieve CSF. Even timely orientation and training programs must be arranged to the banking staffs and the account holders so that they should know about the ethical and legal issues of biometrics.

6.2.5 Bio-Sec Project Development

A biometric-enabled security project known as Bio-Sec was specially developed for the banking sectors. This mechanism has Bio-Sec access controls. Authors of [39] have investigated and discussed this project. In this project, highest priority is given to the security against illegal admission, which could be internal or external. At initial level, Bio-Sec project has included the integration of biometrics with access card in order to secure the identification of internal members of the bank. Figure 21.9 shows the every segment of Bio-Sec project in details.

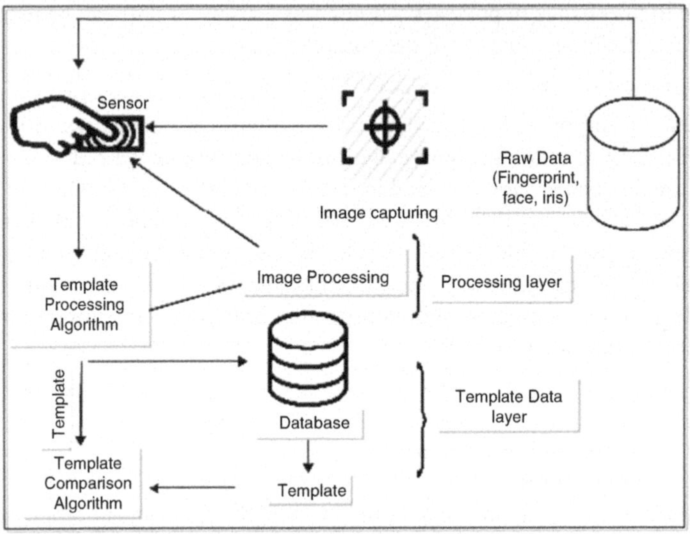

Fig. 21.9 Detailed biometric system in banking sector

Bio-Sec project has deeply classified the roles and responsibilities of individual component. In addition to this access, provisions and authentication processes are also defined properly and assigned the respective duties. The template data layer was used to store the database by using the template comparison algorithm, whereas processing layer was used to perform the image processing; template processing algorithm was used in this layer. This composite segment is known as Bio-Sec project for the security of banking sector.

6.2.6 Outcome from the Case Study

This case study was focused on the biometric-based secure technology implementation on banking sectors.

Several challenges have been identified in this work, and CSF was discussed that really enhances the adaptation rate of biometric technology in banking sectors. Benefit of this case study is that possible challenges in the banking sectors can be estimated before the implementation of biometric technology to the banking sector as there is always a significant amount of risk in the banking domain due to financial transaction involvement. Hence, this case study supports very well especially for technology factors, monetary factors, management factors, and legal and ethical factors. The success rate of the banking system will depend on the successful execution and regular monitoring of the abovementioned factors. CSF helps the banking administration to prepare their business plan, make the system flexible in terms the rectification of problems of clients, and arrange the proper healthy environment to their account holders for a successful biometric security implementation. The architecture of Bio-Sec was given so that it can be used to implement the biometric-based security to the banks.

7 Conclusion

Nowadays the adoption rate of biometric technology is rapidly increasing in all applications. Biometric technology is to be considered as an effective measure for the protection against crime. However, there is always the concern that it violates the privacy and rights of the individuals. These factors may include the possibility of fraud, identity theft, civil liberty violations, and inaccuracy of data. As a result factors may create the conflicts between service provider and public as they may be accused of a crime or may become a victim of discrimination. In these situations, persons may put up the question mark on the storage of biometric database. They can further point out the issue that their personal information may be shared with entities that are not supposed to know them. We must plan for short-term security and long-term security of biometric database separately. This process will reduce the probability of biometric data tracking. A couple of case studies has been covered in this chapter: UID developed by the government of India to provide several facilities to the RoI and inclusion of biometric-based secure technology to existing banking systems. Considering the facts discussed in the second case study, we can conclude that by proper planning, inclusion of relevant blueprint with an appropriate, flexible, and stable biometric scheme that should be focused on ethical, legal, social, and technological issues of the system can build a helpful and secure biometric-based banking system. Further, perfect planning of administration, sufficient homework on fault-tolerant schemes, policy making to sustain the security, and excellent database management will ensure the flawless biometric-based security systems that meet the future's requirements.

References

1. D. Wright et al., Ethical dilemma scenarios and emerging technologies. Technol. Forecast. Soc. Chang. **87**, 325–336 (2014)
2. E. Maria, M. Gameiro, Security, privacy and freedom and the EU legal and policy framework for biometrics. Comput. Law Secur. Rev. **28**, 320–327 (2012)
3. V. Diaz, Legal challenges of biometric immigration control systems. Mexican Law Rev. **7**(1), 1–28 (2015)
4. J. Catherine, Biometric standards–an overview. Inf. Secur. Tech. Rep. **7**(4), 36–48 (2002)
5. M. Baca, J. Cosic, Z. Cosic, Forensic analysis of social networks (Case Study), Proc. of the ITI 2013 35th Int. Conf. on Information Technology Interfaces, 219–224 (2013)
6. M. A. Kowtko, "Biometric authentication for older adults," IEEE Long Island Systems, Applications and Technology (LISAT) Conference 2014, Farmingdale, NY, 2014, pp. 1-6.

7. G. Paul, J. Irvine, IEDs on the road to fingerprint authentication. IEEE Consum. Electron. Mag. **5**, 79–86 (2016)
8. M. Krlic, "Social costs of surveillance and the case of biometrics," 2014 37th International Convention on Information and Communication Technology, Electronics and Microelectronics (MIPRO), Opatija, 2014, pp. 1278-1282.
9. S. Thavalengal, P. Corcoran, Iris recognition on consumer devices-challenges and progress. IEEE Int. Symp. on Technology in Society (ISTAS) Proc., 1–4 (2015)
10. K. Michael, The legal, social and ethical controversy of the collection and storage of fingerprint profiles and DNA samples in forensic science. IEEE Int. Symp. on Technology and Society, 48–60 (2010)
11. A. Krupp, C. Rathgeb and C. Busch, "Social acceptance of biometric technologies in Germany: A survey," 2013 International Conference of the BIOSIG Special Interest Group (BIOSIG), Darmstadt, 2013, pp. 1-5.
12. G. Hornung, M. Desoi, M. Pocs, Biometric systems in future preventive scenarios – legal issues and challenges, 83–94 (2009)
13. C. Sullivan, Digital citizenship and the right to digital identity under Int. law. Comput. Law Secur. Rev. **32**, 474–481 (2016)
14. R. Clarke, Privacy impact assessments as a control mechanism for Australian counter-terrorism initiatives. Comput. Law Secur. Rev., Volume 32, Issue 3, 1–16 (2016)
15. K. Stoychev, T. Georgiev, An alternative approach and attempt to come up with a standard for biometric user authentication in a network based environment. Procedia. Soc. Behav. Sci. **47**, 74–78 (2012)
16. P. Li et al., An effective biometric cryptosystem combining fingerprints with error correction codes. Expert Syst. Appl. **39**, 6562–6574 (2012)
17. C. Tal, M.H. Shiang, An efficient biometrics-based remote user authentication scheme using smart cards. J. Netw. Comput. Appl. **33**, 1–5 (2010)
18. T. Caldwell, Web service-based standards for biometrics interoperability. Biom. Technol. Today **2013**, 9–11 (2013)
19. T. Caldwell, Market report: border biometrics. Biom. Technol. Today **2015**, 5–11 (2015)
20. K.L. Fors, Monitoring migrants or making migrants 'misfit'? Data protection and human rights perspectives on Dutch identity management practices regarding migrants. Comput. Law Secur. Rev. **32**, 443–449, 2016
21. Y. Liu, Scenario study of biometric systems at borders. Comput. Law Secur. Rev. **27**, 36–44 (2011). 2016
22. S.M. Matyas, J. Stapleton, A biometric standard for information management and security. Comput. Secur. **19**, 428–441 (2000)
23. C. Roberts, Biometric attack vectors and defenses. Comput. Law Secur. Rev. **26**, 14–25 (2007)
24. V. Smejkal, J. Kodl and J. Kodl, "Implementing trustworthy dynamic biometric signature according to the electronic signature regulations," 2013 47th International Carnahan Conference on Security Technology (ICCST), Medellin, 2013, pp. 1-6.
25. Y. Sun, M. Zhang, Z. Sun, T. Tan, Demographic analysis from biometric data: achievements, challenges, and New Frontiers. IEEE Trans. Pattern Anal. Mach. Intell., 1–20 (2017). https://doi.org/10.1109/TPAMI.2017.2669035
26. A. S. Munalih, L. Mat Nen, A. Goh, L. K. Win, K. S. Ng and L. Ching Ow Tiong, "Challenge response interaction for biometric liveness establishment and template protection," 2016 14th Annual Conference on Privacy, Security and Trust (PST), Auckland, 2016, pp. 698-701.
27. M. Cehic, M Quigley, Ethical issues associated with biometric technologies, Proceedings of the 2005 Information e-sources Management Association Int. Conf., 1–5 (2005)
28. J.D. Woodward et al., *Army Biometric Applications: Identifying and Addressing Socio Cultural Concerns* (RAND, Santa Monica, 2003). [Online]. Available: http://www.rand.org/pubs/monograph_reports/MR1237
29. BIOVISION, Roadmap for biometrics in Europe to 2010. [Online]. Available: http://www.eubiometricsforum.com/dmdocuments/BIOVISION_Roadmap.pdf

30. Data Protection Working Party of the European commission, *Biometrics* (EC, Brussels, 2003). [Online]. Available: http://www.europa.eu.int/comm/justice_home/fsj/privacy/docs/wpdocs/2003/wp80_en.pdf; Last visited 27 Mar 2007
31. Organization for Economic Co-operation and Development (OECD), *Committee for Information, Computer and communication Policy – Working Party on Information Security and Privacy*, Biometric-based technologies, OECD (2004). [Online]. Available: http://www.oecd.org/sti/security-privacy
32. European Commission Joint Research Center, *Biometrics at the Frontiers: Assessing the Impact on Society* (2005). [Online]. Available: http://www.jrc.cec.eu
33. National Science and Technology Council, *The National Biometrics Challenge*. [Online]. Available: http://www.biometrics.gov/NSTC/pubs/biochallengedoc.pdf
34. K. Yang, E.D. Yingzi, Z. Zhou, Consent biometrics. Neurocomputing **100**, 153–162 (2013)
35. F.Zelazny, The evolution of India's UID program, Center for Global Development, 1–44 (2012)
36. Unique Identification Authority of India, *Ensuring Uniqueness: Collecting Iris Biometrics for the Unique ID Mission* (2006). [Online]. Available: http://uidai.gov.in/UID_PDF/Working_Papers/UID_and_iris_paper_final.pdf
37. Unique Identification Authority of India, *Envisioning a Role for Aadhaar in the Public Distribution System* (2006). [Online]. Available: http://uidai.gov.in/UID_PDF/Working_Papers/Circulated_Aadhaar_PDS_Note.pdf
38. T.S. Siang et al., Ethical implications of digested medical and biometric data. IIMC Int. Conf. Mgt. Corp., 1–9 (2010)
39. S. Venkatraman, I. Delpachitra, Biometrics in banking security: A case study. Inf. Manag. Comput. Secur. **16**(4), 415–430 (2008)
40. Int. Telecommunication Union, ICT Facts and Figures (2011). [Online]. Available: http://www.itu.int/ITU-D/ict/facts/2011/material/ICTFactsFigures2011.pdf
41. A. Rebera, B. Guihen, Biometrics for an Ageing Society Societal and Ethical Factors in Biometrics and Ageing. Int. Conf. of the Biometrics Special Interest Group (BIOSIG), 1–4 (2012)
42. S. Prabhakar, S. Pankanti, A.K. Jain, Biometric recognition: Security and privacy concerns. IEEE Secur. Priv. **2**, 33–42 (2003)
43. E. Mordini, C. Petrini, Ethical and social implications of biometric identification technology. Ann Ist Super Sanità **43**(1), 5–11 (2007)
44. S.C. Dass, Y. Zhu, A.K. Jain, Validating a biometric authentication system: Sample size requirements. IEEE Trans. Pattern Anal. Mach. Intell. **28**(12), 1902–1913 (2006)
45. C. Costanzo, Suddenly, biometric ID doesn't seem like science fiction. Am. Bank. **171**(107), 6–11 (2006)
46. L. Hunter, A. Orr, B. White, Towards a framework for promoting financial stability. Reserve Bank N. Z. Res. Bull. **69**(1), 5–17 (2006)
47. M.S. Obaidat, N. Boudriga, *Security of E-Systems and Computer Networks* (Cambridge University Press, Cambridge, 2007)

Index

A

Aadhaar card, 127, 272
Aadhaar registration process, India, 545
Accelerometer sensor, 236
Activity trackers, 357, 362
Adaptive chosen ciphertext attack (CCA2), 345
Age-related macular degeneration (ARMD)
  druses, 93, 94
  lesions, 92
  symptoms, 92
Algorithms, 19, 20
Alphanumeric passwords, 211
Analytical biosensor, 340
Application programming interfaces (APIs), 371
Artifacts, 291, 293, 299–301
Artificial neural network (ANN), 182, 303, 306
Attentional blink, 300
Attribute based encryption (ABE) architecture, 465
ATVS dataset, 246–248
Auditory verification, 263
Augmented reality (AR), 363
Authentication
  active, 6
  beat-to-beat heartbeat interval, 350
  biometric template, 339
  biometric traits, 402, 403, 405, 406
  body sensors, 348–350
  CA (*see* Continuous authentication (CA))
  cloud computing
    facial scanning, 446
    fingerprint scanning, 442–444
    iris scanning, 444, 445
    security, 442
    vein recognition, 444
  continuous, 6, 8
  could computing (*see* Cloud authentication)
  EEG, 311
  enrollment process, 404
  features, 313
  fingerprint, 350–352
  fuzzy extractors, 342
  human, 401
  identification, 308, 404
  invariant moment fusion (*see* Invariant moment fusion authentication system)
  IoT
    biometric (*see* Biometric security in IoT)
    cyber-attacks, 481
    5 network, 483
    pharmaceutical, 482
    staff secure, 479
  key/smart-cards, 401
  mobile devices, 132
  non-event-related, 312
  pass-thoughts, 370
  personal, 401, 402
  phase, 272, 342
  procedure in medical system, 343
  protocol, 347
  resistant, 310
  scheme, 308
  shoulder-surfing resistance/continuous, 313
  social engineering, 441
  static, 6
  survey of online-signature (*see* Online signature-based biometric recognition system)

M. S. Obaidat et al. (eds.), *Biometric-Based Physical and Cybersecurity Systems*,
https://doi.org/10.1007/978-3-319-98734-7

Authentication (*cont.*)
user, 211
verification, 404
wavelet-domain HMM, 349
wearable (*see* Wearable technologies)
Authorship attribution, 215
Automated attacks, 214
Automatic fingerprint identification systems (AFIS), 5
Autoregressive models, 305

B
Backpack-mounted head camera, 355
Back propagations (BP) NNs, 260
Banking sector
accuracy, 562
Bio-Sec access controls, 565
challenges, 566
client privacy, 562
CSFs, 562
flexibility, 563
implementation, 566
information security models, 562
investigation, 561
legal and ethical factors, 565
management factors, 564
monetary factors, 564
New Zealand, 562
objectives, 561
security and privacy, 563
security models, 561
social and psychological issues, 565
Battery technology, 378
BCI2000, 304
Beat-to-beat heartbeat interval, 350
Behavioral biometrics, 170, 174, 213, 217, 265, 360
HCI (*see* Human-computer interaction (HCI) devices)
Behavioral modalities, 366
Behavioral theory, 265
Bifurcation localization, 82, 83
Biohashing, 273, 274
Bimodal eye biometric system
advantage, 70
aligning system, 71
combined optical system, 71
Cyclops, 71
feature extraction and matching, 72
fused score, 70
image acquisition, 71, 72
Biological traits, 125, 126
Biometric attacks
CCA, 345, 346
CPA, 346, 347
impersonation, 347
man-in-the-middle, 347
software module, 347
template database, 347
Biometric authentication
cloud computing (*see* Cloud computing)
EEG-based, 312
established, 5
fingerprint, 28
online signature-based, 277
wearable (*see* Wearable technologies)
Biometric-based communication protocols
beat-to-beat heartbeat interval, 350
fingerprint-based authentication, 350–352
health monitoring, 348
HTER, 348
IPIs, 348
MIT PhysioBank database, 348
psychological data, 348
secure data transmission, 350–352
security approaches in body-centric network, 348
telemedicine, 348
wavelet-domain HMM, 348–350
Biometric-based security systems
attacks, 345–347
circumvention, 345
coercion, 345
collusion, 345
covert acquisition, 345
DoS, 345
in healthcare facilities, 339
lightweight mechanism, 344
personal medical data, 344
repudiation, 345
types, 346
Biometric fusion
classification, 68
decision-level, 69
feature-level, 68
invariant moment (*see* Invariant moment fusion authentication system)
model, 68
rank-level, 69
score-level, 69
sensor-level, 68
Biometric identification (BioID)
bio template, 450
design, 448
facial and voice recognition, 448

interface, 450
pictures, 450
sign in GUI, 449
user verification, 449
verify prompt, 450
Biometric identifiers, 13, 14
Biometric identity based encryption (BIO-IBE), 469–471
Biometric performance, 6–8
Biometrics
authentication (*see* Authentication)
behavioral (*see* Behavioral biometrics)
biological traits, 125, 126
border control, 127, 128
cognitive (*see* Cognitive biometrics)
EEG (*see* Electroencephalography (EEG)-based biometrics)
Gait recognition (*see* Gait biometrics recognition)
healthcare sensors (*see* Healthcare sensors)
multimodal (*see* Multimodal biometrics)
physiological, 126
3D hand geometry (*see* 3D hand geometry recognition)
2D hand geometry (*see* 2D hand geometry recognition)
Biometrics as a Service (BaaS)
BioID (*see* Biometric identification (BioID))
cloud computing technology, 447
definition, 447
FingerCheck, 452
IaaS, 439
identifiers, 447
ImageWare Systems, 448
IriTech, 451
MorphoTrust, 452, 453
PaaS, 440
SaaS, 440, 447
Biometric security in IoT
advantages, 488–490
applications, 487
authentication approaches, 491
communication, 492
daily activities, 492
eHealth systems (*see* E-health systems)
end users and application areas, 490, 491
extraction
applications, 492
components, 492
effectiveness, biometric scheme, 494
framework, 494
FT, 495
Gabor filtering, 497, 498
local binary pattern generation, 496, 497
RZP, 497, 499
SIFT, 499, 500
grounds, 484
identity, 488
list of, 505
password based/PIN based security, 488
revenue market, 504
safety, security and access plane, 492
technologies, 488
tetrahedron shape security system, 493
type of attack, 504
types, 488
voice biometrics, 484
working procedure, 488
Biometric system
architecture, 375, 376, 378
characteristics, 3
classes, 2
cognitive (*see* Cognitive biometrics)
healthcare sensors
authentication phase, 342
EER, 344
FAR, 344
faulty enrollment, 344
FRR, 344
fuzzy extractor, 342, 343
registration phase, 341, 342
human identification, 2
IoT (*see* Internet of Things (IoT))
operation modes, 4, 5
performance, 6–8, 408, 410
requirements, 3
usage modes, 5, 6
uses, 3, 4
Biometric technology
adoption rate, 567
DNA testing, 537
enrollment and matching process, 536
ethical implications (*see* Ethics and biometrics)
fingerprint and facial scanning, 537
identification, 536
identities, 535
iris scanning, 537
legal implications (*see* Legal implications of biometrics)
perception (*see* Perception of biometrics)
security, Banking sector (*see* Banking sector)
schemes, 536

Biometric technology (*cont.*)
social implications (*see* Social implications of biometrics)
UIDAI
applications, 560
enrollment process, 558
impact, 560
outcomes, 560, 561
program, 555
system architecture, 556
update and maintenance, database, 559
Biometric traits, 2, 5, 9
BioSemi ActiveTwo, 296
BioThenticate, 127
BIOWATCH, 368
Blind source separation (BSS), 303
Blood-glucose biosensor, 340
Blood vessels detection, 80, 81
Blood vessels segmentation
and detection, 80, 81
image enhancement, 80
matched filter, 79
thinning, 81
thresholding, 79
Body-centric communications, 340
Body sensors
biometric-based authentication, 348–350
healthcare applications, 351
in healthcare system, 341
psychological parameters, 338
Border control, 127, 128
Brain-computer interfaces (BCIs), 288
acceptance, 313
components, 298, 299
data acquisition method, 289
extracted, classified and translated, 298
input data, 298
medical tool, 312
P300 speller, 298
passwords, 304
without feedback, 298
Brain's processing, 288
Brain's structure, 288
Bronstein and Drahansky system, 114

C
Cancelable biometrics
concept, 268, 269
filters, 269, 270
hybrid methods, 270, 271
knowledge signatures, 270
non-invertible geometric transforms, 268
salting methods, 270
template protection schemes, 271
Card-based documents, 272
CASIA dataset, 241, 242
Channel frequency response (CFR), 190
Channel state information (CSI)
characteristics, 189, 195
characterizing human identity profile, 201
classifying, SVM, 194
constructing pool, 197
COTS Wi-Fi NICs, 193
CSIR, 190
CSIT, 190
device-free motion recognition techniques, 193
E-eyes recognizes, 193
feature, 197
network adapter, 198
preprocessing method, 201
profile pool, 196, 197
and RSS, 191, 193
signals, 195–197, 199, 200
storage overload, 194
time series, 196, 205
training datasets, 203
Ubuntu 12.04, 195, 201
and Wi-Fi signals, 191, 206
Chosen ciphertext attack (CCA), 345, 346
Chosen plaintext attack (CPA), 346, 347
Ciphertext-policy attribute-based encryption (CP-ABE), 467
Cloud authentication
biometric-based, 459, 460
biometric enrolment stage, 459, 460
challenges
data access, 462
security package, 462
traditional approach, 463, 464
identity, 459
knowledge-based, 459
object-based, 459
privacy-preserving cloud-based, 473
remote (*see* Remote biometric authentication)
Cloud-based solution, 376–378
Cloud computing
Amazon's Elastic Compute Cloud, 437
authentication (*see* Authentication)
characteristics, 438
definition, 437, 455
IaaS, 439
mid-2000s, 437
mobiles, 439, 453

PaaS, 440
SaaS, 440
service (*see* Biometrics as a Service (BaaS))
Cloud service providers (CSPs), 455
Cognitive biometrics
accuracy, 397
behavioral sensing devices, 388
cycle of actions and reactions, 388
detection, human characteristics, 387
eye tracking (*see* Eye tracking biometrics)
human, 397
hybrid model, authentication, 396, 397
keystroke dynamics (*see* Keystroke dynamics biometrics)
mouse dynamics (*see* Mouse dynamics biometrics)
peripheral devices, 388
user behavior, 387
Commercial off-the-shelf (COTS) Wi-Fi devices, 195, 201
Communication system security, 8
cyber-physical ecosystem, 480
ICT, 478
service provider, 479
smart city, 481
Compound annual growth rate (CAGR), 359
Computer security, 167, 184
Computer vision-based identification, 190
Continuous authentication (CA)
accuracies, 212
actively/passively, 212
alphanumeric passwords, 211
architecture, 216, 217
authorship
analysis, 214, 216
attribution, 215
verification, 215, 216, 218
circumvention, 211
data preprocessing, 224, 226
datasets, 216
e-mail, 223
impostors, 224, 225
micro messages, 223, 224
deep learning, 227, 228
enrolment, 217
evaluation method, 226
feature space
application specific features, 220
description, 218, 219
lexical features, 218, 220
N-gram model, 220
selection, 221, 222
syntactic features, 220
writing styles, 218
initial (static), 212
keystroke dynamics, 212
machine learning models, 217
mouse dynamics, 212
n-best, 217
principle, 216
recognition process, 216
shallow classifiers, 226, 227
stylometric (*see* Stylometry)
suspicious behavior, 216
user authentication, 211
verification process, 217
withstand forgery, 228, 229
work on authorship characterization, 215
writing style, 214
Correlation-based matching, 25, 26
Correlation-invariant random filtering (CIRF), 270
Could computing
ABE, 465
BIO-IBE, 469–471
biometric sensors, 472
challenges, 472
characteristics, 456
CP-ABE, 467
CSPs, 455
F-IBE, 468, 469
IaaS, 456
IBE, 467, 468
identification, 457
KP-ABE, 466, 467
layers, services, 458
PaaS, 456
privacy, 473
regulation, 474
remote (*see* Remote biometric authentication)
SaaS, 456
security challenges, 457, 458
Counting human numbers, 205
Criminal justice system, 373
Crossing number (CN), 22, 23
Cryptographic technique, 343
Cumulative match curve (CMC), 129, 156
Cycle rotation metric (CRM), 366
Cyclops, 70

D

Data acquisition, 360
Data communication, 343
Data storage, 360
Daugman's algorithm, 46, 47
Daugman's rubber sheet model, 47, 48, 71

Decision-level fusion techniques, 323
Deep learning, 217, 226–230
Denial-of-service (DoS), 345
Depth of field (DOF), 135
Device-based methods, 192
Device-free human activity recognition, 193
Device-free indoor sensing systems, 207
Device-free methods, 192
Device-free motion recognition techniques, 193
Diabetic retinopathy (DR), 94–96
Diffraction grating, 115
Digital cosine transform (DCT) coefficient, 333
Digital cosine transform (DCT) features, 322
Digital signatures, 266
Digital wavelet transform (DWT) features, 322
Dimensionality reduction, 203
Discriminative multinomial Naïve Bayes (DMNB) classifiers, 215
Dynamic identification recognition, 200
Dynamic signature recognition
  onwards and upwards, 265
  slow and steady, 266
  SWOT analysis, 266
  time information, 265
  transducer, 265
Dynamic time warping (DTW), 257, 366

E

E-commerce
  authentication systems, 275–278
  financial
    and commercial activities, 274
    transactions, 274
  high-risk zone, 275
  insecurities, 275
  objectives, 274
  outcome, 279
  signature-based biometrics, 275
  threats
    random forgery, 278
    simple forgery, 278
    skilled forgery, 279
EEGLAB/BCILAB, 303
EEG recording devices
  BioSemi ActiveTwo, 296
  EMOTIV Epoc, 296, 297
e-Election
  architecture, election process, 515
  electoral authority controls, 515
  entities, 515
  registration, 516
  security mechanisms, e-voting (*see* e-Voting)
e-Government, 531
eHealth systems
  authentication and key agreement scheme, 501
  communication, medical systems, 501
  Maitra and Giri's scheme
    authentication, 503
    log-in phase, 503
    network structure, 502
    registration phase, 502
    setup phase, 501
Election fraud, 513, 514, 520, 521
Election integrity
  biometrics
    registration process, 518
    technologies, 519
  digital ID card, 518
  digital voter ID card, 531
  election process (*see* e-Election)
  e-voting (*see* e-voting)
  properties, 520
  security attacks, 530, 531
  security protocols, electoral system, 528, 529
  security requirements, 520, 521
  traditional election systems, 514, 515
  voters, 518
  voting systems, 513
Electrocardiogram (ECG), 370
Electrode placement, 291, 292
Electrodes, 288, 289, 291–294, 296, 297, 301, 305, 306, 311
Electroencephalogram (EEG), 370
Electroencephalography (EEG)-based biometrics
  acceptance, 312, 313
  artifacts, 291, 293
  attacks, 312
  BCIs, 298, 299
  biofeedback systems, 289
  brain's processing, 288
  classification, 303
  data acquisition method, 289
  definition, 288
  in diagnostics, 289
  electrodes, 291, 292
  emerging biometrics, 370
  EMOTIV Epoc, 289
  ERP, 292–295
  ethics, 312, 313
  event-related paradigms (*see* Event-related paradigms)
  experiments, 299, 300
  features, 299

fMRI, 288
frequency bands, 289
internal organ, 288
and MEG, 288
near-infrared spectroscopy, 288
neurons, 288
non-event-related paradigms (*see* Non-event-related paradigms)
performance, 311, 312
perspectives, 313, 314
privacy, 312, 313
recording devices, 296, 297
resting state with eyes open, 289
RSVP, 299, 300
sampling rates, 289
selecting and extracting features, 301–303
signal-to-noise ratio, 300–302
software
BCI2000, 304
EEGLAB/BCILAB, 303
FieldTrip, 303
Psychtoolbox, 304
spectral density, 289, 290
SSEP, 310
SSVEP, 295
time-frequency representation, 289, 290
visual stimuli, 299
voltages, 288
ElectroMechanical Film (EMFi), 238
Electromyogram (EMG), 301
Electronic signatures, 266
Electrooculogram (EOG), 301
E-mail dataset, 223
Embedded method, 180
Embedded pressure sensor, 236
e-medical system, 337
Emerging biometrics
accelerometer data, 370
accuracy and recognition rates, 369
advanced sensors and associated software, 371
authentication, 370, 371
ECG, 370
EEG, 370
heart rate, 369
human daily activity information, 370
maturity and richness, 371
pass-thoughts, 370
sensors, 369
system architectures
authentication, 374, 375
big data, 374
biomedical information, 374
cloud-based solutions, 376–378
data/template storage, 375, 376
description, 374
template matching, 376
training, 374
verification, 374, 375
wearable, 374
3D signature verification, 371
user conditions, 369
wearable, 371–373
wristband, 371
EMOTIV Epoc, 289, 292, 296, 297
Enrollment phase, 272, 459, 460
Enrolment, 5, 16, 17, 30, 127, 132, 133, 217, 226, 241, 323, 331, 374, 376
Equal error rate (EER), 51, 134, 156, 213, 333, 344
Error-correcting output code (ECOC), 370
Error equal rate (EER), 366
e-security, 165
Ethics, 154
Ethics and biometrics
disclosure, personal identity, 552
identification techniques, 554, 555
illegal immigrants, 552
issues, 552
RAND report, 553, 554
threats, 552
Euclidean distance, 366
Event-related paradigms
analysis, 305
audition and vision, 304
biometric system, 305
components, 305
experimental protocol, 305
internal, 304
secret components, 307, 308
VEPs (*see* Visually evoked potentials (VEPs))
Event-related potentials (ERP)
amplitude, 293
components, 294
latencies, 294
N400, 295
P300, 294, 295
topographies, 294
Evoked potentials
SSEP, 295, 310
VEPs (*see* Visually evoked potentials (VEPs))
e-voting
adaption, 514
Ahmed and Aborizka's scheme

e-voting (*cont.*)
        architecture, 521
        registration phase, 522, 523
        Tally phase, 524
        VDC, 523, 524
    Alrodhan et al.'s scheme
        architecture, 524, 525
        registration phase, 525
        result announcement phase, 527, 528
        steps, voting phase, 525–527
        verification phase, 525
    authenticity, voters, 513
    biometric-based, 518
    channels, 513
    digital voter ID card, 514
    internet-based e-services, 513
    properties, 520
    registration process, 518
    security attacks (*see* e-voting security attacks)
    security issues and challenges, 513
    security mechanisms
        biometric smart token, 517
        blind signature, 517
        digital certificate, 516
        hashed value, 516, 517
        public key infrastructure, 516
    security requirements, 514, 520, 521
e-voting security attacks
    communication messages, 531
    communications, 530
    designing, 530
    DoS, 531
    insider attack, 530
    software, 530
    template database, 531
    victims, 531
    voter masquerading, 530
Extended Gaussian Image (EGI), 129
Extrinsic verification, 216
Eye anatomy, 39, 40
    cornea, 40
    iris, 43–45
    lens, 41
    light projection, 41
    near and far point, 41
    retina, 42, 43
EyeDentification System 7.5
    optical scanning, 59
    representation, 60
    rotation algorithm, 60
Eye diseases
    ARMD
        druses, 93, 94
        lesions, 92
        symptoms, 92
    biometrics, 88
    DR, 94–96
    retinal examination
        binocular indirect ophthalmoscope, 90
        direct ophthalmoscope, 89
        fundus camera, 90
    toxoplasmosis, 96
Eye liveness detection, 38
    eye movement/winking, 87
    iris, 87
    retina, 87, 88
Eye movement scan, 389, 390
Eye tracking biometrics
    BioEye 2015, 391
    categories, 391
    data, 390
    devices, 390
    equal error rate, 391
    GRBFN, 392
    identification rate, 391
    iris recognition system, 388
    machine learning methods, 391
    movement-based, 389
    recording, eye movement, 390
    retina recognition captures, 389
    visual stimuli types, 389

F

Face fusion system
    architecture, 330
    authentication, 332
    covariance matrix, 331
    data sets, 331
    enrolment and comparison, 331
    mean and difference, 330
    recognition algorithm, 331
    Zϕ invariants, 331
Face recognition, 263
    mobile devices, 132
Facial scanning, 446
False acceptance rate (FAR), 156, 213, 256, 333, 334, 344
False rejection rate (FRR), 156, 213, 256, 333, 344
Falsifications, 256
Faulty enrollment, 344
Features extraction, 17, 239, 360
    classification, 21
    CN method, 22, 23
    Daugman's rubber sheet, 84
    eyelid detection, 85
    Gabor filter, 85
    Hamming distance, 87

LBP, 86
mask generation, 85
minutiae, 17–20
Feature subset selection
embedded method, 180
filter process, 179
wrapper method, 179
Federal Trade Commission (FTC), 281
FieldTrip, 303
Filter process, 179
Fine-grained radio signal measurements, 193
FingerCheck, 452
Finger-knuckle-print based system
agriculture, 411
categories, 411
compcode-based ordinal features, 427
correlation, 427
deep matching, 425
dorsal knuckle print, 411
hand crafted features, 425, 426
image enhancement
algorithms, 423
bi-cubic interpolation, 423
CLAHE, 423
histogram equalization, 423
LGBP, 424
performance analysis, 429–432
PolyU FKP database, 429
quality estimation
assessment factors, 412
data acquisition phase, 411
designing, 411
entropy, 413
focus (F), 412
lightning condition effect, 414
max-min normalization technique, 414
parameter clutter, 413
reflection, 413
uniform distribution, 413
ROI extraction
area, 418
central knuckle line detection, 418–420
convex direction coding, 415, 417
knuckle line extraction, 419, 420
segmentation, 420, 421
soft biometric traits, 430
statistical features, 427, 428
structures, 410
Fingerprint
applications, 321
authorized users access, 322
databases, 29
education environment
biometric, 32
planning and ethics approval, 31
requirement, 31
sensor, 32
enhanced image, 325–327
enrolment, 30
frequency image, 325, 326
image database, 333
image-based methods, 322
image enhancement, 322
invariant moments, 329, 330
matching, 21, 24, 25, 34
Minutiae-based methods, 322
MORLET enhancement and compression, 324
MORLET wavelet, 324, 325
online signature-based biometric recognition system, 262
reference point, 327, 328
ridge orientation, 325, 326
ROI, 327, 328
security, 30
transform-based method, 322
verification, 24, 34, 321
verification system, 322
Zernike moments, 328, 330
Fingerprint based authentication, 350–352
Fingerprint performance, 28
Fingerprint recognition
CN method, 22, 23
enrolment, 16, 17
Henry classification system, 16
patterns, 16
sensors, evolution, 15
touchless swipe sensor, 15
Fingerprint scanning, 338, 442–444
Fingerprint sensor, 15
Fisher's Linear Discriminant Analysis (FLDA), 145
Fisherbrains, 306
Fitbit Zip, 357
Floor sensor (FS) approach
data types, 237
EMFi, 238
feature extraction techniques, 239
feature-rich, 238
footstep frames, 238
frame-based floor sensor capture method, 238
fuzzy membership, 239
gait analysis, 237
gait biometric characteristics, 238
GRF, 238–240
Kistler force plate, 238
LDA, 239
LLSRDTW, 239
LTN, 239

Floor sensor (FS) approach (*cont.*)
pathological conditions, 237
PCA, 239
piezoelectric, 238
signal capture, 239
tiles, 238
and wearable sensor approaches, 237
WPD, 239
FM signal, 357
Forensics, 4, 5, 214, 215, 242, 272, 321, 359
DNA, 486
palm print, 485
Forgery
and hackers, 276
identify, 258
and large intra-class variation, 267
random, 278
signatures, 278
simple, 278
skilled, 279
types, 275
Forget-Me-Not device, 356
Fovea localization
anatomical characteristics, 76
multilevel thresholding, 76, 78
ROI, 76
Frame-based floor sensor capture method, 238
Fraud
biometric database, 551
Functional magnetic resonance imaging (fMRI), 288
Function-based methods, 266
Fundus camera
Canon CR-1, 92
lighting systems and image, 91
optical principle, 91
Fusion, *see* Biometric fusion
Fuzzy extractor, 342, 343
Fuzzy identity based encryption (F-IBE), 468, 469
Fuzzy membership, 239
FVC2002 database, 333

G

Gabor filter, 47, 85, 324, 326, 497, 498
Gait analysis, 236, 237
Gait biometrics recognition
and identification, 246
applications, 242–244
ATVS dataset, 246–248
characteristics, 233
classifiers, 246, 247
components, 234
datasets, 241, 242
facial recognition, 240
FS approach, 237–240
fused data, 240
GRF, 249
KNN, 248
LLSRDTW, 249
LTN, 249
MLP, 248
MV approach, 234–236
obtrusiveness, 240
PCs, 249
piezoelectric sensor-based acquisition method, 246
preprocessors, 246, 247
privacy and security, 244, 245
proposed structure, 246
sample extraction, 246
SVM, 248
WS approach, 236, 237
Gait recognition
accelerometer, 366
advances and installation of sensors, 366
classifiers, 366, 367
data collection, 366
sensors, 366
use of smartphones, 366, 367
Gaussian distribution, 303
Gaussian Mixture Model (GMM), 171
Gaussian radial basis function network (GRBFN), 392
Generative attacks, 214
Genuine acceptance rate (GAR), 333, 334
Gesture recognition, 191, 206
Global feature-based recognition, 137
Google Glass, 361, 367
Government, 1, 4, 9
biometric authentication (*see* Authentication)
e-Voting systems (*see* e-Voting)
power, elections, 528
GPS sports watches, 357
Gross Domestic Product (GDP), 372
Ground reaction force (GRF), 238–240, 243, 246, 247, 249

H

Half total error rate (HTER), 348
Hamming distance, 50, 52, 87
HandKey II, 104, 105, 109
HandPunch GT-400, 105

Handwriting/gesture recognition, 367
Head Mounted Displays (HMDs), 363
Healthcare biometrics, *see* Healthcare sensors
Healthcare sensors
  accuracy of system, 352
  advantages, 337, 338
  analytical biosensor, 340
  authentication, 337
  avoiding errors in medication, 339
  biometric, 338
  biometric features, 352
  biometric-based solution, 338 (*see also* Biometric system)
  biometrics, 339
  communication protocols (*see* Biometric-based communication protocols)
  cooperation of user, 352
  data, 337
  decreasing billing fraud, 339
  fingerprint scans, 338
  fraud the system, 352
  hardware, 338
  health information exchange, 339
  marketing, 339, 340
  monitoring applications, 340
  network architecture, 337, 338
  psychological, 337, 338
  reaction capacity to medical emergencies, 338
  remote care applications, 338
  security system (*see* Biometric-based security system)
  types, 340, 341, 346
  uniqueness and distinctiveness, 352
Henry classification system, 16
Hidden Markov models (HMMs)
  in online signature verification system, 257
  signature verification
    data classification, 267, 268
    in financial transaction, 267
    fusion strategies, 268
    intra-class variation, 267
    random and skilled falsification, 267
    shipping information, 267
  stroke-based sequences, 267
Hough transformation, 73, 74
Human body identification, 191
Human gait, 233
Human identity recognition, 190, 201
Human-computer interaction (HCI) data, 213
Human-computer interaction (HCI) devices
  advantages, 192
  channel frequency, 190
  computer vision-based identification, 190
  CSI (*see* Channel state information (CSI))
  device-based methods, 192
  device-free indoor sensing systems, 207
  device-free methods, 192
  dimensionality reduction, 203
  fast identification, 196
  hardware based approaches, 193
  human action behavior recognition, 190
  human body identification, 191
  human identification recognition, 190
  identity recognition, 190
  limitations and challenges
    accurate and fine-grained localization, 206
    accurate and Fine-Grained Localization, 206
    counting human numbers, 205
    multiple identity/gesture recognition, 206
    privacy leakage, 205, 206
  mobile Wi-Fi sensing, 207, 208
  motion sensing, 207
  open environment dynamics, 196
  personal diversity, 196
  recognition methods (*see* Recognition)
  RFID tags, 191
  RSSI, 191, 192
  sensor-based identification, 190
  sensors, 191
  shadowing, 189
  small-scale fading, 189
  system overview and design
    constructing CSI pool, 197
    CSI profile pool, 197, 198
    human identity recognition, 197
    recognition process, 198, 199
    Wide, 197
  testbed experiments, 195, 196
  touch screen, 207
  ubiquitous recognition, 196
  vision-based identification, 190
  Wi-Fi (*see* Wi-Fi signals)
  wireless signal-based identification, 190, 191
Hybrid methods, 182, 270, 271

I

Identification, 155
  accuracy, 489
  and authentication, 283, 369
  authorship, 215

Identification (*cont.*)
automated, 261
biometric, 373, 489, 490
categorization, 243
data reduces, 335
design biometric-based, 343, 344
eye tracking (*see* Eye tracking biometrics)
face, 372
face identification algorithm, 485
fingerprint, 125, 321
forensics, 5
fused algorithm, 323
gait recognition, 234
hand geometry, 486
healthcare, 479
IoT internetworking systems, 483
iris, 485
Keystroke dynamics, 395
non-intrusive method, 263
one-to-many matching, 5, 359
parameters, 255
patient, 339, 353
process, 488
and recognition accuracies, 361
SVM, 239
technique, 258
technology, 484
2D hand geometry recognition
direct measurement, 106, 107
HandKey II, 109
limitation, 109
NIR, 109
silhouette alignment, 107, 108
spoofing, 111
UIN, 272
uses of biometrics, 3
and verification, 280, 359–361
voting
CA, 522
personal, 514
PID, 523
Identity-Based Encryption (IBE), 467, 468
Identity function, 156
Identity theft, 281
Image acquisition, 71, 72
Image enhancement, 18, 80
Image filtering, 17
Image-based methods, 322
ImageWare Systems, 448
Impersonation attack, 347
Impostors dataset, 224, 225
Independent component analysis (ICA), 303
Information security, 8, 166, 380–382
challenges, 457
cloud computing
outsourcing, 458
cognitive biometrics, 389
could computing
data, 458
multi-tenancy, 458
IoT (*see* Internet of Things (IoT))
Infrastructure as a Service (IaaS, 439
Insider attack, 214
Interclass variance, 327
International Association for Pattern Recognition (IAPR), 171
International Civil Aviation Organisation (ICAO), 127
International Conference on Biometrics (ICB), 172
International Data Corporation (IDC), 358, 359
Internet of Things (IoT), 339
applications
agriculture, 482
health care, 480
healthcare, 479
pharmaceutical industry, 482
SIoT, 481
smart cities, 480
smart environment, 478, 479
transportation and logistics, 481
architecture, 478
cost, biometric technology, 504
devices, 477
extract biometric features (*see* Biometric security in IoT)
security (*see* IoT security)
types of technology, 477
types, IoT security (*see* IoT security)
verification, 504
Inter-pulse intervals (IPIs), 348
Intra-class variance, 327
Intrinsic verification, 216
Invariant moment fusion authentication system
binaries method, 323
brute force attack, 321
DCT coefficient, 333
decision-level fusion techniques, 323
definition, 323
disadvantages, 322
EER, 333
enrolment, 323
error rates, 322, 323
face, 330–332

FAR, 333, 334
FAR *vs.* FRR, 333, 334
fingerprint (*see* Fingerprint)
FRR, 333
GAR, 333, 334
GAR *vs.* FAR, 333
hackers, 321
missing and alarm, 333, 334
mono-biometric, 321
receiver operating characteristic, 333
simple binaries method, 322
single biometric system, 322
single identification decision, 323
storage, 323
WFMT, 322
information transmitted, 321
Invariant moments, 329, 330
IoT security
behavioral features, 486, 487
biometric (*see* Biometric security in IoT)
categories, 484
embedded sensor systems, 483
giant network, 483
identification technology, 484
internetworking systems, 483
physiological features, 484, 486
vulnerable attack, 483
Iris anatomy, 43–45
Iris recognition
advantages, 52
characteristics, 45
acceptability, 50
Hamming distance, 51
permanence, 51
reliability, 51
commercial applications and devices, 54, 55
Daugman's algorithm, 46, 47
Daugman's rubber sheet model, 47, 48
disadvantages, 53
encoding, 48
error rates probabilities, 52
Gabor Filter, 47
Hamming distance, 50, 52
history, 45
standard, 53
Iris scan, 262, 263, 321, 444, 445
IriTech, 451

J
Jewelry, 362, 363

K
Key generation procedure, 351
Key-policy attribute-based encryption (KP-ABE) model, 466, 467
Keystroke dynamics
advantages, 174
applications, 175
appropriateness, 175
authentication services, 167
classification methods, 180, 182, 183
cloud validation, 171
datasets, 183
disadvantages, 175
FAR, 174
features
conventional, 176
editing, 178
non-conventional, 177
subset selection, 178, 180
FRR, 174
Gaussian distribution, 171
hand positions, 172
KDA, 172
touchscreen devices, 171, 172
TUBA, 171
workflow, 173
Keystroke dynamics based user authentication (KDA), 172
Keystroke dynamics biometrics, 394, 395
KINECT based 3D face recognition
benchmarking, 153
database, average models, 152
disparity-DB_3, 147
fiducial regions, 150, 151
image capture, 147
noise removal, 148
pre-processing, 147
query matching, 151
standardisation, 152, 153
upward and downward gradients, 149, 150
Kistler force plate, 238, 239
K-nearest neighbors (KNN), 235, 248, 303
Knowledge signatures, 270

L
Large margin nearest neighbors (LMNN), 114
Least Squares Probabilistic Classification (LSPC), 239
Legal implications of biometrics
airports, 539
business organizations, 538

Legal implications of biometrics (*cont.*)
  collection, information, 538
  data protection, 542
  data protection agency, 539
  enrollment process of biometric schemes, 539
  in Europe, 538
  futuristic vision, 541
  government official documents, 538
  government usage, 542, 544
  large-scale schemes, 539
  personal information, citizens, 539
  police measures, 540
  violation of privacy, 540
Lexical features, 218, 220
Licenses, 4, 452
Line scan camera
  2D hand geometry extraction, 118
  3D feature extraction, 120, 121
  edge detection, 119
  mack filtering, 119
  POC palmprint extraction, 118
  requirement, 118
  touchless system, 120
  VeriFinger, 118, 119
Linear discriminant analysis (LDA), 239, 303
Linear time normalization (LTN), 235, 239, 249
Local binary pattern (LBP), 86
Local feature-based recognition, 137
Localized Least Squares Regression with Dynamic Time Warping (LLSRDTW), 239, 249
Lunchtime attacks, 345

M

Machine learning methods, 181, 182
Machine vision (MV) approach, 234–236
Magnetoencephalography (MEG), 288
Man-in-the-middle attack, 347
Mapping function, 343
Matched filter, 80
Matching
  correlation-based, 24, 26
  minutia-based, 24, 27
  pattern-based, 25, 27, 28
Mel-Frequency Cepstral Coefficients (MFCC), 368
Mental stress, 369
Micro messages dataset, 223, 224
Micropayment, 371
Minimum Average Correlation Energy (MACE) biometric filter, 269, 270
Minutiae
  binarized fingerprint images, 18, 19
  definition of characteristics, 17, 18
  detection, 20
  gray-scale fingerprint images, 19
  location, 18
  orientation estimation, 19
  post-processing, 20
  ridge detection, 20
  thinning/skeletonization, 20
  types, 18
Minutiae-based matching, 25–27
Minutiae-based methods, 322
MIT PhysioBank database, 348
Mobile Wi-Fi sensing, 207, 208
Model-based gait biometric, 235
Model-free technique, 234
Mono-biometric authentication systems, 321
MORLET enhancement and compression, 324
MORLET wavelet, 324, 325
MorphoTrust's, 452
Motion sensing, 207
Mouse dynamics biometrics, 170
  categories, 392
  data collection, 392
  detection process, 394
  factors, 392
  graphical user interface, 392
  MSD curves, 393, 394
  security applications, 392
Multi-biometric system
  algorithm, 407
  biometric trait, 408
  instance, 408
  multisensor biometric system, 407
  unimodal, 407
Multi-factor biometric authentication (MFBA), 463, 464
Multi-layer perceptron (MLP), 248, 367
Multi-layer perceptron neural networks (MLP), 239
Multilevel thresholding, 76, 78
Multimodal biometric system, 4
  advantages, 67
  bimodal eye biometric system, 69, 71, 72
  biometric fusion, 67, 69
  invariant moment fusion (*see* Invariant moment fusion authentication system)
  wearable technologies, 360, 361, 368
Multimodal feature (MMF), 129
Multiple identity recognition, 206
Multiple matching algorithms, 407

Multisensor biometric system, 407
Multisensor smart jewelry, 371
Multi-trait system, 408

N

N400, 301
  ERP, 295
Naive Bayes, 367
Near Field Communication (NFC), 368
Near-infrared spectroscopy, 288
Network interface cards (NICs), 193
Network security, 166, 167, 184
Neural network (NN) stage, 182, 259
N-gram analysis, 216, 220, 229
Nike+ iPod Sports Kit, 357
NIST gait database, 234
NIST-USF dataset, 241
Non-event-related paradigms
  identification/authentication, 308
  tasks
    eye and audio tone, 309
    generate words, 309
    imagine speech, 309
    letter composing, 309
    math, 309
    motor imagery, 309
    object counting, 309
    object rotating, 309
    pass-thoughts, 308–310
    recitation, 309
    resting, 309
    resting with eyes closed, 308, 310
    visual counting, 309
Non-intrusive wearable sensors, 237
Non-invertible geometric transforms, 268
Nymi band, 368, 369
Nymi Companion Application (NCA), 377
Nymi Enabled Applications (NEAs), 377

O

Oddball paradigm, 294
Offline signature verification system
  application areas, 258
  biometric database, 258
  components, 259
  dynamic characteristics, 258
  feature extraction stage, 259
  features, 258
  forgeries, 258
  information unavailability, 258
  intermediate stage, 259
  vs. online, 260
  paper-based working culture, 258
  pen-based, 258
  preprocessing stage, 259
  simple paper, 258
  stages, 259
Olea Sensor Networks, 339
One-handed Keystroke Biometric Identification Competition (OhKBIC), 171
Online signature verification system
  DTW, 257
  dynamic properties, 257
  Fourier transform, 258
  frequency domain, 257
  HMM, 257
  key features, 257
  vs. offline, 260
  stylus, 257
Online signature-based biometric recognition system
  auditory verification, 263
  BioHashing, 273, 274
  biometric template protection, 271–273
  cancelable, 268–271
  captures and stores, 261
  dynamic signature recognition, 265, 266
  in E-commerce, 274
  face recognition, 263
  fingerprint, 262
  function-based methods, 266
  HMM, 267, 268
  iris scan, 262, 263
  operation, 261
  retail industry, 280
  retina scan, 262, 263
  security and expediency, 262
  surveillance mechanisms, 261
  systematic comparison, 263
Operation modes, 4, 5
Optic disc localization
  CLAHE and median filter, 75
  gradient vector flow snakes, 73
  Hough transformation, 73
  principal component analysis, 73
  ROI, 74
  watershed segmentation, 74
Optical scanning system, 59
Otsu method, 327
OU-ISIR group, 241, 242

P

P300, 301
  ERP, 294, 295
  speller, 298
  VEPs, 307
Paper-based working culture, 258
Parzen Windows Classifiers (PWCs), 267
Pass-thoughts, 370
Pattern recognition, 167, 181, 182
Perception of biometrics
  enrollment and matching procedure, 550
  false scanning, sensor, 551
  fraud/identity theft, 551
  function creep (FC), 550
  human factors (HFs), 548
  industry, 550
Phone-based sensors, 236
Physiological based traits, 406
Physiological biometrics, 360
Physiological modalities, 366
Physiological sensors, 340
Physiological-based traits, 406
Piezoelectric sensor-based acquisition method, 246
PillCam Capsule Endoscopy, 340
PKI-based electronic signatures, 266
Plagiarism detection, 215
Platform as a Service (PaaS), 440
Potentiometric biosensor, 340
Preprocessing, 360
Principal component analysis (PCA), 194, 195, 235, 236, 239, 302
Principal components (PCs), 249
Principle component analysis (PCA), 137
Privacy, 154
Privacy-preserving biometric, 473
Pseudorandom number, 273
Psychtoolbox, 304
Pupil and iris localization, 84

R

Radial basis function (RBF) NN layers, 260
Radial polynomial, 330
Radial Zernike polynomials (RZP), 497, 499
Radio frequency-based identification, 191
Random forest, 367
Random forgery, 278
Range image maps based 3D face recognition
  classification and query processing, 145
  image capture, 143
  image normalization, 144
  model representation, 144, 145
  profile signatures, 144
Rapid serial visual presentation (RSVP), 299, 300
RBF-kernel SVM, 370
Real signatures, 270
Real-time information access, 372
Received signal strength (RSS), 191–193
Received signal strength indicator (RSSI), 191, 192
Receiver operating characteristics, 333
Recognition
  accuracy, 361, 370
  and activity monitoring, 368
  BioID, 448, 450
  decentralized, 352
  face, 366
  facial, 346, 372, 446
  fingerprint, 346, 442, 443
  gait, 366, 367
  gait biometric (*see* Gait biometrics recognition)
  HandKey II, 109
  handwriting/gesture, 367
  HCI devices
    accuracy evaluation, 203, 204
    dynamic identification, 200
    gesture, 206
    human identity, 197, 201
    identities, volunteers, 201–203
    multiple identity, 206
    PCA, 194, 195
    static people, 199, 200
    SVM, 194
  identification, 136
  IoT
    biometric security (*see* Biometric security in IoT)
    healthcare, 479
    individuals, 478
    Intel smart home platform supports, 479
  iris, 346
  iris scanning, 444
  keystroke dynamics, 394
  local and global feature, 137
  machine learning method, 389
  pattern system, 341
  rates, 361, 369
  3D (*see* 3D hand geometry recognition)
  TOF cameras, 116
  2D (*see* 2D hand geometry recognition)
  vein, 444
  verification, 136
  voice, 346
Recognition
  eye movement biometric, 391

iris recognition system, 388
retina recognition captures, 389
Reference point, 327, 328
Regions of interest (ROI), 74, 327, 328
Registration phase, 341, 342
Relative received signal strength, 192
Reliability, 2, 8
Remote biometric authentication
advantages, 461
attacks, 461
MFBA, 463, 464
schemes, 461
security, 461
Reproduction function, 343
Retail industry
benefits, 282
challenges, 282
characteristics, 280
credit cards, 280
division of signature biometrics, 280, 281
identity theft, 281
objectives, 280
outcome, 282
signature verification process, 281, 282
statistical review, 280
Retina anatomy, 42, 43
Retina scan, 262, 263
Retinal recognition
advantages, 56, 62, 63
characteristics, 61, 62
commercial applications and devices
EyeDentification System 7.5, 64
handheld retinal scanner, 65, 66
ICAM 2001, 65
disadvantages, 63
EyeDentification System 7.5
optical scanning, 59
optical Scanning, 59
representation, 60
rotation algorithm, 60
history, 57, 58
liveness detection, 38
security, 57
Ridge orientation, 325, 326
Robust system, 256
Rotation algorithm, 60
Rotation forest, 367
Rotation Manhattan, 366
Rotational invariant moments, 329

S

Salting methods, 270
Sample extraction, 246
Scalability
IoT, 489
on-demand, 438
Scale-invariant feature transform (SIFT), 499, 500
Secure data transmission, 350–352
Security attacks
cases, 166
passive and active, 167
Sensor-based identification, 190
Sequential minimum optimization (SMO), 215
Shallow classifiers, 226, 227
Shoe-based sensors, 236
Short message verification, 220
Short-time Fourier transform (STFT), 302
Signal-to-noise ratio, EEG
components, 300
handling artifacts, 301
handling background activity, 301, 302
Signature verification system
banking sectors, 255
decision-making method, 256
falsifications, 256
FAR, 256
FRR, 256
generalization, 256
global features, 256
local features, 256
offline, 258, 259
online, 257, 258
robust system, 256
selection, 256
type of situation, 256
Silhouette-based technique, 234
Simple binaries method, 322
Simple forgery, 278
Single biometric system, 271, 322
Skilled forgery, 279
Smart bands, 357, 362
Smart clothing, 361–363
Smart eyewear, 363
Smartwatches, 357, 361, 362
Social implications of biometrics
cyberattacks, 544
European Union (EU)
digitization of borders, 546
identification, security agencies, 547
INS data, 548
migration, 547
profiling, 547
PROGIS, 548, 550
hardware devices, 544
US *vs*. underdeveloped world
anonymity, 545

Social implications of biometrics (*cont.*)
profiling, 546
tracking and surveillance, 545
Social Internet of Things (SIoT), 481
Software as a Service (SaaS), 440, 447, 456
Software defects, 214
Software module attack, 347
Software, EEG
BCI2000, 304
EEGLAB/BCILAB, 303
FieldTrip, 303
Psychtoolbox, 304
Software-defined radio (SDR), 193
SPOT Watch, 357
Static signature verification system, 258
Statistical methods, 180
Steady state evoked potentials (SSEP), 295
EEG, 310
Steady state visually evoked potential (SSVEP), 295
Stepscan floor sensor system, 244
Stereo vision, 117
classification and query processing, 145
image capture, 143
image normalization, 144
model representation, 144, 145
profile signatures, 144
Stride length, 246, 249
Structured light
diffraction grating, 115
IR pattern projection, 113, 114
lasers, 115
stereo vision, 117
TOF scanning, 116
Stylometry, CA
concept drift, 214
data quality issues, 213
features
application-specific features, 220
lexical features, 218, 220
n-gram model, 220
syntactic features, 220
free actions analysis, 213
performance trade-off, 213, 214
security and privacy risks, 214
Support vector machine (SVM), 194, 198, 203, 215, 235, 236, 248, 249, 303
Support vector network, 194
SWOT analysis, 266

T
Technavio's market research analysis, 339
Template protection, 271–273
Testbed experiments
configurations, 195, 196
installations, 195, 196
Thermometric biosensor, 340
Thinning, 72, 81
Time of flight (TOF) scanning, 116
Time-frequency representation, 293
Time-to-authenticate (TTA), 213
Touch and voice behavioral biometrics, 367
Touch screen, 207
Toxoplasmosis, 96
TP-Link TL WR886N router, 195, 201
Traditional biometrics
wearable technologies
gait recognition, 366, 367
gait Recognition, 366
handwriting/gesture recognition, 367
handwriting/Gesture Recognition, 367
identify and verify, 366
multimodal biometrics, 367–369
smartphone's features and capabilities, 366
Transform-based method, 322
Transmitter-receiver (TX-RX) antenna pair, 190, 201
Tree-Re-Weighted (TRW), 368
True Positive Rate (TPR), 370
TUM-IITKGP dataset, 241
Tweets, 223
Two-dimensional continuous wavelet transform (2D CWT), 324, 325
Typing style, 212

U
Ubiquitous recognition, 196
Unimodal biometric structures, 66, 271
Unique identification number (UIN), 272
Unmasking, 215
Unobtrusive gait analysis, 236
User authentication, 211
behavioral, 170
biometrics, 169
cognitive (*see* Cognitive biometrics)
knowledge-based, 168
object-based, 168
Uses of biometrics, 3, 4

V
Vein recognition, 444
Verification, 156
accuracy, 335
auditory, 263

and authentication, 359
authorship, 215–218, 227, 228, 230
BioID application, 449
biometrics, 371, 448
close-set, 24
database, 323
election process
CA, 522
e-voting terminal, 525, 527
voter, 518, 524
voter ID and biometrics, 525
voting certificate, 522
and enrolment, 217, 374
eye Tracking Biometrics, 389
fingerprint, 321, 322
and identification, 359–361
identity, 359
IoT
algorithm, 492
fingerprint biometric, 492
traditional, 492
iris-based verification system, 262
legitimacy, 348
liveliness, 110
moments, 332
Nymi-enabled device, 369
one-to-one matching, 5, 24
open-set, 24
short message, 220
signature systems (*see* Signature verification systems)
steps, 448
template, NCA, 377
3D signature, 371
uses of biometrics, 3
Virtual reality (VR), 363
Vision-based identification, 190
Visually evoked potentials (VEPs)
early components, 305, 306
P300, 307
recognition and memory, 306, 307
structure, 305
Voice-based cancelable biometric templates, 270

W
Watershed segmentation, 74
Wavelet and Fourier-Mellin transform (WFMT), 322
Wavelet Packet Decomposition (WPD), 239
Wavelet-domain Hidden Markov Model (HMM), 348–350
Wearable biometrics
criminal justice system, 373
healthcare, 371, 372
military, 372, 373
privacy control, 372
real-time information access, 372
Wearable sensor (WS) approach, 236–238, 243
Wearable technologies
abandonment, 358
activity trackers, 362
backpack-mounted head camera, 355
biometrics
behavioral, 360
identification, 359, 360
multimodal, 360, 361
physiological, 360
training mode, 359, 360
verification, 359, 360
capabilities, 358, 359
communication capability, 356
developments, 357–358
device prediction, 355
devices, 356
emerging biometrics (*see* Emerging biometrics)
Forget-Me-Not device, 356
functions/applications, 358, 359
jewelry, 362, 363
limitations
battery life, 378
data/template storage, 379
processing power, 379, 380
security and privacy, 380–382
market and growth forecasts, 359
mobility and flexibility, 356
short-lived, 355
smart bands, 362
smart clothing, 361–363
smart eyewear, 363
smartwatches, 361, 362
specifications, 363–365
tracks/monitors, 356
traditional biometric (*see* Traditional biometrics)
transistors and microswitches, 355
Wi-Fi signals
capture motion, 207
characteristics, 206
commercial, 196
COTS, 193, 195, 201
CSI-based, 191
fingerprint, 191
identification recognition, 190

Wi-Fi signals (*cont.*)
  identity recognition, 190
  network connections, 189
  OFDM, 193
  prevalence, 206
  router sends, 203
  test-bed experimental results, 205
  types, 191
  with machine learning, 196
Wireless sensor network, 337, 339
Wireless signal-based identification, 190, 191
Withstand forgery, 228, 229
Worldwide Quarterly Wearable Device Tracker, 358
Wrapper method, 179
Wristband, 371
Writing style
  CA (*see* Continuous authentication (CA))

Z

Zernike moments, 328, 330
ZJU-GaitAcc database, 242

The manufacturer's authorised representative in the EU is Springer Nature Customer Service Centre GmbH, Europaplatz 3, 69115 Heidelberg, Germany. If you have any concerns regarding our products, please contact ProductSafety@springernature.com

Printed and bound by CPI Group (UK) Ltd, Croydon, CR0 4YY
22/09/2026
02220979-0009